# Optically Stimulated Luminescence

# Optically Stimulated Luminescence

## Fundamentals and Applications

EDUARDO G. YUKIHARA
and
STEPHEN W. S. McKEEVER

*Physics Department, Oklahoma State University*
*Oklahoma, USA*

WILEY
A John Wiley and Sons, Ltd., Publication

This edition first published 2011


*Registered office*
John Wiley & Sons Ltd, The Atrium, Southern Gate, Chichester, West Sussex, PO19 8SQ, United Kingdom

For details of our global editorial offices, for customer services and for information about how to apply for permission to reuse the copyright material in this book please see our website at www.wiley.com.



*Library of Congress Cataloging-in-Publication Data*

Yukihara, Eduardo G.
Optically stimulated luminescence : fundamentals and applications / Eduardo G. Yukihara and Stephen W.S. McKeever.
p. cm.
Includes bibliographical references and index.
ISBN 978-0-470-69725-2 (cloth) – ISBN 978-0-470-97705-7 (ePDF) – ISBN 978-0-470-97706-4 (ebook)
1. Optically stimulated luminescence dating. 2. Radiation dosimetry. I. McKeever, S. W. S., 1950– II. Title.
QE508.Y85 2011
535′.356–dc22

2010037269

A catalogue record for this book is available from the British Library.

Print ISBN: 978-0-470-69725-2
ePDF ISBN: 978-0-470-97705-7
oBook ISBN: 978-0-470-97706-4
ePub ISBN: 978-0-470-97721-7

Typeset in 10/12pt Times by Aptara Inc., New Delhi, India.

FSC www.fsc.org MIX Paper from responsible sources FSC® C013604

To Stefanie, my music, my silence

To Joan and our wonderful daughters, Katie and Alison.

# Contents

# Preface

This book was born from our combined years exploring the use of optically stimulated luminescence (OSL) in different areas of radiation dosimetry with a seemingly unending curiosity about the physics of the process and our aspirations and dreams concerning potential applications. We wished to learn about and understand the challenges presented by the many different areas of dosimetry and how OSL could play a useful role and, of course, we hoped that we would make meaningful contributions to the field. The book is an attempt to share our collective experience and organize the available information concerning the different areas of application of the OSL technique. We made a conscious effort to place the material in context and make it useful for a wide audience. In each chapter we have tried to set the stage for more meaningful discussions of the OSL technique by providing background information (though by no means exhaustive) and relevant key references. On some occasions new illustrations or graphs were created to better illustrate the ideas and explanations.

We recognize that to some readers the background information in each chapter may appear over-simplified, especially for specialists in the respective fields. This brings us to the discussion of the intended audience of this book. We tried to make the book both relevant and accessible: relevant for the specialist by providing topical information, and accessible to those without expertise in these particular areas by including important fundamental aspects. We often thought of our audience as students or postdoctoral fellows that have just joined a research laboratory, not necessarily having worked in these areas beforehand. What would we like them to know before they engage in their research activities? It is obvious that one text cannot provide all answers and details, but we hope our book gives readers a general overview of the current problems and an initial reference point before they submerge themselves in the more specialized literature.

# Acknowledgments

Our gratitude first goes to Safa Kasap, who first mentioned the notion of the book to one of us while we were attending an international conference together. The task of writing a book is not to be taken lightly, involving as it does many unsociable and apparently endless hours of hard work. Safa's kind persistence paid off, however, and there is no doubt that without it the book would never have been started.

We are extremely grateful for the time and attention dedicated by several people in reading the manuscript and for their critical and insightful comments. Gabriel Sawakuchi's expertise in medical and space dosimetry, as well as his unselfish dedication, was extremely helpful and resulted in several revisions to the manuscript, improving it immeasurably. We also received very useful comments and suggestions from Gregoire Denis, Oliver Hanson, Emico Okuno and Matthew Rodriguez.

We also thank colleagues and collaborators who contributed with discussion, data, figures and pre-prints, including: Claus Andersen, Eric Benton, Enver Bulur, Ramona Gaza, Razvan Gaza, Carla da Costa Guimarães, Regina DeWitt, H. Y. Göksu, Elizabeth Inrig, David Klein, Gladys Klemic, Matthew Rodriguez, Gabriel Sawakuchi, Kristina Thomsen and Clemens Woda.

Finally, we thank our collaborators who, throughout our careers, have provided wisdom, direction, support and encouragement. They opened opportunities for collaboration in so many different areas, on so many different problems. It has been a pleasure and a delight to work with such clever and capable people. They are too many to mention without risk of leaving someone out by mistake. They know who they are and so we hope they accept this general acknowledgment as an expression of our gratitude.

*E. G. Yukihara and S. W. S. McKeever*

# Disclaimer

Reference to commercial products do not imply or represent endorsement of those products on the part of the authors. The authors have been funded by Landauer Inc. for different projects over the past decade or more.

# List of Acronyms

| | |
|---|---|
| AAPM | American Association of Physicists in Medicine |
| ALARA | as low as reasonably achievable |
| CCD | charge-coupled device |
| CPE | charged particle equilibrium |
| CT | computed tomography |
| CTDI | computed tomography dose index |
| CW-OSL | continuous wave optically stimulated luminescence |
| DOSL | delayed optically stimulated luminescence |
| EPR | electron paramagnetic resonance |
| ERB | Earth's radiation belts |
| ESTRO | European Society for Therapeutic Radiology and Oncology |
| GCR | galactic cosmic rays |
| HCP | heavy charged particle |
| HTR | high temperature ratio |
| HZE | high-charge (Z)–high-energy |
| IAEA | International Atomic Energy Agency |
| ICCHIBAN | Intercomparison of dosimetric instruments for cosmic radiation with heavy ion beams at NIRS |
| ICRP | International Commission on Radiological Protection |
| IEC | International Electrotechnical Commission |
| IMRT | intensity modulated radiation therapy |
| IMPT | intensity modulated proton therapy |
| IR | infrared |
| IRSL | infrared stimulated luminescence |
| ISO | International Organization for Standardization |
| ISS | International Space Station |
| LED | light emitting diode |
| LET | linear energy transfer |
| LM-OSL | linear modulation optically stimulated luminescence |
| NCRP | National Council on Radiation Protection and Measurements |
| OSL/PSL | optically stimulated luminescence/photostimulated luminescence |

| | |
|---|---|
| OSLD/TLD | OSL/TL dosimeter[1] |
| OTOR | one trap-one recombination center (model) |
| PMT | photomultiplier tube |
| PNTD | plastic nuclear track detector |
| POSL | pulsed optically stimulated luminescence |
| RBE | relative biological effectiveness |
| RL | radioluminescence |
| SOBP | spread out Bragg peak |
| SPE | solar particle event |
| TL/TSL | thermoluminescence/thermally stimulated luminescence |
| UV | ultraviolet |
| WHO | World Health Organization |
| YAG | yttrium aluminum garnet ($Y_3Al_5O_{12}$) |

[1] 'OSLD' has been used in the literature for 'optically stimulated luminescence dosimeter' and 'optically stimulated luminescence detector' indiscriminately. In this book we used OSLD for 'optically stimulated luminescence dosimeter' and tried to be more precise in the use of the terms 'dosimeter' and 'detector' (see definitions in Table 3.1). However, in some cases such distinction is not necessary and OSLD can be used for both OSL 'dosimeter' and 'detector' without detriment to the understanding. The same discussion applies for 'TLD.'

# 1

# Introduction

## 1.1 A Short History of Optically Stimulated Luminescence

> Twelfthly, To satisfie my self, whether the Motion introduc'd into the Stone did generate the Light upon the account of its producing heat there, I held it near the Flame of a Candle, till it was qualify'd to shine pretty well in the Dark . . . .
>
> —Boyle, 1664

The readers of this book may find it unusual to start a text with the word "Twelfthly." The above quotation is from a lively description of an experiment performed by Sir Robert Boyle, and the quoted text describes one (the 12th) of several experiments conducted by this seventeenth-century luminary on a piece of diamond loaned to him for the purpose. This prose, presented by Boyle to the Royal Society of London in October of 1663, concerns the phenomenon of thermoluminescence and Boyle's colorful account is now widely regarded as the first written description of its observation (Boyle, 1664). Thermoluminescence (TL) – also known (perhaps more accurately) as thermally stimulated luminescence – is one of a set of properties collectively known as "thermally stimulated phenomena" (Chen and McKeever, 1997). Boyle (1680), as cited by Bender and Marriman (2005), used the beautifully descriptive term "self-shining" to describe the phenomenon of luminescence, but the modern term (and, indeed, the first use of the word "thermoluminescence") is attributed to Eilhardt Wiedemann in his comprehensive studies of a variety of luminescence phenomena: "I have ventured to employ the term luminescence for all those phenomena of light which are more intense than corresponds to the actual temperature" (Wiedemann, 1889; quote from the Oxford English Dictionary, 1997 edition).

Thermoluminescence (TL) refers to the process of stimulating, using thermal energy, the emission of luminescence from a substance following the absorption of energy from an external source by that substance. The usual source of external energy is ionizing radiation and, as such, TL is closely related to phosphorescence, which is the afterglow emitted from a substance after the absorption of external energy. (See Harvey (1957) for a

*Optically Stimulated Luminescence: Fundamentals and Applications* Eduardo G. Yukihara and Stephen W. S. McKeever

comprehensive review of the early literature on this topic. A more modern discussion of TL and its relationship to phosphorescence can be found in Chen and McKeever (1997). Early studies of these phenomena were closely connected with the discovery of radioactivity and the external energy source in these early studies was invariably some form of ionizing radiation, from X-rays, an electron beam or a radioactive substance.

Optically stimulated luminescence (OSL) is a related phenomenon in which the luminescence is stimulated by the absorption of optical energy, rather than thermal energy. It is difficult to identify when studies of OSL (or, as it is also known, photostimulated luminescence, PSL) were first described in the literature. However, certainly the phenomenon was hinted at when initially Edmond Becquerel (1843) and then Henri Becquerel (1883) observed that the phosphorescence from zinc and calcium sulfides was quenched if these materials were exposed to infrared illumination after exposure to an ionizing radiation source. These and other similar observations around this time (Harvey, 1957) noted that the infrared illumination could either increase or decrease the intensity of the phosphorescence. Harvey (1957) reports that Henri Becquerel clearly observed an initial increase in luminescence output on application of the infrared light. The term "photophosphorescence" first appeared to describe these effects some years later (*viz.* 1889, as cited in *The Century Dictionary*, 1889 edition). Nichols and Merritt (1912) also noted that infrared stimulation can increase the luminescence output before rapidly quenching the phosphorescence and discussed the phenomenon in terms of Wiedemann and Schmidt's "electric dissociation" theory, involving the separation of positive and negative charges induced by the absorption of radiation energy (Wiedemann and Schmidt, 1895). Already one can discern the glimmerings of the modern interpretation, which involves ionization of electrons from their parent atoms, despite the fact that these early ideas were formulated before the advent of quantum mechanics and band structure theory.

At this point it is perhaps important to distinguish these effects from the property of "photoluminescence." The latter phenomenon describes prompt luminescence emission (or *fluorescence*) emitted during absorption of the stimulation light. No prior absorption of energy from an external ionizing source is necessary. A notable property of photoluminescence is that the emitted light is of a longer wavelength than that of the stimulation light. Furthermore, the lifetime of photoluminescence emission is such that it decays promptly upon cessation of the stimulation. Emission wavelengths in photophosphorescence, however, can be either longer *or* shorter than the stimulation wavelength, and the emission generally persists for seconds or minutes after the end of the stimulation period. Many examples of photophosphorescence are referred to in the early literature, but part of the difficulty in determining when reports of the phenomenon first appeared relates to the lack of understanding at that time of the physics of luminescence in general. As pointed out by Marfunin (1979), unlike other physical phenomena being studied in the early centuries, a complete understanding of luminescence requires an understanding of quantum mechanics, a field that was not born until the early decades of the twentieth century. Knowledge of quantized energy levels, band structure, and radiative and non-radiative electronic transitions was yet in the future. As a result critical experiments were perhaps not performed or the descriptions of them were vague such that easy identification of the phenomenon being studied is not always clear from the early literature.

Nevertheless, by the mid-twentieth century the understanding that free electrons in delocalized bands were involved in the phosphorescence process was beginning to emerge.

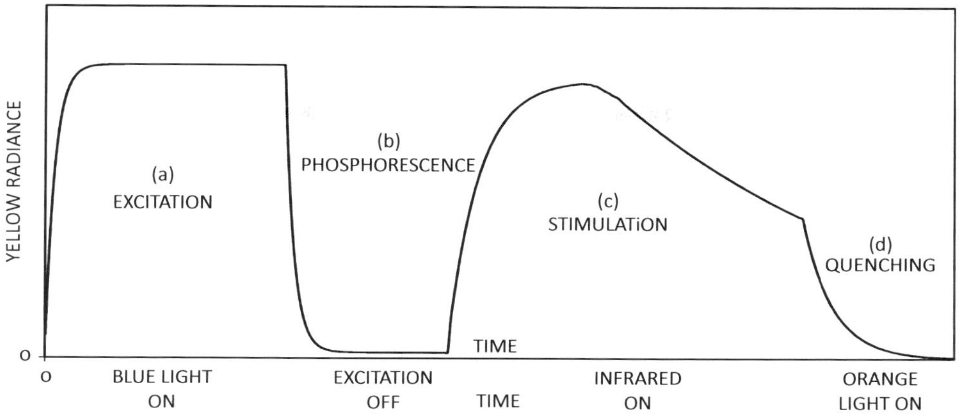


***Figure 1.1*** *The sequence of luminescence emission from Sr(S,Se):$SrSO_4$:$CaF_2$:Sm:Eu. The luminescence during periods (c) and (d) are what we now term optically stimulated luminescence (Schematic redrawing of Figure 3 from Leverenz (1949).)*

As discussed by Leverenz (1950), a debate at that time concerned the connection between photoconductivity and photophosphorescence. Work on sulfide materials (CdS, ZnS) demonstrated that the growth during stimulation and the decay after stimulation of both photoconductivity and photophophorescence were similar in many cases, although other materials seemed to show that a one-to-one connection was not always the case (e.g. Bube, 1951). Nevertheless, a picture emerged that photophosphorescence from those materials for which the luminescence decay was characterized by a $t^{-n}$ law (where $t$ is time and $n$ is usually between ~0.5 and ~2.0) required photostimulated conduction involving free charge carriers in conduction states.

Leverenz (1949) also discusses how infrared light can both quench the phosphorescence or stimulate it. Figure 1.1 illustrates a sequence of possible luminescence events in a complex phosphor made from Sr(S,Se):$SrSO_4$:$CaF_2$:Sm:Eu, as described by Leverenz (1949). Yellow luminescence is emitted during initial excitation with blue light, followed by a rapid decay (fluorescence, or photoluminescence) along with a component with a longer, slower decay (phosphorescence). However, if the material is then subsequently stimulated with infrared light, there is enhanced luminescence (growth and decay), the decay time for which can be rapidly reduced and the luminescence quenched by changing to shorter wavelength illumination (orange).

Even though the same emitting center is being activated in the sequence of luminescence emissions illustrated in Figure 1.1, the observed decay times can vary considerably, depending upon whether or not the sample is being optically stimulated and, if so, the intensity and wavelength(s) chosen. Leverenz (1949) notes: "The emitting center loses control over $\tau$" (the luminescence decay time) "when the energy storage of phosphors consists of trapped excited electrons or metastable states, for then additional activation energy must be supplied to release the trapped electrons." He goes on to state: "This activation energy may be supplied by heat . . . or it may be supplied by additional photons . . ." In these early descriptions of photophosphorescence can be found the essential elements of the phenomenon that we now term optically stimulated luminescence. Namely, after irradiation with the primary ionizing source, energy may be stored in the material in the form of trapped charge carriers (electrons and holes). Release of the trapped charge can then be stimulated by the absorption of optical photons of appropriate wavelength, resulting in luminescence emission. The emission decays with a time constant dictated by the wavelength and intensity of the stimulation light, and the characteristics of the trapping states in the material. This understanding of the processes involved was used by Schulman *et al.* (1951) and Mandeville and Albrecht (1953) to describe the luminescence emitted from alkali halides during optical stimulation following initial gamma irradiation, although OSL was not the term used by these authors. Indeed, nomenclature was still being developed, with Mandeville and Albrecht calling the effect "photostimulation phosphorescence," or alternately "co-stimulation phosphorescence," while Schulman *et al.* preferred the more descriptive (and more accurate) term "radiophotostimulation." Albrecht and Mandeville (1956) used the term "photostimulated emission" when describing what we now know as OSL from X-irradiated BeO. Harvey (1957) discusses the original 1843 observation of E. Becquerel and notes how this effect "could be called photo-stimulation, analogous to thermostimulation, that is thermoluminescence." (See Table 1.1.)

Considering these similar terms used to describe the effect it is perhaps not surprising to discover that the first use of the modern term, optically stimulated luminescence OSL, appeared in the published literature a few years later. Fowler (1963) uses the term when describing a paper that is generally taken to be the first reported use of (what Fowler refers to as) OSL in radiation dosimetry. Antonov-Romanovsky *et al.* (1955) monitored the intensity of infrared-stimulated luminescence from various sulfides, after irradiation, and the intensity of the emitted light was used as a monitor of the dose of initial radiation. Although this is certainly one of the earliest examples of the use of OSL in radiation

***Table 1.1*** *Some early nomenclature, along with the date of first introduction and author(s), for what is now known as optically stimulated luminescence.*

| Description of the Phenomenon | Name of the Phenomenon | Author, and Date of First Introduction |
|---|---|---|
| Light stimulated transfer of electrons from deep traps to shallow traps followed by phosphorescence | Photophosphorescence | Unidentified[a] (1889) |
| | Delayed optically stimulated luminescence | Yoder and Salasky (1997) |
| Light stimulated release of electrons from deep traps followed by radiative recombination and subsequent luminescence | Radiophotostimulation | Schulman *et al.* (1951) |
| | Photostimulation phosphorescence | Mandeville and Albrecht (1953) |
| | Co-stimulation phosphorescence | Mandeville and Albrecht (1953) |
| | Photostimulated emission | Albrecht and Mandeville (1956) |
| | Optically stimulated luminescence | Fowler (1963) |

[a] The term is first cited in *The Century Dictionary*, 1889 edition.

dosimetry, this now-famous 1956 paper refers to work by the same authors (published in Russian) from a few years earlier, in the 1949–1951 era. Nevertheless, despite these early applications, there was a hiatus of more than a decade before this pioneering work was followed by similar studies, notably by Braünlich, Schaffer and Scharmann (1967) and Sanborn and Beard (1967), working with irradiated sulfides. The work of this period even led to a US patent by Wallack (1959) in which was claimed the invention of an OSL gamma radiation dosimeter and system for reading the OSL signal using sulfide materials. Even then, however, OSL still did not catch on as a dosimetry tool; the cause lay in the materials being studied.

The fact that the sulfide materials used by these early pioneers could be stimulated with infrared (IR) light pointed to the fact that the trapped charge was localized in energy levels that were relatively shallow with respect to the delocalized bands, requiring a small de-trapping (activation) energy. This in turn meant that the trapped electrons were unstable at room temperature and decayed through the process of thermal stimulation (and subsequent phosphorescence emission). Thus, the dosimetric signal (i.e. the infrared stimulated luminescence signal) was found to decay with time between the initial absorption of radiation and the time of IR stimulation – a process now commonly referred to as "fading." As a result, OSL dosimetry was slow to be adopted, primarily for lack of suitable materials.

Slowly, however, the published literature began to accumulate descriptions of studies on optically stimulated luminescence effects in a variety of other material types. Most of the studies in this period (the 1970s, 1980s and even into the 1990s) reverted back to the use of photophosphorescence, as practitioners experimented with the optical stimulated transfer of electrons from deep, stable traps, into shallow, unstable traps. The goal was to monitor the subsequent phosphorescence as the charge leaked away from the shallow traps before recombining at the emission sites and to use this as a measure of absorbed dose. The materials used in this period, however, were wide-band-gap insulators, such as BeO (Rhyner and Miller, 1970; Tochilin, Goldstein and Miller, 1969), $CaF_2$ (Bernhardt and Herforth, 1974), $CaSO_4$ (Pradhan and Ayyanger, 1977; Pradhan and Bhatt, 1981) and $Al_2O_3$ (Yoder and Salasky, 1997). (The last work led to coining a new term for photophosphorescence, namely "delayed OSL.") Although photoconducting, narrow-band-gap materials were still studied intensely, focus for dosimetry began to shift away from these materials to wide-band-gap insulating materials with deep, stable traps.

The major breakthrough for use of OSL in dosimetry emerged in a related but quite different area of science – in the world of archeological and geological dating. During the 1980s an effort to establish OSL as a dosimetric tool was taking place in the archaeometric community. Huntley and colleagues (Huntley, Godfrey-Smith and Thewalt, 1985) used OSL from natural quartz to estimate the dose absorbed by this mineral in nature. Through an estimation of the natural dose rate (from natural quantities of uranium, thorium and potassium, as well as cosmic radiation), an estimate of the age of the mineral deposit could be made. This development created a new chronometric tool and at the same time opened an entirely new research field. OSL dating (or "optical dating"; Aitken, 1998) is now an established chronometric method (Bøtter-Jensen, McKeever and Wintle, 2003), but the importance from the point of view of radiation dosimetry is that it demonstrated that OSL could be used with stable, wide-band-gap insulators to determine radiation doses accumulated over millennia. Photophosphorescence was not employed; rather the preferred process being exploited was the direct stimulation of electrons from deep traps, through the conduction band, to recombine with trapped holes at activator sites leading to OSL. As a result, fading was not an issue.

The second major development in establishing OSL as a mainstream dosimetry tool was driven by a further advance in material technology. Anion-deficient $Al_2O_3$, doped with carbon, was developed at the Urals Polytechnical Institute in Russia for use as a sensitive thermoluminescence dosimetry material (Akselrod and Kortov, 1990). However, although $Al_2O_3$:C proved to be an exceptionally sensitive TL detector it also proved to be sensitive to light such that exposure to sunlight or room light caused light-induced fading of the main TL signal. This led to a study by the group at Oklahoma State University of the effects of light stimulation, rather than thermal stimulation, on the luminescence properties of irradiated $Al_2O_3$:C (Markey, Colyott and McKeever, 1995). An added bonus was the fact that performing the optical stimulation at or near room temperature meant that the sample did not have to be heated (as in TL) and thus thermal quenching effects, which limited the TL sensitivity of the material, could be entirely avoided. In parallel, work on OSL from $Al_2O_3$:C was underway at Battelle-Pacific Northwest Laboratories in the USA and several US patents emerged at this time (Miller, 1996, 1998). It was perhaps inevitable that $Al_2O_3$:C should then emerge as the most popular OSL radiation dosimetry material and the only one to date to be successfully commercialized (see www.landauerinc.com).

While OSL dosimetry is unlikely to displace TL dosimetry as the primary luminescence dosimetry method, and while discussions of the pros and cons of each method can be found in the literature (McKeever and Moscovitch, 2003), one major difference between TL and OSL is the fact that a number of different optical stimulation schemes can be adopted for the latter method. The original work of Huntley, Godfrey-Smith and Thewalt (1985) used continuous stimulation of the sample with a constant light intensity – a stimulation scheme now known as continous-wave OSL (CW-OSL). Akselrod, McKeever and colleagues, however, adopted a new scheme using pulsed stimulation, giving rise to pulsed OSL, or POSL (Akselrod and McKeever, 1999; McKeever *et al.*, 1996a). The critical feature of POSL is the need to stimulate the sample with optical pulse widths that are shorter than the luminescence lifetime of the emission center being activated in the luminescence process. The lifetime of luminescence emission from $Al_2O_3$:C is (a long) 35 ms and thus this material lends itself easily to this stimulation scheme. Since the luminescence decays over a period significantly longer than the pulse width (typically 100 ns duration) the luminescence can be monitored between stimulation pulses, not during them. This in turn enables easier separation of the stimulation light from the emission light. Other stimulation schemes include a linear ramp of the stimulation intensity, giving rise to LM-OSL, or linear-modulation OSL (Bulur, 1996). Additional stimulation schemes can easily be envisioned (exponential, sinusoidal, etc.; Bos and Wallinga, 2009).

The twin development of OSL as a tool for radiation dosimetry, primarily based on $Al_2O_3$:C, and OSL dating, primarily (at least initially) based on crystalline natural quartz, has led to enormous growth in the number of OSL publications emerging over the past decade. A simple search on Google Scholar for articles containing the words "optically stimulated luminescence" between 2000 and 2009 reveals 22 270 articles. A search without date restriction shows 25 200 articles. Perhaps of even more interest is the growth in the number of such articles over the past two decades, as shown in Figure 1.2. Finally we may note that a word search on Google finds about 2 850 000 entries containing the phrase "optically stimulated luminescence" (as of the date of this writing). The world of

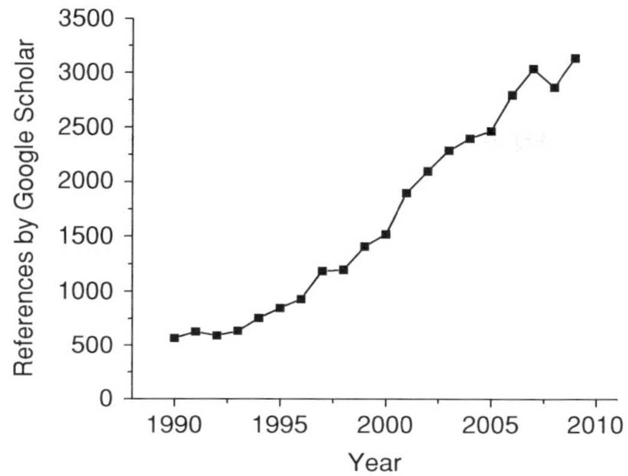


***Figure 1.2*** *The number of references to optically stimulated luminescence found in a web search on Google Scholar, by year.*

stimulated luminescence has come a long way since Boyle's notation of the "self-shining" and "glimmering light" from natural diamond.

## 1.2 Brief Description of Successful Applications

As noted above, one of the key developments in the acceptance of OSL as a tool in radiation dosimetry was the first use of the technique in geological dating of sediments (Huntley, Godfrey-Smith and Thewalt, 1985). Nevertheless, this book does not contain a discussion of applications of OSL in either geological or archeological dating. Our choice to omit this important application of OSL was conditioned by the fact that dating applications cover an enormously large body of literature and represent a topic worthy of a book by itself – quite apart from the fact that neither of the authors considers himself an expert in this field. Indeed, in addition to the wonderfully informative book by Martin Aitken (1998), a recent book by Bøtter-Jensen, McKeever and Wintle (2003) contains a comprehensive description of the modern details of the technique, especially as applied to dating natural quartz and feldspar minerals. Instead, we restrict our attention to other, and newer, applications in the fields of personal, space and medical dosimetry, with additional discourse on the emergence of OSL as a potential technique in security and related fields. For the time being, we give here brief overviews of a few areas on which we will be focusing in the book.

### 1.2.1 Personal

Since the initial dosimetry application described by Antonov-Romanovsky *et al.* (1955) and the development of $Al_2O_3$:C as an OSL material, the application of OSL in personal dosimetry has flourished. Primarily this has been because of the adoption by one of the world's largest radiation dosimetry service providers of the POSL technique, along with the development of a powder-in-plastic form of $Al_2O_3$:C. Landauer Inc., USA (www.landauerinc.com)

designed a personal dosimetry badge known as Luxel[1] as the main dosimeter type in their service-provider network. The measurement technique used by Landauer to read the badges is POSL. Other OSL-based dosimeters now commercialized by the same company include InLight,[1] designed for those users who wish to perform their own dosimetry measurements, and microStar,[1] a field-based version of the InLight system. More recently, a new OSL-based computed tomography quality assurance dosimeter has also become available.

$Al_2O_3$:C currently dominates the commercial OSL-based personal dosimetry market. As we discuss in later pages of this book, however, research on several promising new materials is underway in various laboratories around the world and new applications are beginning to emerge, perhaps leading ultimately to new commercial opportunities.

### 1.2.2 Space

One of the more exciting recent applications of OSL dosimetry is literally "beyond this world." Since the beginning of the human space programs of the USA and the former Soviet Union, thermoluminescence dosimeters have featured strongly in the determination of the dose absorbed by individual astronauts.[2] The complex radiation environment in which the crews operate requires individual dose monitoring for each member of the flight crew. This same complex field, however, also presents the main challenge for dosimetrists. The space radiation field to which the astronauts are exposed originates from three primary sources: galactic cosmic rays (GCR), solar particle events (SPEs) and charged particles trapped in the Earth's magnetic field (Earth's Radiation Belts, ERB). To these sources we can add those secondary particles created via nuclear interactions of the energetic primaries with the surroundings of the astronaut (spacecraft, clothing) and the astronaut's own body. The result is a complex "soup" of energetic particles ranging from electrons and protons to heavy particles (from He to U). Energies vary from tens of MeV/nucleon to relativistic values. The evaluation of total absorbed dose (in Grays, Gy) to each astronaut is a significant challenge, yet it is only half the problem. The ultimate purpose of the dosimetry is to estimate the health burden to each individual caused by the radiation exposure and thus one has to weight the absorbed dose by the relative biological effectiveness (RBE) or quality factor ($Q$) to estimate the weighted dose in Gray-equivalent (Gy-Eq) or Sieverts (Sv), respectively. This, in turn, requires an estimate of the dose carried by each particle type, where the particle types are defined by their energy, charge and mass. For radiation from SPEs, deterministic effects are the primary concern, while for GCR stochastic, cancer-inducing effects are the primary worry.

Because of the complex nature of this challenge the use of a single dosimeter type is not sufficient. This is especially so since thermoluminescence dosimeters (TLDs) are known to have a decreasing efficiency for particles with linear energy transfer (LET, $L$) values greater than approximately 10 keV/μm. Thus, the first requirement is that the luminescence dosimeters be used in conjunction with devices that can monitor doses from heavier particles, with $L$ values up to $\sim 10^3$ keV/μm. The only personal dosimeters so far capable of achieving this goal are plastic nuclear track detectors (PNTDs). However, this alone is not sufficient, since the two dosimeter types (TLDs and PNTDs) have overlapping

[1] "Luxel", "InLight" and "microStar" are all Trade Marks of Landauer Inc., USA.

[2] Although different nations use different terms to describe their space-flight crews, for simplicity the word 'astronaut' will be used throughout this book for all flight crewmembers, irrespective of their national origin.

responses, with TLDs having some sensitivity up to very high $L$ values, and PNTDs having some response down to approximately 5 keV/μm. Thus, detailed calibration of the dosimeters is necessary. Since precise calibration of luminescence dosimeters is non-trivial, the use of additional luminescence dosimeter types is beneficial so that uncertainties in the calibration of each may be offset. Thus, OSL dosimeters (OSLDs) are now being adopted for use in conjunction with TLDs in these applications.

Over and above this general consideration, however, future long-duration missions (to the Moon and Mars, up to perhaps 1000 days in space) may require on-board dosimetry, since the astronauts may not be able to wait until they have returned to Earth before their individual exposures are determined. This in turn will require low-weight, low-energy-consumption, on-board dosimeter readers, and here OSL is again being suggested as the potential dosimeter of choice for this application.

In this book we discuss the physics and challenges associated with the use of OSL dosimeters in these complex radiation fields and highlight future research directions and potential developments, along with several case studies where OSL has been used effectively in space radiation experiments.

### 1.2.3 Medical

The use of OSL in medical dosimetry is embryonic, but growing. The high sensitivity and the all-optical nature of the process are the two properties exploited most in medical dosimetry applications. The high sensitivity means that the dosimeters can be made small, which in turn gives them the property of high spatial resolution, meaning that they have the potential for measurement of dose in regions of severe dose gradients. The all-optical nature of the process means that they can be used with optical fibers to measure doses in difficult-to-access locations, including, potentially, inside the human body. Furthermore, the combination of these two properties allows the use of OSLDs to record dose in near-real-time during exposure, thereby lending an additional capability to the dosimetry system.

Modern advances in radiation medicine – in radiodiagnosis, radiotherapy and interventional radiography – each present dosimetry challenges for the medical physicist that did not exist previously. In radiotherapy the modern movement towards the use of charged particles (protons, and even carbon ions) presents new tests for the dosimetrists when compared with the application of high-energy photons. Even with photons, however, sophisticated intensity modulation techniques create new difficulties over and above the simple goal of "measuring dose." In all of these areas a constant balance has to be made between the treatment necessary to destroy the tumor and the unnecessary exposure of healthy tissue. Innovative applications of OSL dosimetry are now appearing in each of these areas to help the medical physicist and oncologist design the most effective, and least deleterious, treatment for their patients.

In radiodiagnosis OSL has been used with great success in imaging systems – where it has long been known by its alternate name of photostimulated luminescence (PSL). The sensitivity and speed of readout of the stimulated luminescence signal has given radiologists the ability to reduce radiation doses to patients and yet provide high-resolution images to aid diagnosis. However, the use of OSL in this way is not dosimetry. The actual dose to the patient is still determined by conventional OSL (or TL) methods.

In interventional radiography the real-time dose to patients (and surgeons) during surgical procedures can be significant and the ability to provide real-time dose readings while at the same time not producing artifacts in the X-ray image is an attractive goal. Some recent developments in OSL dosimetry may yet lead to the use of fiber-optic-based OSL dosimetry in these applications with significant advantages for both the patient and the doctor.

The advances in, and limitations of, OSL dosimetry in each of these areas will be discussed at length in the book, along with directions for future work.

### 1.2.4 Security

Regrettably, societies around the world find themselves having to consider more and better ways in which they can protect themselves from terrorist attacks, both from inside and outside their communities. The fear of weapons of mass destruction or other high-consequence weapons falling into the hands of terrorists and being used inside our communities is one that is uppermost in the minds of governing officials and security and defense organizations and many state agencies exist around the world for whom their *raison d'être* is the control of the spread of weapons of mass destruction among nations and terrorist groups. Among such weapons are those involving nuclear or radiological devices. The detection of such devices before they can be used is of paramount importance and, as throughout history, defense and security agencies around the world are turning to science and technology to enhance their detection capabilities. The world of science is also being asked to assist in mitigating the possible effects of such weapons, should they be used, so that societies and communities can recover quickly. Science is also being asked to contribute to the realm of investigative forensics in order to assist in the identification of the perpetrators.

In this regard, OSL is being suggested in a number of potential applications – both in the detection of nuclear or radiological devices, in forensics and, more particularly, in the after-the-event triage that must occur if authorities are to cope with the potentially large numbers of injured people in large population areas. Publications on the use of OSL from human teeth, from watch components or jewelry, and from building structures have described how the technique may be able to assist in determining the ranges of radiation doses to which populations may have been exposed during the attack. Other publications have suggested the use of OSL from building and other common materials from our surroundings to identify those rooms in which illicit radioactive materials may have been stored or temporarily held, as part of the normal investigative response to a crime involving such materials – hopefully before their use as a weapon.

Thankfully, such uses have never been tested in practice and, therefore, the effectiveness of OSL in these applications remains to be proved; consequently, discussions of these applications must necessarily be somewhat speculative. It is in this sense that we hope that readers of this text will be interested in the descriptions of these potential applications that we have included in this book. We hope further that they will be inspired to work in these important areas, for the benefit of all of us.

## 1.3 The Future

What of the future? The edging of OSL into so many different and challenging areas presents future scientists with a plethora of opportunities for innovation and invention. We challenge readers to not just satisfy themselves with "dotting 'i's and crossing 't's in tracking down the answer to some minor question, but rather to create novel and interesting ways in which the OSL technique can be applied to important and difficult areas and make incisive in-roads into new fields. The areas we have chosen to highlight in this book (personal dosimetry and the search for new materials, space and medical dosimetry, and security) represent just a few of the areas in which the challenges are difficult and in which new innovations could yield immense rewards. What of other areas? What about optical data storage, information technology, environmental sciences, aerospace, genomics and proteomics, plant sciences and so on? How can OSL be integrated into the world of nanotechnology and biotechnology? A glimpse of what may be possible can perhaps be obtained by perusing the US patent list for all inventions concerning OSL or related phenomena. A search of the US Patent and Trade Mark Office web site using the key words optically stimulated luminescence, photostimulated luminescence, or related terms reveals 195 entries in the almost-25-year period from November 1984 and July 2008. The September 1959 US patent referenced above (Wallack, 1959) is the earliest OSL patent of which the authors are aware and undoubtedly the previous 25-year period, from September 1959 to November 1984, will feature additional invention claims. Of course, a significant number of non-US inventions will also be registered with other international patent offices. Most of the US patented OSL innovations relate to OSL/PSL materials, methods for radiation dose estimation in a variety of applications, or methods for radiation imaging. Thus, although the answers to the above rhetorical questions, and others of similar nature, are unknown at present it is certain that the future holds a rich store of possibilities. Indeed, one might suggest that the possibilities are limitless and that the barriers are only in our imagination.

# 2

# Theory and Practical Aspects

## 2.1 Introduction

Optically stimulated luminescence (OSL) is the transient luminescence observed during illumination of crystalline insulators or semiconductors that were previously excited, typically by exposure to ionizing radiation. The excitation puts the crystal in a metastable state, characterized by electrons and holes (virtual positive charges consisting of empty states in an otherwise full energy band) separately trapped at defects in the crystal lattice. During the OSL process, light stimulates the release of these electrons and holes from these trapping centers, resulting in electron/hole recombination and excitation of luminescence centers in the crystal. OSL consists of the photons emitted when these excited luminescence centers decay to the ground state. (Although OSL processes can, in principle, be observed from any insulating or semiconducting materials, most practical OSL materials are crystals. Therefore we adopt the word *crystal* for the OSL material, while at the same time not implying that only crystals give rise to the phenomenon.)

Figure 2.1 illustrates the process described above, which lays out the basis for a practical use of OSL in ionizing radiation detection and measurement. In the first stage (Figure 2.1a), the OSL detector is exposed to ionizing radiation. The energy deposited by ionizing radiation results in excitations and ionizations: electrons are promoted to the conduction band, where they can move freely throughout the crystal, leaving behind a hole, which can also move freely in the valence band. The electron/hole creation process is represented in Figure 2.1a by the upward arrow with the radiation symbol connecting the valence and conduction bands. There is a probability that these free electrons and holes may become trapped at defects in the crystal lattice, the energy levels for which are represented by the short horizontal lines in the band gap, in between the valence and conduction bands.

After irradiation there is a latency period characterized by a metastable concentration of trapped electrons and holes (Figure 2.1b). If the potential wells associated with the trapping centers are sufficiently deep, then the thermally induced escape probability of the trapped charges is negligible at room temperature. This relatively stable concentration of trapped

*Optically Stimulated Luminescence: Fundamentals and Applications* Eduardo G. Yukihara and Stephen W. S. McKeever

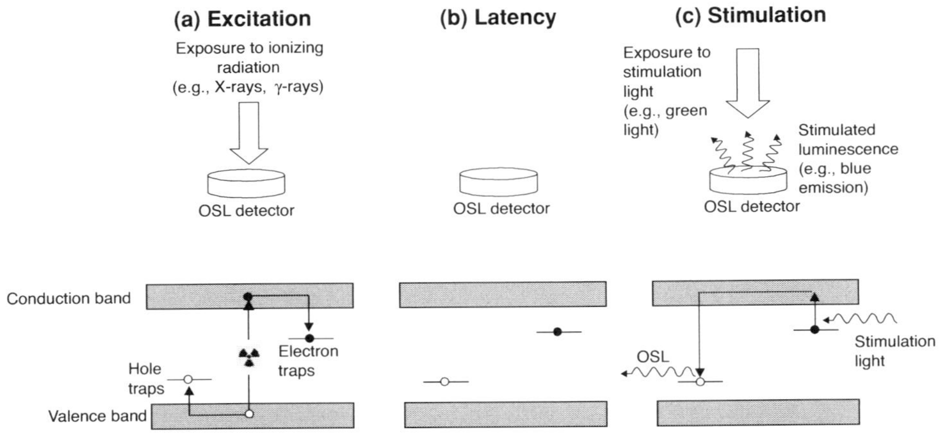


*Figure 2.1* *Different stages involved in the OSL process: (a) excitation of the OSL detector by ionizing radiation creating free electrons (•) and holes (○); (b) latency period characterized by a metastable concentration of electrons and holes captured at defects in the crystal and (c) stimulation of the detector with light, leading to recombination of electron–hole pairs and emission of light (OSL). The upper half represents the interaction of the detector with the ionizing radiation field and stimulation light; the bottom half represents the energy band diagram for the crystal with the available energy levels and corresponding electronic transitions occurring during each stage.*

electrons and holes is related to the energy absorbed by the crystal during the excitation process, that is, to the absorbed dose of radiation; it represents latent information about the radiation field.

The information stored in the OSL detector can be read by light stimulation (Figure 2.1c). In this example, a photon of wavelength $\lambda_{stim}$ (e.g., green light) stimulates the electron to the conduction band. Once in the conduction band, the electron is free to move through the crystal and may reach the trapped hole. The electron/hole recombination process creates a defect in the excited state, which relaxes to the ground state by the emission of a photon of wavelength $\lambda_{OSL}$ (e.g., blue light).

Figure 2.1 outlines how an OSL material can be used as a dosimeter. Figure 2.1a corresponds to the period in which the dosimeter is being worn by the user or exposed to the unknown radiation field that we are interested in characterizing. Figure 2.1b corresponds to the period in which we transport the dosimeter back to the laboratory and store it. Figure 2.1c corresponds to the readout of the information stored in the material, which provides information about the radiation field.

In this simplest model of the OSL process, it can be shown that the OSL intensity decays exponentially during stimulation. For example, let $n$ be the concentration of trapped electrons and $p$ the transition probability per unit time for the trapped electron to escape to the conduction band under optical stimulation. Assuming no possibility of retrapping, $n$ will change according to:

$$\frac{dn}{dt} = -np \tag{2.1}$$

Solution of this equation shows that the concentration of trapped electrons decays exponentially with the stimulation time:

$$n(t) = n_0 \, \mathrm{e}^{-pt}, \tag{2.2}$$

where $n_0 = n(0)$ is the initial concentration of trapped charges. Assuming that all stimulated electrons recombine immediately, the intensity of light emitted is proportional to the rate of electrons escaping the trapping centers $\mathrm{d}n/\mathrm{d}t$, which by Equation (2.1) and Equation (2.2) is:

$$I_{\mathrm{OSL}}(t) \propto \left| \frac{\mathrm{d}n}{\mathrm{d}t} \right| = n_0 p \, \mathrm{e}^{-pt}. \tag{2.3}$$

One should notice that OSL emission requires an initial excitation of the detector (e.g., exposure to ionizing radiation) and involves the transport of charge between different defects in the crystal. In contrast, light emission due to optical absorption and relaxation transitions within the same defect (photoluminescence) is not considered to be OSL for the following reasons: it does not necessarily require exposure to ionizing radiation to be observed, it does not involve charge transport, and it normally involves a single defect.

In the OSL process it is perfectly possible for the emitted photon to have energy higher than the photon involved in the stimulation (i.e., $\lambda_{\mathrm{OSL}} < \lambda_{\mathrm{stim}}$) without violation of energy conservation. The energy of the photon emitted is determined by the nature of the transition between the excited and ground states of the luminescence center, and does not depend on the wavelength of stimulation. Part of the energy emitted as OSL originates from the energy absorbed by the crystal during exposure to ionizing radiation, stored in the form of electrons and holes separately trapped within the crystal.

In fact, one of the advantages of the OSL technique, particularly for low-dose measurements, is the possibility to eliminate confounding signals by monitoring the OSL at wavelengths shorter than the wavelength of the stimulation light. This is often the case in practical OSL applications. At moderate stimulation intensities the probability of non-radiation-induced processes (e.g., multi-photon excitations) leading to photon emissions with $\lambda_{\mathrm{em}} < \lambda_{\mathrm{stim}}$ is very low, and therefore the background signal is minimal. On the other hand, photoluminescence processes producing light with emission wavelengths longer than the excitation wavelengths ($\lambda_{\mathrm{em}} > \lambda_{\mathrm{exc}}$) are quite ubiquitous and may lead to confounding signals if the OSL is detected at wavelengths longer than the stimulation signal.

The above description of the OSL process, although simplistic, captures some of the essential elements of the OSL phenomenon: (i) it shows that there is ultimately a relationship between the intensity of the OSL signal and the absorbed dose of ionizing radiation, the information from the interaction of the detector with the radiation field (i.e., the absorbed dose) being stored as a result of electronic transitions and charge trapping in an imperfect crystal; (ii) it shows how the stored information represented by the trapped charges can be read by stimulating electronic processes in the crystal using light; (iii) it defines OSL as a transient luminescence phenomenon, which decreases over time under light exposure and cannot persist indefinitely because the population of trapped charges is depleted during stimulation.

The OSL process is also related to other luminescence processes in the crystal, which are important not only for a more complete understanding of the OSL process, but also because they lead to complementary techniques to investigate properties of OSL materials.

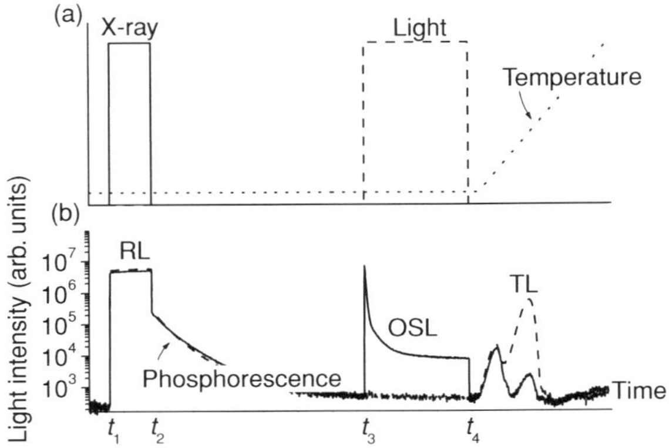


***Figure 2.2*** *(a) Time profile of the X-ray intensity, optical stimulation intensity, and temperature of the material; and (b) logarithmic intensity of light detected by a photomultiplier tube (PMT) of an OSL material that was excited with X-rays and subsequently heated (dashed line) or illuminated and heated (full line). The data show that the optical stimulation can reduce the intensity of the second TL peak in this material when compared to a sample that was not illuminated; this has to do with the optical stimulation of trapped charges that give rise to the second TL peak. Data from multiple experiments were combined for illustrative purposes.*

Of particular interest are the processes of radioluminescence (RL), phosphorescence, and thermoluminescence (TL).

To illustrate these different luminescence processes, Figure 2.2b shows the light emitted by an OSL material during the processes of X-ray irradiation, illumination, or heating represented in Figure 2.2a. The material is exposed to X-rays in the period $t_1 < t < t_2$, optically stimulated in the period $t_3 < t < t_4$, and heated (with the temperature raised linearly) for $t > t_4$. The full line in Figure 2.2b shows the light emitted by a material that was subjected to irradiation, illumination and heating. During X-ray irradiation, RL is observed due to prompt recombination of electron and holes during irradiation. Immediately after irradiation, a transient luminescence related to the thermally stimulated release of charges trapped in shallow trapping centers (e.g., electron traps close to the conduction band or hole traps close to the valence band) at room temperature (phosphorescence) is observed. When the material is illuminated, the light emitted increases rapidly due to *optically stimulated* release of trapped charges resulting in OSL, the signal decaying as the trapped charge population decreases. Then, during the linear increase in temperature, TL is observed due to *thermally stimulated* release of the remaining trapped charges. Two TL peaks are observed, indicating the presence of at least two types of trapping centers. For comparison, Figure 2.2b also shows the light emitted by a material that was irradiated, but was not exposed to light (dashed line). In this case, the trapped charge population is not depleted during the OSL process, resulting in TL peaks of higher intensity than in the case of the sample that was illuminated.

This chapter will now explore in more detail the solid state theory underlying the OSL phenomenon and other theoretical aspects. After providing a theoretical background, we will discuss more applied aspects such as OSL readout and instrumentation, followed by a discussion on complementary techniques used to characterize OSL materials, and an overview of OSL material properties.

The objective here is to provide background information as the basis for discussions contained in subsequent chapters. Some of the background information may seem overwhelming or not relevant for a first-time reader; for those, we recommend revisiting this chapter after reading about the OSL applications. In some topics (e.g. spectroscopic notation), we had to restrict the discussion to a brief outline. We tried to provide key references for readers interested in more detailed explanations.

## 2.2 Basic Aspects of the OSL Phenomenon

The OSL phenomenon and its connection to other luminescence processes (e.g., TL, phosphorescence, RL) and conductivity processes (e.g., thermally stimulated conductivity, photoconductivity) can be explained by considering the available energy levels and transitions of electrons in insulator and semiconductor materials. Although oversimplified, such a phenomenological model goes a long way towards explaining many of the observed phenomena associated with OSL. In this section, we provide a general overview of the relevant aspects.We restrict the discussion mostly to crystalline materials with point defects. However, as noted in Section 2.1, the reader should keep in mind that the OSL phenomenon can also be observed in materials that do not strictly satisfy these conditions, including glasses, plastics, biological materials and so on, or in principle any system in which a metastable state can be created during exposure to radiation and later optically stimulated back to the ground state producing luminescence.

### 2.2.1 Energy Levels in Perfect Crystals

Electrons in a perfect crystal are subjected to the periodic potential associated with the electrical field of the regular arrangement of ions that makes up the crystal. The electronic energy levels can, in principle, be calculated by solving the Schrödinger equation, which for a periodic potential gives rise to bands of allowed energies separated by bands of forbidden energies (energy gaps or band gaps) (Kittel, 1996). To illustrate this result, Figure 2.3 compares the energy levels for a one-dimensional series of potential wells (Figure 2.3d) with the continuous energy levels of the free electron (Figure 2.3a), the discrete energy levels of a simple harmonic oscillator (Figure 2.3b) and the energy levels of a single potential well (Figure 2.3c). The energy bands in a crystal can be considered as being formed by the discrete energy levels of isolated atoms, which when brought together in a crystal split into $(2\ell + 1)N$ closely space levels, where $N$ is the number of atoms in the crystal and $(2\ell + 1)$ is the orbital degeneracy of each level, forming a quasi-continuous distribution of energy levels (Henderson and Imbusch, 1989). These bands are shared by the entire crystal and therefore are also called delocalized energy bands. The energy gaps correspond to Bragg reflection of electron waves in a crystal, for which solutions of the Schödinger equation do not exist (Kittel, 1996).

According to Pauli's exclusion principle, each energy state can be occupied by a single electron only. In the ground state of the crystal, electrons fill the available energy states starting with the lowest energy levels and continuing with levels of increasing energy.

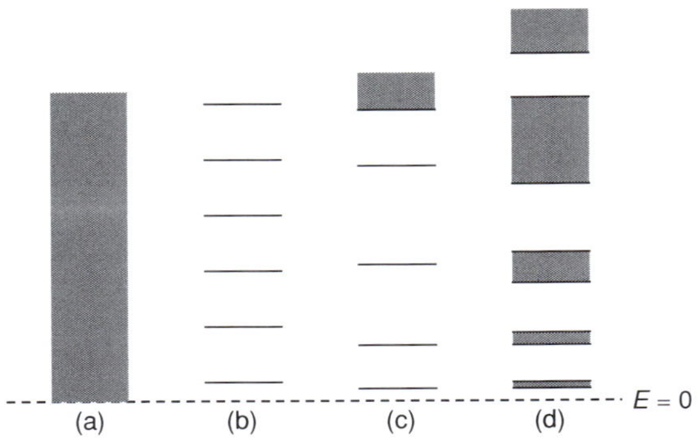


***Figure 2.3*** *Energy levels of: (a) a free electron; (b) a simple harmonic oscillator; (c) a single potential well and (d) a periodic series of potential wells. The latter is similar to the energy levels in a crystal, consisting of energy bands separated by band gaps. Reprinted with permission from Bube, R. H., 1960. Photoconductivity of solids. John Wiley and Sons, Inc., New York.*

When the number of electrons is such as to fill completely all energy bands below a certain energy level, such that the energy bands are completely filled or completely empty, application of an electric field cannot change the total momentum of the electrons to generate an electric current. As a result, the material behaves as an insulator (Kittel, 1996). The highest filled energy band is called the valence band, and the first empty band above that is called the conduction band. In an insulator, the energy gap $E_g$ between the valence band and the conduction band is greater than approximately 2.5 eV, and is therefore much larger than the available thermal energy at room temperature ($E_g \gg kT \approx 0.025$ eV, where $k$ is the Boltzmann constant and $T$ is the absolute temperature). This explains the low electrical conductivity of an insulator. However, when electrons are excited to the conduction band with energy larger than $E_g$, conductivity is observed due to electrons moving in the conduction band and holes moving in the valence band. Once the excitation source is removed, electrons and holes recombine and the crystal returns to its initial state.

### 2.2.2 Defects in the Crystal

Although electron–hole pairs can be created by excitation of a perfect crystal, such systems provide no mechanism for energy to become stored in the form of electrons and holes spatially separated in the crystal, which could later be released to produce OSL. As with many other technological applications of solid-state materials, the OSL process exists because of energy levels introduced in the band gap by defects in the crystal. These defects can trap electrons and holes and effectively store energy.

Defects in a crystal are violations of the periodicity of the crystal lattice (Kittel, 1996; Stoneham, 1975). Examples of defects are missing atoms at lattice points (vacancies), extra atoms in between lattice points (interstitials), foreign atoms replacing lattice atoms (impurities), as well as other extended defects such as dislocations, larger aggregates of simple (point) defects and termination of the crystal (surface). Defects formed by displacements of atoms that form the crystal are called intrinsic defects, while defects formed by foreign atoms are called extrinsic or impurity defects.

Defects can be introduced as a result of thermodynamic equilibrium, doping during the crystal growth process, and post-growth annealing or irradiation with energetic particles or photons. For example, for a crystal in thermodynamic equilibrium at a given temperature there is a probability that a regular atom in the crystal structure will be displaced to an interstitial position, resulting in an intrinsic defect consisting of a vacancy plus an interstitial, called a Frenkel defect. The concentration of this type of defect increases with temperature with a factor $e^{-W/2kT}$, where $W$ is the energy required to take an atom from a lattice position and place it in an interstitial one (Mott and Gurney, 1940). A crystal that is rapidly cooled from close to its melting temperature to room temperature may have a larger concentration of intrinsic defects than would be expected from the thermodynamic equilibrium at room temperature because fast cooling prevents the defects from recombining, therefore freezing them around their high-temperature positions.

A defect that has been extensively studied in alkali halide crystals (e.g., LiF, NaCl, KCl, etc.) and which will be of interest in some OSL materials is the $F$-center. The $F$-center receives its name from the word Farbzentren (color center in German), because this defect is responsible for the coloration of high purity alkali halide crystals heated in the vapor of its cation metal (Pohl, 1939). The $F$-center is an intrinsic defect consisting of an electron captured at an anion vacancy. In a neutral crystal, the number of cation and anion vacancies is the same. However, when the crystal is heated in the vapor of its cation metal, an excess of anion vacancies is formed and the crystal becomes neutral by capture of an electron at these sites (Mott and Gurney, 1940). $F$-centers may also form in other classes of materials. In oxides, $F$-centers consist of oxygen vacancies with two captured electrons (Henderson and Imbusch, 1989).

Optical absorption measurements can be useful in determining the concentration of $F$-centers and other defects in crystals. For example, the concentration of $F$-centers can be obtained from the intensity of the optical absorption bands by Smakula's formula (Henderson and Imbusch, 1989; Marfunin, 1979; Stoneham, 1975). $F$-centers in alkali halides are paramagnetic (consisting of an unpaired electron) and can also be observed by electron paramagnetic resonance (EPR) spectroscopy. EPR measurements confirmed the model of the $F$-center as an electron shared by the six nearest-neighbor cations (Marfunin, 1979), showing the importance of EPR in elucidating the nature of defects in crystals. In oxides, however, the $F$-center has two paired electrons and therefore is not paramagnetic; thus it cannot be observed by EPR.

### 2.2.3 Excitation of the Crystal by Ionizing Radiation

Excitation of the crystal is not only an essential part of the OSL process, responsible for the creation of electron and hole pairs that can become trapped in defects of the crystal, but usually the effect we want to quantify in dosimetry applications. In this sense, we are often interested in the excitation produced by ionizing radiation (e.g., X-rays, $\gamma$-rays, energetic charged particles). Excitation of the crystal can be produced by any radiation with sufficient energy to promote an electron from the valence band to the conduction band (e.g., ultraviolet if the band gap is sufficiently small). Illumination with photons of energy less than the band gap is not able to promote band-to-band transitions, but may still cause photoionization of defects in the crystal giving rise to free charges in the delocalized bands. We restrict our discussion to ionizing radiation.

Ionizing radiation deposits energy in matter primarily by transferring energy to secondary energetic electrons called $\delta$-rays, which in turn continue to distribute the initial energy through subsequent excitation and ionizations until all energy transferred by the primary radiation is dissipated. Charged particles are considered directly ionizing radiation because they mainly transfer energy directly through coulomb interactions with electrons in the medium. Photons and neutrons are considered indirectly ionizing radiation because they first create one or more energetic charged particles (e.g., one energetic electron in the case of a photon interacting by photoelectric effect, or a recoil proton in the case of neutron scattering with a hydrogen nucleus) in a single event, and then these particles produce $\delta$-rays through coulomb interactions.

Although all charged particles transfer energy to matter mostly through excitations and ionizations produced by coulomb interactions, the pattern of energy deposition is very different, depending on the type and energy of the particle. Electrons are easily scattered from their original trajectory. As a result, ionizations occur in a relatively large volume in the material resulting in low ionization densities. Heavy charged particles, on the other hand, follow a straight trajectory except for rare collisions with other heavy ions in the material. In their trajectory they transfer energy to electrons through excitation and ionizations, creating regions of high ionization density around the trajectory path forming what is called a track. Figure 2.4 shows an example of the patterns of energy deposition by a He ion of energy of 200 MeV/u in water.

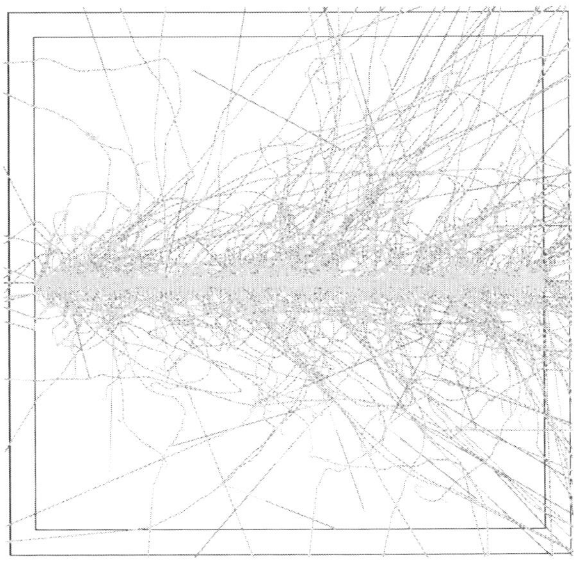

***Figure 2.4*** *Illustration of the pattern of ionizations and secondary electrons created when a He ion with energy of 200 MeV/u crosses a water volume of 100 $\mu$m × 100 $\mu$m × 100 $\mu$m horizontally from left to right, calculated by the Monte Carlo method (courtesy of Dr. Gabriel O. Sawakuchi, Carleton University, Canada). The figure shows that the ionization density is very high close to the particle track, decreasing as the distance from the track increases.*

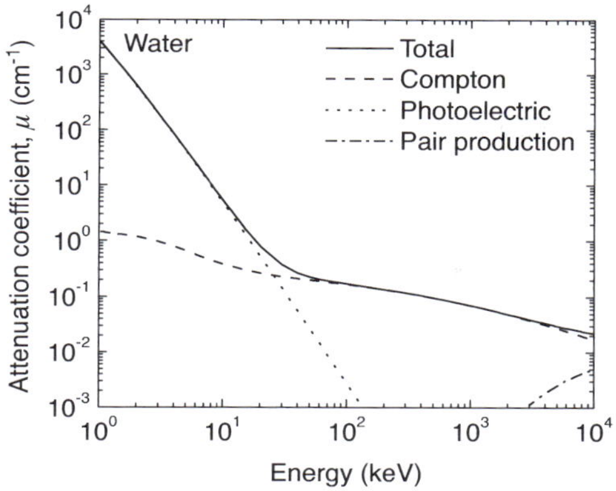


***Figure 2.5*** *Total photon linear attenuation coefficient, defined as the fractional change in the number of photons per unit thickness of material transversed (ICRU, 1998d), and attenuation coefficient due to various interaction processes in water calculated using the database from the US National Institute of Standards and Technology (Hubbell and Seltzer, 2004). This graph shows that photoelectric effect dominates in the low photon energy region, whereas the Compton effect dominates in the intermediate photon energy range. The probability of pair production increases for photon energies larger than 1.022 MeV.*

High-energy photons (X-rays and gamma rays) create charged particles mainly by three processes: the photoelectric effect, Compton scattering and electron–positron pair production. In the photoelectric effect the photon is absorbed by an electron from the inner shells of the atom. The electron is ejected with a kinetic energy equal to the difference between the photon energy and the binding energy of the electron. In Compton scattering the photon of energy $h\nu$ (where $h$ is Planck's constant and $\nu$ is the frequency of the light) interacts with a weakly bound electron of an atom, resulting in a secondary photon with energy $h\nu' < h\nu$ and an electron with kinetic energy equal to the energy difference between the incident and scattered photon, minus the negligible binding energy. Electron–positron pair production can occur in the electric field of a nucleus when photons with energy higher than the rest mass of the particles produced ($h\nu \geq 1.022$ MeV) interact, part of the excess energy being carried as kinetic energy by the particles produced. Each of these processes is dominant in a different energy range, as shown in Figure 2.5, and have different dependences with the atomic number of the constituent elements of the material.

Neutrons create charged particles by elastic scattering (e.g., elastic scattering with a hydrogen atom, creating a recoil proton), inelastic scattering and nuclear interactions (e.g., neutron capture reactions in which the neutron is captured by a nucleus such as that of $^6$Li, which splits into secondary particles such as $\alpha$-particles and tritium ions with energies of the order of 1–3 MeV). These processes are the primary mechanisms through which neutrons are detected.

Regardless of the initial energy transfer mechanism, the original energy from ionizing radiation is dispersed in the material by the $\delta$-rays, primarily through scattering with other electrons and resultant excitations and ionizations, until the energy transferred in each interaction is insufficient to cause further ionizations. As a result of these numerous ionizations, a large number of electrons and holes are created in the crystal. These electrons and holes further lose energy by interaction with crystal vibrations in electron–phonon relaxation or thermalization. In this process, electrons migrate to the bottom of the conduction band and holes to the top of the valence band. This entire process occurs in a time scale of the order of $10^{-11}$ to $10^{-12}$ s (Rodnyi, 1997).

At this point the electron–hole pairs can recombine by different processes. In a perfect crystal this may occur by localization of the electron around the field of the hole followed by subsequent relaxation. The electron localized around a hole forms an exciton, which can recombine, resulting in luminescence with energy close to the energy of the band gap. However, this mechanism is not predominant in large band gap materials or materials with a large number of defects in the crystal lattice. Here, recombination usually occurs when one of the defects first captures a hole and then an electron, or vice versa. If the recombination process is radiative, the luminescence produced by this prompt recombination of electron–hole pairs created by ionizing radiation is the radioluminescence indicated in Figure 2.2. Radioluminescence persists only while the material is exposed to radiation.

### 2.2.4 Trapping and Recombination at Defect Levels

In a perfect crystal, the electrons in the conduction band can be considered as behaving as free charges with an appropriate effective mass, free to move throughout the entire crystal. Defects introduce a local electric field which may capture a free electron from the conduction band. Examples of such fields include the one around an interstitial positive ion, a missing negative ion or a hole (Mott and Gurney, 1940). Therefore, these defects introduce a series of bound states for the charge around the defect, which can be represented as localized energy levels in the upper part of the band gap just below the conduction band, as depicted in Figure 2.6. The discussion is analogous for holes, with the exception that in

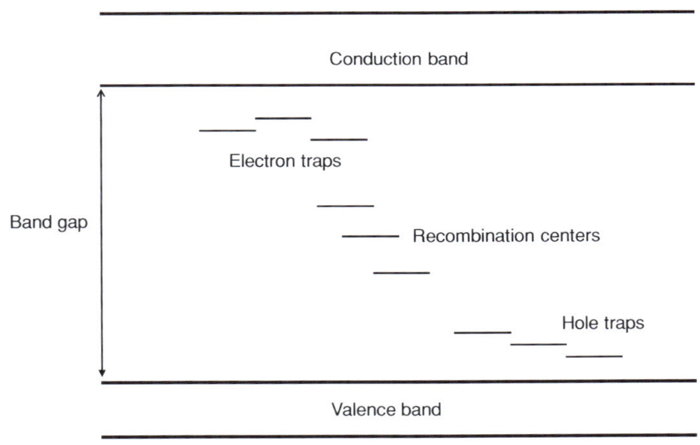


*Figure 2.6 Representation of energy bands in an insulating crystal and localized energy levels introduced by defects. The defect levels close to the conduction band act as electron traps, whereas the defect levels close to the valence band act as hole traps. Defects in the intermediate region of the band gap have a higher probability of acting as a recombinantion center by first localizing and electron and subsequently a hole, or vice versa.*

this case the trapping states appear in the bottom part of the band gap above the valence band.

The localized energy levels introduced by defects, or defect levels, can therefore be responsible for the trapping of electrons and holes that would otherwise recombine. In the absence of thermal or optical stimuli, this concentration of trapped electron and holes can be stable for long periods of time, even geological times, effectively representing a metastable state of the crystal.

Defects may also be responsible for recombination by first capturing a hole from the valence band and subsequently capturing an electron from the conduction band, or vice versa. The defect in which this process takes place is called a recombination center. A trapping center can be distinguished from a recombination center based on the probability of the trapped charge (e.g., electron) being stimulated to the delocalized bands compared to the probability of a charge of opposite sign (e.g., hole) being captured at the same site of the trapped charge resulting in recombination (Bube, 1960). If, under certain conditions of temperature and optical stimulation, an electron trapped at a specific defect is more likely to escape to the conduction band before a hole is captured by the same defect causing recombination, the defect can be classified as a trapping center. If, on the contrary, it is more likely that a hole will be captured by the defect causing recombination before the electron can escape, the defect can be classified as a recombination center. Thus, the energy levels for recombination centers reside in the mid-gap region and a center that is a recombination site at low temperatures can become a trap at higher temperatures.

Considering the processes discussed above, Figure 2.7 illustrates the creation of electron /hole pairs by ionizing radiation, which is followed by trapping or recombination. In this illustration, recombination is radiative, giving rise to radioluminescence.

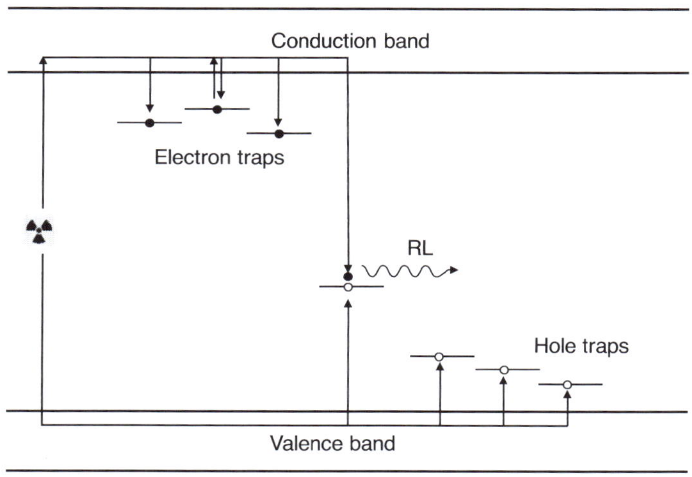


***Figure 2.7*** *Illustration of trapping and recombination of electron/hole pairs created during irradiation. In this example radioluminescence is generated due to the radiative recombination of the electron/hole pairs at recombindation centers. Note that depending on the irradiation temperature, charges trapped at shallow levels may escape during irradiation.*

### 2.2.5 Thermal Stimulation of Trapped Charges

Trapped charges can be stimulated back to the delocalized bands if sufficient energy is provided in the form of heat. Phosphorescence is the luminescence emitted due to recombination of thermally stimulated charge carriers, occurring, for example, after the crystals have been irradiated. Thermoluminescence (TL) is the luminescence emitted due to a similar process, but when the crystal temperature is elevated in a controlled way. Phosphorescence is caused by shallow traps (defect energy levels close to the conduction or valence bands), whereas TL peaks are due to trapping centers at different depths below the conduction band in the case of electron traps, or above the valence band in the case of hole traps. Some of the these trapping centers may be the source of charge carriers, giving rise to the OSL signal during optical stimulation, in which case the TL peaks and the OSL signal are correlated.

The probability of thermal stimulation of the electron to the conduction band (or hole to the valence band) is described by the Boltzmann factor $e^{-E/kT}$, since the defect can be considered in equilibrium with the thermal reservoir represented by the crystal. This probability is given by:

$$p = s\,e^{-\frac{E}{kT}}. \tag{2.4}$$

where $s$ is a proportionality constant called frequency factor, pre-exponential factor, or attempt-to-escape frequency, and $E$ is the activation energy for the process (related to the energy of the defect within the band gap), $k$ is the Boltzmann constant and $T$ is the temperature (Bube, 1960; McKeever, 1985).

In the case of phosphorescence, the probability $p$ is constant and the concentration of trapped charges decays exponentially, as in Equation (2.2). The phosphorescence intensity is, therefore, also described by an exponential decay as in Equation (2.3), but with $p = s\,e^{-\frac{E}{kT}}$.

In the case of TL the crystal is typically subjected to a heating of the type $T(t) = T_0 + \beta t$, where $T_0$ is the initial temperature and $\beta$ is a constant heating rate, and the probability of detrapping $p$ is not constant with time. Combining Equation (2.2) with Equation (2.4), we obtain:

$$\frac{dn}{dt} = -np = -ns\,e^{-\frac{E}{kT}}, \tag{2.5}$$

This equation can be written as:

$$\frac{1}{n}dn = -s\,e^{-\frac{E}{kT}}\frac{dt}{dT}dT, \tag{2.6}$$

where $dt/dT = \beta^{-1}$. This leads to the solution:

$$n(t) = n_0 \exp\left[-\frac{s}{\beta}\int_{T_0}^{T(t)} e^{-\frac{E}{k\theta}}\,d\theta\right], \tag{2.7}$$

where $\theta$ is a dummy variable (i.e., it disappears upon integration). Assuming again that the light intensity is simply proportional to the rate of electrons escaping the trapping centers,

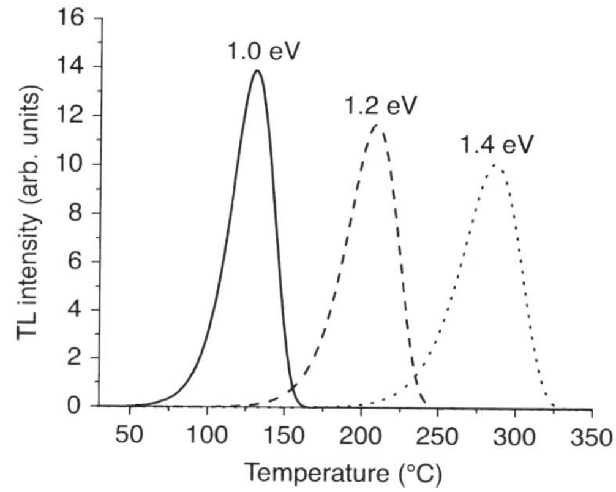


***Figure 2.8*** *TL curves (intensity versus temperature) corresponding to a first-order process characterized by Equation (2.8) with parameters $s = 10^{12}\ s^{-1}$, $\beta = 5\ °C/s$, and various values of the activation energy E (indicated in the figure), and an arbitrary value of $n_0$ (which in this model is just a scaling factor). In the first-order model the peak position is a function of only E and s. The presence of several TL peaks in a material's TL curve indicates the presence of several trapping centers, each with its characteristic thermal stability (i.e., values of E and s).*

the TL intensity can be written as:

$$I_{\mathrm{TL}}(t) \propto \left|\frac{\mathrm{d}n}{\mathrm{d}t}\right| = n_0 s\,\mathrm{e}^{-\frac{E}{kT(t)}} \exp\left[-\frac{s}{\beta}\int_{T_0}^{T(t)} \mathrm{e}^{-\frac{E}{k\theta}}\,\mathrm{d}\theta\right], \tag{2.8}$$

This is a peak-shaped function whose peak position depends on the activation energy $E$, frequency factor $s$ and heating rate $\beta$, and whose peak intensity and total area scale with $n_0$ (see Figure 2.8). Equation (2.8) is the so-called first-order or Randall–Wilkins equation describing the TL phenomenon (Randall and Wilkins, 1945). Other models and advanced theories of the TL phenomenon are discussed in McKeever (1985) and Chen and McKeever (1997).

### 2.2.6 Optical Stimulation of Trapped Charges

Optical transitions can occur between defect levels and the delocalized band, providing the basis for optical stimulation of the radiation-induced trapped charges and resultant OSL. This process is illustrated in Figure 2.9, where we have also included the possibility of the charge being recaptured by other trapping centers. Charges captured by shallow traps can escape to the conduction band by thermal stimulation.

An important aspect to discuss here is the interaction of the defects with the optical radiation field, particularly the effect of the stimulation wavelength and intensity on the transition probability $p$ in Equation (2.1).

The optical transition probability $p$ can be expressed in terms of a photoionization cross-section $\sigma$, defined as the photoionization probability per unit time per atom (or optical

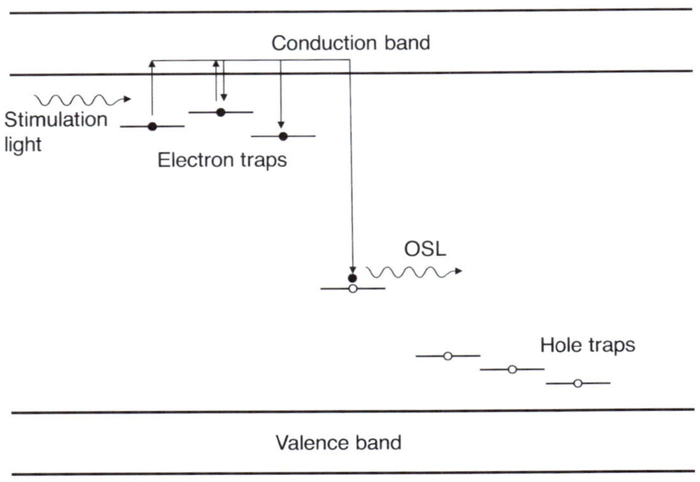


***Figure 2.9*** *Possible electronic transitions following the optical stimulation of electrons from an "optically active" trapping center. The electrons stimulated to the conduction band can undergo recombination with the trapped hole giving rise to OSL, or be recaptured at shallow traps and deep traps. Electrons captured by shallow traps will escape to the conduction band by thermal stimulation. Electrons trapped at deep traps are removed from the process, at least until sufficient energy is provided, allowing them to escape (e.g., by elevating the temperature).*

center) divided by the incident photon flux $\phi$, that is:

$$p = \sigma\phi, \tag{2.9}$$

Photoionization here refers to electron transition from a discrete bound defect state to the quasi-continuum density of states in the conduction band (electron emission) or to electron transition from the valence band to a localized level (hole emission) (Inkson, 1981; Pantelides, 1978). Equation (2.9) shows that the transition probability is proportional to the incident photon flux $\phi(\lambda)$ and, therefore, the higher the photon flux (or light intensity), the higher the rate of trapped charges stimulated. For some materials (e.g. quartz, $Al_2O_3$:C and BeO) the detrapping probability varies linearly with stimulation intensity (Figure 2.10a). For others (e.g. NaCl), a non-linear behavior is observed (Figure 2.10b).

The photoionization cross-section $\sigma$ has a threshold energy $E_I$ below which its value is zero (i.e., the photon energy is smaller than the energy required for the optical transition). This threshold may not be well-defined in experimental measurements due to the possibility of a two-step process in which the charge is optically stimulated to an excited state and then thermally stimulated to the condution band (Pantelides, 1978). Above this threshold energy, the photoionization cross-section is expected to increase due to the contribution of transitions to the quasi-continuum density of states of the delocalized band, and then decrease as the photon energy becomes larger than the energy corresponding to the possible optical transitions.

The theoretical framework for the calculation of the photoionization cross-section $\sigma$ of deep impurity centers is extremely unsatisfactory, basically due to the problem of

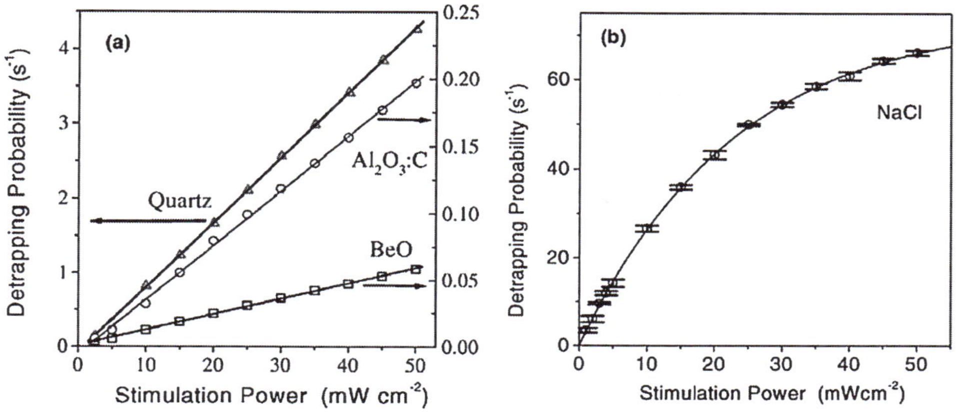


***Figure 2.10*** *Dependence of the detrapping probability $p = \sigma\phi$ with the stimulation power for various materials (Bulur, Bøtter-Jensen and Murray, 2001). Figure (a) shows materials in which the detrapping probability is linear with stimulation power (the detrapping probability is represented in two different scales depending on the material, as indicated by the arrows), whereas (b) shows an example in which this dependence is non-linear. Reprinted from Radiation Measurements, E. Bulur, L. Bøtter-Jensen, A. S. Murray, LM-OSL signals from some insulators: an analysis of the dependency of the detrapping probability on stimulation light intensity, Vol 33, 5, Copyright (2001) with permission from Elsevier.*

defining the initial and final wave functions for the quantum mechanical calculation of the transition probability (Pantelides, 1978). The various assumptions and resultant expressions for the photoionization cross-section were discussed by Bøtter-Jensen, McKeever and Wintle (2003). The main expressions used to express the wavelength dependence of the photoionization cross-section in OSL materials are the expressions by Lukovsky (1965):

$$\sigma(h\nu) \propto \frac{(h\nu - E_I)^{3/2}}{(h\nu)^3}, \tag{2.10}$$

and by Grimmeiss and Ledebo (1975a, b):

$$\sigma(h\nu) \propto \frac{(h\nu - E_I)^{3/2}}{h\nu\,[h\nu + E_I(m_0/m^* - 1)]^2}, \tag{2.11}$$

where $m_0$ is the free electron mass and $m^*$ is effective mass in the band. These expressions have been used usually in a descriptive manner, given the general lack of knowledge on specific defects responsible for the trapping centers and the current status of the theoretical framework.

### 2.2.7 The Luminescence Process

In semiconductors and insulators the probability of electron–hole recombination across the band gap (for example by exciton formation) is low (Blasse and Grabmaier, 1994). In materials of practical interest for OSL dosimetry, luminescence is produced by recombination of electron–hole pairs occurring at defects in the crystal. As discussed in Section 2.2.4, the defect can first capture a hole and subsequently capture an electron (e.g., released from a trapping center), or vice versa. In this recombination process, the electron is first captured in the orbital of an excited state of the defect. Luminescence is produced when the defect returns to the ground state by photon emission, that is, by a radiative transition. Important parameters describing the luminescence process are the emission wavelength, and the lifetime and quantum efficiency of the luminescence process. The latter is related to the relation between radiative and non-radiative processes.

Knowledge of the OSL emission wavelength is important when choosing the appropriate light detection system and optical filters, which in turn is also an important factor when choosing the stimulation wavelength. As discussed in Section 2.1, to take full advantage of the OSL process the stimulation wavelength should be longer than the emission wavelength ($\lambda_{\mathrm{OSL}} < \lambda_{\mathrm{stim}}$) to avoid the detection of photoluminescence from other sources.

The lifetime of the luminescence center can be important if one is interested in using pulsed stimulation in the OSL measurement to improve the signal-to-noise ratio or for time-resolved measurements, as will be discussed later. If a very short light pulse releases electrons from trapping centers, which are then captured at a recombination center creating $N_e$ defects in the excited state, the number of centers in the excited state will decay exponentially according to:

$$N_e(t) = N_e(0)\,\mathrm{e}^{-\frac{t}{\tau_R}}, \tag{2.12}$$

where $t$ is the time since the stimulation pulse and $\tau_{\mathrm{R}}$ is the radiative decay time or lifetime of the luminescence center. The luminescence intensity is proportional to the rate of decay of centers in the excite state, $\mathrm{d}N_e/\mathrm{d}t$, which will also decay exponentially according to $\mathrm{e}^{-t/\tau_R}$.

The lifetime associated with optical transitions within defects in crystals can vary from values less than a nanosecond up to $\sim 10^{-3}$ s (Blasse and Grabmaier, 1994). This is determined by the relative probability of such transitions, which depends strongly on quantum mechanics selection rules that determine whether a transition is "allowed" or "forbidden." In a first approximation in which the wavelength of the electromagnetic radiation (light absorbed or emitted) is large compared to the size of the atom or ion (electric dipole approximation), quantum mechanics considerations lead to conditions that must be satisfied between the initial and final states for the transition to be allowed (Bransden and Joachain, 1983).

Two selection rules that are valid for single and multi-electron centers are the parity and the spin selection rules. The parity selection rule (Laporte selection rule) states that only transitions between states of opposite parity are allowed. For example, for a one-electron center, transition between energy levels within the $f$ shell ($l = 3$) are forbidden, because the parity is given by $(-1)^l$ and the wavefunctions of both initial and final states have the same parity (both are odd, i.e., the signal of the wavefunction is changed under reversal of spatial coordinates, e.g., $x \to -x$, $y \to -y$, $z \to -z$). The spin selection rule states that transitions between initial and final states characterized by different values of total spin angular momentum ($\Delta S \neq 0$) are said to be spin forbidden. However, in solids some forbidden transitions are still possible due to influence of the crystalline field or due to high-order term effects (e.g., magnetic dipole and electric quadrupole), which are not considered in the electric dipole approximation. As a result, "forbidden" transitions can still be observed, but usually with low probability, that is, they are weak and have relatively long lifetimes when compared to "allowed" transitions (Blasse and Grabmaier, 1994; Henderson and Imbusch, 1989).

The quantum efficiency for the luminescence process is associated with the relative number of centers that return to the ground state radiatively in comparison with the number of centers that return to the ground state non-radiatively. In the fluorescence process, in which the optical center is brought to the excited state by light absorption and can return to the ground state by light emission, the quantum efficiency can be defined in terms of number of photons emitted by the center per unit time divided by the number of photons absorbed per unit time. In our case the luminescence center is brought to the excited state by electron–hole recombination, but the concept of quantum efficiency defined in terms of radiative and non-radiative transitions still applies. If the radiative transition rate (probability of radiative transition) is $\Gamma_R = \tau_R^{-1}$, where $\tau_R$ is again the radiative lifetime, and the non-radiative transition rate is $\Gamma_{NR} = \tau_{NR}^{-1}$ where $\tau_{NR}$ is the lifetime for non-radiative transitions, the luminescence efficiency $\eta$ will be given by the probability of radiative transition divided by the total transition probability (i.e., including radiative and non-radiative) (Di Bartolo, 1968):

$$\eta = \frac{\Gamma_R}{\Gamma_R + \Gamma_{NR}} = \frac{\tau_R^{-1}}{\tau_R^{-1} + \tau_{NR}^{-1}}. \tag{2.13}$$

Several aspects of the luminescence process related to the interaction of the electrons with the vibrating environment can be understood with the help of the configurational coordinate diagram (Figure 2.11), including the possible dependence of the quantum efficiency with temperature. In this diagram, one assumes the defect at a distance $R_0$ from its neighboring

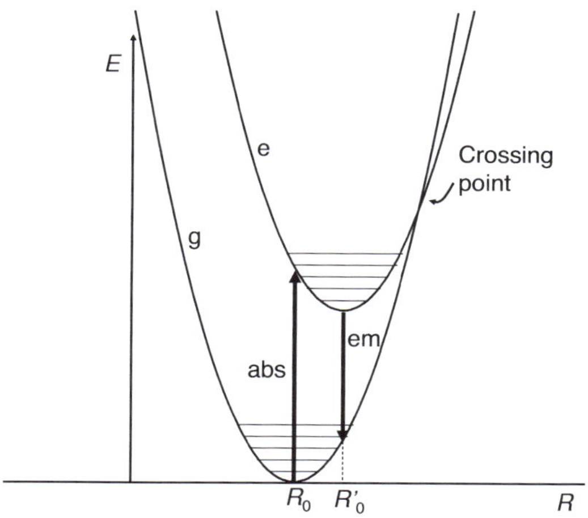


*Figure 2.11* Configurational coordinate diagram representing the potential energy curves of the ground state (g) and excited state (e) of a luminescence center as a function of the separation R between the center and neighboring ions. $R_0$ represents the equilibrium position when the center is in the ground state, whereas $R'_0$ represents the equilibrium position of the center in the excited state. The energies for a process of photon absorption (abs) and emission (em) are represented by the vertical arrows. The horizontal lines represent vibrational energy levels of the defect. Diagram based on Blasse and Grabmeir (1994).

ions and approximates the vibrations in the crystal by harmonic oscillation of the ions around the equilibrium position. The potential energy curve for the ground state is a parabola and the vibrational states of the harmonic oscillator are represented by the horizontal lines within the parabola. The excited state is represented in the same way, except that it can have an equilibrium position at a distance $R_0'$ different than $R_0$. Optical transitions are represented by vertical arrows in this diagram, because the electronic transitions occur much faster than the crystal relaxation around the new equilibrium position (Franck–Condon principle) (Mott and Gurney, 1940). For the center represented in Figure 2.11, a transition can occur from the lowest vibrational states of the ground state to high vibrational states of the excited state (vertical arrow labeled "abs" to indicate absorption). The system can then relax to a new equilibrium position in the diagram by phonon emission, achieving the lowest vibrational state of the excited state (relaxed excited state). From this point in the diagram a radiative transition can occur to a high vibrational state of the ground state (vertical arrow labeled "em" to indicate emission), after which it again relaxes to the bottom of the ground state by phonon emission.

The configurational coordinate diagram explains the difference in energy or wavelength of the absorption bands and emission bands of a luminescence center. As can be observed in Figure 2.11, the emitted photon has lower energy than the absorbed photon and therefore the emission band is shifted to longer wavelengths when compared to the absorption band (Stokes shift). The degree of the Stokes shift is related to how strongly the electrons are coupled to the vibrating environment.

The configurational coordinate diagram also shows that there is a probability that the system will gain sufficient energy by phonon interaction to reach vibrational levels where the two parabolas cross. If this happens, the system can relax to the ground state by phonon emission through the vibrational states of the ground state. This represents a non-radiative transition. Since the probability of the system occupying the higher vibrational states increases with temperature, the probability of non-radiative transitions increases. Considering that the probability of non-radiative transition is given by $\Gamma_{NR} = \Gamma_0 \, e^{-E/kT}$, where $E$ is related to the energy required for the system to reach the crossing point in the configurational diagram and $\Gamma_0$ is a constant, Equation (2.13) can be re-written as:

$$\eta(T) = \frac{\Gamma_R}{\Gamma_R + \Gamma_0 \, e^{-E/kT}} = \frac{1}{1 + C \, e^{-E/kT}}. \tag{2.14}$$

where $C = \Gamma_0/\Gamma_R$ is a constant. Equation (2.14) shows that at low temperatures the quantum efficiency approaches 1, since the available thermal energy is insufficient for the system to reach the crossing point. As the temperature increases and $kT \sim E$, the efficiency is reduced, approaching zero at high temperatures when $C \gg 1$. This behavior is illustrated in Figure 2.12.

Another consequence of non-radiative transitions is that the lifetime of the luminescence process becomes shorter than the radiative lifetime as the temperature increases. Since the excited centers will return to the ground state with a total probability given by $\Gamma = \Gamma_R + \Gamma_{NR}$, the lifetime of the luminescence is given by (Di Bartolo, 1968):

$$\frac{1}{\tau} = \frac{1}{\tau_R} + \frac{1}{\tau_{NR}}. \tag{2.15}$$

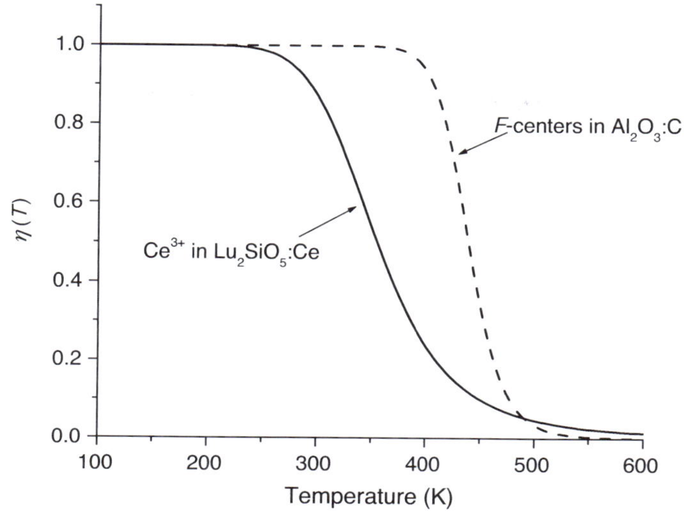


***Figure 2.12*** *Quantum efficiency $\eta(T)$ according to Equation (2.14) for $Ce^{3+}$ in $Lu_2SiO_5$:Ce scintillators ($C = 3.45 \times 10^5$ and $E = 0.32$ eV) (Dorenbos et al., 1994a) and F-centers in $Al_2O_3$: C ($C = 8.5 \times 10^{11}$ and $E = 1.04$ eV) (Akselrod et al., 1998b). The figure indicates that the probability of radiative transitions decreases as the temperature increases, causing a reduction in the luminescence efficiency of the recombination centers.*

Assuming again that $\Gamma_{NR} = \Gamma_0\,e^{-E/kT}$, this equation can also be written as:

$$\tau(T) = \frac{\tau_R}{1 + \tau_R \Gamma_0\,e^{-E/kT}}. \tag{2.16}$$

This equation shows that the parameters $E$ and $\Gamma_0$ can also be obtained by a measurement of the luminescence lifetime as a function of the temperature (e.g. Akselrod *et al.*, 1998b).

The characteristics of the luminescence process (e.g., the width, number and position of emission bands) are determined by the type of defect, its site in the crystal, interaction with the vibrating environment of the crystal and the resultant separation of the energy levels. For example, the crystal field can lift the degeneracies of the energy levels of a free ion (i.e., ion "free in space") differently according to the symmetry of the site and the strength of the crystal field. An excellent introduction to this topic can be found in Blasse and Grabmeir (1994), whereas more in-depth discussions can be found, for example, in Henderson and Imbush (1989) and in Di Bartolo (1968). Here we restrict our discussion to a few general aspects.

The luminescence bands are labeled by indicating the initial and final energy levels of the associated optical transitions. However, a general approach to the calculation of the energy levels is not possible due to differences in the nature of the centers and their interaction with surrounding atoms (Henderson and Imbusch, 1989). It is instructive here to consider the case of optical transitions between deeply bound states ($F$-centers, dopant transition and rare-earth ions), in which the influence of the crystal field may often be treated as a perturbation to the energy levels of the free ion.

Neglecting the magnetic interaction between the electrons' spin and the magnetic field due to the motion of electrons around the nucleus (spin–orbit coupling), the energy levels of multi-electron atoms can be labeled according to their values of the total orbital angular momentum $L$ and the total spin angular momentum $S$ by the so-called $LS$ terms $^{2S+1}L$. The symbol $L$ = S, P, D, F, G, H and so on indicates the total orbital angular momentum ($L$ = 0, 1, 2, 3, etc., respectively) (Bransden and Joachain, 1983). The superscript $2S + 1$ indicates the multiplicity associated with $S$ (number of degenerate energy levels comprising the term). Spin–orbit coupling may lift some of the degeneracy of the $LS$ term according to their total angular momentum $J$. For example, for two $np$ electrons the $^3$P state can be split into the states $^3P_0$, $^3P_1$ and $^3P_2$ according to the different values of $J$ ($J$ = 0, 1 or 2), whereas the $^1$D and $^1$S have multiplicity 1 and therefore are not affected and are labeled $^1D_2$ and $^1S_0$, respectively (Henderson and Imbusch, 1989).

The energy level scheme represented above can be applied to trivalent rare-earth ions in which $n$ electrons in the incomplete 4$f$ shell (represented by $4f^n$) are shielded from the crystalline environment by the electrons in the filled 5d and 6s orbitals and transitions occur between the 4$f$ orbitals. Because of this shielding from the crystalline environment, the minima of the excited and ground state in the configurational coordinate diagram in Figure 2.11 coincide ($\Delta R = 0$) and the optical transitions give rise to the sharp emission lines. Because the transitions are within the 4$f$ orbitals, they are parity forbidden. As a result, the lifetime of the luminescence can be relatively long ($10^{-3}$ s) (Blasse and Grabmaier, 1994). In the case of rare-earth ions in which the optical transitions occur between energy levels that are affected by the crystalline field, such as $5d \rightarrow 4f$ transitions in $Ce^{3+}$ and $Eu^{2+}$, the emission can be broad (band emission) and the lifetime considerably shorter because the transitions are allowed. For example, the lifetime of $Ce^{3+}$ emission is of the order of nanoseconds. The lifetime of $Eu^{2+}$emission is of the order of microseconds because the emission is slowed down by spin selection rules (Blasse and Grabmaier, 1994).

Up to this point, we have not considered the influence of the crystalline field in the energy levels. The crystalline field can split the energy levels given by the $LS$ terms of the free ion according to how they are affected by the local symmetry of the defect site. Therefore, by analyzing the symmetry of a particular site, it is possible to derive and classify the new energy levels according to the symmetry type. This is done using group theory, which is discussed in more detail, for example, in Henderson and Imbush (1989). In general, these new energy levels are labeled as A, B, E and T, where A and B indicate singlet levels (single energy states), E indicates doublets (two degenerate energy states) and T indicates triplets (three degenerate energy states).

For example, consider a defect consisting of a $Cr^{3+}$ ion substituting for $Al^{3+}$ in aluminum oxide, surrounded by six oxygen ions in an octahedral configuration. According to the symmetry of the octahedral configuration, the $LS$ terms of the $Cr^{3+}$ free ion ($^4$F, $^4$P, $^2$G, $^2$F, etc.) are split into five energy levels labeled $A_1$, $A_2$, E, $T_1$, $T_2$. The energy of each level varies according to the strength of the crystal field and are represented by Tanabe–Sugano diagrams (Henderson and Imbusch, 1989; Sugano, Tanabe and Kamikura, 1970). The spin multiplicity is preserved in the label of the crystal field energy levels (e.g., the $^4$F level is split into $^4A_2$, $^4T_2$ and $^4T_1$ energy levels).

In summary, radiative processes associated with the OSL emission are usually related to optical transitions within specific defects in the crystal. The main characteristics (wavelength, lifetime, efficiency) are determined by the interaction with the crystalline field (e.g., symmetry, crystal field strength, coupling with vibrations in the crystal, etc.). These defects can also be characterized by other techniques such as photoluminescence and radioluminescence and, therefore, more information is usually available about the recombination and luminescence centers than the trapping centers involved in the OSL process.

### 2.2.8 Rate Equations for OSL and TL Processes

A general statistical description for thermally stimulated processes and an arbitrary distribution of defect levels is given by Bräunlich (1979) and Chen and McKeever (1997), but the resultant equations are mathematically intractable. In simplified conditions the formalism leads to simultaneous, non-linear differential equations describing the rates of the electronic transitions among the different energy levels in the crystal. These have been useful in modeling and explaining several aspects of the OSL and TL processes.

Some underlying assumptions to set up and numerically solve these rate equations are: (i) charge recombination occurs by charge transport through the delocalized energy bands and (ii) the process is quasi-static, that is, at any instant the system is in quasi-equilibrium. Quasi-equilibrium means that the process proceeds sufficiently slowly so that the system is at any instant close to equilibrium and the concentrations of charges in the conduction and valence band remain relatively small.

Before presenting a more general formulation of the rate equations, let us start with a simplified TL/OSL model consisting of the delocalized energy bands (valence and conduction bands) and two types of discrete defect levels, one representing trapping centers and one representing recombination centers. This model, usually referred to as the OTOR model (one-trap–one-recombination center) (Chen and McKeever, 1997), is depicted in Figure 2.13. The meaning of the symbols used in the formalism are indicated in the figure caption.

Based on this simplified model and conservation of charge, we can write the rate equations. The first equation states that the rate of variation in the concentration of trapped electrons $\mathrm{d}n/\mathrm{d}t$ is equal to the rate of electrons being captured $[N - n(t)]A_n n_c(t)$ minus the rate of electrons escaping the trapping centers $n(t)p$, where $p$ is the transition probability rate per unit time ($s^{-1}$) of electrons stimulated out of the traps:

$$\frac{\mathrm{d}n(t)}{\mathrm{d}t} = [N - n(t)]A_n n_c(t) - n(t)p. \tag{2.17}$$

A similar equation can be written for the rate of variation in the concentration of trapped holes $\mathrm{d}m/\mathrm{d}t$ decreasing due to capture of electrons from the conduction band and increasing due to the capture of holes from the valence band:

$$\frac{\mathrm{d}m(t)}{\mathrm{d}t} = [M - m(t)]A_h n_v(t) - m(t)A_m n_c(t). \tag{2.18}$$

The equations for the change in concentration of electrons in the conduction band $\mathrm{d}n_c/\mathrm{d}t$ and holes in the conduction band $\mathrm{d}n_v/\mathrm{d}t$ can be written as:

$$\frac{\mathrm{d}n_c(t)}{\mathrm{d}t} = f - \frac{\mathrm{d}n(t)}{\mathrm{d}t} - m(t)A_m n_c(t), \tag{2.19}$$

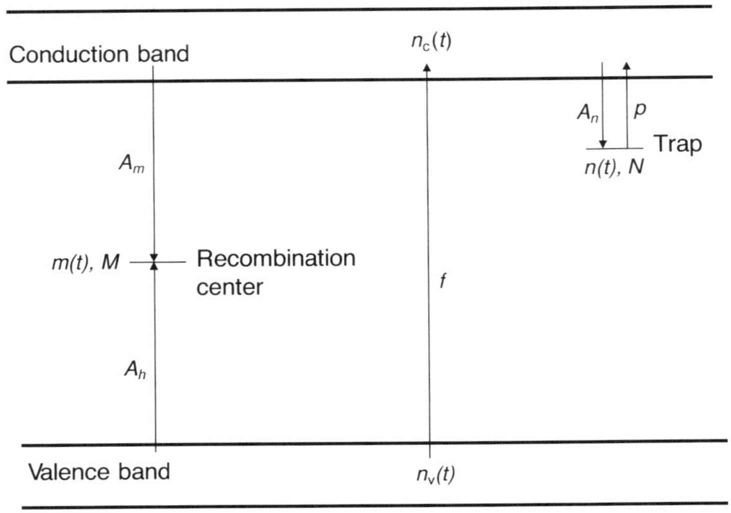


***Figure 2.13*** *Simple OSL model containing one trap and one recombination center (OTOR model). In this diagram, n(t) is the concentration of trapped electrons, m(t) the concentration of trapped holes, $n_c(t)$ the concentration of electrons in the conduction band, and $n_v(t)$ the concentration of holes in the valence band. The probability for optical or thermal stimulation is p. N and M are the total concentration of trapping and recombination centers, respectively. The excitation by ionizing radiation is represented by the rate of electron–hole pairs created per unit volume by ionizing radiation given by f. An electron in the conduction band can be captured by empty traps with a probability $A_n$ or recombine with a trapped hole with a probability $A_m$. Holes in the valence band can be trapped by the recombination center with a probability $A_h$. In simplified version of the OTOR model, the probabilities of hole recombination at the trapping center or hole release from the recombination center are considered negligible.*

and

$$\frac{dn_v(t)}{dt} = f - [M - m(t)]A_h n_v(t). \tag{2.20}$$

This set of equations can be used to represent processes taking place in the crystal during irradiation, optical stimulation or thermal stimulation. During irradiation $f \neq 0$, whereas in the absence of radiation $f = 0$. Optical stimulation of electrons from the trapping centers to the conduction band can be represented by Equation (2.9), whereas thermal stimulation can be represented by Equation (2.4). Here, the temperature can be held constant (e.g., during irradiation) or increased linearly as during a TL readout. The set of differential equations can be solved numerically using mathematical software packages or other user-specific codes (Bailey, 2001; McKeever *et al.*, 1997).

In quasi-equilibrium all holes are assumed to be immediately captured by recombination centers and all electrons to be immediately captured by recombination or trapping centers, such that:

$$dn_c/dt \cong 0, \tag{2.21}$$

and

$$\mathrm{d}n_{\mathrm{v}}/\mathrm{d}t \cong 0. \tag{2.22}$$

Under these assumptions, Equation (2.19) can be written as:

$$n_{\mathrm{c}}(t) = \frac{f + n(t)p}{[N - n(t)]A_n + m(t)A_m}, \tag{2.23}$$

Equation (2.18) can then be written as:

$$\frac{\mathrm{d}m(t)}{\mathrm{d}t} = f - m(t)A_m n_{\mathrm{c}}(t) \tag{2.24}$$

One limitation of the OTOR model is that the concentration of available recombination centers (i.e., trapped holes) is zero before irradiation ($m = 0$). This means that during irradiation the RL signal given by $m(t)A_m n_{\mathrm{c}}(t)$ is initially zero, increasing with dose as the concentration of trapped holes $m$ increases. This is often not the behavior observed in RL measurements, which usually exhibit a sharp increase to a fixed level when the irradiation is turned on and then may change slowly as the irradiation continues (see Figure 2.2). This behavior indicates a pre-existing concentration of recombination centers ($m \neq 0$) allowing for immediate recombination from the moment the radiation source is turned on, which is not surprising in real crystals containing many types of defects, or direct transfer of energy from excitons to the luminescence centers. Numerical calculations have shown that this pre-existing concentration of recombination centers is necessary to explain the RL signal of $Al_2O_3$:C, for example (Pagonis *et al.*, 2009).

Another problem with the OTOR model is that at the end of irradiation, $n = m$. When the OSL curves are calculated numerically using the rate equations shown above, the OSL decay curve deviates significantly from the simple exponential decay, being better described as a stretched exponential (Chen and Leung, 2003). If an existing concentration of recombination centers existed, such that $m \gg n$, the shape of the OSL curve would be significantly different. This is illustrated in Figure 2.14.

For the reasons discussed above, it is more realistic to consider an initial concentration of holes $m_0$ in the recombination center before irradiation. These trapped holes can be considered to be compensated by an identical concentration of trapped electrons in deep defects (not shown in Figure 2.6). In a real crystal the number of defects is always large, providing various trapping and recombination sites for the electron–hole pairs and preventing the situation described in the previous paragraph from happening.

Figure 2.15 shows the numerical solution of the rate equations for a specific set of parameters (indicated in the caption) during irradiation ($f \neq 0$) in the absence of stimulation ($p = 0$). Figure 2.15a shows that the concentration of trapped charges $n$ increases linearly with absorbed dose of radiation until saturation, whereas the concentration of trapped holes $m$ remains approximately constant, increasing slightly at high doses. Figure 2.15b shows the rates for trapping and recombination as a function of the dose. This result explains the fact that the radioluminescence intensity, which is proportional to the rate of recombination, can increase as a function of the dose delivered to the crystal even at a constant irradiation dose rate (Andersen *et al.*, 2006; Pagonis *et al.*, 2009).

Interesting results can also be obtained by solving the same set of equations for the OSL readout ($f = 0$; $p \neq 0$). In this case, the defect occupancy represented in Figure 2.15a

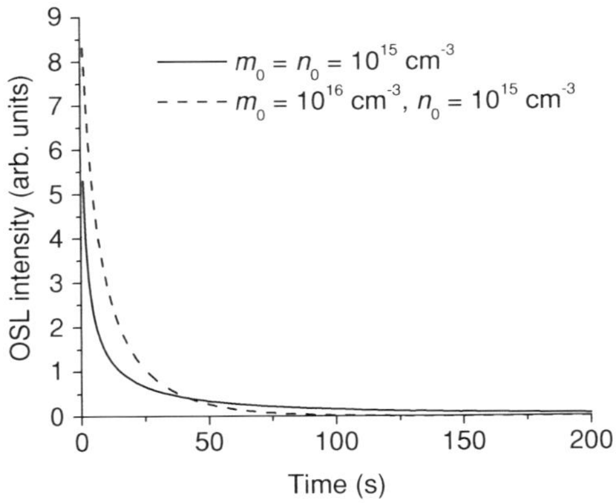


***Figure 2.14*** *OSL decay curves obtained by the numerical solution of the rate equations for the OTOR model for the readout process with $m_0 = n_0 = 10^{15}$ cm$^{-3}$, or for a model with $m_0 = 10^{16}$ cm$^{-3}$ and $n_0 = 10^{15}$ cm$^{-3}$. This initial concentration of trapped holes is assumed to be compensated by electrons trapped in deep traps which do not participate in the process. Other parameters in the calculation are: total concentration of traps $N = 10^{15}$ cm$^{-3}$, electron trapping probability $A_n = 10^{-7}$ cm$^3$ s$^{-1}$, recombination probability $A_m = 10^{-8}$ cm$^3$ s$^{-1}$ and optical detrapping probability $p = \sigma\phi = 0.1$ s$^{-1}$. Quasi-equilibrium conditions were not assumed in this calculation. The results show that the curves can be considerably different depending on whether $m = n$ or $m > n$. In a real crystal the latter is more likely, due to the large number of pre-existing defects that can act as recombination centers.*

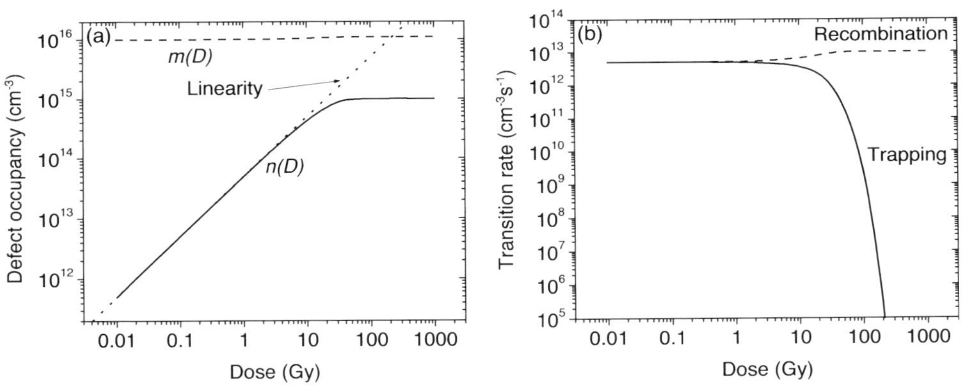


***Figure 2.15*** *Numerical solution of the rate equations during irradiation for the following set of parameters: total concentration of traps $N = 10^{15}$ cm$^{-3}$, electron trapping probability $A_n = 10^{-7}$ cm$^3$ s$^{-1}$, initial concentration of recombination centers $m_0 = 10^{16}$ cm$^{-3}$, recombination probability $A_m = 10^{-8}$ cm$^3$ s$^{-1}$, rate of ion-pair production $f = 10^{-13}$ cm$^{-3}$ s$^{-1}$ (electron–hole pair creation efficiency of $10^{-14}$ pairs Gy$^{-1}$). Graph (a) shows the concentration of trapped electrons $n(D)$ and the concentration of recombination centers $m(D)$ along with the linearity curve. Graph (b) shows the trapping rate $[N - n]A_n n_c$ and recombination rate $mA_m n_c$. In this case, stimulation is assumed to be absent ($p = 0$).*

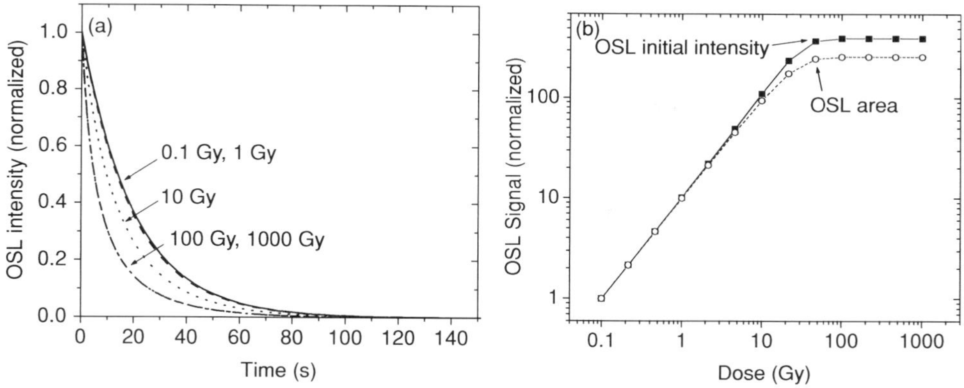


***Figure 2.16*** *(a) Normalized OSL curves and (b) OSL area (total OSL intensity) and initial OSL intensity versus dose, obtained solving the numerical rate equations with the parameters shown in Figure 2.15 and with a probability rate for optical stimulation* $p = \sigma\phi = 0.1\ s^{-1}$*. The initial defect occupancy as a function of the dose is taken from the data in Figure 2.15a. Graph (a) shows that the OSL curve shape may depend on the dose and, as a result, the dose response in graph (b) may be different for the OSL area or initial OSL intensity.*

for a given dose is used as the initial concentration $n_0$ and $m_0$ in the calculation. Figure 2.16a shows the OSL curves obtained from the numerical solution of the rate equations for various doses, that is, various initial defect occupancies, normalized to the initial intensity for comparison of the curve shape. The OSL curve shapes are observed to depend on the dose, the decay becoming faster as the dose increases. In addition, Figure 2.16b shows that the initial OSL intensity and total OSL area increase linearly with dose for low doses, but they may not be proportional to each other at high doses. Both these phenomena (change in the OSL curves with dose and non-proportionality of initial OSL intensity and OSL area) have been observed experimentally, and the model presented here was used as a plausible explanation for such observations (Yukihara *et al.*, 2004a).

More complex models, such as the one illustrated in Figure 2.17, are useful to explain the influence of shallow and deep traps in the OSL process. The model in Figure 2.17 includes shallow traps (STs), main dosimetric traps (MDTs) and deep traps (DTs). Shallow traps can capture electrons and release them by thermal stimulation at room temperature (we assume here no optical stimulation from shallow traps). Charges trapped in the main dosimetric traps are thermally stable and can escape only by optical stimulation. Deep traps only compete for charges released to the conduction band.

Figure 2.18a shows possible OSL curves obtained when shallow traps or deep traps are introduced. Deep traps can cause an overall reduction in the OSL intensity by capturing charges that would otherwise recombine producing OSL, without affecting the shape of the OSL curve. Shallow traps also cause an initially lower OSL intensity, but an equilibrium is eventually reached between the rate of capture and thermal stimulation to and from the shallow traps. From that point on, the OSL decreases due to the depletion of the MDT. The resultant OSL curve is characterized by an initial increase in the OSL intensity followed by a decrease. Note that in this simulation the OSL intensity for the system with shallow

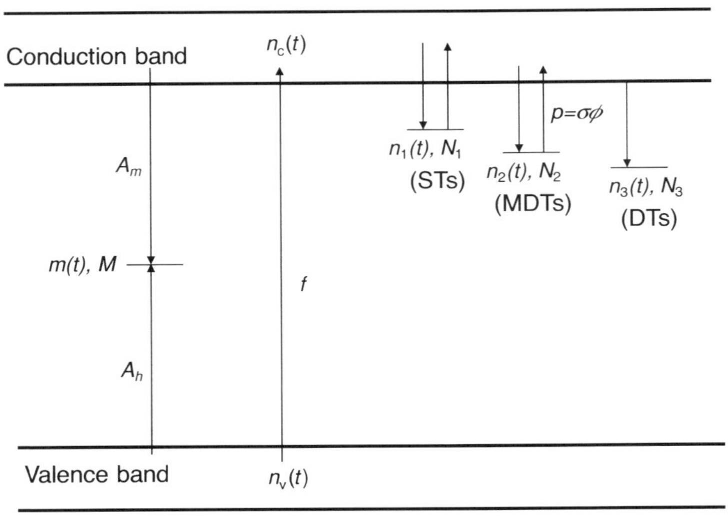


***Figure 2.17*** *A more realistic model of the OSL process including a shallow traps (STs), main dosimetric traps (MDTs) and deep traps (DTs), in addition to recombination centers.*

traps is higher than the OSL intensity for the system without shallow traps for $t > 60$ s, showing the additional contribution of the thermally stimulated charges from the shallow traps to the detected signal. In other words, the main effect of shallow traps is to delay the luminescence process.

A more general set of rate equations can be written for an arbitrary number $q$ of traps and an arbitrary number $r$ of recombination centers, including possible thermal stimulation

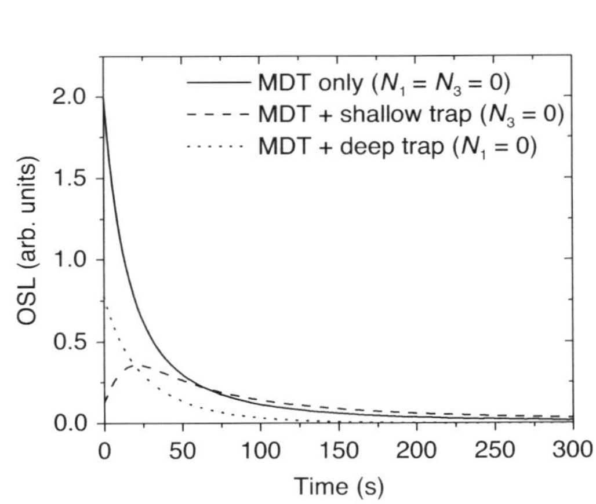


***Figure 2.18*** *OSL curves obtained using a set of rate equations for the model in Figure 2.17 and the following parameters:* $A_1 = 10^{-6}$ cm$^3$ s$^{-1}$, $A_m = A_2 = A_3 = 10^{-7}$ cm$^3$ s$^{-1}$, $E_1 = 0.75$ eV, $E_2 = 1.2$ eV, $s_1 = s_2 = 10^{12}$ s$^{-1}$, $p = \sigma\phi = 0.05$ s$^{-1}$. *The graph shows the effect of shallow traps and deep traps in the OSL curve, in comparison to a system which only contains main dosimetric traps.*

of holes from the recombination centers and valid for simultaneous optical and thermal stimulated release of electrons and holes (Bailey, 2001). For each trapping center we can write an equation for the change in concentration of trapped charges, where the equation for the *i*th trapping center is:

$$\frac{\mathrm{d}n_i(t)}{\mathrm{d}t} = [N_i - n_i(t)]\, A_i n_\mathrm{c}(t) - n_i(t)\sigma_i\phi - n_i(t)s_i e^{-E_i/kT}. \tag{2.25}$$

In the same way, the equation for the *j*th recombination center is:

$$\frac{\mathrm{d}m_j(t)}{\mathrm{d}t} = \left[M_j - m_j(t)\right] A_{h,j} n_\mathrm{v}(t) - m_j(t) A_{m,j} n_\mathrm{c}(t) - m_j(t) s_j e^{-E_j/kT}, \tag{2.26}$$

where the last term on the right side of the equation corresponds to the thermal release of holes from the recombination centers. The equations for the conduction and valence band can be written as:

$$\frac{\mathrm{d}n_\mathrm{c}(t)}{\mathrm{d}t} = f - \sum_{i=1}^{q} \frac{\mathrm{d}n_i(t)}{\mathrm{d}t} - \sum_{j=1}^{r} m_j(t) A_{m,j} n_\mathrm{c}(t) \tag{2.27}$$

and

$$\frac{\mathrm{d}n_\mathrm{v}(t)}{\mathrm{d}t} = f - \sum_{j=1}^{r} \left[M_j - m_j(t)\right] A_{h,j} n_\mathrm{v}(t) + \sum_{j=1}^{r} m_j(t) s_j e^{-E_j/kT}. \tag{2.28}$$

As one can infer, the large number of free parameters in such a system of equations does not lead to an unequivocal interpretation. The situation becomes worse as the number of trapping centers and recombination centers increases. However, numerical calculations using rate equations can be used with a high degree of sophistication if the model is constrained by several, independent experimental observations. To avoid or at least minimize such ambiguities, one should consider: (i) simulating both irradiation and OSL readout stages for a variety of doses; (ii) comparing the results from the simulations during irradiation and readout stages with the experimental OSL dose response; (iii) checking whether the simulation results simultaneously reproduce a variety of other experimental observations.

For example, a similar set of rate equations was used by Bailey (2001) to reproduce several experimental data on the RL, TL and OSL of quartz considering a model including five types of trapping centers and four recombination centers (Bailey, 2001). Numerical solution of rate equations has been extensively used to explain a variety of phenomena, including the effect of bleaching and phototransfer on the TL and OSL signals (McKeever and Morris, 1994), sensitivity changes in quartz (McKeever *et al.*, 1996, 1997), the temperature dependence of OSL decay curves (McKeever *et al.*, 1997), OSL from optical fiber probes under simultaneous irradiation and pulsed stimulation (Polf *et al.*, 2004), radioluminescence of $Al_2O_3$:C (Pagonis *et al.*, 2009) and the possible effects that martian temperature cycles may have on the OSL of sediments on Mars (McKeever *et al.*, 2009).

It is important to keep in mind a comment by Bräunlich regarding thermally stimulated processes that also applies here: "An acceptable fit of experimental data with solutions of the rate equations, obtained with experimentally determined values of activation energies and some of the reaction rate constants, yields the conclusion that the absolute rate theory is consistent with experiment. But this should not be mistaken as proof" (Bräunlich, 1979).

### 2.2.9 Temperature Dependence of the OSL Signal

Up to this point the influence of temperature on the OSL signal has not been considered explicitly. The OSL may be influenced by the detector's temperature due to thermal quenching of the luminescence, competition by shallow traps and thermally assisted optical stimulation. Although one of the processes may be dominant in a particular system or under restricted conditions, it is not always possible to separate them.

#### *2.2.9.1 Thermal Quenching*

Thermal quenching is the reduction in luminescence intensity with increasing temperatures. There are at least two explanations for this phenomenon: a decrease in the luminescence quantum efficiency with increasing temperatures (Mott–Seitz model), and a reduction in the concentration of recombination centers during heating (Schön–Klasens model) (Bräunlich, 1979; Chen and McKeever, 1997; McKeever, 1985).

In the Mott–Seitz model, the luminescence efficiency decreases with increasing temperature due to a reduction in the quantum efficiency of the luminescence centers, as discussed in Section 2.2.7 and illustrated in Figure 2.12. The process can be described by Equation (2.14) and is accompanied by a reduction in the luminescence lifetime, as shown by Equation (2.16). Thermal quenching of luminescence of $F$-centers in $Al_2O_3$:C was shown to be consistent with this type of process (Akselrod *et al.*, 1998b). However, this process is not a concern for OSL measurements carried out in a temperature range in which the luminescence efficiency is $\sim$1. This is one of advantages of the OSL process when compared to TL, since in the latter the sample must be heated and therefore the luminescence efficiency can vary during the TL readout. Moreover, it can be inferred from the configurational coordinate diagram in Figure 2.11 that the probability of a non-radiative transition is related to the parameter $\Delta R = R_0 - R_0'$, that is, to the offset between the minimum of the two parabolas representing the excited and ground states of the defect. Therefore, centers with small Stokes shift (small $\Delta R$) should be less affected by thermal quenching (Blasse and Grabmaier, 1994).

Another source of thermal quenching is thermally stimulated change in the concentration of recombination centers (Schön–Klasens model) (Chen and McKeever, 1997; McKeever, 1985). As discussed earlier in connection with the recombination process, luminescence occurs when the recombination center first traps a hole and subsequently an electron (or vice versa). If the hole escapes the defect by thermal stimulation before recombination with the electron takes place, the decrease in the concentration of recombination centers will result in a reduction of luminescence.

When multiple optically and thermally stimulated processes are present, the situation becomes very complicated, requiring an investigation using the rate equation described in Section 2.2.8, sometimes in conjunction with the Mott–Seitz model of thermal quenching. For example, Akselrod *et al.* (1998b) have shown that the degree of thermal quenching in

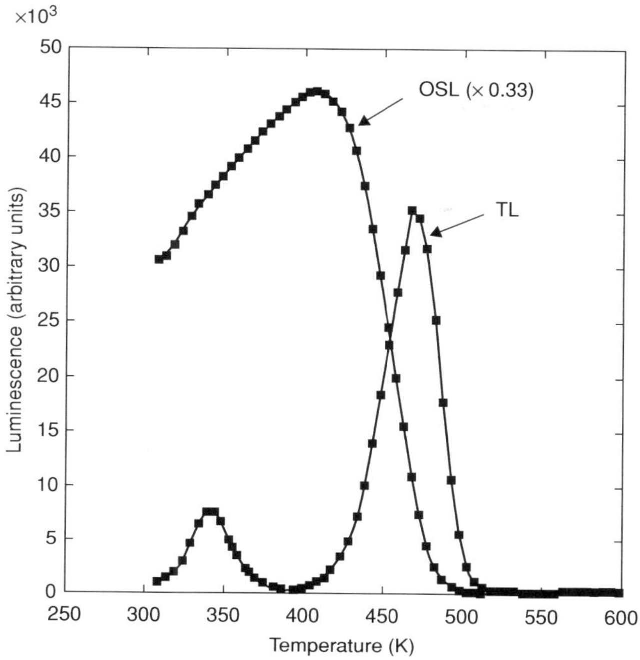


*Figure 2.19* *OSL signal as a function of readout temperature in comparison with the TL signal from the same sample. As the temperature increases, the apparent OSL intensity increases due to a quick release of charges captured at the shallow traps by thermal stimulation. Reprinted from Radiation Protection Dosimetry, B. G. Markey et al., The temperature dependence of optically stimulated luminescence from alpha-$Al_2O_3$:C, Vol. 65, Copyright (1996) with permission from Oxford University Press.*

TL processes may be affected by the occupancy of deep traps, as a result of a combination of the kinetics of the process with the thermal quenching of the *F*-centers in $Al_2O_3$:C.

### 2.2.9.2 Competition by Shallow Traps

In some cases, it has been observed that the OSL intensity may increase when the temperature is raised in a limited temperature range due to a reduction in the competition by shallow traps. This is illustrated for $Al_2O_3$:C in Figure 2.19. When the OSL is measured at a low temperature, the electrons can be captured by shallow traps instead of recombining with the trapped holes. Therefore, the OSL intensity is lower than if the shallow traps were absent or full. When the temperature increases, the electrons captured by the shallow traps are quickly released back to the conduction band by thermal stimulation, recombining with the trapped holes and resulting in an apparent increase in the OSL intensity.

The influence of shallow traps can also be seen in the shape of the OSL curve as an initial increase followed by the typical exponential-like decrease, as shown in Figure 2.18. The temperature dependence of the OSL curve is also affected by the stimulation intensity. McKeever *et al.* (1997) investigated the temperature dependence introduced

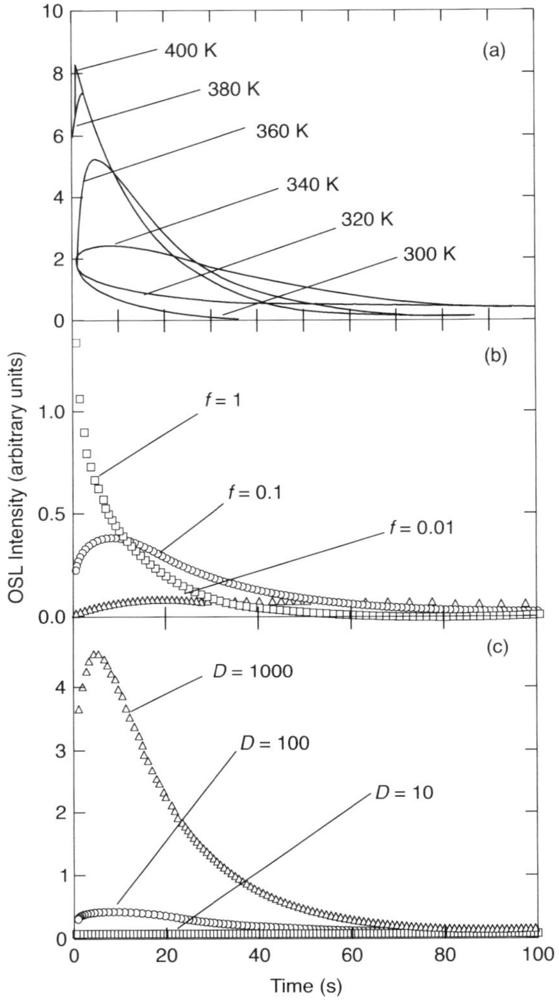


***Figure 2.20*** *Simulated OSL curves for a model containing three traps and two recombination centers at (a) various temperatures, (b) various stimulation rates ($f = \sigma\phi$) and (c) various doses D in arbitrary units (McKeever et al., 1997). Reprinted from Radiation Measurements, S.W.S. McKeever, L. Bøtter-Jensen, N. Agersnap Larsen, G. A. T. Duller, Temperature dependence of OSL decay curves: Experimental and theoretical aspects. Vol 27, Copyright (1997) with permission from Elsevier.*

by shallow traps, thermally assisted optical stimulation, thermal quenching and localized donor–acceptor recombination. Figure 2.20 shows examples of OSL curves obtained for a system consisting of three traps and two recombination centers in various conditions. This figure exemplifies the variety of OSL curves that can be observed when the readout is performed under different conditions of temperature and stimulation intensity, or when the OSL materials are irradiated with different doses.

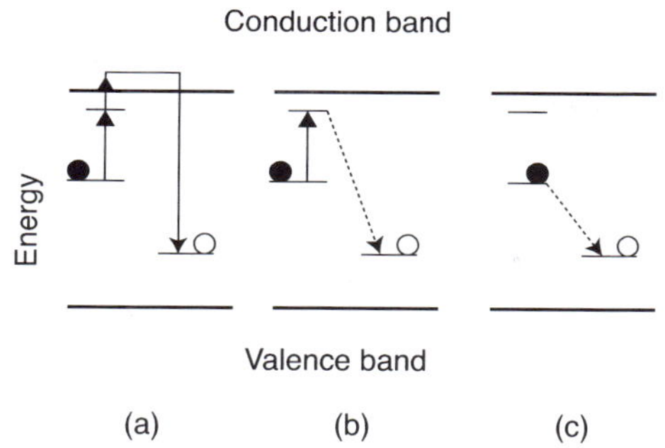


***Figure 2.21*** *Alternative recombination process involving thermally assisted stimulation of electrons to the conduction band (a) and quantum tunneling from the excited (b) or ground (c) states. Reprinted from Physics and Chemistry of Minerals, N. R. J. Poolton et al., Electrons in feldspar IL on the wavefunction of electrons trapped at simple lattice defects. Vol 29, Iss. 3. Copyright (2002) with permission from Springer.*

McKeever *et al.* (1997) pointed out that the OSL curves at low and high temperatures in Figure 2.20a are identical if normalized to the initial intensity. They conclude that the influence of shallow traps does not affect the rate of OSL decay, which according to Equations (2.3) and (2.9) is dictated by the factor $\sigma\phi$, when the temperature is low enough such that the shallow traps are stable, or high enough such that the electrons trapped at shallow traps escape promptly. This is important to distinguish between the effect of shallow traps and the effect of thermally assisted optical stimulation. In the latter, the rate of decay increases with temperature.

### *2.2.9.3 Thermally Assisted Optical Stimulation*

The infrared stimulated luminescence (IRSL) from feldspars shows both an increase in the total OSL intensity and an increase in the OSL decay rate (McKeever *et al.*, 1997). The mechanism initially put forward to explain such observation in feldspars is that infrared (IR) excites the trapped charge defect from the ground state ~2.0 eV below the conduction band to an excited state ~0.8 eV below the conduction band (Hütt, Jaek and Tchonka, 1988). Once in the excited state, the trapped electron can be released to the delocalized bands by a thermally stimulated process (see Figure 2.21a). Therefore, an increase in the readout temperature increases the probability of the electrons escaping the trapping center, resulting in an OSL signal. This hypothesis was supported by the fact that the stimulation spectrum of the IR stimulated signal is characterized by a resonance peak at ~1.4 eV, indicating an optical transition to an energy level with well-defined energy instead of a transition to a band. More recently, other models have been invoked to explain the OSL in feldspars, including the possibility of tunneling from the ground or excited states to the recombination centers (Figure 2.21b and c) and optically stimulated excitation of electrons to "band-tail" states below the conduction band, where the electrons can move with reduced mobility. In this model, electrons in the band-tail states undergo a thermally stimulated random walk process, which obviously introduces a temperature dependence in the resultant OSL process (Poolton *et al.*, 2002a, 2002b, 2009).

### 2.2.10 Other OSL Models

#### 2.2.10.1 Non-First-Order Kinetics

In TL theory, the first-order kinetic process described by Equations (2.5) and (2.7) has been complemented by other models that consider the possibility of retrapping of the stimulated charges back in the same defect. Under simplified conditions, these models lead to non-first-order equations for the TL process (McKeever, 1985).

By analogy with TL equations, the rate equations for a one-trap–one-recombination center model during OSL readout can be written as (Chen and McKeever, 1997):

$$\frac{\mathrm{d}n}{\mathrm{d}t} = -n\sigma\phi + (N-n)A_n n_\mathrm{c}, \tag{2.29}$$

$$\frac{\mathrm{d}m}{\mathrm{d}t} = -mA_m n_\mathrm{c}, \tag{2.30}$$

and

$$\frac{\mathrm{d}n}{\mathrm{d}t} + \frac{\mathrm{d}n_\mathrm{c}}{\mathrm{d}t} = \frac{\mathrm{d}m}{\mathrm{d}t}, \tag{2.31}$$

where $n = n(t)$ is the concentration of trapped electrons, $m = m(t)$ is the concentration of trapped holes, $N$ is the total concentration of trapping centers, $A_n$ is the retrapping probability, $A_m$ is the recombination probability, $\sigma$ is the photoionization cross-section and $\phi$ is the photon flux. In quasi-equilibrium conditions ($\mathrm{d}n/\mathrm{d}t \approx \mathrm{d}m/\mathrm{d}t$ and $n_\mathrm{c} \ll n, m$), these equations can be simplified to obtain:

$$n_\mathrm{c} = \frac{n\sigma\phi}{(N-n)A_n + mA_m}, \tag{2.32}$$

which leads to:

$$\frac{\mathrm{d}n}{\mathrm{d}t} = -n\sigma\phi\left[\frac{mA_m}{(N-n)A_n + mA_m}\right]. \tag{2.33}$$

For the condition of slow retrapping, $(N-n)A_n \ll mA_m$, Equation (2.33) reduces to the first-order rate equation, as in Equation (2.1). For the condition of strong retrapping expressed by $A_n/A_m = 1$ and with $n - m$, this equation reduces to the rate equation for the second-order model:

$$\frac{\mathrm{d}n}{\mathrm{d}t} = -n^2\frac{\sigma\phi}{N}. \tag{2.34}$$

This leads to an OSL intensity $I \propto |\mathrm{d}n/\mathrm{d}t|$ equal to:

$$I(t) \propto \frac{n_0^2\sigma\phi}{N}\left(1 + \frac{n_0\sigma\phi t}{N}\right)^{-2}, \tag{2.35}$$

which is the second-order equation for the OSL curve.

Comparing the second-order rate Equation (2.34) with the first-order rate Equation (2.1), a general-order kinetics equation similar to the one used in TL (Chen and McKeever, 1997)

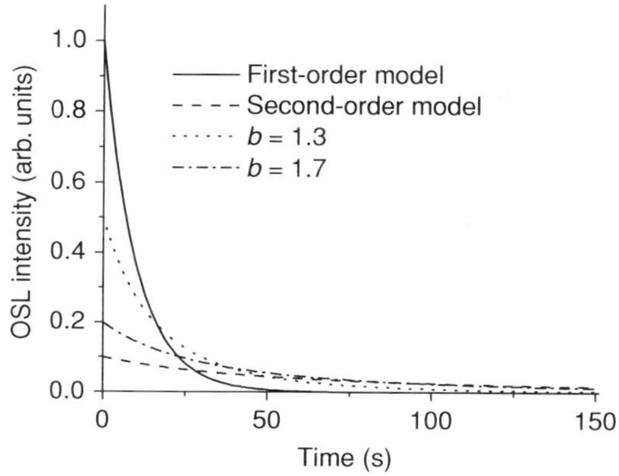


***Figure 2.22*** *OSL curves in the first-order model, second-order model and general-order model with* $b = 1.3$ *and* $1.7$. *In the calculations we used* $n_0 = 10^{15}$ $cm^{-3}$, $N = 10^{16}$ $cm^{-3}$ *and* $p = \sigma\phi = 0.1$ $s^{-1}$.

could be proposed:

$$\frac{dn}{dt} = -\frac{n^b}{N^{b-1}} p, \tag{2.36}$$

where $b$ is a dimensionless kinetic parameter equal to 1 for first-order kinetics or 2 for second-order kinetics, and $p = \sigma\phi$. This leads to the solution:

$$I_{\mathrm{OSL}}(t) = \frac{n_0^b p}{N^{b-1}} \left[ 1 + (b-1) \left( \frac{n_0}{N} \right)^{b-1} pt \right]^{-\frac{b}{b-1}}, \tag{2.37}$$

This equation reproduces the first-order model when $b \to 1$ and the second-order model when $b = 2$. A comparison between the OSL curves for these different models is shown in Figure 2.22.

Discussions of the OSL process in terms of non-first-order kinetic equations have been limited (e.g., Bulur, 1996; Kitis, Furetta and Pagonis, 2009). There are several reasons for that. First of all, OSL curves are often the result of a superposition of multiple trapping centers with similar values of photoionization cross-section or a distribution of trapping centers, therefore resulting in OSL curves that may have first-order kinetics (i.e., shape independent of the dose), but which consist of a superposition of a large number of exponential decays. Secondly, it has been demonstrated that the kinetic order is not constant throughout the measurement, except for first-order kinetics, because the relative rate of retrapping and recombination is the result of a dynamic competition process that changes during readout (Sunta *et al.*, 2005). Thirdly, analytical expressions are mostly useful for deconvolution of different processes; however, in a multi-component system a non-first-order process implies interaction between trapping centers and, therefore the principle of superposition is not valid anymore. Deconvolution of an OSL curve with multiple non-first-order processes is therefore questionable. Although non-first-order equations have been used to describe the shape of the OSL curves, this approach precludes a deeper understanding of the physics behind the OSL process and the discussions have been, in general, sterile. The system of rate equations discussed earlier already includes the possibility of retrapping and, therefore, can be used to investigate the physical behavior of OSL systems without assuming non-first-order models.

Chen and Leung (2003) have shown that with a certain range of parameters, the OSL process for the one-trap–one-recombination center model leads to an OSL decay that can be described by a stretched-exponential function of the type:

$$I = I_0 \exp\left[-(t/\tau)^{\beta}\right], \tag{2.38}$$

with $0 < \beta < 1$. As discussed in Section 2.2.8, this is related to a condition ($m_0 = n_0$), which leads to results that are not typical of real crystals.

### 2.2.10.2 Spatially Correlated Defects and Localized Transitions

Up to this point, we have relied on an energy band diagram with simple isolated defects (e.g., Figure 2.6) and recombination by charge transport through the delocalized bands to describe the OSL process and the underlying kinetics of the OSL process. Such a description, coupled with an appropriate set of rate equations, tends to be clear and capable of explaining at least qualitatively most of the experimental observations in OSL. Nevertheless, the quantity of experimental data that are difficult to explain with such elementary models is accumulating in the literature, indicating a need for models considering spatially correlated defects and localized transitions.

In TL it has long been recognized that "the portrayal of an idealized band picture and isolated traps and recombination sites seems totally inappropriate for most radiation dosemeter materials" (Townsend and Rowlands, 1999). Characteristics such as the dependence of the TL peak on a particular emission band, or on the ionic radius of dopants, and "athermal phosphorescence" have been considered as evidence of a close association between trapping and recombination centers in extended defect models, in some cases including the possibility of recombination by quantum tunneling between defects instead of charge transport through the delocalized bands. In LiF:Mg,Ti, a widely used TL dosimeter, the concept of spatially correlated trapping and luminescence centers has been central to the explanation of the material's dose response and the influence of ionization density on the TL properties of the material (e.g., Horowitz, 2001; Livingstone *et al.*, 2009).

Possibly because OSL is more recent than TL as a widespread dosimetry technique, the concept of spatially correlated defects is not often considered in OSL models. Models of OSL dose response and dose-dependence of the OSL curve have been based on competition processes between isolated defects and recombination through the delocalized bands (Chen and Leung, 2001a, b; Chen, Pagonis and Lawless, 2006; Lawless, Chen and Pagonis 2009; Pagonis, Chen and Lawless, 2006; Pagonis *et al.* 2009). Two notable exceptions are the OSL models for feldspar and BaFBr:$Eu^{2+}$.

The IRSL in feldspar has been described in terms of recombination through quantum tunneling between spatially correlated defects and in terms of thermally assisted charge hopping through band-tail states below the conduction band (Poolton *et al.*, 2002a, 2002b). The involvement of band-tail states and tunneling in the OSL process in feldspars is supported by detailed OSL measurements carried out at low temperature (Poolton *et al.*, 2009). Tunneling between closely associated defects is probably responsible for anomalous fading in feldspars (Wintle, 1973), which is a decrease in the TL and OSL signal that cannot be explained by thermal stimulation of trapped charges. The evidence for spatially correlated defects in BaFBr:$Eu^{2+}$ is discussed in Section 2.7.1.3.

There are some experimental results in OSL dosimetry mostly related to ionization density effects that have been only explained in an ad hoc fashion or not fully understood at all. These include:

(i) The OSL from different emission bands in $Al_2O_3$:C (UV and $F$-center emission bands) seems to be related to trapping centers of similar thermal stability, but the OSL decay curve associated with one recombination center decays faster than the OSL from the other (Yukihara and McKeever, 2006b).
(ii) The OSL associated with the UV emission band in $Al_2O_3$:C increases with time over the period of 1–2 weeks, whereas the OSL associated with the $F$-center emission remains constant (Yukihara and McKeever, 2006b).
(iii) The rate of decay of the OSL curve and the relative intensity between the UV and $F$-center emission bands show strong dependence on ionization density (i.e., absorbed dose or type of particle) (Yukihara and McKeever, 2006b; Yukihara *et al.*, 2006).

Similar changes in the rate of the decay of OSL curve were observed for different X-ray energies, again emphasizing the role of ionization density in determining the OSL of $Al_2O_3$:C (Jain, Bøtter-Jensen and Thomsen, 2007).

A satisfactory physical explanation for these behaviors is still lacking and may require more detailed experiments coupled with a model that includes spatially correlated defects. As pointed out by Townsend and Rowlands (1999) in TL dosimetry, "rather than consider large scale complexes as unusual it may be that these are the norm, and truly isolated trapping and recombination centres are the exception."

## 2.3 OSL Readout

Having discussed the underlying theory of the OSL process, we turn now to more practical aspects of OSL dosimetry, starting with basic elements of an OSL reader and readout modalities.

### 2.3.1 Basic Elements of an OSL Reader

An OSL reader consists of two essential elements: a light source to stimulate the OSL detector and a light transducer to detect the luminescence from the detector (Figure 2.23). A stimulation filter is usually placed between the light source and the OSL detector to select the stimulation wavelength or band that is optimum for a specific OSL material. The OSL is detected by a light transducer of high sensitivity and low noise, typically a photomultiplier tube (PMT). Detection filters are placed between the PMT and the OSL detector to block the scattered stimulation light and isolate specific OSL emission bands characteristic of the OSL material. The optical elements and detector holder are secured in place in a light-tight enclosure (not shown in Figure 2.23).

The essential problem in designing an OSL reader is to discriminate the OSL from the stimulation light, which have intensities many orders of magnitude apart, while optimizing the collection of OSL by the light transducer. As an example, for a typical stimulation

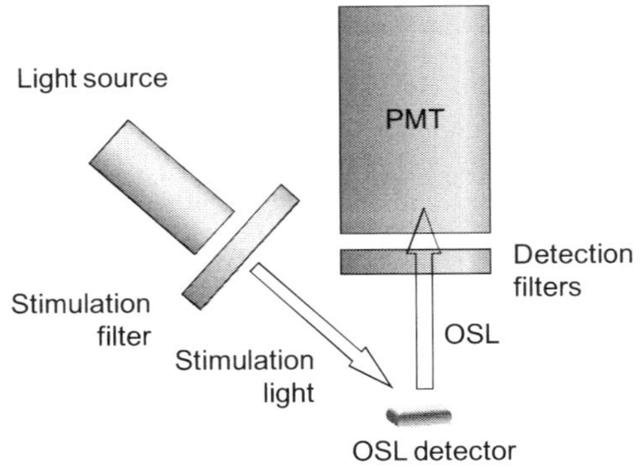


***Figure 2.23*** *Schematics of an OSL reader with the essential elements for stimulation and detection of the OSL signal from a detector. In this diagram, the light transducer is a photomultiplier tube (PMT). Stimulation filters are used to block short wavelength components of the light source which could be transmitted by the detection filter. The combination of stimulation and detection filter should prevent the stimulation light from reaching the PMT, while transmitting most of the OSL from the detector.*

intensity of 10 mW cm$^{-2}$ of green light (525 nm) there are more than $10^{16}$ photons cm$^{-2}$ s$^{-1}$ reaching the detector position. In contrast, the number of OSL photons detected by the PMT can be of the order of just a few hundred photons per second for very low doses. Therefore, this is a significant signal-to-noise problem and OSL reader design involves the judicious choice of stimulation light, stimulation filters, light transducer and detection filters appropriate for each OSL material. Additional optical elements (mirrors or lenses) may also be used to increase the photon collection and the intensity of the detected OSL signal, while decreasing the intensity of the scattered light from the stimulation source.

### 2.3.2 Stimulation Modalities

The OSL technique provides fine control over the stimulation process, the degree of control depending on the particular light source. Optimization of the readout process involves not only choice of stimulation wavelength, but also determination of the best modulation and duration of the stimulation intensity. Several approaches have been proposed for the OSL readout to take advantage of this degree of control.

#### *2.3.2.1 Continuous-Wave OSL (CW-OSL)*

The simplest OSL readout approach consists of stimulating the detector with light of constant intensity, a modality known as continuous-wave OSL (CW-OSL). In this case, discrimination between OSL and stimulation light is based only on wavelength separation ($\lambda_{OSL} \neq \lambda_{stim}$). Therefore, it is essential that the transmission band of the detection filters does not overlap with the spectrum of the stimulation light, which is determined by the light source and optical filters used in front of it. CW-OSL is the most widely used stimulation method in dosimetry and luminescence dating due to its simplicity and frequently satisfactory performance.

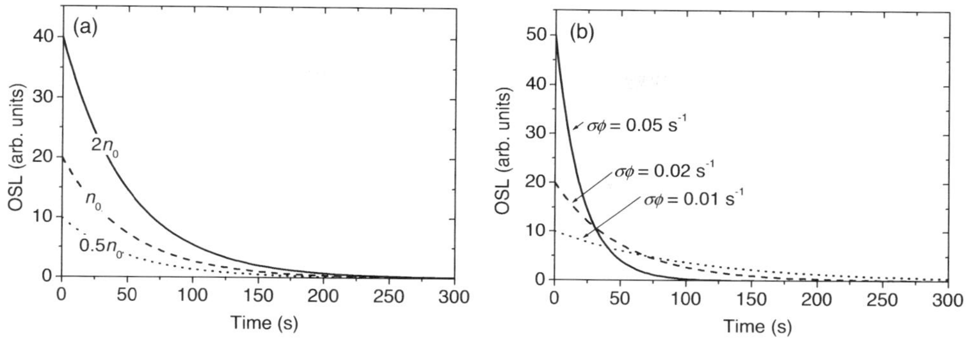


***Figure 2.24*** *(a) OSL curves for $\sigma\phi = 0.02\ s^{-1}$ and various values of $n_0$ and (b) for fixed $n_0$ and various values of $\sigma\phi$ in the first-order model described by Equation (2.3). Figure (a) shows that the initial concentration of trapped charges $n_0$ is a scaling parameter; since $n_0$ is determined by the previous exposure to ionizing radiation, the absorbed dose of radiation affects the OSL intensity, but not the curve shape. Figure (b) shows that increasing the stimulation power (i.e., the photon flux $\phi$) makes the OSL signal decay faster, but the total OSL area remains constant.*

A simple description of the OSL curve obtained using CW-OSL for a first-order recombination process is given by Equation (2.3) with $p$ given by Equation (2.9). Examples of OSL curves for different values of $n_0$ and $\sigma\phi$ are given in Figure 2.24.

In this simple model, the rate of decay of the OSL curve depends only on $\sigma$ and $\phi$, but not on $n_0$, which means that the shape of the OSL curve does not change with the dose (see Figure 2.24a). On the other hand, the total area under the OSL curve depends only on $n_0$, but not on $\sigma$ or $\phi$, because

$$\int_0^\infty I_{\mathrm{OSL}}(t)\mathrm{d}t \propto \int_0^\infty n_0\sigma\phi e^{-\sigma\phi t}\mathrm{d}t = n_0. \tag{2.39}$$

This mean that the total area under the OSL curve, or the total number of photons emitted, is not affected by changes in the stimulation intensity. The areas under the OSL curves in Figure 2.24b are all the same.

The observations above have practical implications for dosimetry. First of all, for an ideal OSL detector stimulated with constant intensity the absorbed dose will always be related to the OSL curve integrated for any fixed period of time. For example, if one chooses the OSL signal $S$ to be the total counts over the first 1 s of stimulation, $S$ will be proportional to the absorbed dose. If the OSL signal during this first 1 s of stimulation provides sufficient sensitivity and precision, the stimulation can be interrupted, leaving the remaining trapped charges available for future readout. Therefore, depending on the stimulation intensity and duration, a significant part of the information stored in the detector can be preserved for re-evaluation of the absorbed dose at a later date, making the OSL detectors re-readable.

However, as one can see from Figure 2.24b, the initial intensity of the OSL curve does depend on the stimulation intensity $\phi$, and is therefore affected by fluctuations in the stimulation source. In this case, stability of the stimulation source is essential to obtain reproducible results. The total OSL area, however, is independent of the stimulation intensity.

Therefore, if the OSL signal is chosen to be the total area under the OSL curve, the results will be less dependent on fluctuations in the intensity of the stimulation light.

The OSL curve of a real material is rarely described by a single exponential decay. There are many factors contributing to deviations from this ideal behavior, including: (i) contribution from more than one trapping center or trapping centers giving rise to a distribution of $\sigma(\lambda)$ values; (ii) departure from first-order kinetics caused, for example, by re-trapping of the charge carriers after stimulation and (iii) non-uniform illumination of the material. Some of these aspects have already been discussed in Sections 2.2.8 and 2.2.10.

### 2.3.2.2 Pulsed OSL (POSL)

Despite of the simplicity of the CW-OSL technique, the need to completely remove the scattered stimulation light often requires the use of optical filtration that ends up reducing significantly the intensity of the measured OSL signal. This problem is more critical the smaller the wavelength separation between the stimulation light and the OSL emission band becomes.

In some situations, particularly low-dose measurements, it may be advantageous to use additional temporal discrimination between the stimulation light and the OSL emission instead of extra optical filters. This can be done using a pulsed light source and a gated detection system that detects the OSL signal only in the period between the stimulation pulses, when there is no scattered light. If the intensity of the scattered stimulation light is too high, it may be necessary to gate the PMT, that is, to switch the PMT high voltage off during stimulation and back on in the period between the stimulation pulses. However, the most practical approach is to use optical filtration to reduce the scattered light to a safe level, avoiding PMT damage, while gating only the photon counter to detect only the pulses arriving in between the stimulation pulses. This approach avoids the drastic measure of turning the PMT high-voltage on and off, improving the stability (and probably the durability) of the reader. This OSL readout modality is called pulsed OSL (POSL). The applicability of POSL for dosimetry using $Al_2O_3$:C detectors was first explored by Markey, Colyott and McKeever (1995), and further demonstrated in subsequent publications (Akselrod and McKeever, 1999; McKeever and Akselrod, 1999).

Figure 2.25 illustrates a POSL scheme in which short laser pulses are used to stimulate the detector at a frequency of 4000 Hz, that is with one laser pulse every 250 μs (Figure 2.25a). The photon counter gate is open during a 220 μs interval in between the laser pulses (Figure 2.25b), which means that only the PMT counts arriving in this time interval are counted.

Figure 2.26 shows an example of the improved discrimination achieved using the POSL technique. Figure 2.26a shows the PMT signal obtained using only optical filtration with a specific filter combination, the PMT signal being monitored continuously during stimulation. The PMT counts during readout of the un-irradiated detector are around 1750 cps and the OSL signal for a dose of 1 mGy is slightly above the background. Figure 2.26b shows the readout of detectors in similar conditions, except that in this case only the PMT counts in the period between the stimulation pulses were monitored. The PMT counts due to scattered stimulation light were reduced by almost two orders of magnitude, down to ~25 cps, without a significant reduction in the radiation-induced OSL signal (the difference

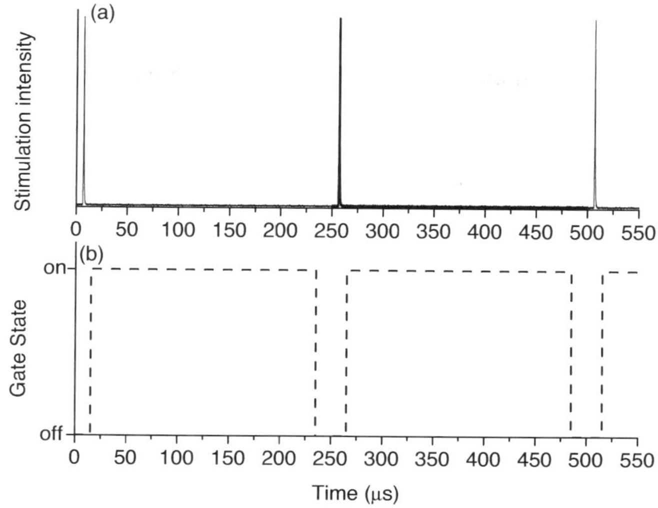


***Figure 2.25*** *Schematics of the POSL readout scheme, showing (a) the optical stimulation pulses and (b) the gate state of the photon counter used to count the photons arriving at the PMT only between the stimulation pulses. The pulse frequency in this example is 4 kHz.*

between the 1 mGy and the 0 mGy curves). The OSL signal for the detector irradiated with 1 mGy is now more pronounced against the background.

This temporal discrimination depends of course on the lifetime of the luminescence centers responsible for the OSL signal, which is the characteristic time scale in which the excited centers relax to the ground state once recombination takes place. If the luminescence

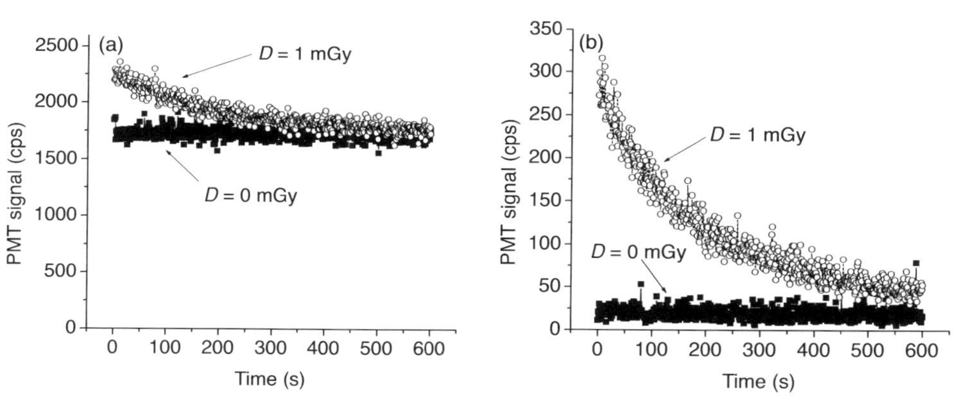


***Figure 2.26*** *Illustration of an OSL measurement of $Al_2O_3$:C using a pulsed light source with (a) a continuous photon counter and (b) a gated photon counter (only PMT pulses in between stimulation pulses are counted). The measurement was also carried out for a detector that was not irradiated; in this case, the PMT signal is due to PMT dark counts and scattered stimulation light reaching the PMT and counted by the photon counter.*

lifetime is shorter than the stimulation pulse duration, the luminescence centers decay immediately after recombination and, consequently, the OSL emission occurs only during stimulation. In this case, the OSL intensity in the periods between laser pulses is negligible and the POSL technique is not effective. However, when the luminescence lifetime is significantly longer than the stimulation pulses, a significant fraction of the luminescence centers decays after the stimulation pulse is over, producing OSL in the period between stimulation pulses.

This point can be illustrated by finding an expression for the OSL intensity when the stimulation is on and off. This expression can be obtained writing the rate equations for the concentration of luminescence centers in the excited state $m^*(t)$ during stimulation:

$$\frac{\mathrm{d}m^*(t)}{\mathrm{d}t} = -\frac{m^*(t)}{\tau} - \frac{\mathrm{d}n(t)}{\mathrm{d}t}. \tag{2.40}$$

This equation states that the concentration of excited luminescence centers $m^*$ decreases due to relaxation, which occurs with a probability equal to $\tau^{-1}$, $\tau$ being the luminescence lifetime, and increases at the same rate as the concentration of trapped electrons $n$ decreases due to recombination. The rate of decrease in $n$ with optical stimulation is given by Equation (2.1), which has a solution given by Equation (2.2). Equation (2.40) can then be rewritten as:

$$\frac{\mathrm{d}m^*(t)}{\mathrm{d}t} + \frac{m^*(t)}{\tau} = n_0 p e^{-pt}. \tag{2.41}$$

Here $p = \sigma\phi$ is the probability of photoionization of the trapping centers. The solution for this equation is (Boas, 2005):

$$m^*(t) = m_0^* e^{-t/\tau} + \frac{n_0 p}{-p + \tau^{-1}} \left(e^{-pt} - e^{-t/\tau}\right), \tag{2.42}$$

and the OSL intensity is given by $I_{OSL}(t) \propto m^*(t)/\tau$. In this equation, $n_0$ and $m_0^*$ are the initial concentrations of trapped electrons and excited luminescence centers at the beginning of each stimulation pulse, respectively. In between the stimulation pulses, $p = 0$ and the OSL intensity is simply:

$$I_{\mathrm{OSL}}(t) \propto m^*(\Delta t) e^{-t/\tau}/\tau, \tag{2.43}$$

where $m^*(\Delta t)$ is the concentration of excited centers at the end of a stimulation pulse of duration $\Delta t$.

Figure 2.27 shows the OSL intensity during and after a 0.1 s stimulation pulse for a material having luminescence centers characterized by a luminescence lifetime $\tau = 35$ ms. During the stimulation pulses, electrons escaping the trapping centers recombine, increasing the concentration of excited luminescence centers, leading to an increase in the OSL signal. Once the stimulation pulse is over, the OSL decays with a characteristic lifetime $\tau$, the intensity being reduced to ~4% after three lifetimes.

If an OSL detector having an initial concentration of trapped electrons is stimulated with a light pulse identical to the one shown in Figure 2.27 every second (frequency of 1 Hz), the OSL intensity will behave as shown in Figure 2.28a, with the luminescence decaying to zero in the period between stimulation pulses. The fraction of the OSL signal emitted after the stimulation pulses is ~32% of the total OSL signal, implying that ~68% of the signal is lost. However, if the stimulation frequency is increased, keeping the stimulation pulse

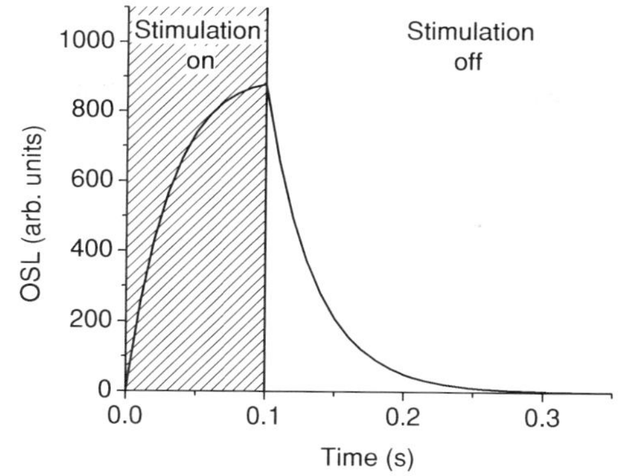


***Figure 2.27*** *OSL signal calculated using Equations (2.42) and (2.43) for a 0.1 s stimulation pulse. In the calculations the lifetime of the luminescence center was $\tau = 35$ ms and the optical stimulation probability rate $p = \sigma\phi = 1\ s^{-1}$. The increase in OSL during stimulation pulse is caused by electrons escaping the trapping center and recombining, creating excited luminescence centers. When the stimulation light is switched off, the concentration of excited luminescence centers and the OSL intensity decays exponentially with lifetime $\tau$.*

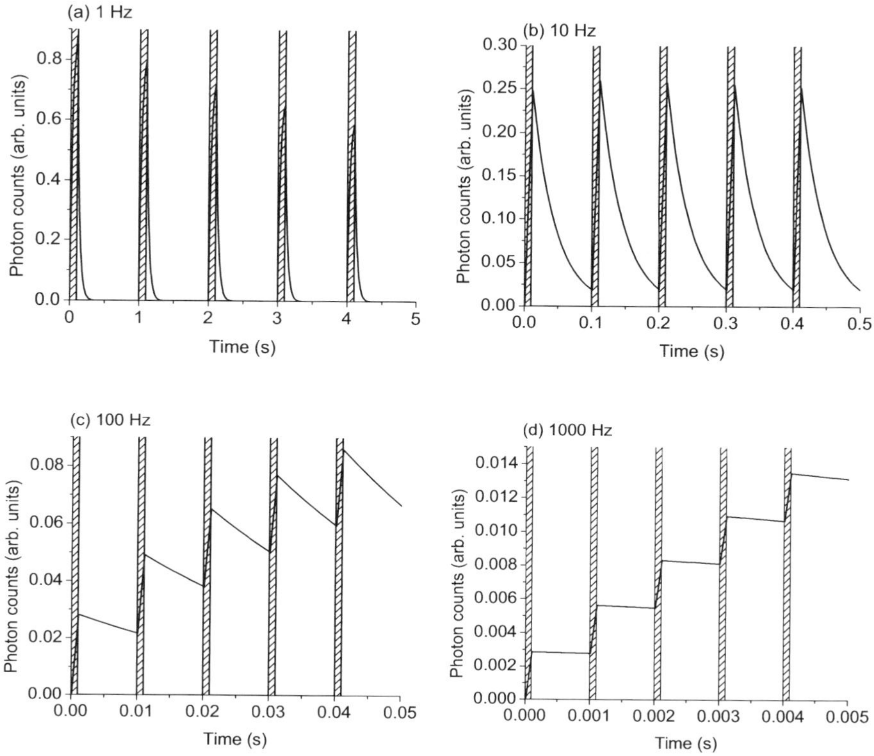


***Figure 2.28*** *OSL signal calculated using Equations (2.42) and (2.43) for different pulse frequencies. In all cases the pulse width is 10% of the pulse period, the luminescence lifetime is $\tau = 35$ ms, and the optical stimulation probability is $p = \sigma\phi = 1\ s^{-1}$.*

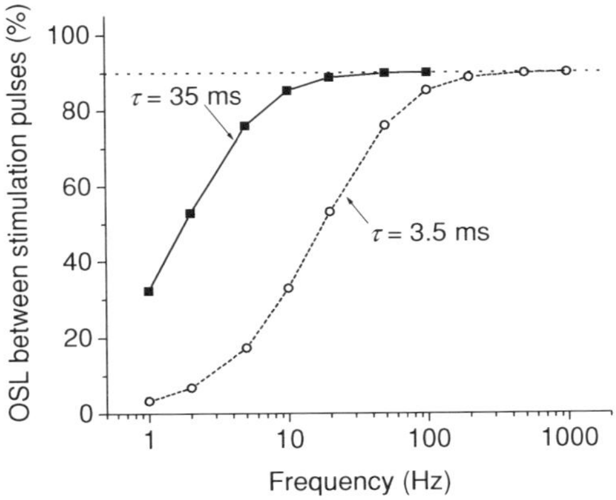


***Figure 2.29*** *OSL emitted in between stimulation pulses (averaged over 1 s of stimulation) as a function of pulse frequency for a material with a luminescence center characterized by a luminescence lifetime $\tau = 35$ ms and another with a luminescence lifetime $\tau = 3.5$ ms. In both cases the pulse width is equal to 10% of the pulse period and the optical stimulation probability rate is $p = \sigma\phi = 1\ s^{-1}$. As the pulse frequency increases, the OSL emitted in between the stimulation pulses approaches the limiting case of 90%.*

duration at 10% of the period between pulses, the OSL intensity will have the characteristics shown in Figure 2.28b–d. The higher the frequency, the higher the fraction of OSL emitted in the period between stimulation pulses, therefore increasing the number of photons that can be detected using the POSL technique.

This is illustrated in Figure 2.29 for hypothetical materials having luminescence lifetimes equal to 35 and 3.5 ms. This figure shows the fraction of the total OSL signal emitted in between stimulation pulses as a function of stimulation frequency (the duty cycle, i.e., the stimulation period with respect to the interval between the stimulation pulses, kept at 10% in all cases). In the case of $\tau = 35$ ms, a stimulation frequency of 100 Hz is sufficient to achieve maximum efficiency for this duty cycle. (The efficiency can be increased even more by decreasing the duty cycle.)

In practice, each data point in the OSL curve corresponds to the total PMT counts in between stimulation pulses for a fixed number of pulses or integration period, as seen in Figure 2.26. The general equations for the OSL process discussed with respect to the CW-OSL still apply, except for the fact that the intensity may be lower due to the OSL lost (not detected) during the stimulation pulses.

Pulsed stimulation can also be used to separate overlapping emission bands originating from luminescence centers with distinct luminescence lifetimes. Because the POSL is more efficient in detecting luminescence from centers with long luminescence lifetimes, the appropriate choice of the timing parameters and gate window can isolate one particular emission band, which sometimes is difficult to achieve on the basis of wavelength discrimination, particularly for emission bands that overlap to a great extent. Detection of the signal during and in between light pulses can be used to separate the OSL emission from different emission bands or even from different materials, provided the luminescence centers have different luminescence lifetimes. For example, the OSL from feldspar and quartz can be separated using this approach (Thomsen *et al.*, 2006). For materials with fast luminescence lifetimes, it has also been suggested that OSL detection during stimulation can be used for better discrimination against background signal (Niedermayer and Bayer, 2010).

#### *2.3.2.3 Linear Modulation OSL (LM-OSL)*

Whereas CW-OSL and POSL are the predominant stimulation schemes in dosimetry, the high degree of control in the stimulation intensity (for example by controlling the LED current) opens the possibility of other stimulation modes to be devised.

Bulur (1996) proposed recording the OSL curves using stimulation with an intensity that increases linearly with time, instead of remaining constant, an approach later referred to as linear modulation OSL (LM-OSL). Using the LM-OSL technique, the OSL intensity as a function of time shows an initial linear increase as the intensity increases followed by a decrease due to the depletion of the trapped charge concentration. Therefore, the LM-OSL intensity graphs are peak-shaped curves, instead of exponential-like decays.

Consider, for example, a first-order OSL process in which the stimulation photon fluence rate $\phi$ is increased linearly from zero until a maximum value $\phi_{\max}$ during a period $T$. In this case, Equation (2.1) has to be replaced with:

$$\frac{\mathrm{d}n}{\mathrm{d}t} = -n\sigma\frac{\phi_{\max}}{T}t, \tag{2.44}$$

where we omitted the time and wavelength dependence of the variables to simplify the notation. The solution for this equation leads to the following expression for the LM-OSL intensity:

$$I_{\mathrm{LM-OSL}}(t) \propto n_0\sigma\frac{\phi_{\max}}{T}te^{-\frac{1}{2}\sigma\frac{\phi_{\max}}{T}t^2}, \tag{2.45}$$

This function is plotted in Figure 2.30 in comparison to an equivalent CW-OSL curve. The LM-OSL curve reaches a maximum at an instant $t_{\max}$ given by:

$$t_{\max} = \sqrt{\frac{T}{\sigma\phi_{\max}}}, \tag{2.46}$$

which is related to the photoionization cross-section $\sigma$. The LM-OSL maximum intensity is proportional to

$$I_{\max} \propto n_0\sqrt{\frac{\sigma\phi_{\max}}{T}}e^{-\frac{1}{2}} = \frac{n_0}{t_{\max}}e^{-\frac{1}{2}}. \tag{2.47}$$

As shown in Figure 2.30, the LM-OSL curve has a reduced intensity when compared to an equivalent CW-OSL curve, which is one of the disadvantages of this approach (Bulur, 1996). The areas under the curves are the same, however.

One proposed advantage of the LM-OSL technique, in comparison with CW-OSL, is the easy visualization of OSL curves containing more than one component, due to the contribution from different trapping centers, each characterized by a different photoionization cross-section. This is exemplified in Figure 2.31, which shows calculated CW-OSL and LM-OSL curves for a system with two types of trapping center. In the CW-OSL curves the

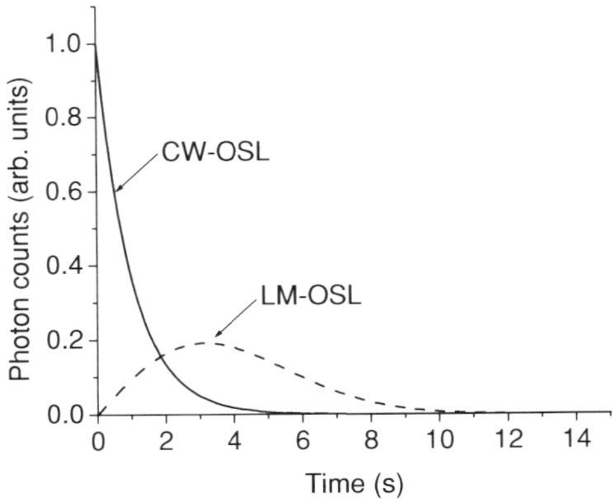


*Figure 2.30* Comparison between CW-OSL calculated using Equation (2.3) ($\sigma\phi = 1\ s^{-1}$, $n_0 = 1\ cm^{-3}$) and LM-OSL calculated using Equation (2.45) ($T = 10\ s$, $p = \sigma\phi_{max} = 1\ s^{-1}$, $n_0 = 1\ cm^{-3}$).

two components appear as superimposed exponential decays (Figure 2.31a), whereas in the LM-OSL curves they appear as superimposed peaks (Figure 2.31b) and are more easily distinguished by eye.

The disadvantage of the LM-OSL technique is the need to modulate the stimulation intensity, which may introduce complications related to: (i) non-linearity of the stimulation ramp; (ii) wavelength shift of the stimulation spectrum due to operation of the stimulation source at a non-constant power and (iii) variable PMT background associated with leakage of stimulation light.

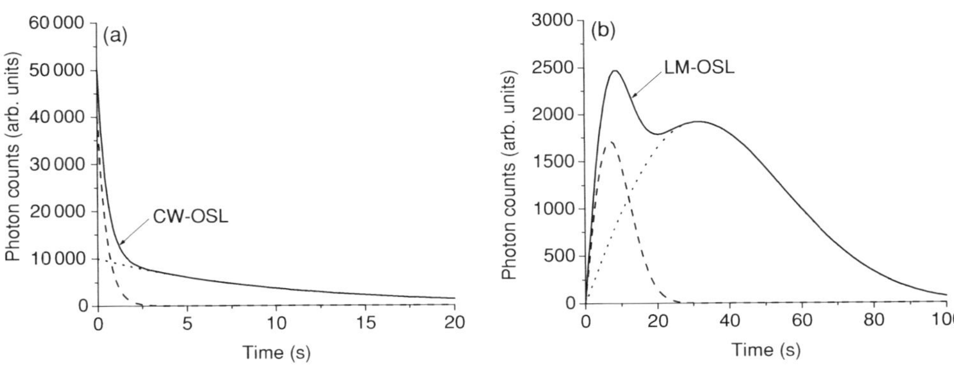


*Figure 2.31* (a) CW-OSL and (b) LM-OSL curves for a system with two trapping centers characterized by the parameters: $n_{0,1} = 10^5\ cm^{-3}$, $\sigma_1 = 0.001\ cm^{-2}\ s^{-1}$, $n_{0,2} = 2\times10^4\ cm^{-3}$, $\sigma_2 = 0.02\ cm^{-2}\ s^{-1}$, $\phi_{max} = 100\ cm^{-2}\ s^{-1}$, $T = 100\ s$, following Bulur (1996). The figure shows that the two OSL components are better visualized in the LM-OSL measurement.

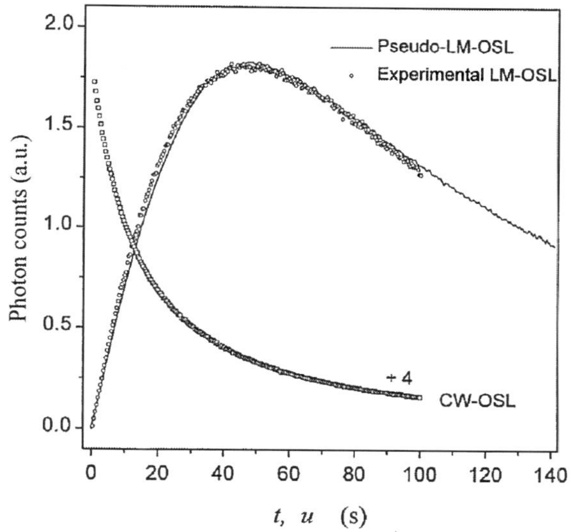


***Figure 2.32*** *Experimental and pseudo-LM-OSL curves of a Na-feldspar sample; the pseudo-LM-OSL curve was obtained from the CW-OSL curve also shown in the figure. Reprinted from Radiation Measurements, Bulur, E., A simple transformation for converting CW-OSL curves to LM-OSL curves. Vol. 32, 141–145. Copyright (2000) with permission from Elsevier.*

Bulur (2000) showed that it is possible to convert CW-OSL curves to LM-OSL curves using a simple mathematical transformation. Adopting the variable transformation:

$$u = \sqrt{2tT} \text{ or } t = \frac{u^2}{2T}, \tag{2.48}$$

the OSL intensity as a function of $u$ is given by:

$$I(u) \propto \left|\frac{\mathrm{d}n}{\mathrm{d}u}\right| = \left|\frac{\mathrm{d}n}{\mathrm{d}t}\right|\left|\frac{\mathrm{d}t}{\mathrm{d}u}\right| = n_0\sigma\frac{\phi}{T}ue^{-\frac{1}{2}\sigma\frac{\phi}{T}u^2}, \tag{2.49}$$

where we have used Equation (2.1) with $p = \sigma\phi$ and replaced $t$ with $u^2/(2T)$. This equation is identical to Equation (2.45), with $\phi_{\max} = \phi$ and $u$ instead of $t$. Therefore, the LM-OSL curves can be calculated from the CW-OSL data by applying the variable substitution given by Equation (2.48) and multiplying each data point by $u/T$. The curves obtained are called "pseudo-LM-OSL curves." The value of $T$ is arbitrary, but Bulur (2000) proposes using $T = 2T_{\mathrm{CW}}$, where $T_{\mathrm{CW}}$ is the total measurement time of the CW-OSL curve.

An example of the equivalence between an experimental LM-OSL curve and a pseudo-LM-OSL curve is shown in Figure 2.32. This figure presents the CW-OSL, LM-OSL and pseudo-LM-OSL curves for a feldspar sample. The experimental and pseudo-LM-OSL curves agree reasonably well, demonstrating the validity of this approach. Bulur (2000) also showed that the transformation is applicable to other OSL models (e.g., second-order and general-order kinetics).

Since proposed, LM-OSL has been used in numerous studies, having been applied to quartz (Agersnap Larsen *et al.*, 2000; Bulur, Bøtter-Jensen and Murray, 2000, 2001; Bulur *et al.* 2002; Demuro *et al.*, 2008; Jacobs, Wintle and Duller, 2006; Jain *et al.*, 2005; Kitis, Polymeris and Kiyak, 2007; Kiyak, Polymeris and Kitis, 2007; Kuhns, Agersnap Larsen and McKeever, 2000; Li and Li, 2006a, 2006b; Packman *et al.*, 2007; Polymeris *et al.*, 2009; Rittenour *et al.*, 2005; Singarayer and Bailey, 2003, 2004; Wang, Lu and Zhao, 2006), $Al_2O_3$:C (Bulur, Bøtter-Jensen and Murray, 2001; Whitley and McKeever, 2001, 2002), BeO (Bulur, Bøtter-Jensen and Murray, 2001; Bulur and Yeltik, 2010b), NaCl (Bulur, Bøtter-Jensen and Murray, 2001), $ZrO_2$ (Rivera *et al.*, 2003), ZnS and SrS based phosphors (Bulur and Göksu, 1997), and feldspars (Fattahi *et al.*, 2006). Countering claims that LM-OSL facilitates the separation of overlapping OSL components, Bos and Wallinga (2009) argued that there is no clear advantage in this regard; instead, CW-OSL measurements are recommended to avoid the complications associated with ramping the stimulation light.

#### 2.3.2.4 Other Stimulation Modalities

Based on concepts similar to those described in the previous section, other modulation schemes are also possible. For example, Mishra *et al.* (2008) proposed the use of a non-linear modulation technique, in which the stimulation intensity is varied with $t^l$ where $l$ is an arbitrary number. This approach can lead to OSL peaks that are more symmetric than LM-OSL peaks. An exponential increase in the stimulation intensity would reduce the difference in intensity between the OSL components more than the LM-OSL technique, but it would require variation in stimulation intensity by many orders of magnitude to be an effective experimental tool. For example, the intensity of a laser light can be modulated by several orders of magnitude using combinations of step-motor controlled graduated neutral density filters. The optical density of these types of filters typically changes linearly with position, allowing an exponential increase in the intensity by a linear motion of the filter.

Various stimulation modes (e.g., linear, exponential, hyperbolic) were investigated systematically by Bos and Wallinga (2009) for a first-order OSL process. The authors concluded that the different modes can be useful for visualization because they highlight different aspects of the OSL curves with multiple components, but provide no advantage in terms of a better separation of the components.

## 2.4 Instrumentation

One of the attractive points of the OSL technique is that it is technically readily accessible. An OSL reader having the essential parts shown in Figure 2.23 with sensitivity comparable to commercial readers can be built with readily available off-the-shelf components.

The overall design considerations involve: (i) choice of a light source for stimulation and appropriate optical filters to select the stimulation wavelengths; (ii) choice of a light transducer to detect the OSL signal and convert it to an electronic signal that can be digitized, as well as associated optical filters to select the OSL emission band and reject the stimulation light; (iii) choice of stimulation modality (e.g., CW-OSL, POSL, LM-OSL, etc.); and (iv) considerations regarding optimization of light collection for improved signal detection. These considerations are intertwined and depend highly on the material properties. Although some configurations can be used for general purpose and basic characterization of OSL materials, improved performance can be achieved by optimizating the OSL reader for a specific material and application.

We now explore in more detail some of the elements of an OSL reader.

### 2.4.1 Light Sources

The choice of light source for the OSL measurement needs to consider the optimum stimulation wavelength and modality (e.g., CW-OSL versus POSL, etc.) for the particular material. This choice cannot be made independently of the selection of OSL emission band. For most materials, the shorter the stimulation wavelength, the higher the photoionization cross-section of the trapped charges (see Section 2.2.6). However, the smaller the wavelength separation between stimulation light and OSL signal, the more difficult it is to block the scattered stimulation light from reaching the PMT. Therefore, the choice of stimulation wavelength is restricted by the choice of the detection wavelength. The OSL emission characteristics also govern the choice of the best stimulation modality, since POSL is only applicable for luminescence centers that have a relatively long luminescence lifetime.

Taking the material properties and OSL emission band into account, factors to consider when choosing a stimulation source are the spectral distribution of the source (monochromatic versus broadband emission), the stimulation intensity that can obtained, whether the source should be continuous or pulsed (and the time scale of the pulsing scheme), and other factors, such as stability, lifetime, cost and how convenient it is to change the wavelength (if necessary).

#### *2.4.1.1 Lamps*

Tungsten halogen lamps, arc lamps and gas discharge lamps can be used to stimulate the OSL signal, provided optical filters or monochromators are used to select the appropriate wavelength. From these, halogen lamps are probably the most cost-effective solution to stimulate the OSL materials in the visible and near-IR (250–2700 nm), providing a continuous spectrum, high light output and easy operation. The main drawbacks are the relatively short operating life (thousands of hours) in comparison with LEDs (hundreds of thousands of hours), the presence of UV light in the case of arc lamps and gas discharge lamps, and the low efficiency (most of the light output is filtered when selecting the desired stimulation band). Microsecond pulses can be obtained using pulsed arc lamps, but for halogen lamps pulsed stimulation requires the use of optical shutters, which are inherently slow. Figure 2.33 shows examples of a stimulation unit based on a halogen lamp.

#### *2.4.1.2 Light-emitting diodes (LEDs)*

The light from LEDs originates from the recombination of electron–hole pairs in a forward biased p-n diode junction. This recombination process results in a broad emission band with a peak that depends on the semiconductor material used in the LED. LEDs are now fabricated covering the IR to the ultraviolet range of the spectrum. Figure 2.34 illustrates the design of the stimulation module used in the Risø TL/OSL readers (Section 2.5.2).

Advantages of using LEDs for OSL measurements include the small size, efficiency, long operating life and possibility of pulsing them in the microsecond range for POSL measurements. Due to these advantages, IR, green or blue LEDs are the stimulation source of choice in commercial OSL readers such as the Risø TL/OSL readers and Landauer's InLight system.

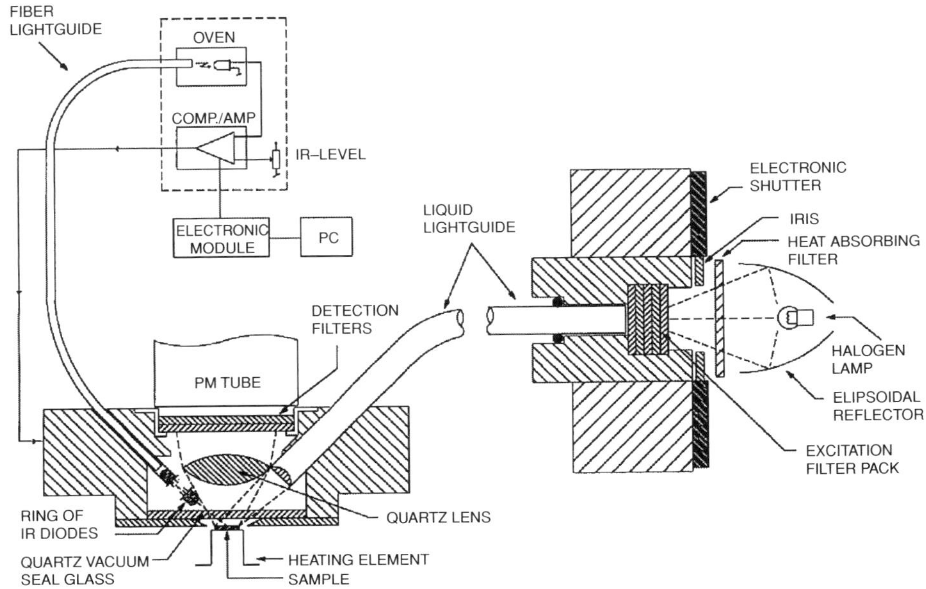


***Figure 2.33*** *Examples of stimulation modules using halogen lamps and a liquid lightguide to carry the stimulation light to the sample. Reprinted from Radiation Protection Dosimetry, Bøtter-Jensen, L., Murray, A.S., Developments in optically stimulated luminescence techniques for dating and retrospective dosimetry. 84 (1–4), 307–315. Copyright (1999) with permission from Oxford University Press.*

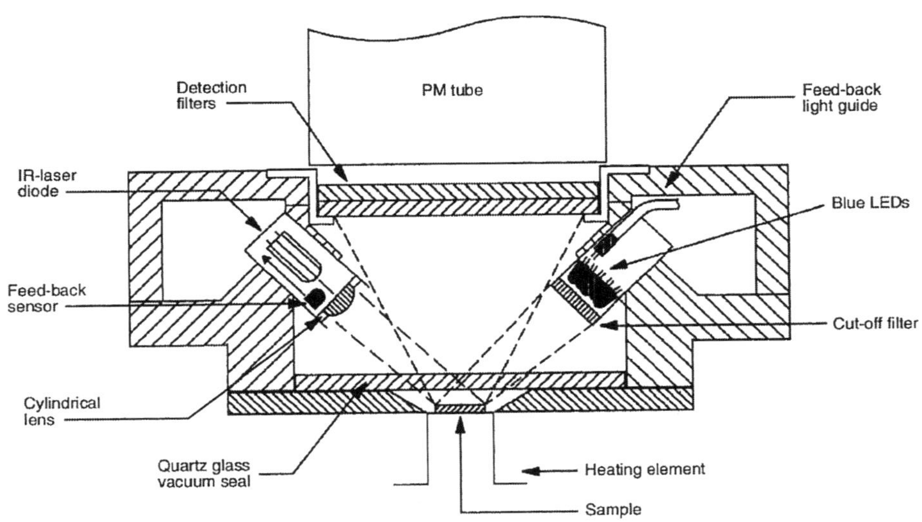


***Figure 2.34*** *Stimulation unit used in the Risø TL/OSL readers, based on IR and blue LEDs. Reprinted from Radiation Protection Dosimetry, Bøtter-Jensen, L., Bulur, E., Murray, A.S. and Poolton, N.R.J., Enhancements in luminescence measurement techniques. 101, 119–124. Copyright (2002) with permission from Oxford University Press.*

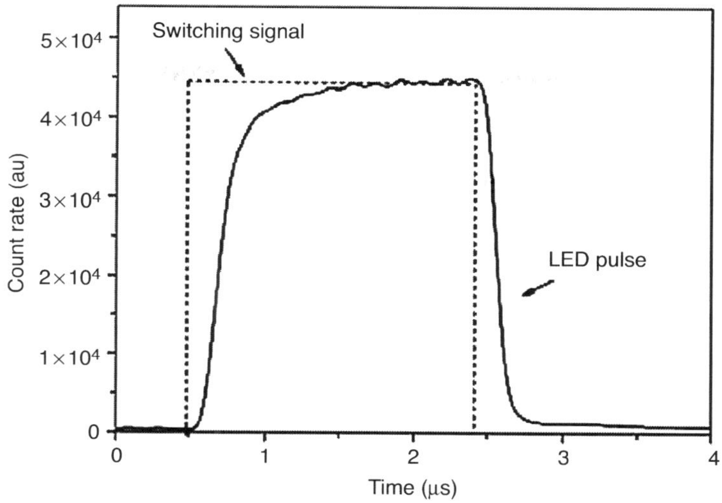


***Figure 2.35*** *A ~2-μs pulse from a standard Risø TL/OSL reader using a diode pulsing unit. The time profile of the signal used to switch the LED is shown by the dashed line. This figure shows that POSL measurements with relatively short light pulses can be performed using LEDs. Repritned from Radiation Measurements, Denby, P.M., Bøtter-Jensen, L., Murray, A.S., Thomsen, K.J., Moska, P., Application of pulsed OSL to the separation of the luminescence components from a mixed quartz/feldspar sample. Vol. 41, 774–779. Copyright (2006) with permission from Elsevier.*

POSL measurements have been performed using LEDs in the microsecond range (Chithambo and Galloway, 2000a; Denby *et al.*, 2006). As an example, Figure 2.35 shows LED pulses obtained using a driver designed for the Risø TL/OSL reader (Denby *et al.*, 2006).

### *2.4.1.3 Lasers*

Laser light is produced in a medium containing fluorescent species (e.g., impurity in a crystal, dye molecule in a solvent, atom in a gas) that can be excited to a metastable excited state (e.g., optically or by an electrical current). The medium is placed in an optical resonant cavity formed by mirrors at opposite ends of the medium to amplify the light wave and select a preferential direction of propagation. One of the mirrors is only partly reflective, allowing a fraction of the laser light in the cavity to be transmitted giving rise to the laser beam.

Laser light is highly monochromatic, directional and can be pulsed with high frequency and short pulse duration for POSL measurements. Lasers are available in a variety of media and species, providing wavelengths from the UV to the IR. In addition to the fundamental laser line, wavelengths equal to 1/2, 1/3 and 1/4 of the fundamental wavelength can be obtained using second-harmonic generation and third-harmonic generation. Thus, a Nd:YAG laser producing a 1064 nm laser line can be used to obtain 532 nm, 355 nm or 266 nm lines. Other types of lasers, such as argon ion lasers, are capable of multiple line operation (514.5 nm, 457.9 nm, 488.0 nm, 351 nm). The line can be selected by introducing a prism in the laser resonant cavity, but the laser intensity will depend on the selected line. Fluorescent dye lasers based on dyes (fluoresceins, coumarins and rhodamines) in a liquid medium (water, ethanol, methanol) can produce light ranging from the UV to the IR and can be tuned continuously over a wavelength range of tens of nanometers. The line is selected, for example, by using a diffraction grating at the end of the resonant cavity. Semiconductor (diode) lasers are also important because they are compact, relatively inexpensive and can be used directly or for pumping other solid-state lasers (e.g., diode-pumped Nd:YAG laser).

The small wavelength bandwidth makes it relatively easier to block the stimulation light from lasers. Laser line mirrors can be used in combination with glass bandpass filters to effectively prevent a specific laser line from reaching the PMT. Although lasers are highly monochromatic, optical filters are sometimes necessary to prevent other light components originating from the laser system (e.g., UV from the plasma in some gas lasers) or light leakage from the room from reaching the sample and the PMT.

The narrow-beam, highly directional light produced by the laser can also be focused to stimulate the OSL from narrow regions of a material. For example, 532 nm light from a Nd:$YVO_4$ solid-state diode-pumped laser or a 830 nm IR laser have been focused down to a spot of 20 μm in diameter for OSL stimulation of single grains for luminescence dating applications, achieving stimulation powers (photon flux) more than three orders of magnitude higher than standard OSL stimulation sources (Bøtter-Jensen *et al.*, 2002, 2003). The disadvantage of laser light is that a beam expander needs to be used for illumination of samples of larger diameter, such as detectors of a few millimeters in diameter. With an expanded beam, variation in the light intensity across the laser beam may result in a photon flux that is higher in parts of the sample than in others.

Lasers can be pulsed at high frequencies and with very short pulse durations (down to femptoseconds). The advantage of lasers over LEDs is that lasers can be Q-switched to produce light pulses of very high intensity. In a Q-switched laser the cavity is prevented from resonating, for example by blocking, removing, or displacing one of the cavity mirrors. Since laser light is not being produced in this condition, the metastable excited state is not depleted and population inversion continues to increase under continuous pumping. When the cavity is again allowed to resonate, the excited state is suddenly depleted by stimulated emission, giving rise to a pulse of light of high intensity. The situation is analogous to a flood gate opening and closing: when closed, the amount of water accumulates on one side of the gate; when opened, the water is allowed to flow creating a large wave.

The disadvantage of high stimulation intensities is the possibility of inducing luminescence from multi-photon processes in the material (e.g., two-photon absorption), which may contribute to the measurement background and not be of interest for a particular application. The OSL intensity increases linearly with stimulation power, whereas two-photon absorption processes contributing to the background increase quadratically. Investigations on the laser-induced background in POSL measurements of $Al_2O_3$:C have shown the importance of restricting the peak power delivered to the sample in order to decrease the background (Akselrod and McKeever, 1999).

For more information on laser systems, see for example Hecht (2001), Svelto (1998) and Sze and Ng (2007).

### 2.4.2 Light Detectors

All available commercial OSL readers and most prototype readers use PMTs for OSL detection because of their high sensitivity, fast response and low noise levels. Although disadvantages of PMTs include the need for high voltage and associated electronics, compact PMT packages powered by low voltages (e.g., 5 V) and containing the high-voltage power supply and signal-conditioning electronics are now available. Charge-coupled devices (CCDs) and other light transducers have been used for OSL detection, but only in laboratory prototypes. This situation may change in the future due to the fast-paced progress in optoelectronics.

#### *2.4.2.1 Photomultiplier Tubes*

A PMT consists of a vacuum tube containing a photo-sensitive cathode (photocathode) and a series of electrodes held at increasingly higher electric potential, capable of generating up to $10^8$ electrons at the final anode for each photon incident in the photocathode.

The principle of operation of a PMT can be broken down into three stages: light conversion, electron multiplication and charge collection (Figure 2.36). Light conversion is the first step in the detection process and consists of the conversion of photons incident on the photocathode to electrons emitted into the vacuum. This only occurs when electrons produced by the photoelectric effect have sufficient energy to escape the photocathode. This process can be characterized by a quantum efficiency, which is the ratio of the number of photoelectrons emitted to the number of incident photons. The photoelectrons generated at the photocathode are then focused and accelerated towards a sequence of electrodes (dynodes) for multiplication. Each dynode is coated with materials that produce more than

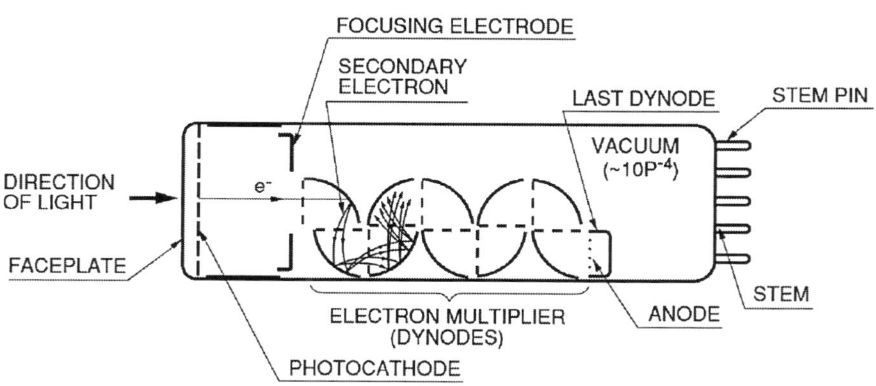


***Figure 2.36*** *Diagram illustrating the operation of a photomultiplier tube (PMT). Photons incident in the photocathode create electrons that are continuously accelerated inside a vacuum tube towards electrodes held at increasingly higher electrical potentials. The number of electrons is multiplied every time the accelerated electrons collide with the electrodes, generating a signal that can be detected as an electric current or voltage pulse in the anode. Reprinted from Photomultiplier Tubes: Basics and Applications (Third Edition) Hamamatsu. Copyright (2006) with permission from Hamamatsu Photonics K. K.*

one secondary electron for each incident electron, which is accelerated by the electrical potential difference between the dynodes during the transit period. For a PMT containing $n$ dynodes, each characterized by the same ratio between secondary electrons to incident electrons (secondary emission ratio) $\delta$, the multiplication factor is in principle $\delta^n$. Finally, the electrons are collected at the anode, generating an electrical current or a sudden change (pulse) in electrical potential.

The PMT is usually connected to a socket containing a voltage-divider circuit and connections for the signal and high voltage. The PMT and socket are enclosed in a light-tight housing. A high-voltage power supply is used in conjunction with the voltage-divider circuit to establish the appropriate electric potential differences between the electrodes. Voltages ranging from 500 to 3000 V are used, the polarity and value depending on the PMT design. Because of the influence of electric and magnetic fields in the PMT performance, electrostatic and magnetic shielding may be required for stable performance. If the housing is made of a metal, sufficient electrostatic shielding is achieved by grounding it. Magnetic shielding can be achieved by wrapping the PMT with a material of high magnetic permeability such as Permalloy (Hamamatsu Photonics K. K., 2007).

There are also PMT packages available integrating the PMT, voltage-divider circuit, high-voltage power supply and signal conditioning in one package. Models are available in which the package is powered by a low voltage (e.g., +5 V) and the signal output is a transistor–transistor logic (TTL) pulse (>2 V) for each photon detected.

The spectral response of the PMT is described by the quantum efficiency, which depends on the photocathode material and type (transmission or reflection mode), window and the photon energy. PMTs commonly used in OSL applications have a bialkali photocathode, a spectral response in the 160–650 nm range and a peak radiant sensitivity around ~400 nm. Sensitivity in the UV range requires windows that transmit in this range, typically fused silica glass.

There are two PMT operating modes: current mode and photon counting mode. In the first mode, the anode current is used as an indication of the detected light intensity, whereas in the second mode voltage pulses associated with electrons generated by a single photon detection are counted. The anode current can be as low as a few nanoamperes, corresponding to the PMT current in the dark, up to maximum rated currents of the order of hundreds of microamperes, depending on the PMT type. This provides good linearity and wide dynamic range. However, photon counting results in higher signal-to-noise ratios and therefore is more suitable for detection of the low light levels associated with the OSL from insensitive materials or materials exposed to low doses.

There are several sources of noise (dark current or dark counts) in the PMT, including thermionic emission from the photocathode and dynodes, leakage current, ionization current from residual gases and environmental radiation. Therefore, appropriate operation of the PMT involves the determination of the voltage to be applied and, in photon counting mode, of the discrimination voltage level below which pulses are not counted to achieve optimum signal-to-noise ratio. The choice of the voltage can be made based on measurements of the number of counts versus voltage in the presence of a light source (signal) compared to the same measurement in the dark (noise). Figure 2.37 shows an example of such measurement, where the signal ($S$), noise ($N$) and signal-to-noise ratio ($S/N$) are plotted as a function of the applied voltage, indicating an optimum signal-to-noise ratio for voltages in the 1050–1150 V range. If a multi-channel pulse height analyzer is available, the discrimination level

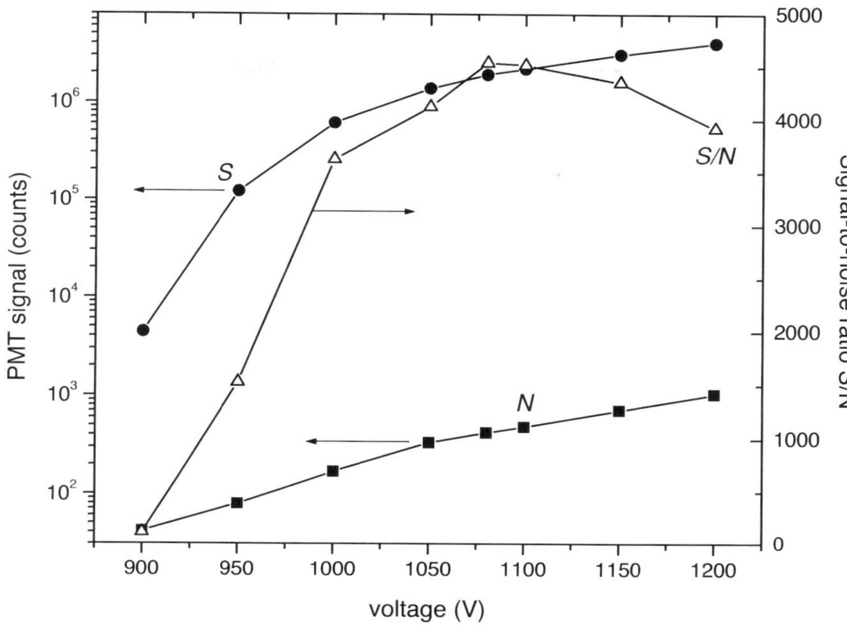


*Figure 2.37* Example of PMT signal and noise as a function of voltage for a particular PMT model and discrimination voltage (−10 mV). The figure shows that the ratio between signal (S) and noise (N) is optimum for voltages in the 1050–1150 V range.

can be based on the pulse height distribution for a particular applied voltage. Alternatively, signal-to-noise curves similar to Figure 2.37 can be measured for various discrimination voltage levels to determine the optimum combination of applied voltage and discrimination level. Because of the temperature dependence of thermoionic emission, cooling of the PMT can be effective in reducing the dark count rate particularly for photocathodes with small potential barriers for electron emission, used for increased quantum efficiency at long wavelengths and enhanced red and IR sensitivity.

In photon counting mode the PMT signal linearity is excellent over a wide range, being limited by pulse overlapping for high count rates. The upper limit of linearity is therefore related to the pulse-pair resolution, which is the minimum period between pulses that can be separately detected. Typical rise and fall times and pulse widths are in the nanosecond to tens of nanoseconds range. For a pulse-pair resolution of 18 ns, the deviation from linearity is less than 5% at $10^6$ cps, but increases rapidly for higher count rates. Correction for the PMT non-linearity can be made using the approximation in Equation (2.50):

$$N = \frac{N'}{1 - N't}, \tag{2.50}$$

where $N$ is the actual count rate ($s^{-1}$), $N'$ is the measured count rate ($s^{-1}$), and $t$ is the pulse-pair resolution (Hamamatsu Photonics K. K., 2007). The PMT sensitivity drifts with time and for that reason a warm-up period of 30–60 min is recommended.

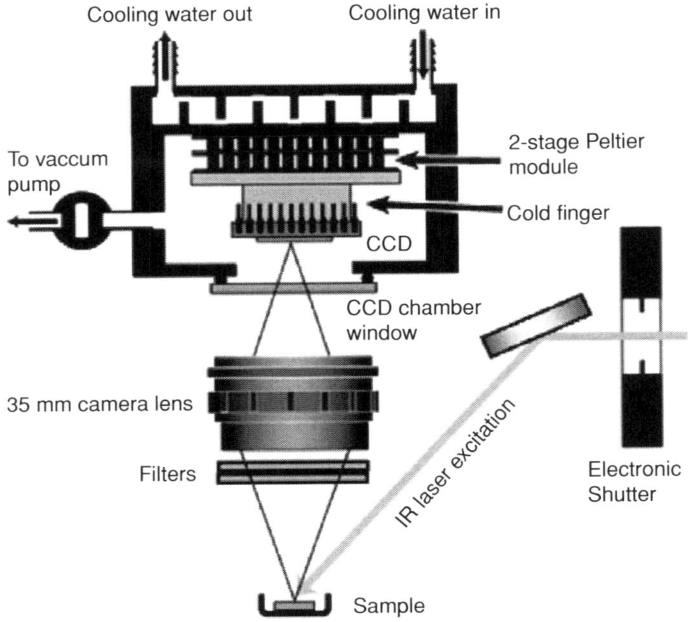


***Figure 2.38*** *Example of CCD-based OSL system for imaging. Reprinted from Radiation Measurements, Baril, M.R., CCD imaging of the infra-red stimulated luminescence of feldspars. Vol. 38, 81–86. Copyright (2004) with permission from Elsevier.*

Comprehensive reviews on the basics and applications of PMT can be found in online resources available from PMT manufacturers (e.g., Hamamatsu Photonics K. K., 2007).

#### *2.4.2.2 Charge-Coupled Devices (CCDs)*

CCDs are position-sensitive light detectors described as offering "low dark current, low noise, low-voltage operation, good linearity and good dynamic range" (Sze and Ng, 2007). CCDs have been used on a limited scale for OSL measurements with spatial resolution (OSL imaging) or wavelength resolution (OSL emission spectroscopy).

In OSL imaging, the objective is to map the OSL intensity as a function of position on the sample or detector. Figure 2.38 shows an example of a setup for OSL imaging. Applications include mapping dose gradients across the sample or simultaneously measuring the OSL from multiple bright grains within the sample (Baril, 2004; Duller, Bøtter-Jensen and Markey, 1997; Greilich, Glasmacher and Wagner, 2002; Greilich and Wagner, 2006). In principle, the OSL image can also be compared with other imaging techniques, such as optical microscopy or elemental mapping, to better characterize the source of the OSL signal (Greilich and Wagner, 2006). Other interesting applications of OSL imaging include the possibility of distinguishing between static and dynamic exposures in personal dosimetry (Akselrod, Agersnap Larsen and McKeever, 2000) and of dose mapping for medical applications (see Chapter 5). Spatially resolved measurements can also be achieved using PMTs and a laser or sample scanning system (Bailiff and Mikhailik, 2003; Bøtter-Jensen, 1997; Idri *et al.*, 2004).

In OSL emission spectroscopy, the OSL is collected and focused into a diffraction grating to disperse the light spatially according to its wavelength. The dispersed light is focused onto the CCD with different wavelengths falling onto different groups of pixels on the CCD. The CCD output per pixel, or group of pixels, is then a measure of the intensity of the light as a function of wavelength. An example of a setup for OSL emission spectroscopy can be found in Rieser, Habermann and Wagner (1999). Compact spectrometers containing the diffraction grating and the CCD are currently available and have been used for OSL spectroscopy (Bos *et al.*, 2002).

#### 2.4.2.3 *Other Detectors*

Other semiconductor light detectors such as photodiodes and avalanche photodiodes can in principle be used for detection of OSL. The advantage is the possibility of integrating the OSL material with the semiconductor light source (LED) and light detector (photodiode) in a single packaged unit.

Such an integrated sensor based on the OSL from CaS:Ce,Sm, an IR GaAs LED and a GaAsP photodiode with sensitivity below the mGy level has been described and characterized for space dosimetry and real-time monitoring of radiation beams (Dusseau *et al.*, 2000; Garcia *et al.*, 2006; Vaillé *et al.*, 2003, 2005). A similar approach has also been proposed, in this case using $Al_2O_3$:C dosimetric material as a substrate to grow a high quantum-efficiency GaN-based p-i-n photodiode (Chen, Poochinda and Stoebe, 2006). The authors estimated that such a device could in principle detect doses as low as 0.06 mGy, but this has not been demonstrated experimentally.

### 2.4.3 Optical Filters

Optical filters must be judiciously chosen to prevent stimulation light from reaching the light detector and to optimize the signal-to-noise ratio. The choice must be based on a combination of factors, such as OSL emission wavelengths of interest, optimum stimulation wavelength, choice of stimulation modality, required sensitivity and so on. There is no unique answer for a particular problem, but some filter combinations have been popularly adopted by different research groups and in different equipment. Table 2.1 lists some of the filter combinations reported in the literature and the materials with which these combinations have been used.

Figure 2.39 shows the emission from blue LEDs, centered at ~470 nm, and the transmission spectra of the Schott GG-420 (Schott AG) and Hoya U-340 (Hoya Corporation, USA) filters used in the Risø TL/OSL readers. The Schott GG-420 filter is used in front of the LEDs to cut the tail of the LED emission extending into the detection region, as can be observed in the figure. Some of the bandpass glass filters such as Hoya U-340 and Schott UG-1 transmit in the IR region (>700 nm). To solve this problem, some filters are coated with a metal oxide layer (e.g., Schott DUG11X) or are used in combination with another bandpass filter (e.g., Schott BG-39) to block the IR component.

**Table 2.1** *Examples of light sources and optical filters used for OSL measurements of different materials, as reported in the literature*

| Stimulation | Stimulation filters[a] | Detection filters[a] | Transmission window[b] | Stimulation mode[c] | Suitable material[d] | Available instruments/References |
|---|---|---|---|---|---|---|
| 532 nm (Nd:YAG laser) | Schott GG-495 | Corning 5-58 | 340–480 nm | POSL | $Al_2O_3$:C | Luxel systems (Akselrod and McKeever, 1999) |
| Green LEDs (525 nm) | Schott GG-495 | Hoya U-340 (7.5 mm) | 270–380 nm | CW-OSL | $Al_2O_3$:C[f] | Risø readers (Bøtter-Jensen, Thomsen and Jain, 2010) |
| | Schott OG-515 | Hoya B-370 | 300–450 nm | CW-OSL | $Al_2O_3$:C | InLight readers (Jursinic, 2007) |
| | Schott GG-495 | Corning 5-58 | 340–480 nm | POSL | $Al_2O_3$:C | (Yukihara *et al.*, 2009) |
| Blue LEDs (470 nm) | Schott GG-455 | Hoya U-340 (7.5 mm) | 270–380 nm | CW-OSL | Quartz, feldspars, $Al_2O_3$:C | Risø reader (Bøtter-Jensen, Thomsen and Jain, 2010) |
| | Schott GG-455 | Schott UG-1 + Schott BG-39 | 330–410 nm | CW-OSL | | (Smetana *et al.*, 2008) |
| | UV protection film | Schott DUG11X[e] | 290–380 nm | CW-OSL | BeO | (Sommer, Jahn and Henniger, 2008) |
| IR LEDs (830 nm) | | Schott BG-39 | 330–620 nm | CW-OSL | Feldspars | Risø reader (Bøtter-Jensen, Thomsen and Jain, 2010) |

[a] Filter designations are mostly based on published literature, but equivalent filters are sometimes available from different manufacturers. For example, Kopp 5113 filters have similar characteristics to Corning 5-58.
[b] The lower and upper transmission wavelengths were approximated based on the wavelength corresponding to 5% of the maximum transmission.
[c] "POSL" indicates the filter combinations in which background can be high if used in CW-OSL mode. For filter combinations indicated as CW-OSL, pulsed stimulation is sometimes used for time-resolved discrimination between the OSL from different materials or luminescence centers (see Section 2.3.2.2).
[d] This column indicates most used OSL materials that have been used with the particular stimulation source and filter combination. It does not mean that other materials cannot be used with satisfactory performance, depending on the intended application.
[e] The DUG11X is a Schott UG 11 coated with metal oxide layer to block the secondary transmission band around 720 nm.
[f] When using Hoya U-340 filters with $Al_2O_3$:C, most of the signal is from the main luminescence centers (*F*-centers), which consists of a broad emission band centered at 415 nm. However, a UV emission band centered at 335 nm is also detected. See Section 2.6.1.

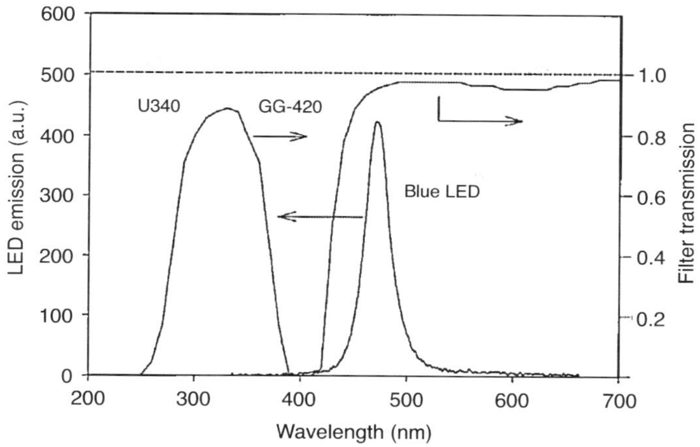


***Figure 2.39*** *Emission spectrum of blue LEDs and transmission spectrum of optical filters used in front of the PMT (Hoya U-340) and in front of the LEDs (Schott GG-420) in the commercial Risø TL/OSL readers. Reprinted from Radiation Physics and Chemistry, Bøtter-Jensen, L., Murray, A.S., Optically stimulated luminescence techniques in retrospective dosimetry. Vol. 61, 181–190. Copyright (2001) with permission from Elsevier.*

### 2.4.4 Light Collection

The need for optical filters and additional optical ports in the design of an OSL reader often imposes a limit on the distance between the OSL detector and the light detector (say, PMT), to the detriment of light collection. Additional light-collection optics can be used to improve the OSL signal detected by the PMT, particularly for samples that are extremely insensitive to ionizing radiation (i.e., which have weak OSL signal) or for light detection at very low doses. Markey *et al.* (1996a) showed that the use of ellipsoidal mirrors, with the sample placed in one of the focal points (Figure 2.40), can increase the sensitivity of the instrument by a factor of between 2 and 3.5. A similar approach has been used by Rieser, Habermann and Wagner (1999) for OSL emission spectroscopy. Other authors have achieved an improvement of up to six times with a combination of a convex lens and an ellipsoidal mirror (Yukihara *et al.*, 2007). If lenses are used for improving the light collection, care must be taken to make sure the glass material transmits the wavelength associated with the OSL emission. For UV detection, UV-grade fused silica lenses are necessary. Anti-reflective coatings can be applied to reduce transmission losses (Fischer, Tadic-Galeb and Yoder, 2008).

### 2.4.5 Sample Heaters

OSL measurements do not necessarily require a sample heater. However, it may be advantageous to perform the OSL stimulation at elevated temperatures. Raising the sample temperature may increase the OSL intensity due to: (i) an increase in stimulation efficiency caused by the presence of thermally assisted processes, or (ii) a decrease in the influence of shallow traps (at higher temperature the charges in the shallow traps are quickly released back to the conduction band and become available for recombination). However, elevated temperatures may also cause a decrease in the OSL intensity due to thermal quenching of the

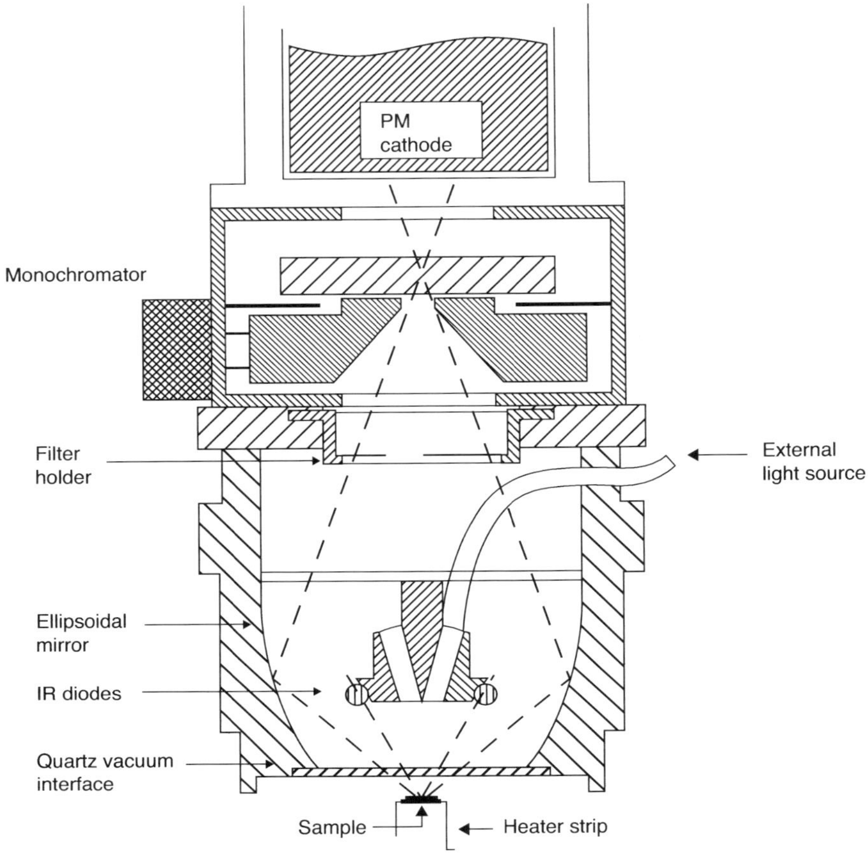


***Figure 2.40*** *Example of an OSL reader design incorporating an ellipsoidal mirror to improve the OSL collection. Reprinted from Radiation Protection Dosimetry, Markey, B.G., Botter Jensen, L., Poolton, N.R.J., et al. A new sensitive system for measurement of thermally and optically stimulated luminescence. 66 (1–4), 413–418. Copyright (1996) with permission from Oxford University Press.*

luminescence. OSL measurements at elevated temperatures have typically been performed using combined TL/OSL readers in which the sample is heated by passing a controlled electrical current through a heating element made of a low mass, high-resistance alloy strip. The temperature is monitored using a thermocouple connected to the heater.

## 2.5 Available OSL Readers

### 2.5.1 Experimental Arrangements

In previous sections we discussed a variety of approaches by which the general instrument layout shown in Figure 2.23 has been implemented by different research groups (e.g., Figures 2.33, 2.34 and 2.38). In addition to those, it is illustrative to include here a description of

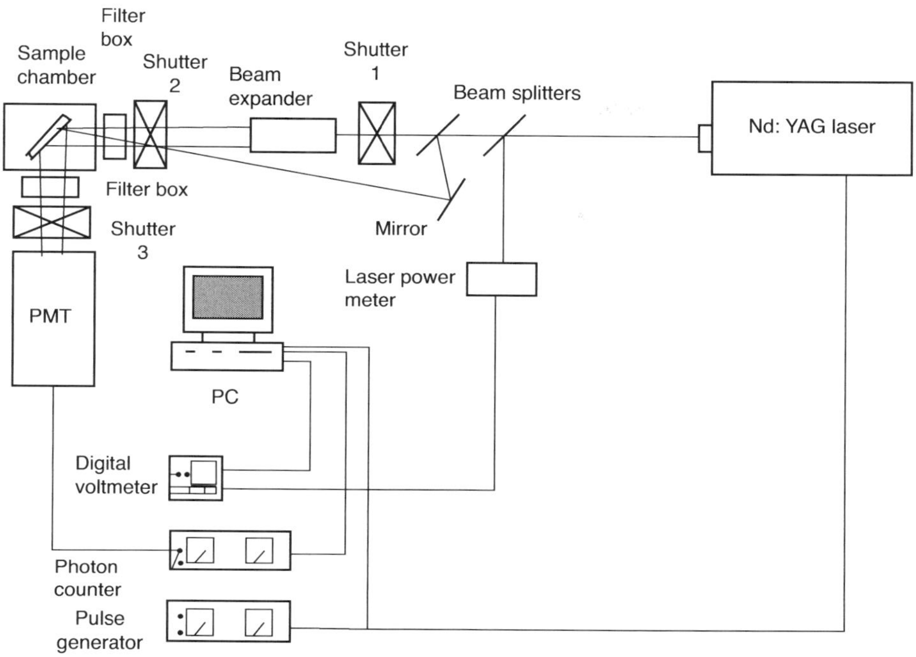


***Figure 2.41*** *Schematic representation of an OSL system which uses a pulsed laser for stimulation of the OSL signal. Reprinted from Radiation Protection Dosimetry, Akselrod, M.S., McKeever, S.W.S., A radiation dosimetry method using pulsed optically stimulated luminescence. Vol. 81, 167–176. Copyright (1999) with permission from Oxford University Press.*

the laser-based system shown in Figure 2.41, which was used to develop the POSL dosimetry system currently used by Landauer (Luxel dosimetry system, Landauer Inc., USA).

The system shown in Figure 2.41 uses the 532 nm laser line from a frequency-doubled Nd:YAG laser to stimulate the OSL from a material placed inside a sample chamber, while the OSL is detected by a PMT. Beam splitters are used to separate the beam into a weak intensity beam and a strong beam. The weak beam is used to stimulate samples irradiated with high doses, whereas the strong beam is used to stimulate samples irradiated with low doses. The choice of stimulation intensity is controlled by a computer using shutters placed in the beam path. A beam expander is used to provide uniform illumination of the sample. In this setup a glass filter which transmits the stimulation light, but blocks light with shorter wavelength (e.g., Schott OG-515) is used in the stimulation window of the sample chamber. A combination of 532 nm laser line mirror and a bandpass glass filter centered at 420 nm (e.g., Corning 5-58) is used in the detection window of the sample chamber.

### 2.5.2 Automated Research Readers

Research on OSL materials has benefited tremendously from the development of automated readers capable of carrying out sequences of OSL measurements and irradiations,

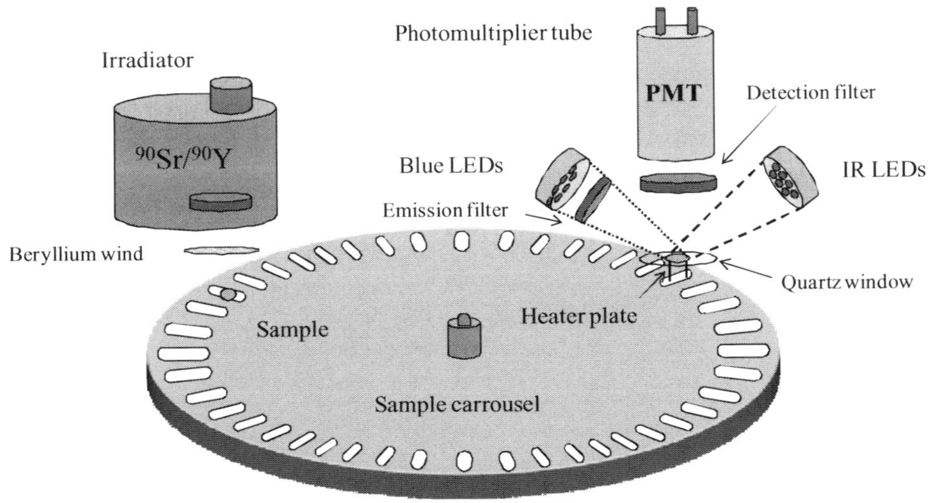


*Figure 2.42* Schematics of the Risø TL/OSL reader (Risø National Laboratory, Denmark). Figure courtesy of Dr. Kristina J. Thomsen, Risø National Laboratory.

as well as TL measurements and heating procedures, of multiple samples. There are two such systems currently available commercially, the Risø TL/OSL reader (Risø National Laboratory, Denmark) and the Daybreak TL/OSL readers (Daybreak Nuclear and Medical Systems, Inc.). Detailed information can be found elsewhere (Bortolot, 2000; Bøtter-Jensen, 2000; Bøtter-Jensen, McKeever and Wintle, 2003; Bøtter-Jensen et al., 2002, 2003, 2010; Thomsen *et al.*, 2006).

The basic stimulation and detection scheme of the Risø TL/OSL reader was shown in Figure 2.34. LED clusters (blue or IR) are used for CW-OSL and LM-OSL measurements, while the OSL is detected by a PMT. A longpass glass filter is used in front of the blue LED to remove short wavelengths that could reach the PMT (see Figure 2.39). The filter pack used in front of the PMT can be changed according to the stimulation type (see Table 2.1). For readout, the sample is placed in contact with a heating element that allows OSL measurements at elevated temperatures, pre-irradiation or post-irradiation heating procedures and TL measurements. The entire system is mounted in a sample handling system with capacity for up to 48 samples (Figure 2.42). The system can perform long sequences of irradiation, readout and heating stages for all samples. Irradiation is typically performed using a $^{90}$Sr/$^{90}$Y beta source, although the use of X-ray generators is being investigated (Andersen, Bøtter-Jensen and Murray, 2003). One needs to be aware of the possibility of irradiation or illumination of adjacent sample positions (Bray, Bailey and Stokes, 2002). For this reason, the user may consider spacing the samples in the sample turntable.

Recent developments of the Risø TL/OSL reader include the possibility of performing time-resolved OSL measurements using pulsed stimulation (Denby *et al.*, 2006), CW and LM-OSL measurements of single grains using green and IR lasers, and RL measurements using a light guide connecting the sample position below the irradiator to the light detection system (Bøtter-Jensen *et al.*, 2003).

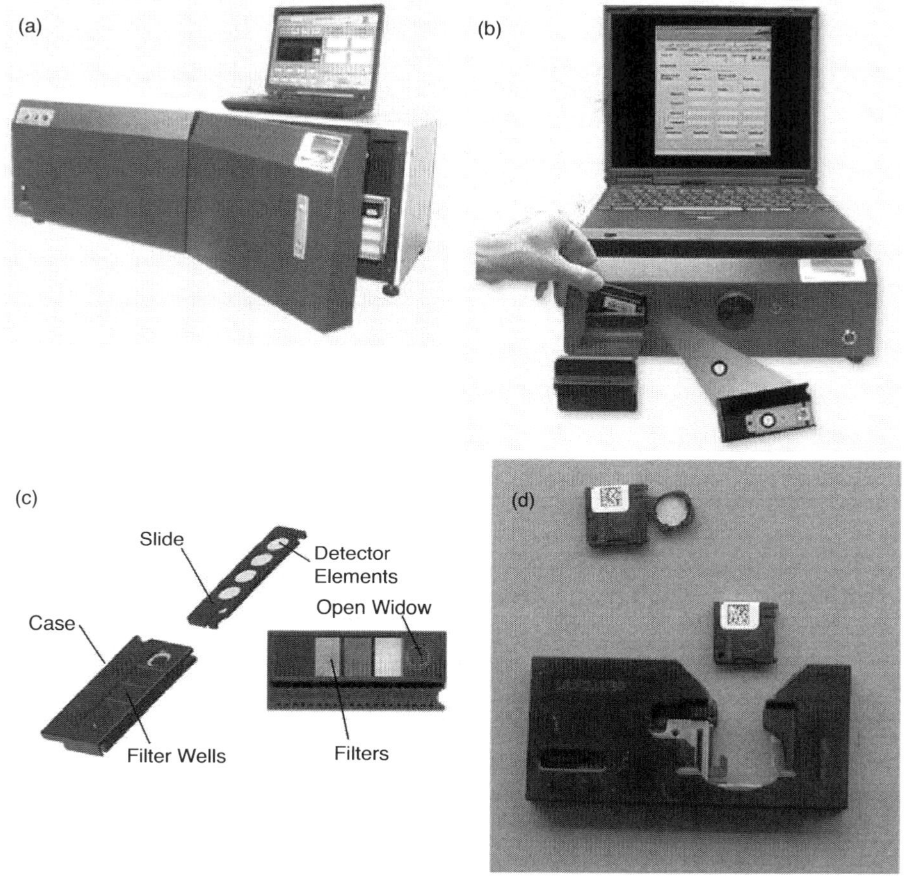


***Figure 2.43*** *InLight dosimetry system: (a) 200 unit automatic reader; Courtesy of Landauer (b) microStar reader; Courtesy of Landauer (c) InLight dosemeter; Reprinted from Radiation Light Dosimetry, Perks, C.A., Le Roy, G. and Prugnaud, B., Introduction of the InLight Monitoring Service, Vol. 125, 220–223. Copyright (2007) with permission from Oxford University Press and (d) nanoDot dosimeter and adapter. Courtesy of Landauer.*

### 2.5.3 Commercial Dosimetry Readers

At the time of writing there were only two dosimetry systems commercially available: the Luxel and the InLight dosimetry systems (Landauer Inc., USA). Luxel dosimeters are read in-house using readers similar to the one shown in Figure 2.41. InLight readers (including the portable microStar) are available for readout onsite. Figure 2.43 shows examples of InLight readers and dosimeters.

The InLight readers use green light from an array of LEDs for OSL stimulation. The LEDs can be operated in "weak beam" mode (six LEDs only) for readout of detectors irradiated with high doses, or "strong beam" mode (38 LEDs) for readout of detectors

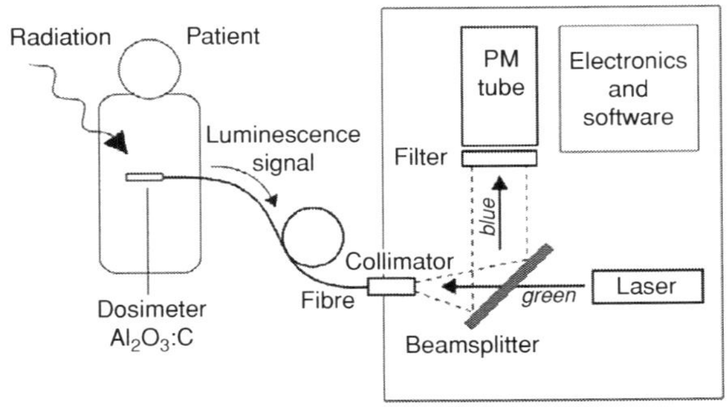


***Figure 2.44*** *Conceptual diagram of the optical fiber systems developed by the research group at Risø National Laboratory. The OSL detector is placed at the tip of an optical fiber, which is connected to a reader containing the stimulation and detection components. The stimulation light from the laser is injected into the optical fiber to stimulate the OSL detector on the other end of the optical fiber. The OSL signal travels back through the optical fiber and is refelected by the beamsplitter towards the PMT. Optical filters are used in front of the PMT to reject the stimulation light while transmitting the OSL. Reprinted from Physics in Medicine and Biology, Aznar, M.C. et al., Real-time optical-fibre luminescence dosimetry for radiotherapy: physical characteristics and applications in photon beams. Copyright (2004) with permission from IOP Publishing Ltd.*

irradiated with low doses. The OSL is detected by a PMT using Hoya B-370 filters (see Table 2.1). Each detector is stimulated for a period of ~1 s.

The InLight readers are designed for readout of InLight dosimeters exclusively, which are based on the same type of OSL detector used in the Luxel dosimetry system. Figure 2.43c shows in detail the case, detector slide and detector elements of the InLight dosimeter badge. Figure 2.43d shows a case containing a single OSL element used for medical applications (nanoDot) and the adapter used to read the nanoDot dosimeter in the microStar reader.

### 2.5.4 Optical Fiber Systems

Optical fiber systems having an OSL material as the probe's detector have been proposed for real-time medical dosimetry (Andersen *et al.*, 2003; Polf *et al.*, 2002; Ranchoux *et al.*, 2002) and environmental monitoring applications (Klein *et al.*, 2006).

Figure 2.44 shows one example of an optical fiber OSL system. The reader consists of a green laser coupled to an optical fiber that carries the stimulation light to the OSL probe ($Al_2O_3$:C detector). The OSL emitted by the detector is carried by the optical fiber to the reader and reflected by the beamsplitter (dichroic mirror) to the PMT. In addition to the OSL signal measured under optical stimulation, the system also allows real-time measurements of the dose rate using the RL signal observed during irradiation, or a combination of RL and OSL measurements carried out during irradiation. The probe itself does not require electrical connections and is immune to electromagnetic interference, but light generated in the optical fiber during irradiation may be an issue. For environmental measurements, the only limitation on the distance between the reader and the probe is attenuation by the optical fiber. (See more on optical fiber systems in Chapters 5 and 6.)

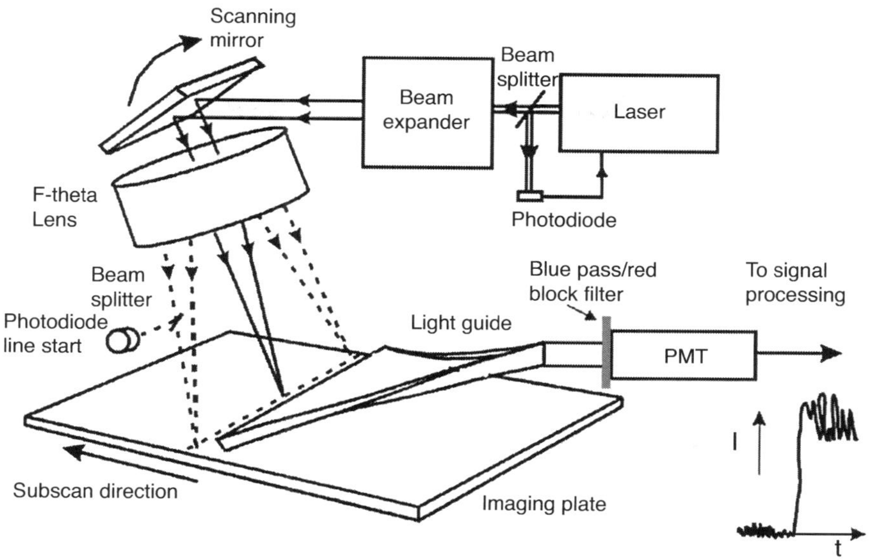


*Figure 2.45 Schematics of a phosphor screen or image plate reader. The image is obtained by scanning a laser spot over the surface of a screen or imaging plate containing the OSL material (photostimulable phosphor); the OSL is detected by a PMT through the light guide. Reprinted from Physics in Medicine and Biology, Rowlands, J.A., The physics of computed radiography, Vol. 47, R123–R166. Vol. 47, R123–R166. Copyright (2002) with permission from IOP Publishing.*

### 2.5.5 Imaging Systems

A major area of application of the OSL technique is X-ray imaging for digital radiography (Bushberg *et al.*, 1994; Rowlands, 2002; Sonoda *et al.*, 1983). In this area, the term photostimulable luminescence (PSL) is often used instead of OSL, and the materials are referred to as photostimulable phosphors or X-ray storage phosphors. In this application, the traditional X-ray photographic film is replaced with a polymeric film (phosphor screen or image plate) containing OSL material mixed with an organic binder. The image is obtained by scanning the phosphor (known in this application as a "screen") with a laser spot (usually a HeNe or solid-state red laser) and detecting the resultant OSL signal, which is then converted to a digital output and processed to obtain the final image. Figure 2.45 shows the schematics of a reader, showing how the OSL from the phosphor screen is read using a laser spot scanned over the screen surface. Advantages of these systems in comparison to photographic films include high dynamic range, linearity and reusability (Thoms, Von Seggern and Winnacker, 1991), but they are not typically used for dosimetry because the phosphors commonly used have high effective atomic number and an OSL signal that fades with time.

There is currently no commercial reader for dose mapping (2D dosimetry) using OSL detectors, although dose mapping has been used routinely in commercial OSL dosimetry for discrimination between static and dynamic exposures (Akselrod *et al.*, 2000) and in TL dosimetry (Olko *et al.*, 2006a). Dose mapping has been performed by using either position-sensitive light detectors (see Section 2.4.2.2) or by scanning the detector with respect to the detection system (Bailiff and Mikhailik, 2003; Bøtter-Jensen, 1997; Idri *et al.*, 2004). Landauer recently introduced a dosimeter capable of measuring dose profiles in X-ray computed tomography (CT) based on the OSL from $Al_2O_3$:C. The system uses a laser to stimulate OSL measurements every 0.1 mm along a 15 cm OSL strip (Landauer, 2006). An equivalent system based on LEDs and a motorized translational stage has also been described for dose profile measurements in X-ray CT and in proton therapy (Yukihara *et al.*, 2009, 2010). An OSL spot-scanning system with 1 mm spatial resolution based on BeO was described by Jahn, Sommer and Henniger (2009).

Gaza *et al.* (2006a) used a confocal microscope system and a moving sample stage (*x*–*y* plane) to map the OSL emission from beta-irradiated ($^{90}$Sr/$^{90}$Y source) $Al_2O_3$:C single crystals as a function of (*x*, *y*) position and depth into the chip (*z*-direction). The OSL was stimulated with a diode-pumped Nd-YAG laser at 532 nm and the OSL emission was detected with a PMT. This arrangement allowed measurement of both the depth–dose profile $D(z)$ for the beta particles and the two-dimensional variation of the OSL over the *x*–*y* planes at various depths, *z*.

### 2.5.6 Portable OSL Readers

In addition to Landauer's InLight microStar portable reader, portable OSL readers have been described by Thomsen *et al.* (2008) and Smetana *et al.* (2008). In both cases the instrument design is similar to the one shown in Figure 2.34, except for the sample handling system. Portable readers can be easily deployed for applications in security or prompt readout in personal or medical dosimetry.

## 2.6 Complementary Techniques

Complementary techniques can be used to characterize the properties of OSL detectors for optimization of the readout system and understanding of the underlying physical processes. Essential measurements include the determination of the OSL emission and stimulation specta, which informs the choice of stimulation source and detection window. Correlation between OSL and TL signals can assist in determining the presence and influence of shallow traps, identification of optically active or inactive traps, the presence of more than one trap contributing to the OSL signal, and the influence of deep traps. Here we provide some examples of measurements that are helpful to understand the OSL process.

### 2.6.1 OSL Emission and Stimulation Spectrum

Determination of the OSL emission spectrum of a particular material is essential to choose the appropriate detection window for the OSL measurements. Since the choice of detection window depends on the separation between stimulation and detection wavelengths, it is also important to determine how the OSL intensity varies with the stimulation wavelength, that is, to determine the OSL stimulation spectrum.

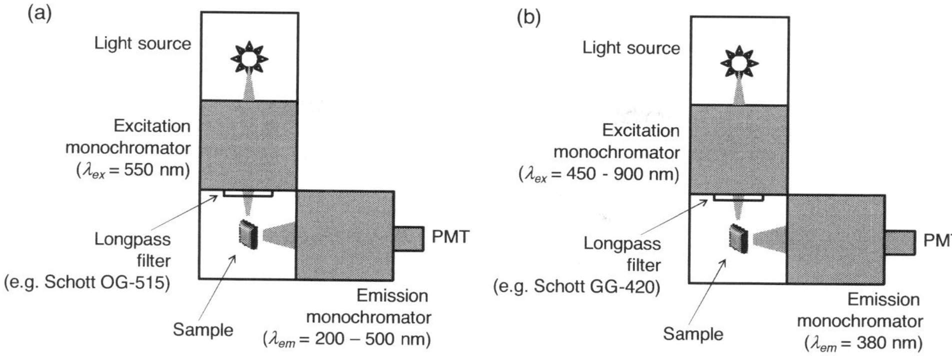


***Figure 2.46*** *Schematic setup for measurement of: (a) the OSL emission spectrum, and (b) the OSL stimulation spectrum using a spectrofluorometer.*

Systems with various designs have been used to determine the OSL emission and stimulation spectrum (e.g., Bos *et al.*, 2002; Bøtter-Jensen *et al.*, 1994; Rieser, Habermann and Wagner, 1999). Here we will illustrate this type of measurement using a spectrofluorometer because of its flexibility and wide availability in research laboratories. A spectrofluorometer has a light source and two monochromators, one to select a particular wavelength of the light source to excite or stimulate the sample, and the second one to select the wavelength of the luminescence to be detected. Both monochromators can be scanned independently to vary the wavelength of the excitation/stimulation light and of the detected luminescence. Additional optical ports may be available for stimulation of the sample with other light sources (e.g., pulsed LED or laser).

Figure 2.46a shows the schematic layout for the measurement of an OSL emission spectrum. The excitation monochromator is fixed at a constant wavelength $\lambda_{ex}$ (e.g., 550 nm) to stimulate the sample. The OSL emission spectrum is determined by scanning the wavelength ($\lambda_{em}$) of the emission monochromator between 200 nm and 500 nm, while monitoring the PMT signal. A longpass filter (e.g., Schott OG-515) is shown in the stimulation light path to prevent the second harmonic of the stimulation light (225 nm in this case) from reaching the sample and the detection system. The measured signal also needs to be corrected for the spectral response of the detection system.

Figure 2.46b shows the schematic layout for the measurement of an OSL stimulation spectrum. In this case, the wavelength ($\lambda_{em}$) of the emission monochromator is fixed at the emission band of interest (e.g., 380 nm). The excitation wavelength ($\lambda_{ex}$) is scanned from 450 and 900 nm, while monitoring the PMT signal to obtain the OSL stimulation spectrum. The longpass filter has been changed to allow for a wider stimulation range. The OSL intensity must be corrected for the variable intensity of the stimulation light, which can be done using the signal from a photodiode usually included in commercial systems to monitor the intensity of the light at the exit of the excitation monochromator.

Examples of OSL emission and stimulation spectra of $Al_2O_3$:C measured using the system described above are shown in Figure 2.47. The main OSL emission band of $Al_2O_3$:C is observed at ~415 nm (Figure 2.47a). The stimulation spectrum shows that the OSL intensity increases with decreasing stimulation wavelength (Figure 2.47b).

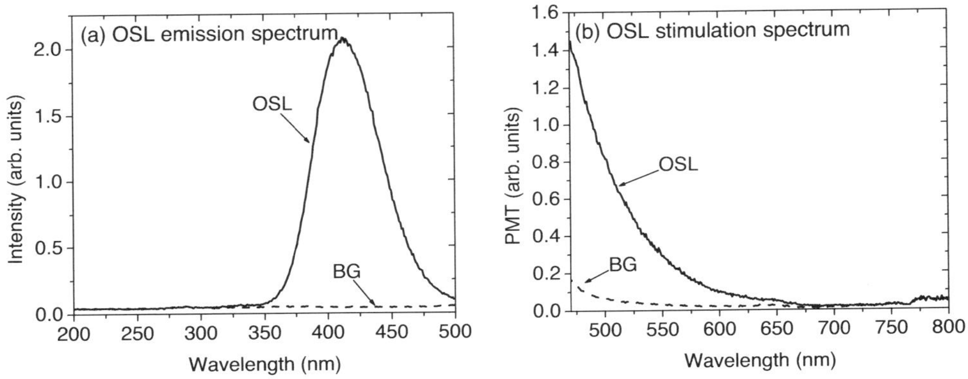


*Figure 2.47* (a) Emission and (b) excitation spectrum of the OSL of $Al_2O_3$:C irradiated with 1 Gy. BG indicates the background signal obtained after the OSL signal was erased by optical stimulation. The emission spectrum shows a dominant band at ~415 nm, whereas the stimulation spectrum shows a continuous increase in the OSL intensity with decreasing wavelengths.

### 2.6.2 Lifetime and Time-Resolved OSL Measurements

Information on the lifetime of the luminescence centers is important to determine the timing parameters used in the POSL technique (see Section 2.3.2.2). The lifetime can be measured using a pulsed excitation source (X-ray or UV) and a multi-channel scaler, in which the PMT pulses are classified into different channels according to the time elapsed (delay) since the excitation pulse. More simply, the OSL lifetime can also be directly measured by stimulating an irradiated sample with a pulsed light source. In both cases, the excitation/stimulation source pulses and the multi-channel scaler must be synchronized. Figure 2.48 shows an example of an OSL lifetime measurement of the main emission band of $Al_2O_3$:C, attributed to $F$-center emission, which has a lifetime of 35 ms (Akselrod *et al.*, 1998b; Evans, 1995).

Pulsed stimulation can also be used for time-resolved OSL measurements, allowing for the separation between emission bands of different lifetimes. For example, in the CW-OSL emission spectrum of $Al_2O_3$:C (Figure 2.47a) the $F$-center emission dominates. However, using pulsed stimulation and monitoring the luminescence during and in between the stimulation pulses, the spectrum shown in Figure 2.49 is obtained. A UV emission band centered at ~335 nm can be seen during the laser pulses, whereas in between the laser pulses only $F$-center emission is observed. As discussed in Section 2.3.2.2, time-resolved measurements have also been used to separate the OSL emissions from different minerals.

### 2.6.3 Correlations Between OSL and TL

The OSL phenomenon is inextricably related to the thermoluminescence (TL) phenomenon, except that in the case of TL the stimulation of the trapped charge population that ultimately leads to recombination and luminescence is by thermal energy, provided to the crystal by heating. The relationships between OSL and TL signals are important not only from a theoretical point of view, but they also help identify how deep, with respect to the conduction

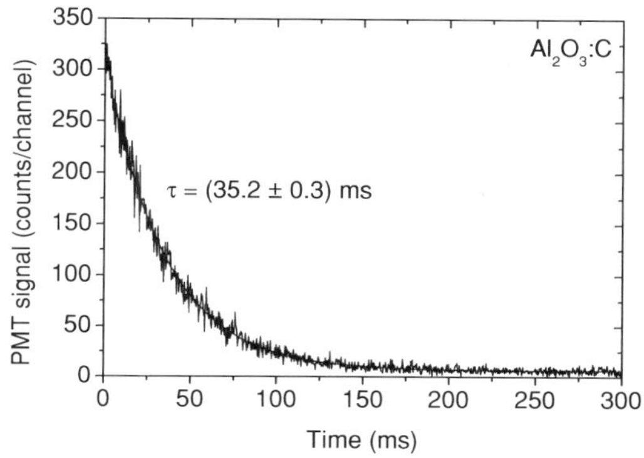


***Figure 2.48*** *Lifetime measurement of the OSL from $Al_2O_3$:C, in which the photons are counted as a function of the delay following a 10 ns green laser stimulation pulse. The OSL intensity decreases exponentially with a lifetime of ~35 ms, characteristic of F-center luminescence in $Al_2O_3$:C. The signal shown in the figure corresponds to the accumulated PMT counts after several stimulation pulses.*

band, are the trapping centers that contribute to the OSL signal. This in turn leads to an estimate of the traps' thermal stability. The TL/OSL correlations also aid in understanding the influence of other trapping centers (e.g., shallow traps or deep traps) on the OSL process.

Thus, there is a series of combined TL/OSL experiments that can provide useful insight into the OSL process and its relationship with trapping centers in the material. These include

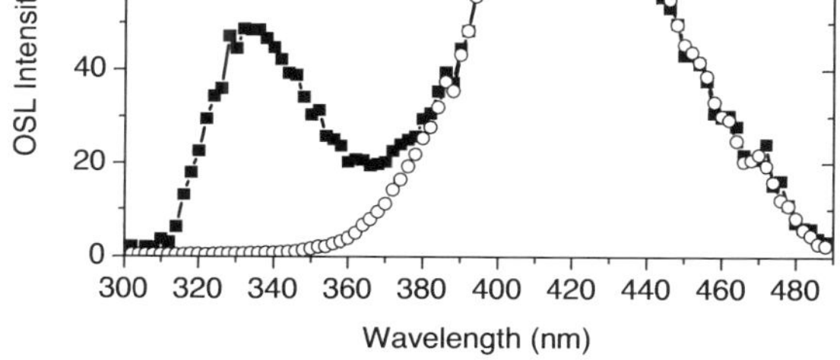


***Figure 2.49*** *Time-resolved OSL emission from $Al_2O_3$:C measured using pulsed stimulation and monitoring the OSL signal during and in between laser pulses. The laser pulses have a duration of ~100 ns and the pulse-to-pulse interval is 250 μs. The relatively high intensity of the UV emission band in this graph is a result of the short laser pulses: the UV emission persists only during the laser pulse, whereas the F-center luminescence persists during the entire 250 μs cycle. For more experimental details, please see Yukihara and McKeever (2006b).*

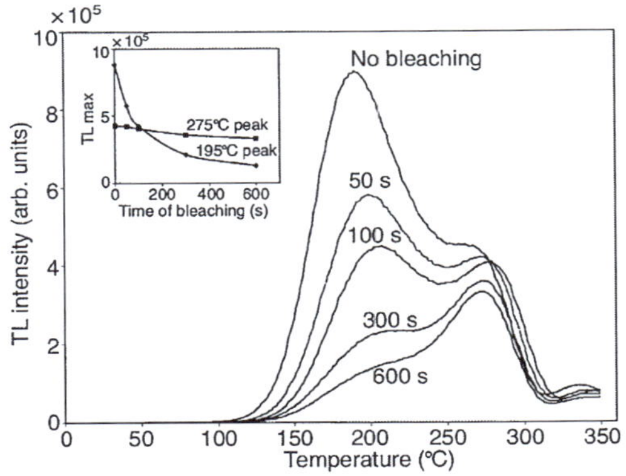


***Figure 2.50*** *TL curves of $Al_2O_3$:C following irradiation with 0.01 Gy and bleaching for various periods of time. The figure shows that the trapping centers responsible for the low temperature part of this double-peak structure are the most affected by optical stimulation, and likely the ones contributing the mostly to the OSL signal. Reprinted from Radiation Light Dosimetry, Akselrod, A.E., Akselrod, M.S., Correlation between OSL and the distribution of TL traps in $Al_2O_3$:C. Vol. 100 (1–4), 217–220. Copyright (2002) with permission from Oxford University Press.*

monitoring the effects of optical bleaching on the TL curves, and the effects of thermal annealing on the OSL curves.

### 2.6.3.1 Comparison Between TL Curves Before and After Bleaching

It is possible to investigate which TL peaks are affected by optical stimulation by comparing the TL curves of the material obtained immediately after irradiation with those obtained after irradiation and optical stimulation (bleaching), provided that the TL curves are reproducible.

As an example, Figure 2.50 compares the TL curve of $Al_2O_3$:C after irradiation (no bleaching) with the TL curves after irradiation and bleaching for different periods of time. This sample is a particular type of $Al_2O_3$:C having a double-peak TL structure (Akselrod and Akselrod, 2002). The figure shows that the optical stimulation preferentially bleaches the low-temperature portion of this composite TL curve, leaving a "hard-to-bleach" component on the high-temperature side almost intact, indicating that the low part of the TL peak is likely to be the one contributing to the OSL signal during optical stimulation.

One problem with such types of measurement is the possibility that the decrease in the TL signal is not caused directly by optical stimulation of the associated trapped charge population, but by the decrease in the concentration of recombination centers due to charges stimulated from other traps (Chruścińska, 2006; McKeever, 1991). Chruścińska (2009) stresses the importance of repeating the experiments at different doses, particularly at low doses, to test whether or not the concentration of recombination centers is affecting the results.

### 2.6.3.2 Partial Cleaning and Step-Annealing Experiments

An alternative method to investigate the influence of different trapping centers in the OSL process is to heat the sample to a temperature $T_{max}$ with a controlled heating rate and

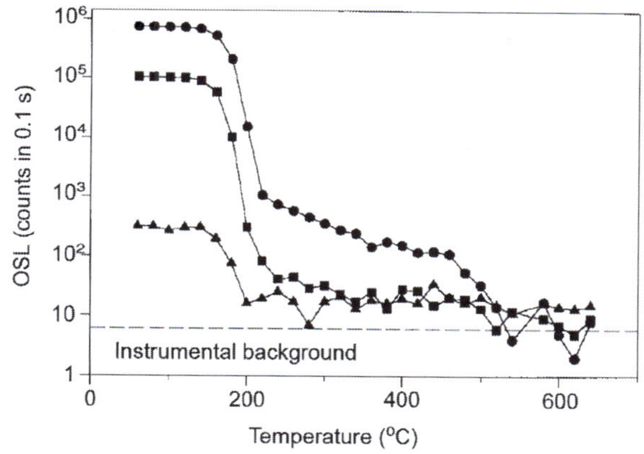


*Figure 2.51* OSL signal from $Al_2O_3$:C irradiated with three different dose levels (● = 1 Gy, ■ = 110 mGy and ▲ = 100 μGy) and pre-heated to the indicated temperature. The data indicate the temperature ranges responsible for depletion of the trapping centers associated with the OSL signal. Reprinted from Radiation Protection Dosimetry, Bøtter-Jensen, L., et al., Retrospective assessment of environmental dose rates using optically stimulated luminescence from $Al_2O_3$:C and quartz. Vol. 84 (1–4), 537–542. Copyright (1999) with permission from Oxford University Press.

duration prior to the OSL measurements, cool the sample to room temperature and observe the influence of this "pre-heat" on either the total OSL signal or shape of the OSL curve measured at room temperature. The purpose of the pre-heat is to empty the trapping centers related to the TL peaks at temperatures below $T_{max}$.

Step-annealing experiments refer to those in which the process above is applied repeatedly such that $T_{max}$ is increased in increments, the samples being pre-heated to increasingly higher temperatures before each OSL measurement. The sample is first irradiated, pre-heated to a temperature $T_1$, cooled to room temperature and probed with a short stimulation pulse (e.g., 0.1 s). The duration and intensity of the light pulses are chosen to not significantly deplete the OSL signal. The sample is then pre-heated again to a temperature $T_2 > T_1$, cooled to room temperature and the OSL is once more probed with a short stimulation. By repeating this process for increasing temperatures, it is possible to determine the temperature range over which the trapping centers responsible for the OSL signal are emptied.

Figure 2.51 shows a typical result of a step-annealing procedure applied to three $Al_2O_3$:C samples, the pre-heating being applied from 60 to 640 °C in 20 °C increments. We notice that the OSL signal is essentially constant for pre-heat temperatures in the 60–140 °C range. This suggests that: (i) there are no shallow traps contributing to the OSL signal and (ii) that the stimulation pulses are not depleting the signal. The main drop in OSL signal occurs around 150–220 °C, indicating that this is the temperature range in which the trapping centers responsible for the OSL signal are thermally emptied. This result agrees with the conclusion from the data shown in Figure 2.50, although the $Al_2O_3$:C samples used in the two studies are different. For pre-heating to temperatures above 250 °C, the OSL signal continues to decrease gradually, indicating the contribution from trapping centers located deeper in the energy band gap.

The degree of OSL depletion due to the stimulation pulses should also be tested by repeatedly probing the OSL signal with short pulses, but without pre-heating. If necessary, these data can be used to correct the data obtained with the step-annealing procedure. Also, instead of short pulses the step-annealing can be performed by measuring the complete OSL curve. In this case, the sample needs to be re-irradiated prior to each pre-heating and tested for sensitivity changes after each cycle. To avoid accumulation of charges at deep traps, it may be necessary to heat the sample to empty all trapping centers.

### 2.6.4 Other Phenomena

OSL can be included in a larger class of stimulated relaxation phenomena, along with thermally stimulated conductivity, thermally stimulated depolarization current, photostimulated conductivity, and so on. (Bräunlich, 1979; Chen and McKeever, 1997). In these phenomena, two conditions are required: (i) first the system must be excited from equilibrium to a non-equilibrium steady-state situation (metastable state); (ii) then the system must be stimulated, that is, provided with sufficient energy to overcome the potential barrier between the metastable and equilibrium states. The nomenclature for the different phenomena typically stipulates the stimulation source and the particular signal measured during the relaxation process. For instance, photostimulated conductivity is the transient conductivity observed when the material is irradiated and subsequently stimulated with light. For historical reasons the term "thermoluminescence" has survived as the equivalent to "thermally stimulated luminescence" and both are still used in the literature.

## 2.7 Overview of OSL Materials

OSL is observed in a wide variety of natural and artificial materials. Well-known OSL materials include quartz, feldspar and carbon-doped aluminum oxide, but OSL has also been observed in topaz, ceramics, dental enamel, common salt, washing powder, water softener, monocrystalline silicon carbide and a variety of doped and undoped materials such as magnesium silicate, spinel, glasses and so on. In spite of the wide variety of material exhibiting OSL, just a few of them have been widely used in OSL applications. In fact, the availability of materials has been one of the disadvantages of the OSL technique when compared to the TL technique.

In this section we will describe some of the general properties of OSL materials of interest for the applications discussed in this book. In particular we discuss the OSL stimulation and emission spectra, luminescence lifetime, OSL mechanisms and relationship with TL. Table 2.2 summarizes some of the material properties discussed in subsequent sections. Properties relevant to particular applications will be discussed in the respective chapters. A more detailed material description organized by type of material can be found in Bøtter-Jensen, McKeever and Wintle (2003), and an excellent overview on the properties of various OSL materials can be found in Pradhan, Lee and Kim (2008). TL properties of some of these materials can also be found in McKeever, Moscovitch and Townsend (1995).

***Table 2.2*** *Summary of OSL materials, relevant properties and key references*

| OSL material | Application or potential application | Optimum stimulation wavelength | Emission bands and lifetimes | $Z_{eff}$[a] | References |
|---|---|---|---|---|---|
| $Al_2O_3$:C | Personal, environmental, medical and space dosimetry | 525 nm (green) | 335 nm (<7 ns), 420 nm (35 ms) | 11.3 | (Akselrod and McKeever, 1999; McKeever and Akselrod, 1999) |
| BeO | Personal dosimetry | 435 nm (blue) | 335 nm (TL), 380 nm (OSL) | 7.2 | (Bulur and Göksu, 1998; Sommer and Henniger, 2006; Sommer, Freudenberg and Henniger, 2007; Sommer, Jahn and Henniger, 2008) |
| MgO:Tb | Personal dosimetry | 500–570 nm (green) | 350–700 nm ($Tb^{3+}$ lines) | 10.8 | (Bos, Prokić and Brouwer, 2006) |
| BaFX:$Eu^{2+}$ (X = Br, Cl, I) | X-ray imaging | 495 nm, 590 nm | 490 nm (750 ns) | ~50 | (Rowlands, 2002; Sonoda *et al.*, 1983; von Seggern, 1999) |
| $Cu^+$-doped fused quartz | Optical fiber dosimetry | 790 nm (red) | 540 nm | 11.8 | (Justus *et al.*, 1999a; Justus *et al.*, 1999b) |
| Quartz | Luminescence dating, retrospective dosimetry | Blue | 365 nm (~30–40 μs) | 11.8 | (Denby *et al.*, 2006; Krbetschek *et al.*, 1997) |
| Feldspar | Luminescence dating, retrospective dosimetry | IR | Visible (Schott BG-39) | 11.5 (albite)<br>13.7 (K-feldspar) | (Clark and Bailiff, 1998) |
| MgS:Ce,Sm SrS:Ce, Sm CaS:Ce,Sm | Optical fiber dosimetry, integrated sensor | IR | 490 nm (36 ns, 165 ns) | 14.6 (MgS)<br>34.6 (SrS)<br>18.5 (CaS) | (Dusseau *et al.*, 2000; Lapraz *et al.*, 2006; Ravotti *et al.*, 2007) |
| KCl:Eu and KBr:Eu | X-ray imaging, optical fiber dosimetry | 560 nm (KCl) and 620 nm (KBr) | 420 nm (~1 μs) | 18.1 (KCl)<br>31.5 (KBr) | (Douguchi *et al.*, 1999; Gaza and McKeever, 2006; Klein and McKeever, 2008; Merckle, Powell and Wilson, 1978; Nanto *et al.*, 1993) |

*(Continued)*

***Table 2.2*** (Continued)

| OSL material | Application or potential application | Optimum stimulation wavelength | Emission bands and lifetimes | $Z_{eff}$[a] | References |
|---|---|---|---|---|---|
| YSO | Near-real time detection | Green | 320–475 nm | 33.6 | (Hazelton *et al.*, 2010) |
| Porcelain | Retrospective dosimetry | Green | 410 nm | | (Bøtter-Jensen *et al.*, 1996; Poolton, Bøtter-Jensen and Jungner, 1995) |
| Common salt, washing powder, water softener | Retrospective (accident) dosimetry | Blue | UV (Hoya U-340) | | (Larsson *et al.*, 2005; Thomsen, Bøtter-Jensen and Murray, 2002) |
| Electronic components and substrates | Retrospective (accident) dosimetry | IR and blue | Visible (Schott BG-39) or UV (Hoya U-340) | | (Godfrey-Smith, 2006; Göksu, 2003; Inrig, Godfrey-Smith and Khanna, 2008; Inrig, Godfrey-Smith and Larsson, 2010; Larsson *et al.*, 2005; Mathur *et al.*, 2007; Woda and Spöttl, 2009; Woda, Greilich and Beerten, 2010) |
| Dental enamel | Retrospective (accident) dosimetry | IR and green | Visible (Schott BG-39) or UV (Hoya U-340) | 15.6 (Shimano *et al.*, 1989) | (DeWitt *et al.*, 2010; Godfrey-Smith and Pass, 1997; Godfrey-Smith, 2008; Yukihara *et al.*, 2007) |
| Dental ceramics | Retrospective (accident) dosimetry | IR (140 °C) and blue | Visible (Schott BG-38) or UV (Hoya U-340) | | (Bailiff *et al.*, 2002; Veronese *et al.*, 2010) |

[a] Calculated for materials of well-established composition as discussed in Chapter 3 (see Section 3.4.1.1) with $m = 3.5$ (Bos, 2001). Otherwise, references are given.

### 2.7.1 Artificial Materials

#### 2.7.1.1 $Al_2O_3$:C

Carbon-doped aluminum oxide ($Al_2O_3$:C) is the most widely used OSL material in X-ray, gamma and beta radiation dosimetry applications as the basis for the Luxel and InLight dosimetry systems (Landauer Inc., USA). OSL measurements are typically carried out with green stimulation, and detection of the main emission band is in the blue region of the spectrum, centered at ~415 nm. The emission and stimulation spectra are shown in Figure 2.47.

$Al_2O_3$:C was introduced in the 1990s as a high sensitivity TL material with a reported sensitivity 40–60 times that of LiF:Mg,Ti TL detectors (Akselrod *et al.*, 1990; Akselrod and Kortov, 1990; Akselrod, Kortov and Gorelova, 1993). The crystal is grown in a reducing atmosphere in the presence of carbon to increase the concentration of oxygen vacancies responsible for the main emission band centered at 415 nm. It has been suggested that the substitution of $O^{2-}$ by $C^{4-}$ promotes the formation of oxygen vacancies by providing charge compensation (Yang *et al.*, 2008). The material was then developed for OSL applications (Akselrod *et al.*, 1998a; Akselrod and McKeever, 1999; Markey, Colyott and McKeever, 1995; Markey *et al.*, 1996b; McKeever *et al.*, 1996). The detector response is linear for several orders of magnitude, up to ~10 Gy.

$Al_2O_3$:C is also used as a probe in optical fiber dosimetry systems for medical and environmental applications (see Chapters 5 and 6). In addition, $Al_2O_3$:C-based dosimeters were recently made sensitive to neutrons with the addition of neutron converters in the dosimeter formulation (see Chapter 3). $Al_2O_3$ doped with carbon and magnesium ($Al_2O_3$:C,Mg) has been recently developed for volumetric optical memory storage and nuclear track detection (Akselrod *et al.*, 2003; Akselrod and Akselrod, 2006) and is currently being investigated for OSL applications.

Point defects in $Al_2O_3$ that have been extensively investigated include $F^+$- and $F$-centers, which are one- and two-electron centers associated with oxygen vacancies (Evans, 1995). Absorption and emission bands associated with these defects are illustrated in Figure 2.52. The OSL, TL, RL and photoluminescence emission spectra of $Al_2O_3$:C are dominated by a broad emission band centered at ~415 nm (~3.0 eV) (Akselrod *et al.*, 1990; Erfurt *et al.*, 2000; Markey, Bøtter-Jensen and Duller, 1997) (see Figure 2.47a). This band has been attributed to radiative transition from the $^3$P excited state to the $^1$S ground state of the $F$-center, represented by:

$$F^*\text{-center} \rightarrow F\text{-center} + h\nu_{3\,\text{eV}}, \tag{2.51}$$

where $F^*$-center represents the excited state of the $F$-center. Because this transition is spin-forbidden, the luminescence lifetime is long, 35 ms at room temperature (Akselrod *et al.*, 1998b). $F$-centers also give rise to an absorption band at ~205 nm (6.0 eV) associated with transition from the $^1$S ground state to the $^1$P excited state (see Figure 2.52). Because of the small energy difference between the $^1$P excited state and the bottom of the conduction band, excitation with 205 nm can efficiently ionize $F$-centers, producing $F^+$-centers and free electrons that may be captured at other defects in the material. The $F$-center emission shows thermal quenching above ~400 K (see Figure 2.12) accompanied by reduction in luminescence lifetime with temperature (Akselrod *et al.*, 1998b).

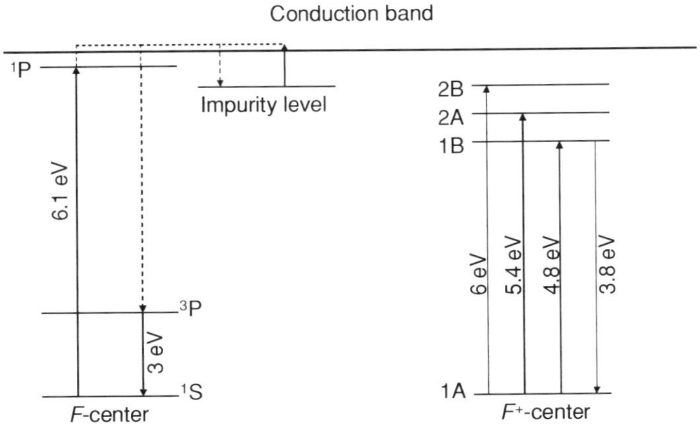


***Figure 2.52*** *Optical transitions of F- and $F^+$-centers in in $Al_2O_3$ (Lee and Crawford, 1979). The electronic structure of the F-center can be treated as a helium atom, the 6.1 eV absorption band being assigned to the $^1S \rightarrow {}^1P$ transition, and the 3.0 eV emission band to the $^3P \rightarrow {}^1S$ transition (Lee and Crawford, 1979). Electrons excited to the $^1P$ energy level may escape to the conduction band and be trapped at other impurity levels in the crystal. The $^1S$ state of the $F^+$-center in a spherical symmetry corresponds to the 1A state in the $C_2$ symmetry of the oxygen site, whereas the $^2P$ state is split into the states 1B, 2A and 2B by the anisotropic potential (La, Bartram and Cox, 1973). Reprinted from Physical Review B, Lee, K.H., Crawford Jr., J.H., Luminescence of the F center in sapphire. Vol. 19 (6), 3217–3221. Copyright (1979) by the American Physical Society.*

A weak UV emission band at 330 nm (3.75 eV) has been observed in the TL emission spectrum of $\alpha$-$Al_2O_3$ (Summers, 1984). This band is attributed to radiative transitions from the 1B excited state to the 1A ground state of $F^+$-centers based on known optical transitions of the $F^+$-centers, which shows a 3.8 eV emission band and 4.85 eV and 5.41 eV absorption bands (see Figure 2.52). The lifetime of this luminescence band is $<7$ ns, in agreement with predicted values for the $F^+$-center luminescence (Evans, 1995). The $F^+$-center emission band shifts to longer wavelengths with increasing temperatures and exhibits thermal quenching around 400 K (Evans, 1995). This emission band is more prominent in $Al_2O_3$:C,Mg because an increased concentration of $F^+$-center is promoted by Mg-doping and charge compensation (McKeever *et al.*, 1999).

The UV emission band is typically difficult to observe in the TL or OSL of $Al_2O_3$:C because of the dominant $F$-center luminescence. However, the difference in lifetime between the $F$-center and the UV emission band allows the bands to be separated using time-resolved OSL techniques, for example by using the POSL technique and monitoring the OSL signal from $Al_2O_3$:C both during and in between the stimulation light pulses (Yukihara and McKeever, 2006b). Figure 2.49 shows the OSL emission spectrum during and in between laser pulses. During the light pulses, the OSL emission comes from both UV and $F$-center luminescence, whereas in between the light pulses the OSL emission comes only from $F$-center luminescence. Because of the long lifetime of the $F$-center luminescence (35 ms) compared to the time interval between laser pulses in this measurement (250 μs), the $F$-center intensity is essentially constant in this time scale and as a result the

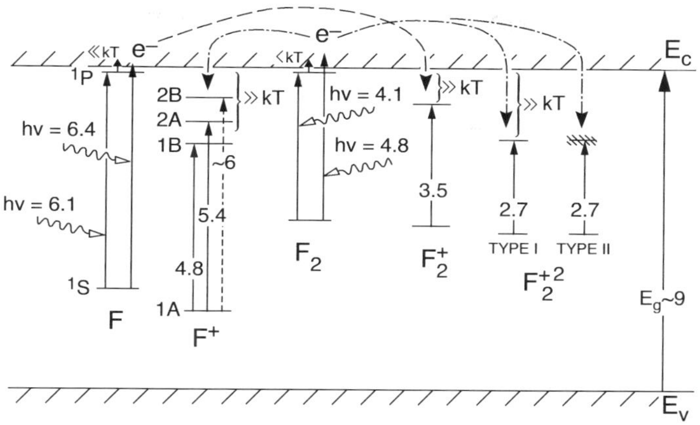


*Figure 2.53* *Proposed energy band diagram of the single and aggregate defects in α-$Al_2O_3$ (Evans, Pogatshnik and Chen, 1994; Pogatshnik, Chen and Evans, 1987). Reprinted from IEEE Transactions on Nuclear Science, Pogatshnik, G.J., Chen, Y., Evans, B.D., A model of lattice defects in sapphire. Vol. NS–34 (6), 1709–1711, 1987. © 2007 IEEE.*

415 nm emission band is identical in both curves in Figure 2.49. During the laser pulses, the UV emission band centered at ~335 nm (3.70 eV) is observed.

Other defects proposed to account for other optical absorption bands and photo-conversion observations in $\alpha$-$Al_2O_3$ include aggregate vacancies ($F_2$, $F_2{}^+$ and $F_2{}^{2+}$ centers) (see Figure 2.53). As in many other materials, $Cr^{3+}$ emission at 695 nm has been observed in $Al_2O_3$ samples (Summers, 1984). To a certain extent, identification of the defects is based on indirect arguments, such as the reciprocal relationship between the 6.1 and 4.8 eV absorption bands observed in bleaching experiments (Lee and Crawford, 1977). Therefore, the identifications are not completely free from uncertainties in interpretation, principally due to the influence of other trapping centers during photo-conversion, irradiation and annealing, as well as the overlap of optical absorption and emission bands.

Several trapping centers have been observed in $Al_2O_3$:C by TL measurements (Figure 2.54). The main trapping center used in TL and OSL dosimetry is associated with a broad TL peak at ~450 K (~180 °C) at 1 °C/s (peak C in Figure 2.54). The TL peak position is independent of the dose for doses <10 Gy (Walker *et al.*, 1996; Yukihara *et al.*, 2003), suggesting that the broad TL peak is a result of a superposition of a distribution of first-order TL peaks. Analysis of the TL curve shape is complicated because of distortion of the TL curve due to thermal quenching of the *F*-center luminescence. The main dosimetric TL peak at ~450 K is light sensitive; the corresponding trapping center is associated with most of the OSL signal in $Al_2O_3$:C. Step-annealing measurements show that the OSL signal decays in the same temperature range as the main TL peak (Markey *et al.*, 1996b). Moreover, the low-temperature part of the TL peak seems to be more sensitive to light than the high temperature part; as a result, the TL peak shifts to high temperatures with illumination (Figure 2.50).

Low-temperature TL peaks associated with shallow traps are observed at ~260 and ~310 K at a heating rate of 0.3 K/s (see peaks A and B in Figure 2.54a), varying with the sample and the heating rate. These shallow traps are responsible for a

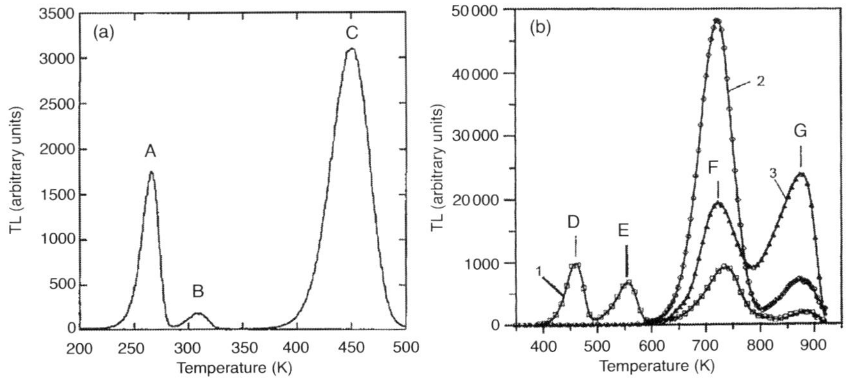


***Figure 2.54*** *TL curves of $Al_2O_3$:C in (a) the 200–500 K range (heating rate of 0.3 K/s) and (b) the 400–900 K range (heating rate of 2 K/s). The TL curves in (a) were obtained after 10 mGy $^{137}Cs$ gamma irradiation, whereas the TL curves in (b) were obtained after UV illumination of the crystal at elevated temperatures (460 K for curve 1, 620 K for curve 2 and 720 K for curve 3). Peaks labels were changed from the original publications. Figure (a) was reprinted Radiation Protection Dosimetry, Colyott, L.E., Akselrod, M.S., McKeever, S.W.S., Phototransferred thermoluminescence in α-$Al_2O_3$:C. Vol. 64 (1–4), 263–266). Copyright (1996) with permission from Oxford University Press; Figure (b) was reprinted from Radiation Protection Dosimetry, Kortov, V.S., Milman, I.I., Nikiforov, S.V., The effect of deep traps on the main features of thermoluminescence in dosimetric α-$Al_2O_3$ crystals. Vol. 84 (1–4), 35–38. Copyright (1999) with permission from Oxford University Press.*

temperature-dependent delayed luminescence during room temperature OSL measurement (Markey *et al.*, 1996b). These shallow traps have been explored in dosimetry systems utilizing the "delayed" OSL (DOSL) readout, in which charges are optically transferred to shallow traps and the phosphorescence is then monitored at room temperature following the end of stimulation (Yoder and Salasky, 1997). However, it has been shown that most of the charges optically stimulated from the traps undergo prompt recombination, which makes the POSL technique more advantageous than DOSL (Akselrod *et al.*, 1998a).

Deep traps are difficult to observe in $Al_2O_3$:C using luminescence techniques because of thermal quenching of the luminescence centers. The TL peaks observed in Figure 2.54b were obtained by illuminating the $Al_2O_3$:C samples with UV at different elevated temperatures. However, a large population of charges trapped at deep trapping centers can be inferred indirectly from their influence on the concentration of $F$ and $F^+$-centers and on the TL and OSL sensitivities after high dose irradiation using step-annealing investigations (Akselrod and Gorelova, 1993; Yukihara *et al.*, 2003; Yukihara *et al.*, 2004a). Furthermore, deep traps can be observed in thermally stimulated conductivity measurements, which do not suffer from thermal quenching problems (Agersnap Larsen, Bøtter-Jensen and McKeever, 1999). The most important of the deep traps include a deep hole trap which becomes unstable around 800–875 K and a deep electron trap that becomes unstable at ~1100–1200 K. These have been shown to be important in dictating the high-dose characteristics of the OSL from this material and these, in turn, are important in governing the response of the material to heavy charged particles (see Chapter 4).

### 2.7.1.2 BeO

Ceramic beryllium oxide (BeO) has long been known to be capable of storing energy upon exposure to ionizing radiation (Albrecht and Mandeville, 1956; Mandeville and Albrecht, 1954; Rhyner and Miller, 1970; Tochilin, Goldstein and Miller, 1969). The material has been investigated as a TL dosimeter, but the TL signal was shown to be sensitive to light (McKeever, Moscovitch and Townsend, 1995). Taking advantage of the material's light sensitivity, an OSL dosimetry system based on BeO ceramics (Thermalox™ 995, Brush Wellman Inc., USA) has been developed as an alternative to $Al_2O_3$:C, particularly for personal and medical dosimetry applications (Sommer and Henniger, 2006; Sommer, Freudenberg and Henniger, 2007; Sommer, Jahn and Henniger, 2008). Favorable dosimetric characteristics of BeO include a high sensitivity to ionizing radiation, a dose response linear over six orders of magnitude (from ~5 μGy to ~5 Gy), a low effective atomic number ($Z_{eff} = 7.2$), and relatively low cost.

The OSL curve seems to be composed of several components, but the analysis is complicated due to phototransfer of charges from deep traps and the influence of shallow traps. Figure 2.55 shows typical OSL curves of BeO ceramics fitted with two OSL decay components and one component due to phototransfer of charges to a shallow trap.

The TL curve of BeO shows three main peaks at ~80 °C, ~200 °C and ~340 °C, the exact temperature depending on the heating rate used to record the TL curve. Although TL recorded before and after OSL readout (Figure 2.56) indicates that the trapping centers associated with the first two TL peaks are affected by light, step-annealing experiments show that the OSL signal is actually associated with a trapping center that becomes unstable

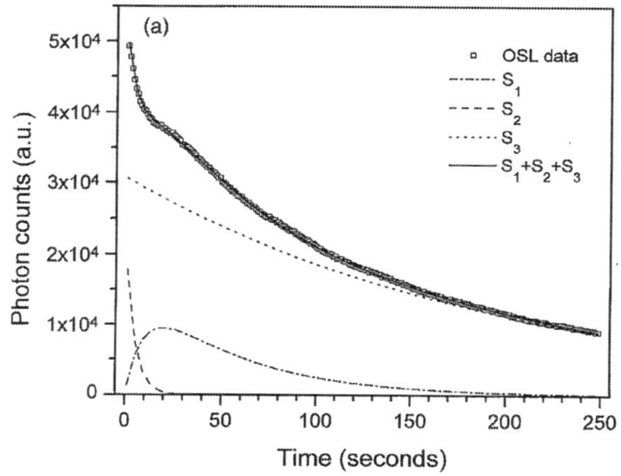


***Figure 2.55*** *OSL from BeO ceramics following irradiation with 125 mGy at 50 °C and preheating to 125 °C to remove unstable components. The curve was decomposed in three components, one corresponding to a quickly decaying OSL signal ($S_2$), one corresponding to a slowly decaying OSL signal ($S_3$) and one corresponding to phototransfer to shallow traps ($S_1$) (Bulur and Göksu, 1998). Reprinted from Radiation Measurements, Bulur, E., Göksu, H.Y., OSL from BeO ceramics: new observations from an old material. Vol. 29 (6) 639–650. Copyright (1998) with permission from Elsevier.*

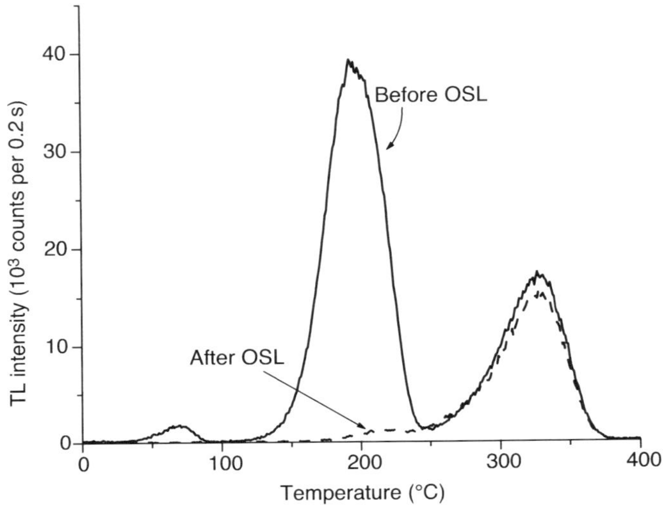


***Figure 2.56*** *TL curve of BeO ceramics following irradiation, before and after OSL readout. The TL readout was at a heating rate of 5 °C/s and obtained using Hoya U-340 filters in front of the PMT.*

at 340 °C. Figure 2.57 shows the residual OSL signal after pre-heating to different temperatures (indicated in the abscissa). These data show that the OSL signal is relatively unaffected by pre-heating temperatures up to ~250 °C.

It has been assumed that the OSL emission spectrum of BeO is similar to its TL emission spectrum (Sommer, Jahn and Henniger, 2008), which has an emission band at 335 nm (McKeever, Moscovitch and Townsend, 1995). However, recent results (unpublished) indicate that the OSL emission spectrum is composed of a dominant band at ~370 nm with a shoulder at ~310 nm (Figure 2.58a). In additively colored BeO, an emission band at 3.4 eV (~365 nm) has been attributed to $F$-centers (oxygen vacancy with two electrons captured), while an emission band at 3.9 eV (320 nm) has been attributed to $F^+$-centers (Pustovarov *et al.*, 2001). However, currently there are no data offering a direct correlation between the OSL emission bands and specific defects in BeO. The OSL stimulation spectrum of BeO has been reported to have a broad peak centered at 435 nm (Bulur and Göksu, 1998), but more recent results do not confirm this observation. The data in Figure 2.58b indicate a continuous increase in the OSL intensity with decreasing stimulation wavelengths.

The luminescence lifetime associated with the OSL signal has also been measured for various temperatures, as shown in Figure 2.59. The lifetime at room temperature is ~27 μs, but it decreases with increasing temperatures to ~800 ns at 140 °C. The decrease in luminescence lifetime as a function of temperature (Figure 2.60a) can be described by Equation (2.16) with parameters $\tau_R = (28.6 \pm 0.03)$ μs, $\Gamma_0 = (9.5 \pm 7.2) \times 10^{12}\ s^{-1}$, and activation energy $E = (0.568 \pm 0.023)$ eV. The decrease in lifetime is accompanied by a reduction in luminescence efficiency (Figure 2.60b), as expected from the Mott–Seitz model of thermal quenching expressed in Equation (2.14). Similar results were obtained by Bulur and Yeltik (2010a, b), who reported a lifetime of 26 μs at 30 °C independent of

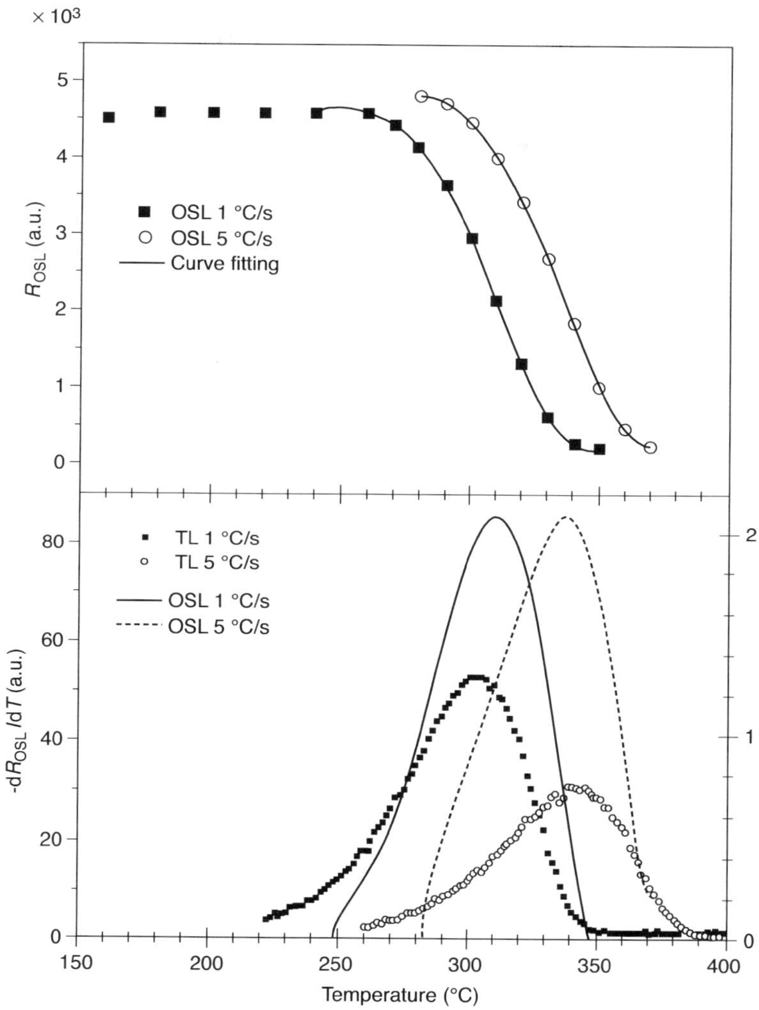


***Figure 2.57*** *Residual OSL signal $R_{OSL}$ and $dR_{OSL}/dT$ as a function of pre-heat temperature for two heating rates, compared to the TL curves obtained at the same heating rates (Bulur and Göksu, 1998). The data in this figure show that the trapping centers responsible for the OSL signal are stable for temperatures up to 250 °C, indicating that the trapping centers responsible for the TL peak at ~200 °C are not contributing to the OSL signal. Reprinted from Radiation Measurements, Bulur, E., Göksu, H.Y., OSL from BeO ceramics: new observations from an old material. Vol. 29 (6) 639–650. Copyright (1998) with permission from Elsevier.*

the dose between 0.1 and 500 Gy, and also characterized by thermal quenching with an activation energy of ~0.56 eV. The *F*-center emission band in BeO at 365 nm (3.4 eV) is reported to have a lifetime of 700 ms at 77 K and to be characterized by thermal quenching above 100 K with an activation energy of 0.070 eV (Gorbunov, Kruzhalov and Springis,

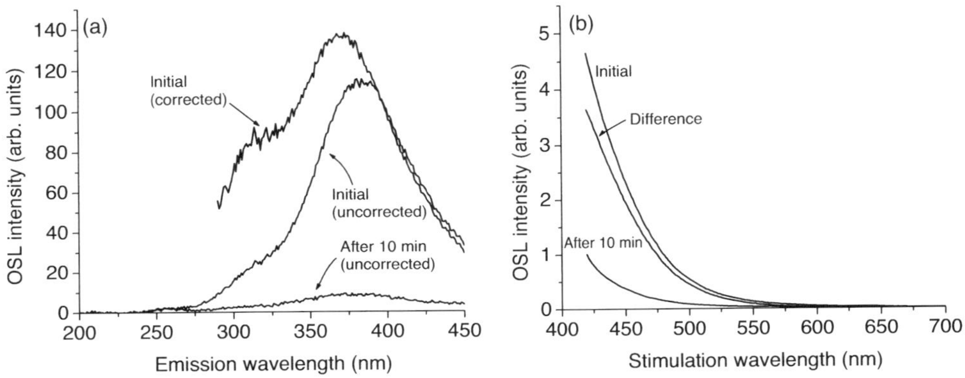


***Figure 2.58*** *(a) OSL emission spectrum (uncorrected and corrected by the instrument response) and (b) OSL stimulation spectrum of BeO. The OSL emission spectrum was obtained with stimulation at 550 nm, while the OSL stimulation spectrum was obtained monitoring the OSL at 380 nm. The spectrum after 10 min illumination with blue light (470 nm) is also shown in the figures. Unpublished results from the authors.*

1987). Therefore, it is unlikely that the OSL emission band of BeO shown in Figure 2.58 is the same emission reported by Gorbunov *et al.*, which they attributed to $F$-centers.

### 2.7.1.3 X-Ray Storage Phosphors

The commercial X-ray storage phosphors used in digital radiography are mostly based on BaFX:$Eu^{2+}$ where X = Br, Cl, I or a mixture of them (Rowlands, 2002; Sonoda *et al.*, 1983). Important types of defects in these crystals are $Eu^{2+}$ ions substituting for Ba, and $F^+$-centers consisting of either an F vacancy or an X vacancy. The OSL process is explained

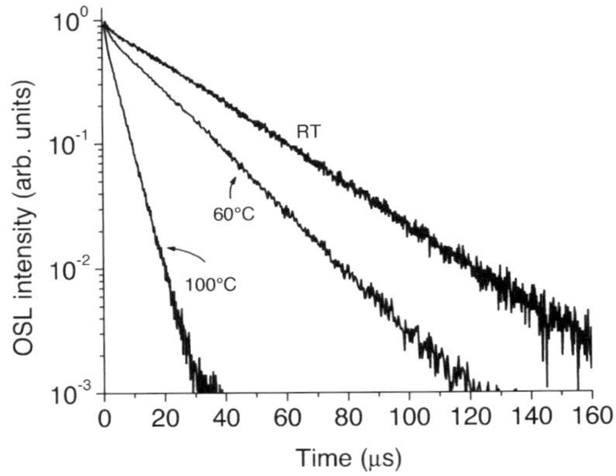


***Figure 2.59*** *Luminescence decay of BeO after 10 ns laser pulses measured at room temperature (RT), 60 and 100 °C. Unpublished results from the authors.*

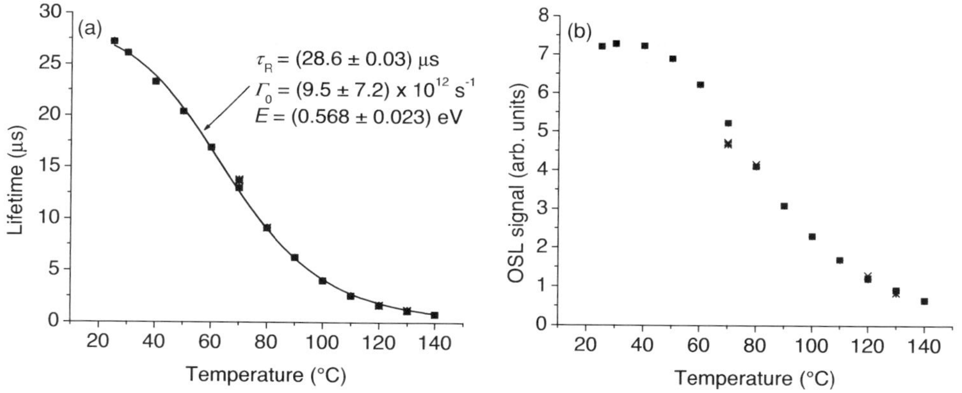


***Figure 2.60*** *(a) Luminescence lifetime and (b) OSL signal produced by 10 ns laser pulses as a function of temperature in BeO. Unpublished results from the authors.*

by trapping of electrons at $F^+$-centers (which are converted to $F$-centers) and holes at $Eu^{2+}$ sites forming a $Eu^{2+}$/hole complex (von Seggern, 1992) (see Figure 2.61). The OSL stimulation spectrum shows two peaks at 495 and 590 nm assigned to $F$-centers (Thoms, von Seggern and Winnacker, 1991) associated with the Br sites. Stimulation at the $F$-center absorption band liberates the electrons to the conduction band, which then recombine with the $Eu^{2+}$/hole complex (von Seggern, 1992). Recombination results in excited $Eu^{2+}$ centers

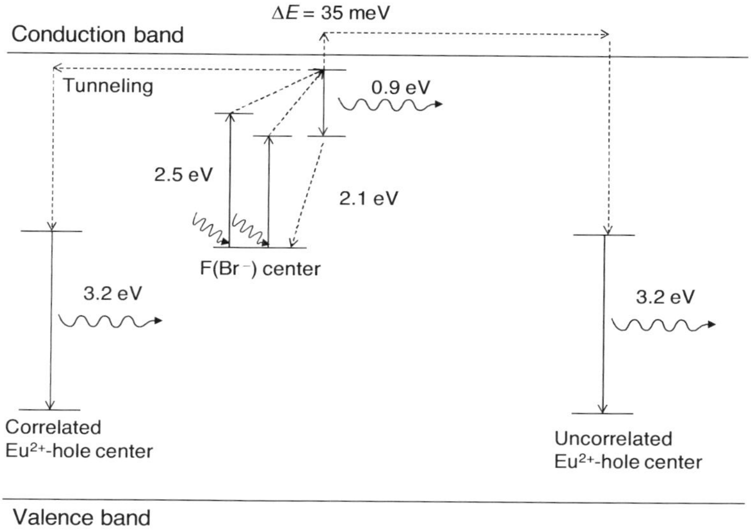


***Figure 2.61*** *Model for OSL in BaFBr:Eu$^{2+}$ X-ray storage phosphors, redrawn from von Seggern (1992). Reprinted from Nuclear Instruments & Methods in Physics Research A, von Seggern, H., X-ray imaging with photostimulable phosphors. Vol. 322, 467–471. Copyright (1992) with permission from Elsevier.*

which relax, giving rise to a broad emission band at ∼390 nm with a luminescence lifetime of ∼750 ns, corresponding to the $4f^65d \rightarrow 4f^7$ transition of $Eu^{2+}$ (Takahashi, Kohda and Miyahara, 1984; von Seggern, 1992, 1999). This process is thermally assisted, meaning that thermal stimulation is required for the electron to escape from the $F$-center excited state to the conduction band, and its intensity is therefore reduced at low temperatures. The existence of OSL emission at low temperatures is interpreted as being due to quantum tunneling from the $F$-center excited state to the trapped hole center in spatially correlated defects (Thoms, von Seggern and Winnacker, 1991).

The OSL signal in photostimulable phosphors exhibits fading at room temperature, resulting in a loss of latent image by ∼25% in a period of 8 h (Rowlands, 2002). In spite of that, and the relatively high effective atomic number ($Z_{eff} \sim 50$), investigations have been carried out to use phosphor screens in radiation dosimetry (Ariga *et al.*, 2007; Boukhair *et al.*, 2001; Ohuchi and Yamadera, 2002; Ohuchi, Yamadera and Baba, 2003).

Photostimulable phosphors have also been used for neutron imaging by combining them with neutron converters (Fröhlich-Schlapp *et al.*, 2005; Masalovich *et al.*, 2005). Other materials investigated as X-ray storage phosphors include RbBr:$Tl^+$ (Rowlands, 2002), KCl:$Eu^{2+}$ (Fröhlich-Schlapp *et al.*, 2005; Nanto *et al.*, 1993), $BaCl_2$:$Eu^{2+}$ (Appleby *et al.*, 2006), CsBr:$Eu^{2+}$ (Appleby *et al.*, 2009a, b) and $RbCdF_3$:$Mn^{2+}$ (Dotzler *et al.*, 2007).

#### 2.7.1.4 $Cu^+$-Doped Fused Quartz

$Cu^+$-doped fused quartz was proposed for use as an OSL probe in optical fiber systems (Huston *et al.*, 2002; Justus *et al.*, 1997). The material is produced by thermal diffusion of $Cu^+$ ions into clear, fused quartz at a temperature of 1100 °C (Justus *et al.*, 1999a, 1999b). The OSL can be stimulated with 790 nm light, resulting in a luminescence peak at 540 nm attributed to $Cu^+$ ions. The half-life of the trapped charge population (i.e., of the OSL signal) was observed to be ∼1.5 s in the conditions of the experiment. The OSL signal is related to TL peaks at ∼150 °C, ∼220 °C and ∼320 °C. The low-temperature peak decays at room temperature, indicating trapping centers that are thermally unstable. The OSL signal also exhibits fading of ∼30% in a 10 month period following irradiation (Justus *et al.*, 1999b). The luminescence of the recombination centers following a short laser pulse can be described by two exponential decays with lifetimes of ∼50 and ∼100 μs (Justus *et al.*, 2004). The OSL signal is reported to be linear in the dose range from 0.01 to 10 Gy (Huston *et al.*, 2002).

#### 2.7.1.5 KCl and KBr

KCl:Eu has been proposed as a photostimulable phosphor for digital radiography applications (Nanto *et al.*, 1993), whereas KBr:Eu has been proposed as an OSL probe for optical fiber dosimetry and applications in medicine (Gaza and McKeever, 2006; Klein and McKeever, 2008). The main OSL emission band in KCl:Eu and KBr:Eu is centered at 420 nm and has a lifetime of ∼1 μs, attributed to $4f^65d \rightarrow 4f^7$ transitions of the $Eu^{2+}$ ions occupying cation sites (Merckle, Powell and Wilson, 1978; Nanto *et al.*, 1993). The OSL stimulation spectrum of KCl:Eu shows a peak at 560 nm coinciding with optical absorption by $F$-center defects in this crystal, whereas the stimulation peak of KBr:Eu is shifted to 620 nm (Douguchi *et al.*, 1999; Nanto *et al.*, 1993). For crystals of intermediate composition ($KCl_xBr_{1-x}$:Eu), the OSL stimulation peak shifts gradually between 560 and 620 nm, according to the Cl content relative to Br (Douguchi *et al.*, 1999).

One of the interesting properties of KBr:Eu is the fast decay of the OSL signal. The entire OSL curve can be read in a period of ~30 ms (Klein and McKeever, 2008). The main disadvantages are the fading of the OSL signal (Douguchi *et al.*, 1999) and the relatively high effective atomic number of KCl ($Z_{eff} = 18.1$) and KBr ($Z_{eff} = 31.5$). Similar properties were also observed in KCl:$Cu^+$ and KBr:$Cu^+$, except that the emission band from $Eu^{2+}$ is replaced by two emission bands at 400 nm and a smaller peak appearing as a shoulder at 450 nm. The first is attributed to $Cu^+$ emission. The OSL stimulation spectrum in the Cu-doped samples also coincide with the *F*-center absorption bands in this crystal (Bandyopadhyay, Russell and Chakrabarti, 1999).

#### *2.7.1.6 MgS:Ce,Sm, SrS:Ce,Sm, and CaS:Ce,Sm*

This class of materials have been investigated for dose mapping (Dusseau *et al.*, 1998; Dusseau *et al.*, 1999; Idri *et al.*, 2004; Ranchoux *et al.*, 2001), optical fiber dosimetry (Benoit *et al.*, 2008b), thermal neutron detection (Benoit *et al.*, 2008a) and as the radiation-sensitive material in an integrated OSL sensor (Dusseau *et al.*, 2000; Garcia *et al.*, 2006; Plattard *et al.*, 2002; Vaillé *et al.*, 2003, 2005). The OSL can be measured with IR stimulation, resulting in emission at 490 nm characteristic of $Ce^{3+}$ in this host material, with a lifetime of 36 ns and 165 ns (Lapraz *et al.*, 2006; Ravotti *et al.*, 2007). The OSL is related to trapping centers introduced by Sm that are emptied by heating to 75 °C and, therefore, are unstable at room temperature.

### 2.7.2 Natural Materials

Naturally occurring materials are important for luminescence dating and retrospective dosimetry using OSL. A simple description of the properties of natural materials is complicated due to the wide variability in composition and impurities. This is pronounced in the case of feldspar, which is an aluminosilicate with variable composition described by a ternary diagram representing the solid solutions of $K[AlSi_3O_8]$, $Na[AlSi_3O_8]$ and $Ca[Al_2Si_2O_8]$. Greater complexity is introduced by the occurrence of more than one phase and the presence of inclusions within the minerals. Even quartz, however, can show wide variability in properties from one geological site to another. A comprehensive overview of the physical properties of several minerals relevant for luminescence dating can be found in Krbetschek *et al.* (1997).

#### *2.7.2.1 Quartz*

Quartz and feldspars are the most important natural OSL materials for luminescence dating and retrospective dosimetry due to their abundance in the Earth's crust and consequent ubiquitous presence in sedimentary deposits, archeological artifacts, building materials, ceramics and so on. Due to its simpler crystal structure, suitable OSL properties and the development of appropriate protocols, quartz has become the most reliable material for those applications (Wintle, 1997, 2008).

Despite its importance, knowledge of the specific defects involved in the OSL process in quartz is still lacking. Several defects associated with Al, Ge and Ti impurities (substituting for $Si^{4+}$ and charge compensated), along with $OH^-$ ions and oxygen vacancies have been studied, but clear correlations between TL/OSL signals and specific defects, such as the one between the 110 °C TL peak and a $(GeO_4)^-$ electron trap, are quite rare (Bøtter-Jensen, McKeever and Wintle, 2003; McKeever, Chen and Halliburton, 1985).

The OSL stimulation spectrum of quartz shows a continuous increase in OSL intensity with decreasing wavelengths (Bøtter-Jensen *et al.*, 1994). For this reason, the OSL from quartz is stimulated using blue light in most applications. Emission bands have been reported over a wide range of wavelengths, from 175 to 700 nm (Krbetschek *et al.*, 1997). The dominant band in the UV region is detected at 365 nm (Huntley, Godfrey-Smith and Haskell, 1991) and is characterized by thermal quenching in the temperature range from 100 to 300 °C (Nanjundaswamy, Lepper and McKeever, 2002; Wintle, 1975). The principal lifetime of this OSL signal is of the order of 30–40 μs (Chithambo and Galloway, 2000b; Denby *et al.*, 2006), although other lifetime components in this wavelength range have been detected (Chithambo and Ogundare, 2009; Galloway, 2002).

There are several peaks in the TL curve of quartz and several investigations have shown that the OSL signal is related to the same trapping center responsible for a TL peak at 325 °C (Smith *et al.*, 1986; Spooner, 1994; Wintle and Murray, 1997).

For a more detailed review on the OSL properties of quartz, we refer the reader to Bøtter-Jensen, McKeever and Wintle (2003).

#### 2.7.2.2 *Feldspar*

A significant body of information regarding the phenomenology of TL and OSL of feldspars can be found in the literature, mainly due to its importance for luminescence dating and retrospective dosimetry applications. This stems from its presence in many sedimentary deposits, high luminescence yield, ability to stimulate OSL with IR radiation and saturation only at very high doses of ionizing radiation (>1000 Gy).

OSL of feldspars can be stimulated with either visible or IR radiation (Hütt, Jaek and Tchonka, 1988). The initial mechanism for IR stimulation proposed by Hütt *et al.* states that trapped charges are promoted to an excited state of the trapping center by absorption of ~1.43 eV photons, from which they can escape to the conduction band by thermal assistance. Stimulation with visible light occurs by direct transitions from the ground state to the conduction band by absorption of ~2.5 eV photons. More recent detailed experiments at low temperatures support a different model in which recombination occurs by charge transport through band-tail states below the conduction band and tunneling from the excited states to the recombination centers (Poolton *et al.*, 2002a, 2002b, 2009).

The main OSL emission bands are reported at 280, 320–340, 400 and 560–570 nm and in the red/IR range (Clarke and Rendell, 1997; Huntley, Godfrey-Smith and Haskell, 1991; Krbetschek and Rieser, 1995). Clark and Bailiff (1998) presented evidence of strong interaction between UV and visible emission bands, suggesting that the UV emission can stimulate some of the visible luminescence. Time-resolved measurements of IRSL from various types of feldspars and in various detection windows indicate luminescence processes with lifetimes that range from tens of nanoseconds to microseconds (Clark and Bailiff, 1998).

In laboratory-irradiated samples, the TL curve is characterized by an almost continuous emission from room temperature to 450–500 °C with a maximum around 180 °C. In natural samples, the emission starts around 200 °C and extends up to 450–500 °C with one or two peaks (Duller, Bøtter-Jensen and Markey, 1997). The TL curve shape and the number of discernable maxima depend on the region of the emission spectrum detected, that is, on the choice of filters (Duller and Bøtter-Jensen, 1997).

Duller (1997) noticed that IR bleaches the TL peak between 100 and 200 °C, but pulse annealing through this temperature range does not significantly affect the IR-stimulated luminescence signal. On the other hand, IR stimulation causes a reduction in the entire TL curve. Therefore, it has been suggested that IR stimulation is able to stimulate charges from trapping centers in a wide range of depths below the conduction band, but only those trapping centers associated with high temperature TL peaks contribute to the IRSL signal from feldspars (Duller, Bøtter-Jensen and Markey, 1997). As discussed earlier, in BeO it has also also been observed that optical stimulation reduces the intensity of a TL peak at ~200 °C that does not seem to give rise to the OSL signal, since the OSL signal does not start to decrease until the material is heated above 250 °C. There is currently no model supported by detailed investigations to explain this disconnection between the TL and OSL signals in these materials, only speculations that the charges optically stimulated from some of the trapping centers recombine non-radiatively or by emission in a wavelength outside the detection window of the OSL system (Bulur and Göksu, 1998; Duller, 1997).

Previous studies on the LM-OSL of Na and K feldspar performed by Bulur and Göksu (1999) and Bulur (2000) using IR stimulation showed similar curves for both samples, described using a linear combination of three first-order kinetic components. First-order kinetics was assumed based on results showing that the LM-OSL peak positions are independent of the doses used, ranging from ~3 to 180 Gy (Bulur, 2000). LM-OSL measurements at different temperatures from 30 to 150 °C showed that the intensities are increased and that the peaks occur earlier with increasing measurement temperatures.

The problem that continues to be a challenge for practical applications of OSL from feldspars is anomalous fading, which is the decrease in the OSL signal that cannot be explained by a loss in the trapped charges population by thermally stimulated processes (Huntley and Lamothe, 2001; Wintle, 1973). The phenomenon may be associated a loss in the trapped charge population due to recombination via tunneling. For more complete information we refer the reader to the reviews by Duller (1997) and Krbetschek *et al.* (1997), as well as Bøtter-Jensen, McKeever and Wintle (2003).

### 2.7.2.3 *Ceramic Materials*

Ceramic materials found in the environment (bricks, tiles, roof tiles, porcelain fittings, etc.) have been an important source of information for dose reconstruction studies. In most cases, quartz and feldspars are the main minerals contributing to the OSL signal from ceramic materials.

In the case of porcelain, minerals include mullite, cristobalite, aluminum oxide and quartz (Bailiff, 1997). The OSL stimulation spectrum shows a broad peak at 500 nm. The emission spectrum shows emission in the red ($Cr^{3+}$), at 410 nm (*F*-centers from $Al_2O_3$) and, in some cases, peaks at 490 nm and 580 nm attributed to $Dy^{3+}$. These results indicate that $Al_2O_3$ is one of the main contributors to the OSL from porcelain (Bøtter-Jensen *et al.*, 1996; Poolton, Bøtter-Jensen and Jungner, 1995).

#### 2.7.2.4 Dental Enamel

Dental enamel (hydroxyapatite) was suggested as a potential dosimeter for triage in case of a radiological or nuclear accident by Godfrey-Smith and Pass (1997). Although the material does not exhibit high sensitivity, OSL has been detected in the visible using IR stimulation, or in the UV range (Hoya U-340 filters) using green or blue stimulation. Unfortunately, due to the low OSL sensitivity there is a lack of understanding of the OSL mechanism and related stimulation and emission spectra (Yukihara *et al.*, 2007). In addition to the low sensitivity, some of the challenges are the rapid fading of the OSL and the large sample-to-sample variability (DeWitt *et al.*, 2010; Godfrey-Smith, 2008).

### 2.7.3 Electronic Components

The OSL signals from electronic components, which have been called "fortuitous dosimeters," have been investigated as a complementary dosimetry technique to biodosimetry assays for triage and retrospective dosimetry in the case of radiological or nuclear accidents or terrorist attacks. Göksu (2003) have shown that it is possible to detect the OSL signal from telephone chip-cards with IR stimulation (800–950 nm) and broadband detection in the visible (320–650 nm). Subsequently, other investigations also demonstrated the possibility of OSL detection with blue stimulation (470 nm) and UV detection (270–380 nm) (Mathur *et al.*, 2007). This kind of study was extended to electronic components of a variety of devices, other types of chip cards, and SiC substrates (Bassinet, Trompier and Clairand, 2010; Godfrey-Smith, 2006; Inrig, Godfrey-Smith and Khanna, 2008; Woda and Spöttl, 2009; Woda *et al.*, 2009; Woda, Greilich and Beerten, 2010). For example, Inrig, Godfrey-Smith and Khanna (2008) demonstrated that OSL of surface-mounted resistors with alumina porcelain substrates can be consistently detected with blue stimulation and UV detection. More recent studies have focused in correcting for fading of the OSL signal, which is one of the main problems with this type of material (Inrig, Godfrey-Smith and Larsson, 2010).

### 2.7.4 Other OSL Materials and Material Needs

A variety of other materials have been investigated for various OSL applications. These include $MgO:Tb^{3+}$ (Bos, Prokić and Brouwer 2006), magnesium aluminate spinel ($MgAl_2O_4$) (Alagu Raja *et al.*, 2009; Yoshimura and Yukihara, 2006a, b), barium aluminoborate glasses (Yoshimura and Yukihara, 2006a, b), terbium-activated silicates and aluminates (Mittani, Prokić and Yukihara, 2008), thalium-doped ammonium salts (Le Masson *et al.*, 2004b), hydrated magnesium sulfates (Le Masson, Bos and van Eijk, 2001), $KMgF_3:Ce^{3+}$ and $NaMgF_3:Ce^{3+}$ (Le Masson *et al.*, 2001, 2002), and $Al_2O_3$ with other dopants (Yoshimura, 2007) or prepared by alternative synthesis processes (Blair *et al.*, 2009, 2010), to name a few. It has also been shown that OSL can be measured in commercial scintillators such as YSO (Hazelton *et al.*, 2010).

It is often the case that one of the main problems observed with these and other materials discussed earlier is fading of the OSL signal. This highlights one of the disadvantages of the OSL technique, the fact that the OSL technique simultaneously releases charges from multiple trapping centers, some that are thermally stable and some that are not. Consequently the OSL signal may fade over different time scales, depending on the depth of the multiple traps contributing to the OSL process. Although pre-heating can be used to empty shallow traps associated with unstable OSL components, this is not always practical for commercial applications. Even when shallow traps are emptied prior to the OSL readout it may still be necessary to perform the OSL at elevated temperatures to eliminate or minimize the influence of shallow traps as competitors.

In summary, there is still a scarcity of OSL materials with demonstrated properties suited to dosimetry applications. Progress on the development of OSL materials with engineered properties has been slow and most research has focused on the OSL characterization of existing materials. For some applications (e.g., retrospective dosimetry) this is the only possible approach. However, the development of synthetic materials with optimal characteristics, such as was the case with $Al_2O_3$:C, will require a more focused effort and one in which material synthesis and characterization of luminescence and physical properties are carried out hand-in-hand.

# 3
# Personal Dosimetry

## 3.1 Introduction

The use of radiation in modern society is based on a comprehensive radiological protection system that has been continuously evolving since the recognition of the deleterious effects of radiation and the establishment of the International Commission on Radiation Units and Measurements (ICRU) in 1925 (originally the International X-ray Unit Committee) and the International Commission on Radiological Protection (ICRP) in 1928 (originally the International X-ray and Radium Protection Committee). This radiological protection system and the expertise of the scientific and technical communities are embodied in the recommendations of ICRP and ICRU and other international and national organizations, as well as in established international and national standards. These recommendations serve as the basis for legislation and implementation of radiological protection worldwide, although with variations from country to country. A very informative compilation and discussion of implementation of standards for individual monitoring in Europe can be found in Fantuzzi *et al.* (2004).

The ICRP system of radiological protection, recently revised in the ICRP Publication 103 (ICRP, 2007), is based on three principles: justification, optimization and application of dose limits. The principle of justification states "any decision that alters the radiation exposure situation should do more good than harm." The principle of optimization, widely known as the ALARA principle (As Low As Reasonably Achievable), states "the likelihood of incurring exposure, the number of people exposed, and the magnitude of their individual doses should all be kept as low as reasonably achievable, taking into account economic and societal factors." The third principle states "the total dose to any individual from regulated sources in planned exposure situations other than medical exposure of patients should not exceed the appropriate limits specified by the Commission." The latter applies only to situations involving the planned introduction and operation of radiation sources, defined by the ICRP as "planned exposure situations" (ICRP, 2007).

*Optically Stimulated Luminescence: Fundamentals and Applications* Eduardo G. Yukihara and Stephen W. S. McKeever

One of the fundamental requirements in this system of radiological protection is, of course, the measurement or assessment of radiation doses. Personal dosimetry, a term here broadly used to indicate the assessment of doses in individual and area monitoring, is necessary to demonstrate compliance with dose limits and meet regulatory requirements. Area monitoring refers to the use of dosimeters or survey meters in the workplace to identify abnormal conditions, whereas individual monitoring refers to the use of dosimeters worn by the workers. Objectives of individual monitoring for external exposures include (ICRP, 1997):

(i) Assessment of effective dose and, if appropriate, equivalent doses for compliance purposes
(ii) To contribute to the control of operation and design of facilities
(iii) To provide information for health surveillance and treatment in case of accidental exposures.

Solid-state passive dosimeters, including TL dosimeters and more recently OSL dosimeters, have been widely used as cost-effective personal dosimetry techniques because of the small detector size, high sensitivity, freedom from electromagnetic, mechanical and environmental interference (as much as possible), and the ability to register the absorbed dose accumulated in the detector over extended periods of time. International standards exist for the general characteristics and performance requirements of passive dosimetry systems (IEC, 2007) and of TL dosimetry systems (IEC, 2006). International standards covering specifically OSL dosimetry systems are currently not established, but are planned to be included in future parts of the IEC 62387 (IEC, 2006).

In this chapter we provide the basic framework and information regarding the application of OSLDs as personal and area dosimeters, including definition of quantities of interest, establishment of a terminology and formalism, discussion of calibration procedures, OSL detectors and, finally, OSL dosimetry systems currently in use. The last part of this chapter is dedicated to neutron-sensitive OSL detectors, which require special attention because of recent developments and the potential for future advances.

We should mention that individual and area monitoring programs involve a series of activities including management and technical procedures (e.g., accreditation, procedures for document control, quality control, record keeping, etc.), which will not be discussed here. Except for the definition of quantities of interest that will be necessary for our discussion, we focus specifically on OSL dosimetry systems, dosimetric characteristics of OSL detectors and methods of analysis. We also focus only on assessment of external exposures to ionizing radiation. Here, it is important to keep in mind the meaning of terms such as material, detector, dosimeter and dosimetry system, as shown in Table 3.1. Also, some standards and other documents of relevance for personal dosimetry are collected in Table 3.2.

## 3.2 Quantities of Interest

In the course of the discussion on OSL applications in radiation dosimetry included in the present and subsequent chapters, different quantities related to radiological protection

*Table 3.1* Definitions adapted from the international standard for thermoluminescence dosimetry systems for personal and environmental monitoring (IEC, 2006)

| Term | Definition | Examples |
|---|---|---|
| OSL material | Substance exhibiting optically stimulated luminescence | $Al_2O_3$:C<br>BeO |
| OSL detector | Specified quantity of OSL material, or such material incorporated with other non-luminescent material into a matrix, being defined by mass, shape or size or the mass of material incorporated in the matrix | $Al_2O_3$:C single crystal<br>$Al_2O_3$:C Luxel detector<br>BeO pellet |
| OSL dosimeter | Passive device consisting of one or more OSL detectors, which may be mounted in a holder intended to work on a person's body or placed in an environment for the purpose of assessing the appropriate dose equivalent at or near the position where it is placed | Luxel dosimeter<br>InLight dosimeter |
| OSL dosimeter reader | Instrument used to measure the light emitted from the detectors in OSL dosimeter, consisting of a stimulation light source, a light-measuring device and the associated electronics | InLight reader |
| OSL dosimetry system | The OSL dosimeter, reader and all associated equipment and procedures used for assessing the indicated value | Luxel dosimetry system<br>InLight dosimetry system |

against external exposures to ionizing radiation will be required, depending on the specific application or task. This section provides a brief overview of these quantities and key bibliographical references.

The system of quantities for radiological protection has evolved over the years into a system of inter-related physical quantities and two sets of radiological protection quantities, that is, protection and operational quantities. Protection quantities have the purpose of specifying exposure (dose) limits "to ensure that the occurrence of stochastic health effects is kept below unacceptable levels and that tissue reactions are avoided" (ICRP, 2007). Operational quantities were defined by the ICRU (1993) for practical measurements, both for area and individual monitoring, for the purpose of demonstrating compliance with those dose limits. These two sets of quantities are related to physical quantities, as illustrated in Figure 3.1. Protection and operational quantities can be related to each other using radiation transport codes and mathematical models, based on which conversion coefficients have been derived (ICRP, 1996).

### 3.2.1 Absorbed Dose and Other Physical Quantities

Absorbed dose is the fundamental dosimetric quantity used to provide a correlation between energy deposition and effects of ionizing radiation. As defined by the ICRP (2007), it is:

***Table 3.2*** *Some standards and documents of relevance for personal dosimetry. See Fantuzzi* et al. *(2004) for a more complete listing*

| Document | Title | Reference |
|---|---|---|
| ICRP Publication 103 | The 2007 recommendations of the International Commission on Radiological Protection | (ICRP, 2007) |
| ICRP Publication 74 | Conversion coefficients for use in radiological protection against external radiation | (ICRP, 1996) |
| ICRU Report 51 | Quantities and units in radiation protection dosimetry | (ICRU, 1993) |
| ICRU Report 60 | Fundamental quantities and units for ionizing radiation | (ICRU, 1998a) |
| ICRU Report 66 | Determination of operational dose equivalent quantities for neutrons | (ICRU, 2001) |
| IEC 61066 | Thermoluminescence dosimetry systems for personal and environmental monitoring | (IEC, 2006) |
| IEC 62387-1 | Radiation protection instrumentation – Passive integrating dosimetry systems for environmental and personal monitoring – Part 1: General characteristics and performance requirements | (IEC, 2007) |
| ISO 4037-1 | X and gamma reference radiation for calibrating dosimeters and doserate meters and for determining their response as a function of photon energy – Part 1: Radiation characteristics and production methods | (ISO, 1996a) |
| ISO 4037-2 | X and gamma reference radiation for calibrating dosimeters and doserate meters and for determining their response as a function of photon energy – Part 2: Dosimetry for radiation protection over the energy ranges 8 keV to 1,3 MeV and 4 MeV to 9 MeV | (ISO, 1997) |
| ISO 4037-3 | X and gamma reference radiation for calibrating dosimeters and doserate meters and for determining their response as a function of photon energy – Part 3: Calibration of area and personal dosimeters and the measurement of their response as a function of energy and angle of incidence | (ISO, 1996b) |
| ISO 4037-4 | X and gamma reference radiation for calibrating dosimeters and doserate meters and for determining their response as a function of photon energy – Part 4: Calibration of area and personal dosimeters in low energy X reference radiation fields | (ISO, 1996c) |
| ISO 6980-1 | Reference beta particle radiations – Part 1: methods of production | (ISO, 2006b) |
| ISO 6980-2 | Reference beta particle radiations – Part 2: calibration fundamentals related to basic quantities characterizing the radiation field | (ISO, 2004) |
| ISO 6980-3 | Reference beta particle radiations – Part 3: calibration of area and personal dosimeters and determination of response as a function of energy and angle of incidence | (ISO, 2006a) |
| ISO 8529-1 | Reference neutron irradiations – Part 1: Characteristics and methods of production | (ISO, 2001) |
| ISO 8529-2 | Reference neutron irradiations – Part 2: Calibration fundamentals related to basic quantities characterizing the radiation field | (ISO, 2000b) |
| ISO 8529-3 | Reference neutron irradiations – Part 3: Calibration of area and personal dosimeters and determination of their response as a function of neutron energy and angle of incidence | (ISO, 2000a) |
| ANSI:HPS N13.11-2001 | Personal Dosimetry Performance – Criteria for Testing | (ANSI/HPS, 2001) |
| ANSI/HPS N13.52-1999 | American National Standard – Personnel Neutron Dosimeters (Neutron Energies Less Than 20 MeV) | (ANSI/HPS, 1999) |

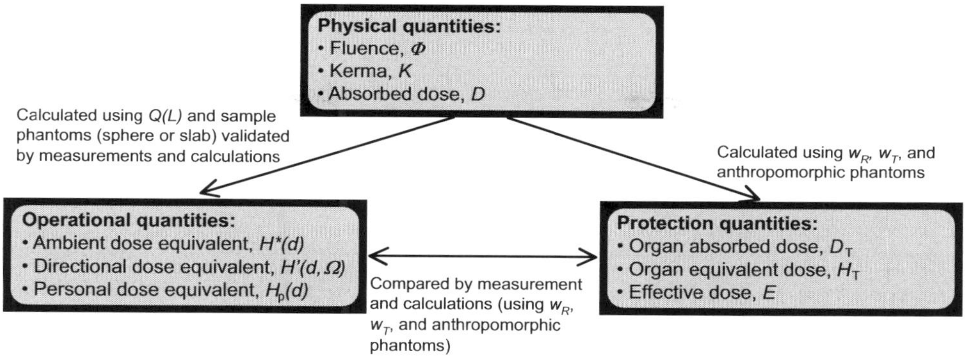


***Figure 3.1*** *Relationship between physical, operational and protection quantities related to radiological protection. Adapted from the ICRP Publication 74 (ICRP, 1996).*

- Measurable
- Well-defined at any point in matter
- Used for all types of radiation and any irradiation geometry
- Primary standards exist to allow its determination by measurement.

The absorbed dose weighted by the quality factor $Q$ (to take into account the biological effectiveness of different types of radiation) is the basis for the definition of dose equivalent quantities (e.g. $H$) and operational quantities for external exposures, such as the personal dose equivalent $H_p(10)$ and ambient dose equivalent $H^*(10)$. The absorbed dose averaged over an extended mass and weighted by radiation weighting factors $w_R$ and tissue weighting factor $w_T$ is the basis for the definition of the protection quantities equivalent dose $H_T$ and effective dose $E$ (see Section 3.2.2).

The absorbed dose $D$ quantifies deposition of energy in matter and is defined as:

$$D = \frac{d\bar{\varepsilon}}{dm}\left[\text{units} : \frac{J}{kg} = Gy\right] \tag{3.1}$$

where $d\bar{\varepsilon}$ is the mean energy imparted to matter of mass $dm$. The quantity $d\bar{\varepsilon}$ is the mean energy that is deposited in the volume of interest of mass $dm$, therefore including not only interaction of the radiation field with $dm$ but also the interaction from secondary charged particles depositing energy in $dm$. The energy imparted $\varepsilon$ is defined more precisely by $\varepsilon = R_{in} - R_{out} + \Sigma Q$, where $R_{in}$ is the energy (excluding rest energies) of all charged and uncharged ionizing particles entering the volume, $R_{out}$ is the energy (excluding rest energies) of charged and uncharged ionizing particles leaving the volume, and $\Sigma Q$ is all changes in rest mass in any interaction occurring inside the volume (ICRU, 1993, 1998a). The absorbed dose is a non-stochastic quantity defined at a point as an average over a quantity of mass $dm$ of a specific material (e.g., air, water, tissue, acrylic, etc.). The units indicated in Equation (3.1) are derived from the International System of Units (SI) and recommended by international bodies such as the ICRU (ICRU, 1998d). Use of the "rad" (1 Gy = 100 rad) is discouraged, but persists. Fluctuations in the energy deposited are taken into account in microdosimetry (e.g. Rossi and Zaider, 1996) and will not be considered here.

Another physical quantity that is often used is the kerma. The kerma quantifies the conversion of energy from uncharged particles (e.g. photons or neutrons) to kinetic energy of charged particles and is defined as:

$$K = \frac{\mathrm{d}E_{\mathrm{tr}}}{\mathrm{d}m} \left[\text{units} : \frac{\mathrm{J}}{\mathrm{kg}} = \mathrm{Gy}\right] \tag{3.2}$$

where $\mathrm{d}E_{\mathrm{tr}}$ is the "initial kinetic energies of all the charged particles liberated by uncharged particles in a mass $\mathrm{d}m$ of material" (ICRU, 1998a). Kerma is also a non-stochastic quantity defined at a point as an average over $\mathrm{d}m$ of a specific material. Since the kerma quantifies conversion of energy, its value includes the entire energy from secondary electrons created in $\mathrm{d}m$, but which will leave the volume and deposit energy elsewhere. The kerma can be divided in two components ($K = K_{\mathrm{c}} + K_{\mathrm{r}}$), depending on how the transferred energy is spent in the medium. The collision kerma $K_{\mathrm{c}}$ refers to the energy transferred to charged particles that is spent in ionizations and excitations, whereas the radiative kerma $K_{\mathrm{r}}$ refers to the energy that is spent in radiative processes, that is, carried away by photons (Attix, 2004).

For a monoenergetic fluence $\Phi(E)$ of uncharged particles (i.e., particles per unit area incident on a sphere of cross-sectional area $\mathrm{d}a$) of energy $E$, the kerma can be calculated using the appropriate mass energy transfer coefficient of the material $\mu_{\mathrm{tr}}/\rho$ for the corresponding particle energy $E$ (Attix, 2004):

$$K = \Phi(E)E\left(\frac{\mu_{\mathrm{tr}}}{\rho}\right)_E \tag{3.3}$$

The collision kerma can be calculated by:

$$K_{\mathrm{c}} = \Phi(E)E\left(\frac{\mu_{\mathrm{en}}}{\rho}\right)_E \tag{3.4}$$

where $\mu_{\mathrm{en}}/\rho$ is the mass energy absorption coefficient of the material for specific uncharged particles of energy $E$.

Another quantity of interest is the exposure $X$, defined as "the quotient $\mathrm{d}Q$ by $\mathrm{d}m$, where $\mathrm{d}Q$ is the absolute value of the total charge of the ions of one sign produced in air when all the electrons and positrons liberated or created by photons in air of mass $\mathrm{d}m$ are completely stopped in air" (ICRU, 1998d), that is:

$$X = \frac{\mathrm{d}Q}{\mathrm{d}m} \quad \left[\text{units} : \frac{\mathrm{C}}{\mathrm{kg}}\right]. \tag{3.5}$$

Exposure is easily measured using ionization chambers. For a monoenergetic fluence $\Phi(E)$ of photons of energy $E$, the exposure can be calculated using the appropriate mass energy absorption coefficient of the material $\mu_{\mathrm{en}}/\rho$ for the corresponding particle energy $E$ (Attix, 2004):

$$X = \Phi E\left(\frac{\mu_{\mathrm{en}}}{\rho}\right)_{E,\mathrm{air}}\left(\frac{e}{W}\right)_{\mathrm{air}} = (K_{\mathrm{c}})_{\mathrm{air}}\left(\frac{e}{W}\right)_{\mathrm{air}} \tag{3.6}$$

where $e$ is the electron charge, and $W$ is the mean energy required to produce an ion pair in air (33.97 eV/pair) (Attix, 2004). This equation shows the relationship between exposure and collision air kerma. In addition to the SI units shown in Equation (3.5), exposure is still often expressed in röntgen, R (1 R $= 2.580 \times 10^{-4}$ C kg$^{-1}$ air).

One should notice that the kerma is related to the process of energy conversion between uncharged and charged particles, whereas the dose is related to the process of energy deposition. It is not possible to write a general equation relating the dose to the radiation field as, for example, in Equation (3.4), because the dose is related to the energy deposited by secondary particles that can be created in other points in the medium (Attix, 2004). However, in conditions of charged particle equilibrium (CPE) the absorbed dose can be approximate by the collision kerma $K_c$ (Attix, 2004):

$$D \overset{\text{CPE}}{=} K_c \tag{3.7}$$

This is because in CPE, for each particle leaving the volume of interest $V$ there is, on average, an identical particle entering $V$. For example, if a photon enters $V$ and transfers an amount of energy $E_{tr}$ to a secondary electron that leaves the volume, causing ionizations and excitations inside and outside $V$, $E_{tr}$ contributes fully to the collision kerma, but only a fraction of $E_{tr}$ (the fraction related to ionizations and excitations inside $V$) contributes to the absorbed dose (assuming no radiative losses). However, in CPE this electron is compensated by an identical electron that will deposit its energy inside $V$ and, therefore, the collision kerma and the absorbed dose are numerically identical.

For a radiation field of monoenergetic charged particles of energy $E$ and fluence $\Phi(E)$, the absorbed dose can be calculated by:

$$D = \Phi \left(\frac{S}{\rho}\right)_c \tag{3.8}$$

where $(S/\rho)_c$ is the collision mass stopping power (Attix, 2004). The mass stopping power of charged particles for a material is defined as "the quotient of d$E$ by $\rho$d$l$, where d$E$ is the energy lost by a charged particle in traversing a distance $dl$ in the material of density $\rho$" (ICRU, 1998d):

$$\frac{S}{\rho} = \frac{1}{\rho}\frac{\mathrm{d}E}{\mathrm{d}l} \tag{3.9}$$

The collision stopping power refers to the fraction of the stopping power due to collisions with electrons in the material. The quantity $S$ is called linear stopping power.

One physical quantity that is important to describe the interaction of charged particles with matter is the linear energy transfer (LET). The LET is one of the quantities used to describe the quality of the radiation in terms of the spatial distribution of ionization events created by the radiation in a certain material. This description of the spatial distribution (or ionization density) is important because it affects the response of both biological systems and solid-state detectors, including TL and OSL detectors. The LET, although an incomplete description of the complex pattern of energy deposition in matter (e.g., see Horowitz, 2003), is often used as a parameter in describing the effects of ionization density in a detector and in the definition of dose equivalent and operational quantities (see Section 3.2.3).

The LET, usually called unrestricted LET, $L_\infty$, or simply $L$, of a charged particle in a specific material is defined as:

$$L = \frac{\mathrm{d}E}{\mathrm{d}l}, \tag{3.10}$$

where d$E$ is the energy lost by the particle due to electronic collisions in traversing a distance d$l$ (ICRU, 1993). The LET is usually expressed in keV/μm in a material.

The LET definition above is identical to the linear electronic stopping power. However, other definitions which exclude the kinetic energy from electrons release in excess of a certain cut-off value Δ also exist (ICRU, 1998a). Track-averaged or dose-averaged LET values can also be calculated based on the energy distribution of secondary charged particles created by indirectly ionizing radiation (e.g., photons and neutrons). For example, the track-averaged LET of $^{60}$Co gamma rays in water is calculated as 0.23 keV/μm (ICRU, 1970).

A more detailed description of radiometric quantities specifying the radiation field, interaction coefficients describing the interaction of radiation with matter and dosimetric quantities can be found in the ICRU Report 60 (ICRU, 1998a) and in Attix (2004).

### 3.2.2 Protection Quantities

For the purpose of limiting exposure to ionizing radiation, the absorbed dose is averaged over specific organs and tissue, weighted by the biological effectiveness of different types of radiation to obtain the equivalent dose to an organ or tissue $H_{\mathrm{T}}$, and compounded into the quantity effective dose $E$, taking into account the radiosensitivity of different organs and tissues.

The mean absorbed dose to the tissue or organ is defined by:

$$\bar{D}_{\mathrm{T}} = \frac{\int_{\mathrm{T}} D(x, y, z)\rho(x, y, z)\mathrm{d}V}{\int_{\mathrm{T}} \rho(x, y, z)\mathrm{d}V} \tag{3.11}$$

where $D(x,y,z)$ and $\rho(x,y,z)$ are, respectively, the absorbed dose and the density at a point in the organ or tissue T over which the integration is carried out. This definition "assumes that, for low doses, the mean value of absorbed dose in a specific organ or tissue can be correlated with radiation detriment from stochastic effects in all parts of that organ or tissue with sufficient accuracy for the purposes of radiological protection" (ICRP, 2007).

The equivalent dose to an organ or tissue T is defined by:

$$H_{\mathrm{T}} = \sum_{\mathrm{R}} w_{\mathrm{R}} D_{\mathrm{T,R}} \quad \left[\text{units} : \frac{\mathrm{J}}{\mathrm{kg}} = \mathrm{Sv}\right] \tag{3.12}$$

where $D_{\mathrm{T,R}}$ is the mean absorbed dose to the tissue or organ for radiation of type R, $w_{\mathrm{R}}$ are radiation weighting factors given in Table 3.3, and the sum is carried over all types of radiation.

The weighting factors for neutrons are given as a function of the neutron energy (in MeV) (ICRP, 2007):

$$w_{\mathrm{R}} = \begin{cases} 2.5 + 18.2\exp\left\{-\left[\ln(E_n)\right]^2/6\right\} & \text{for} \quad E < 1\mathrm{MeV} \\ 5.0 + 17.0\exp\left\{-\left[\ln(2E_n)\right]^2/6\right\} & \text{for } 1\mathrm{MeV} \leq E_n \leq 50\mathrm{MeV} \\ 2.5 + 3.25\exp\left\{-\left[\ln(0.04E_n)\right]^2/6\right\} & \text{for} \quad E_n > 50\mathrm{MeV} \end{cases} . \tag{3.13}$$

This function is plotted in Figure 3.2.

***Table 3.3*** *Radiation weighting factors $w_R$ for calculation of the equivalent dose $H_T$ (ICRP, 2007)*

| Radiation type | $w_R$ |
|---|---|
| Photons | 1 |
| Electrons and muons | 1 |
| Protons and charged pions | 2 |
| Alpha particles, fission fragments, heavy ions | 20 |
| Neutrons | See Figure 3.2 and Equation (3.13) |

The effective dose $E$ can then be defined as the sum of $H_T$ over all tissues:

$$E = \sum_{T} w_T H_T \tag{3.14}$$

where $w_T$ are the tissue weighting factors given in Table 3.4.

Since calculation of $H_T$ and $E$ include weighting factors based on radiobiological and epidemiological data for humans, averaged over both sexes and ages, these quantities refer to a reference person and do not take into account characteristics of an individual.

Based on these definitions, the ICRP currently recommends the dose limits summarized in Table 3.5, where the effective dose should also take into account internal exposures due to intake of radionuclides (not discussed here).

The recommended values for the tissue and radiation weighting factors $w_T$ and $w_R$ used respectively in the definition of effective and equivalent doses take into account variations in stochastic effects in different tissues due to different types of radiation. Stochastic effects are probabilistic, have no threshold, and are considered to increase in frequency with the radiation dose. Therefore, these quantities are intended for radiological protection in the low dose range, in which stochastic effects are the primary concern (ICRP, 2007).

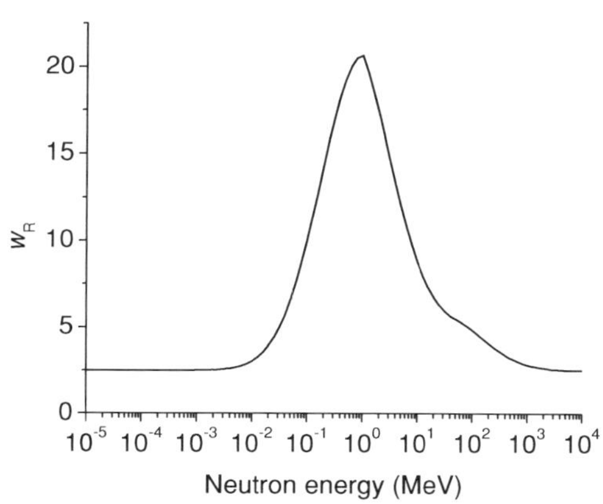


***Figure 3.2*** *Radiation weighting factors $w_R$ for neutrons as a function of neutron energy given by Equation (3.13).*

*Table 3.4* Tissue weighting factors $w_T$ for the calculation of the effective dose $E$ (ICRP, 2007)

| Tissue | $w_T$ | $\sum w_T$ |
|---|---|---|
| Bone-marrow (red), colon, lung, stomach, breast, remainder tissues | 0.12 | 0.72 |
| Gonads | 0.08 | 0.08 |
| Bladder, esophagus, liver, thyroid | 0.04 | 0.16 |
| Bone surface, brain, salivary glands, skin | 0.01 | 0.04 |

For high doses typical of emergency situations, the primary concern is deterministic effects or tissue reactions, which are characterized by a clinical reaction resulting from the depletion of cell population in the tissue. Deterministic effects have a threshold and the severity increases with the dose of radiation. In these situations, the quantities effective and equivalent doses should not be used to determine treatment. Instead, the quantity that should be used is the absorbed dose to the tissue weighted by the relative biological effectiveness (RBE) when appropriate (ICRP, 2007).

### 3.2.3 Operational Quantities

Because the radiation protection quantities cannot be measured, operational quantities have been defined for practical measurements in area and individual monitoring (ICRU, 1993). The operational quantities are defined at an appropriate point (as opposed to protection quantities, which are defined as averages over an extended mass) and are aimed at "providing an estimate or upper limit for the value of the protection quantities related to an exposure, or potential exposure of persons under most irradiation conditions" (ICRP, 2007). They can be calculated using knowledge of the radiation fluence at the point of interest, allowing radiation-monitoring devices (e.g. personal dosimeters) to be calibrated in terms of these quantities. For these reasons, they are normally used in regulations. For individual monitoring the recommended quantity is the personal dose equivalent, $H_p(d)$, whereas for

*Table 3.5* Recommended limits in "planned exposure situations" according to the ICRP Publication 103 (ICRP, 2007)

| Type of limit | Occupational | Public |
|---|---|---|
| Effective dose | 20 mSv per year averaged over defined periods of five years (not to exceed 50 mSv in any single year) | 1 mSv in a year |
| Annual equivalent dose in: | | |
| Lens of eye | 150 mSv | 15 mSv |
| Skin | 500 mSv | 50 mSv |
| Hands and feet | 500 mSv | – |

The values are provided here only to give an idea of the order of magnitude of recommended exposure limits; consult the original reference for complete details.

area monitoring the recommended quantity is the ambient dose equivalent $H^*(d)$. Both are based on the definition of dose equivalent $H$.

The dose equivalent is defined at a point as the product of the absorbed dose $D$ and the quality factor $Q$:

$$H = QD. \tag{3.15}$$

The quality factor $Q$ is defined by the ICRP to take into account the difference in effectiveness of different types of radiation in producing biological effects and is specified as a function of the unrestricted linear energy transfer $L$ of charged particles in water as:

$$Q(L) = \begin{cases} 1 & \text{for} \quad L \leq 10 \\ 0.32L - 2.2 & \text{for } 10 < L < 100 \\ 300/\sqrt{L} & \text{for} \quad L \geq 100 \end{cases} . \tag{3.16}$$

where $L$ is in keV/μm (ICRP, 2007).

The personal dose equivalent $H_p(d)$ is defined by the ICRU for individual monitoring as "the dose equivalent in soft tissue, at an appropriate depth, $d$, below a specified point on the body" (ICRU, 1993). The appropriate depth $d$ is expressed in millimeters and it is chosen depending on whether the radiation is weakly or strongly penetrating. For strongly penetrating radiation, the appropriate depth is $d = 10$ mm and the quantity $H_p(10)$ is used. For weakly penetrating radiation, the appropriate depth $d$ is 0.07 mm for the skin and 3 mm for the lens of the eyes. The units are the same as the dose equivalent, that is, J kg$^{-1}$ = Sv.

The ambient dose equivalent $H^*(d)$ is defined at a point in the radiation field as "the dose equivalent that would be produced by the corresponding expanded and aligned field, in the ICRU sphere at a depth, $d$, on the radius opposing the direction of the aligned field" (ICRU, 1993), where the ICRU sphere is a 30 cm diameter tissue-equivalent sphere intended to approximate a human body. For low-penetrating radiation, the operational quantity directional dose equivalent $H'(d,\Omega)$ is appropriate, where $H'(d,\Omega)$ is defined as "the dose equivalent that would be produced by the corresponding expanded field, in the ICRU sphere at a depth, $d$, on a radius in a specified direction, $\Omega$" (ICRU, 1993). A depth of 0.07 mm is used for the skin, and 3 mm for the lens of the eye.

## 3.3 Dosimetry Considerations

### 3.3.1 Definitions

Given the lack of international standards specific for OSL dosimetry, we have adopted, with a few modifications, the terms and definitions used in relevant international standards (IEC, 2006, 2007), which are consistent with the "ISO/IEC Guide 99: International vocabulary of metrology – Basic and general concepts and associated terms" (ISO/IEC, 2007) and the "ISO/IEC Guide 98-3: Guide to the expression of uncertainty in measurement (GUM:1995)" (ISO/IEC, 2008). A few definitions are collected in Table 3.6.

Figure 3.3 illustrates the relationship between the signal $S$, the indicated value $M$, and the quantity the dosimeter is intended to measure, for example, the personal dose equivalent $H_p(10)$, the ambient dose equivalent $H^*(10)$ and so on, here generically indicated by the

***Table 3.6*** *Some definitions adapted from the International Standard IEC 61066 on thermoluminescence dosimetry systems for personal and environmental monitoring (IEC, 2006)*

| Term[a] | Definition[b] |
|---|---|
| (Reference) calibration factor $N_0$ | Quotient of the conventional true value of a quantity $H_{r,0}$ and the indicated value $M_{r,0}$ at the point of test for a reference radiation under reference conditions: $N_0 = \frac{H_{r,0}}{M_{r,0}}$ |
| Indicated value $M$ | Value of the measurand given directly by a measuring instrument on the basis of its calibration curve |
| Reference response $R_0$ | Response for a reference value $H_{r,0}$ of the quantity to be measured under reference conditions: $R_0 = \frac{M_{r,0}}{H_{r,0}}$ |
| Relative response $r$ | Quotient of the response $R$ and the reference response $R_0$ $r = \frac{R}{R_0}$ |
| (Readout) signal $S$ | Quantity obtained in a reader after readout of a detector from which the indicated value of the dose equivalent is evaluated |
| Response $R$ | Ratio, under specified conditions, given by the relation: $R = \frac{M}{H}$ where $M$ is the indicated value of the quantity measured by the equipment or assembly under test, and $H$ is the conventional true value of this quantity |
| Value of the measurand $H$ | Value that can be obtained from the indicated value $M$ by applying the model function for the measurement |

[a] $H$ is used here to specify generically one of the dose equivalent quantities of interest, for example, $H_p(10)$, $H_p(0.07)$ or $H^*(10)$.
[b] Here the symbols $M$ and $H$ are used in place of the symbols $E$ and $C$ used in the original definitions from the IEC 61066 (IEC, 2006).

value of the measurand $H$. "Measurand" means the quantity intended to be measured (ISO/IEC, 2007).

The starting point of the evaluation is the readout of the OSL signal from the detector or detectors contained in a dosimeter. The result of such a readout is the signal $S$ or signals $S_g$ ($g = 1, 2, 3,\ldots$), where $g$ indicates the detector element in the dosimeter or different signals from the same detector. For example, in OSL dosimetry the signal $S_1$ can be defined as the PMT counts over 1 s of stimulation (initial OSL intensity) for detector element 1. Alternatively, $S_1$ can be defined as the total PMT counts accumulated over a period of stimulation in which most of the trapped charges associated with the OSL signal are depleted, that is, the total area under the OSL decay curve. Except when the entire OSL curve is recorded, this choice is made during the design of the OSL reader, taking into

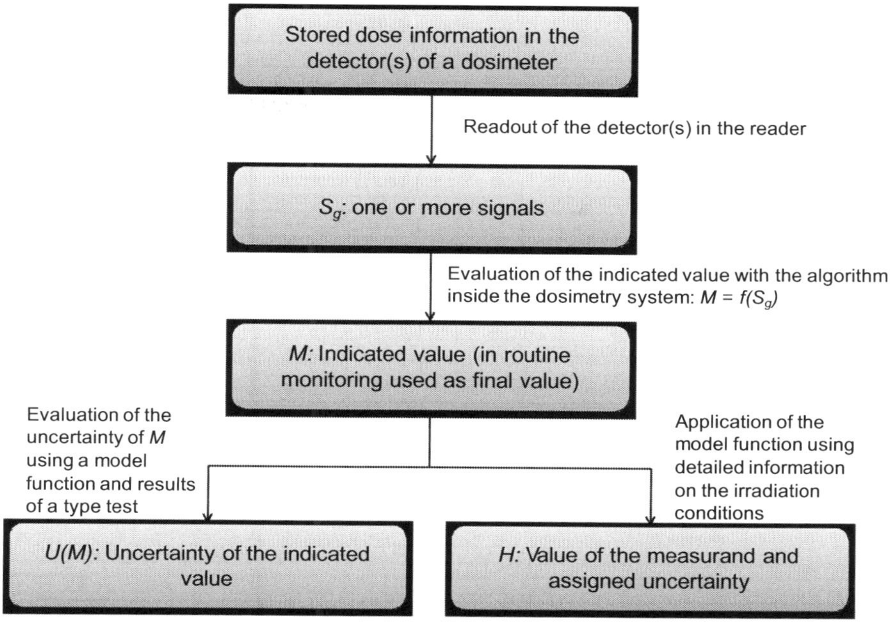


***Figure 3.3*** *Connection between the dose information stored in the detectors of a dosimeter to the indicated value M (used as a final value in routine monitoring) and value of the measurand H (evaluated only when higher precision is required, for example in case of an accident), based on diagram of the International Standard IEC 61066 on thermoluminescence dosimetry systems for personal and environmental monitoring (IEC, 2006).*

account requirements of sensitivity, precision and throughput. However, more than one signal can also be defined based on a single OSL detector. For example, one can define two signals $S_1$ and $S_2$ as the PMT counts from two different emission bands and/or two different integration intervals of the OSL decay curve. In TL dosimetry, the equivalent situation is to define $S_1$ and $S_2$ as the area under two different peaks in the TL curve.

The signal $S$ is the signal corrected for the instrumental background of the reader:

$$S = S_{\text{raw}} - S_{\text{bg}} \tag{3.17}$$

where $S_{\text{raw}}$ is the PMT signal observed for the detector element and $S_{\text{bg}}$ is the PMT signal due to PMT dark counts, electronic noise, stimulation light leakage and so on, which can be evaluated using a blank (annealed or bleached) detector.

The OSL signal $S$ (or signals $S_g$) evaluated from the OSL dosimeter is then used in conjunction with a dose algorithm to obtain the indicated value $M$, which is the value displayed by the instrument. The steps of the algorithm may include subtraction of the instrument background (if not corrected before), application of a calibration factor and detector sensitivity factors, and the combination of the different signals $S_g$. Depending on the algorithm, the indicated value $M$ can be the estimation of the quantity of interest used in routine monitoring as the final value.

Further evaluation can be made to estimate the uncertainty of the indicated value or to correct the indicated value using detailed information on the irradiation conditions (e.g. as in case of accidents) to obtain a more accurate estimate of the quantity of interest. If necessary, the indicated value may be corrected for influencing factors which are independent of the indicate value (type "S," for sum) and for those which affect the response of the detector (type "F," for factor) (IEC, 2007). The fully corrected indicated value $M$ can be obtained by:

$$M = \left( M_{\text{raw}} - \sum_j M_j \right) \prod_i k_i \tag{3.18}$$

where $M_{\text{raw}}$ is the uncorrected indicated value, $M_j$ are the indicated values due to influence quantities of type "S" (e.g. natural background dose, electrical or mechanical disturbances, etc.) and $k_i$ are the correction factors for influence quantities of type "F" (e.g., angle of incidence, energy, etc.). The correction factors $k_i$ are the reciprocal of the relative responses $r$ defined in Table 3.6. The natural background dose is usually estimated using control dosimeters or fixed national or regional averages (Lopez Ponte *et al.*, 2004).

The quantity of interest $H$ can then be evaluated by use of the appropriate calibration factor $N$:

$$H = NM \tag{3.19}$$

The value of $H$ should then be identical to the conventional true value of the quantity of interest if the dosimeter was exposed in conditions identical to those used for obtaining $N$. "Conventional true value" means "the value attributed to a particular quantity and accepted, sometimes by convention, as having an uncertainty appropriate for a given purpose" (ISO/IEC, 2008).

The formalism presented here is very similar to that discussed in the International Standard IEC 61066 (IEC, 2006), except that some symbols were replaced to avoid confusion. For example, the symbol "$E$" used for the indicated value in the IEC 61066 can be too easily confused with "energy" here. Therefore, it was replaced by "$M$," which is also the symbol used for the indicated value in other relevant standards (e.g., ISO, 1999) and in medical dosimetry (see Chapter 5). One should keep in mind the latitude one has in designing the dose algorithm for the calculation of the indicated value $M$ most appropriate for each dosimetry system. Besides the details of the dose calculation algorithm, the formalism should not depend on the application of the OSL detectors, except that the quantity of interest should be chosen appropriately, for example, $H_{\text{p}}(10)$ for external exposure to penetrating radiation, $H'(0.07)$ for area monitoring of low penetrating radiation, and so on. The calibration conditions should, of course, also be chosen appropriately.

### 3.3.2 Dose Calculation Algorithm

The dose calculation algorithm is the evaluation procedure used to obtain the indicated value $M$ from the signal $S$ or signals $S_g$. The dose calculation algorithm may include quality control parameters (e.g. reader stability), calibration factors, corrections due to fading, non-linearity, background subtraction and so on, some of which may already have been determined at an earlier stage (Fantuzzi *et al.*, 2004).

In a dosimeter containing elements that respond differently to variations in the radiation field (e.g., different photon energies or different beta or neutron components), the algorithm may combine the results from the different elements to evaluate the type of radiation field (e.g. mean photon energy, contribution from beta particles, contribution from neutrons) so that the appropriate correction and calibration factors can be applied. The different element responses can be due, for example, to the use of detectors with different compositions (different materials) or detectors of the same material behind different filters. The use of different elements with an appropriate algorithm can extend the applicability of a dosimeter to a wider variety of conditions, which is particularly important for satisfactory performance in mixed fields (e.g., mixed gamma-ray and beta-ray fields, or mixed gamma-ray and neutron fields). Based on the estimation of the radiation field and use of the appropriate correction and calibration factors, a multi-element dosimeter is able to estimate the deep and shallow dose equivalent, that is, $H_p(10)$ and $H_p(0.07)$. The parameters of the algorithm are determined by characterizing the response of the elements in the badge in known radiation fields. The badge design, detector elements and algorithm are therefore integral parts of the dosimetry system.

The algorithms used by different dosimetry systems are usually kept confidential and, for this reason, in this section we will discuss a few concepts without details of any specific algorithm. More discussion on dose calculation algorithms for personal dosimetry can be found in Moscovitch (1993) and Stanford and McCurdy (1990).

#### *3.3.2.1 Reader Calibration Factor and Element Correction Coefficients*

A methodology for calibration of the reader and individual elements is described for TL dosimetry using LiF:Mg,Ti or LiF:Mg,Cu,P by Moscovitch (1993, 1999) and adapted here with minimum modifications. The reader calibration establishes a known relationship between the signal detected and the reference value for the quantity of interest, whereas element correction coefficients are used to account for variations in the detector's sensitivity. For convenience, the reader calibration is established on a regular basis using a local source, for example, a $^{60}$Co or $^{137}$Cs gamma rays.

The reader calibration is established by dividing the dosimeter population into two groups: the calibration dosimeters and the field dosimeters. Calibration dosimeters are used exclusively for reader calibration, whereas the field dosimeters are used for routine individual or area monitoring. The reader calibration factor $\mathrm{RCF}_i$ for a specific dosimeter element $i$ can be defined as:

$$\mathrm{RCF}_i = \frac{\langle S \rangle_i}{L} \tag{3.20}$$

where $\langle S \rangle_i$ is the average of the signal in element $i$ (properly corrected for instrument background) over calibration dosimeters exposed to the local source with a known dose or irradiation time $L$.[1]

[1] $L$ here should not be confused with LET.

Element correction coefficients for the $i$th element in the $j$th dosimeter can be defined for the calibration dosimeters ($ECC_{ij}$) and field dosimeters ($ecc_{ij}$) as:

$$\mathrm{ECC}_{ij} = \frac{\langle S \rangle_i}{S_{ij}}, \tag{3.21}$$

and

$$\mathrm{ecc}_{ij} = \frac{\langle S \rangle_i}{s_{ij}} \tag{3.22}$$

where $S_{ij}$ and $s_{ij}$ are the signals (properly corrected for instrument background) in the $i$th element of the $j$th calibration or field dosimeter, respectively, and $\langle S \rangle_i$ is the average signal of the $i$th element over all calibration dosimeters. When the signal of the $i$th element and $j$th dosimeter is multiplied by the appropriate element correction coefficient the resultant value is numerically identical to the average signal of the $i$th element in the calibration dosimeters. For a dosimeter irradiated in unknown conditions, we can find the dose or exposure time $L_{ij}$ to the local source that would produce a signal in the $i$th element of the dosimeter identical to the signal measured:

$$L_{ij} = \frac{\mathrm{ecc}_{ij} s_{ij}}{\mathrm{RCF}_i} \tag{3.23}$$

To establish the connection between $L_{ij}$ and the quantity of interest, a set of dosimeters needs to be irradiated under reference conditions appropriate for the particular type of dosimeter (see Section 3.3.3). The calibration factor $a_i$ for element $i$ can then be obtained by:

$$a_i = \frac{C}{\langle L \rangle_i} \tag{3.24}$$

where $C$ is the conventional true value of the quantity, for example personal dose equivalent $H_p(10)$, and $\langle L \rangle_i$ is the average of $L_{ij}$ for the $i$th element over all irradiated dosimeters. For a dosimeter irradiated under identical reference conditions with an unknown dose, the quantity of interest can be estimated based on the calibration factors $a_i$ and the values $L_{ij}$. For dosimeters irradiated in conditions other than the reference conditions, the algorithm needs to combine the indicated values of the different elements to assess the mean energy of the photon field, the beta component and the neutron component, and apply appropriate corrections and calibration factors.

#### 3.3.2.2 Determination of Mean Photon Energy

Depending on the effective atomic number of the OSL detector, it may be necessary to correct for dependence of the detector's response on the energy of the photon field (see Section 3.4.1.1). This can be done by using a combination of two elements having different effective atomic numbers (i.e., different photon energy response curves) or by using filters to absorb the low-energy photons and alter the photon energy response of one of the elements.

This is exemplified in Figure 3.4, which shows the ratio between the TL signal from two materials, $CaSO_4$ and $Li_2B_4O_7$ (Figure 3.4a) or the ratio between the TL signal from the same material (LiF:Mg,Ti) in a open window position or behind lead (Pb) filter (Figure 3.4b). In the first case, $Li_2B_4O_7$ is almost tissue-equivalent with an effective atomic number

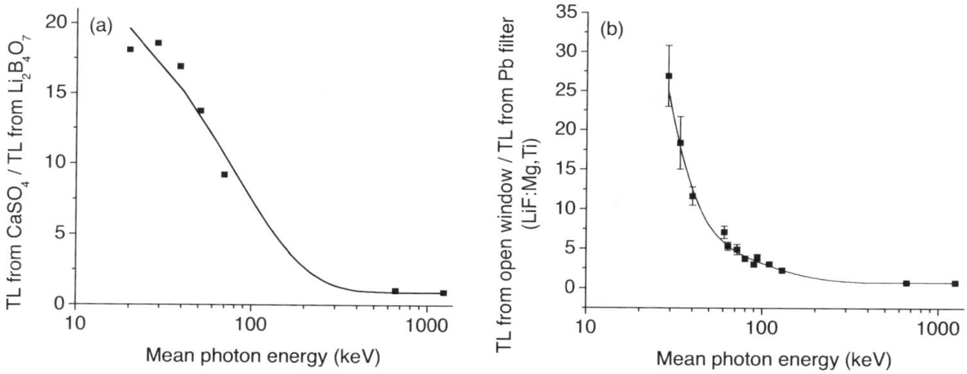


***Figure 3.4*** *(a) Ratio between the TL signal from two different detector materials exposed to X-ray or gamma-ray fields of different mean energies, based on data from Stanford and McCurdy (1990) and (b) ratio between the TL signal from two different detector materials exposed to X-ray or gamma-ray fields of different mean energies, based on data from Guimarães and Okuno (2003).*

$Z_{eff} = 7.32$, whereas $CaSO_4$ has an effective atomic number $Z_{eff} = 15.6$ (Bos, 2001). When both detectors are exposed to the same photon fluence, more energy will be deposited in the $CaSO_4$ than in $Li_2B_4O_7$ due to the higher probability for photoelectric effect in the material and higher mass energy absorption coefficient (see Section 3.4.1.1). Therefore, the response from $CaSO_4$ is higher than $Li_2B_4O_7$ and the ratio between the TL signals from these two detectors define a curve that can be used to determine the mean energy of an unknown photon field. In the case of the same material under different filters, low-energy photons will be more attenuated by the filter than high-energy photons, resulting in a ratio between the TL signals from the detectors in the two positions that is energy dependent. Again, this ratio can be used to determine the mean energy of the photon radiation field.

### *3.3.2.3 Determination of Shallow and Deep Doses*

For radiation protection purposes it may be necessary to evaluate the personal dose equivalent $H_p(10)$ ("deep dose equivalent") for control of effective dose, and the personal dose equivalent $H_p(0.07)$ ("shallow dose equivalent") for control of doses to the skin, hands, and feet. Evaluation of shallow doses is a concern particularly in the presence of beta rays and low-energy photons.

In the case of photons, the deep dose can be evaluated using the detectors behind appropriate filtration that blocks low-penetrating radiation. The shallow dose can be determined using detectors with minimum cover ("open window").

In mixed photon/beta fields it is necessary to separate the photon component from the beta component. One solution is to position the detector used for photon discrimination beyond the range of the most energetic beta-rays in the field. However, in the presence of low-energy photons it is still necessary to account for the fact that low-energy photons will produce a higher signal in the shallow detectors than in the deep detectors (Stanford and McCurdy, 1990). Therefore, first it is necessary to estimate the mean photon energy using a method such as the one discussed in the previous section. Then, this information can be used to estimate the photon contribution in the shallow detectors, allowing a more accurate assessment of the beta contribution. Complications also arise due to the detector thickness, particularly for low-energy beta rays with ranges smaller than the detector thickness. Thin detectors are advantageous in these situations due to the smaller correction factors.

#### 3.3.2.4 *Determination of Neutron Dose*

For dosimeters with neutron capabilities, a common problem is to separate the neutron component from the ever-present gamma component of the radiation fields. One approach commonly adopted in TL dosimetry is to use a pair of detectors, one which has low neutron sensitivity, but is sensitive to gamma rays (e.g., $^{7}$LiF:Mg,Ti) and the other which is sensitive to neutron and gamma rays (e.g., $^{6}$LiF:Mg,Ti) (ICRU, 2001). The net signal due to neutrons can then be found from the difference between the responses of the two detectors, after correction for the sensitivity of the individual elements.

However, it is still necessary to estimate the neutron contribution to the effective dose based on the net signal due to neutrons. Unfortunately, knowledge of calibration factors in realistic neutron fields of unknown energy spectra is one of the most challenging problems in neutron dosimetry. This problem will be discussed in more detail in the next section.

### 3.3.3 Reference Calibration Fields for Personal and Area Dosimeters

Calibration of the personal and area dosimeters involves the determination of the calibration factor $N$ or the response $R$ relating the (corrected) dosimeter's indicated value $M$ to the conventional true value $H$ of the quantity of interest for a specified type of dosimeter and reference conditions (e.g., quality of the beam, angle of incidence, temperature and pressure, etc.):

$$R = \frac{M}{H}, \text{ or } N = \frac{H}{M} \tag{3.25}$$

The calibration process consists in first determining a basic quantity (e.g., air kerma) free-in-air at the point of test in a reference field, followed by calculation of the operational quantity by the application of conversion coefficients, and finally irradiation of the dosimeter at the point of test on a phantom (for personal dosimeters) or free-in-air (for area dosimeters) (ISO, 1996b). The calibration fields, procedures for personal and area dosimeters, and conversion coefficients are described in the ISO 4037 for X-rays and gamma-rays, in the ISO 6980 for beta radiation, and in the ISO 8529 for neutrons (see Table 3.2).

To determine the calibration factor $N$, the dosimeter to be calibrated can be irradiated at the point of test after irradiation of a standard (calibrated) instrument. Depending on the stability of the radiation source, a monitor for the source output may be necessary. An alternative is to irradiate both dosimeter and standard instrument simultaneously. In this case the irradiations should be performed with the dosimeter to be calibrated and standard instrument positioned symmetrically with respect to the axis of the radiation field, at the same distance from the source, and they should not influence each other by more than 2%. Moreover, the measurement should be repeated by exchanging the dosimeter and the standard instrument, the calibration factor being calculated taking into account both

*Table 3.7 Phantoms for calibration of personal dosimeters (ISO, 1996b)*

| Phantom | Use | Type and dimensions |
|---|---|---|
| ISO rod phantom | Finger, for $H_p(0.07)$ | PMMA cylinder<br>19 mm diameter × 300 mm length |
| ISO pillar phantom | Wrist, ankle, for $H_p(0.07)$ | Water-filled cylinder with PMMA walls<br>73 mm outer diameter × 300 mm length |
| ISO water slab phantom | Body, for $H_p(10)$ | Water-filled cylinder with PMMA walls<br>30 cm × 30 cm × 15 cm<br>(front wall 2.5 mm thick, other walls 10 mm thick) |
| ISO solid PMMA slab phantom | May be used in place of ISO water slab phantom for mean energy equal to or above that of the $^{137}$Cs | PMMA slab<br>30 cm × 30 cm × 15 cm |

irradiations, as described for example in the ISO 4037-3 (ISO, 1999). A third possibility is to irradiate the dosimeters in a known gamma radiation field.

Dosimeters for area monitoring should be calibrated in ambient dose equivalent $H^*(10)$ or $H'(0.07)$ by exposure free in air. For personal dosimeters worn on the body, used to measure $H_p(10)$, calibration should be performed by irradiation on the ISO water slab phantom. For dosimeters worn on a finger used to measure $H_p(0.07)$, calibration should be performed using the ISO rod phantom. It is interesting to note that for calibrations on or in a phantom, the combination personal dosimeter *plus* phantom is considered as the dosimeter to be tested. A list of phantom characteristics is given in Table 3.7.

In the case of neutrons, the strong energy dependence of the TL and OSL dosimeters with neutron energy makes it difficult to determine the appropriate calibration factor without more knowledge on the radiation field. For this reason, it is recommended that location- or field-specific calibration factors be established (Burgkhardt and Piesch, 1988; ICRU, 2001).

### 3.3.4 Uncertainty Analysis and Expression of Uncertainty

As stated by the ISO/IEC Guide 98-3 on uncertainty of measurement, "when reporting the result of a measurement of a physical quantity, it is obligatory that some quantitative indication of the quality of the result be given so that those who use it can assess its reliability. Without such an indication, measurement results cannot be compared, either among themselves or with reference values given in a specification or standard" (ISO/IEC, 2008). This section provides rudimentary elements on uncertainty analysis and the expression of uncertainty based on the ISO/IEC Guide 98-3.

### 3.3.4.1 Evaluation of Uncertainty

Uncertainty in the result of a measurement $x_i$ subjected to a random error can be evaluated by statistical methods, that is, based on a frequency distribution of a series of observations (type A evaluation of standard uncertainty), or by a scientific judgment based on prior knowledge or information on the possible variability in $x_i$, for example using previous measurement data, data from calibration or other certificates and so on (type B evaluation of uncertainty). This classification of type A or B evaluation is meant for discussion purposes only. In both cases the uncertainty represents the state of knowledge about the result of the measurement and can be quantified by variances or standard deviations (positive square root of the variance).

In most cases the probability distribution of a continuous random variable $x$ can be approximated by a normal or Gaussian distribution:

$$f(x) = \frac{1}{\sigma\sqrt{2\pi}} e^{-\frac{1}{2}\left(\frac{x-\mu}{\sigma}\right)^2}, \tag{3.26}$$

which is centered at the expected value $\mu$ and has a full-width at half maximum (FWHM) that is related to the standard deviation $\sigma$ by FWHM $= 2\sigma\sqrt{2\ln 2}$. Thus, the frequency distribution of a set of measurements of a quantity $x$ can be represented by the expected value $\mu$ and the standard deviation $\sigma$, where the interval between $\mu - \sigma$ and $\mu + \sigma$ encompasses ~68% of the measured values.

For type A estimation of standard uncertainty, the best estimate for the expected value[1] $\mu_i$ of a quantity $x_i$ based on a number $n$ of observations $x_{i,j}$ ($j = 1,\ldots,n$) is the arithmetic mean or average given by:

$$\mu_i = \bar{x}_i = \frac{1}{n}\sum_{j=1}^{n} x_{i,j} \tag{3.27}$$

The *experimental standard deviation* $\sigma$ can be estimated based on the same dataset by:

$$\sigma(x_i) = \sqrt{\frac{1}{n-1}\sum_{j=1}^{n}\left(x_{i,j} - \bar{x}_i\right)^2} \tag{3.28}$$

The experimental standard deviation represents the variability of $x_{i,j}$ about the mean $\bar{x}_i$ and therefore represents the uncertainty of a single measurement. The uncertainty in the *mean value* calculated from $n$ measurements, the experimental standard deviation *of the mean*, is given by:

$$\sigma(\bar{x}_i) = \frac{\sigma(x_i)}{\sqrt{n}} \tag{3.29}$$

The number of observations $n$ should be large enough to provide a good estimate of the mean and experimental standard deviation of the mean.

As in the ISO/IEC Guide 98-3, the standard uncertainty denoted $u(x_i)$ represents the uncertainty in the estimate of $x_i$, regardless of the type of uncertainty evaluation. In the

[1]For simplicity, here we use the same symbols for the measurand and the estimate of the measurand. The ISO/IEC Guide 98-3 uses different symbols.

***Table 3.8*** *Examples of calculation of combined standard uncertainty $u_c(y)$ from the uncertainties $u_i(x_i)$ of the quantities $x_i$ when $y = f(x_1, x_2, \ldots)$ using Equation (3.32)*

| Function | Combined variance |
|---|---|
| $y(x_1, x_2) = x_1 + x_2$ or $y(x_1, x_2) = x_1 - x_2$ | $u_c^2(y) = u^2(x_1) + u^2(x_2)$ |
| $y(x_1, x_2) = x_1 x_2$ or $y(x_1, x_2) = \dfrac{x_1}{x_2}$ | $\left[\dfrac{u_c(y)}{y}\right]^2 = \left[\dfrac{u(x_1)}{x_1}\right]^2 + \left[\dfrac{u(x_2)}{x_2}\right]^2$ |
| $y(x) = cx$ | $u_c^2(y) = c^2 u^2(x)$ |
| $y(x) = x^c$ | $\left[\dfrac{u_c(y)}{y}\right]^2 = c^2 \left[\dfrac{u(x)}{x}\right]^2$ |

In these examples, the term $c$ is a fixed constant.

case of type A uncertainty evaluation, $u(x_i) = \sigma(\bar{x}_i)$. For type B uncertainty evaluation, see discussion in the ISO/IEC Guide 98-3 (ISO/IEC, 2008).

### 3.3.4.2 *Propagation of Uncertainties*

If $N$ quantities $x_i$ are used as input quantities to obtain a derived quantity $y$, as in:

$$y = f(x_1, x_2, \ldots, x_N), \tag{3.30}$$

the quantity $y$ can be estimated by using the estimates for the input quantities $\bar{x}_i$ given by Equation (3.27) and:

$$y = f(\bar{x}_1, \bar{x}_2, \ldots, \bar{x}_N), \tag{3.31}$$

In this case, the uncertainty in $y$ is given by propagation of uncertainties.

For uncorrelated quantities, the combined uncertainty is the positive square root of the combined variance given in a first approximation by:

$$u_c^2(y) = \left(\frac{\partial f}{\partial x_1}\right)^2 u^2(x_1) + \left(\frac{\partial f}{\partial x_2}\right)^2 u^2(x_2) + \ldots + \left(\frac{\partial f}{\partial x_N}\right)^2 u^2(x_N) \tag{3.32}$$

Table 3.8 gives some examples of functions and evaluation of the combined uncertainty.

In the case of correlated quantities, other terms expressing the covariance between the input quantities $u(x_i,x_j)$ should be considered (see ISO/IEC, 2008):

$$u_c^2(y) = \sum_{i=1}^{N}\sum_{j=1}^{N} \frac{\partial f}{\partial x_i}\frac{\partial f}{\partial x_j} u(x_i, x_j), \tag{3.33}$$

where $u(x_i,x_j) = u(x_j,x_i)$, and $u(x_i,x_j) = 0$ if the quantities $x_i$ and $x_j$ are not correlated. For example, if the quantity $y$ is obtained from a function $y = a + bx$, where $a$ and $b$ are obtained by curve-fitting procedures, the quantities $a$ and $b$ are correlated (an underestimation in the slope $b$ leads to an over-estimation in the intercept $a$, and vice versa) with $u(a,b) < 0$ (anti-correlation). The uncertainty in $y$ can be estimated using Equation (3.33) and $u(x)$, $u(a)$, $u(b)$ and $u(a,b)$. The latter can usually be obtained from the covariance matrix provided by the fitting routines.

It may be preferable in some cases, as, for example, when $y$ is the result of a numerical algorithm, to obtain an estimate of $y$ from the average of independent determinations $y_j$, that is:

$$\bar{y} = \frac{1}{n}\sum_{j=1}^{n} y_j = \frac{1}{n}\sum_{j=1}^{n} f(x_{1,j}, x_{2,j}, \ldots, x_{N,j}). \tag{3.34}$$

In this case, the experimental standard deviation and experimental standard deviation of the mean can be estimated based on the values $y_j$ using Equations (3.28) and (3.29).

#### 3.3.4.3 *Expression of Uncertainty*

To avoid confusion, the ISO/IEC Guide 98-3 (ISO/IEC, 2008) recommends stating the numerical result of a measurement as in one of the following examples, quoted from the ISO/IEC Guide 98-3:

(i) $m_s = 100.021\,47$ g with (a combined standard uncertainty) $u_c = 0.35$ mg.
(ii) $m_s = 100.021\,47(35)$ g, where the number in parentheses is the numerical value of (the combined standard uncertainty) $u_c$ referred to the corresponding last digits of the quoted result.
(iii) $m_s = 100.021\,47(0.000\,35)$ g, where the number in parentheses is the numerical value of (the combined standard uncertainty) $u_c$ expressed in the unit of the quoted result.
(iv) $m_s = (100.021\,47 \pm 0.000\,35)$ g, where the number following the symbol $\pm$ is the numerical value of (the combined standard uncertainty) $u_c$ and not a confidence interval.

It is also recommended that the relative combined standard uncertainty $u_c(y)/|y|$ ($|y| \neq 0$) is also included, when appropriate.

#### 3.3.4.4 *Confidence Interval*

Although the combined standard uncertainty $u_c(y)$ is the recommended parameter for expressing quantitatively the uncertainty of the result of a measurement, the ISO/IEC Guide 98-3 (ISO/IEC, 2008) recognizes that it may be necessary for certain applications to express the uncertainty in a way that defines an interval encompassing a large fraction of the measurement results, that is, associated with higher level of confidence, for example ~95% as opposed to ~68%. If that is the case, the expanded uncertainty defined by $U = ku_c(y)$ can be used, where $k$ is called the coverage factor. For a large number of observations, $k = 1$ corresponds to a level of confidence of ~68%, $k = 2$ to ~95% and $k = 3$ to ~99%.

For a small number of observations, the confidence interval has to be determined based on the $t$-distribution and the number of degrees of freedom (ISO/IEC, 2008). In practice, the confidence interval encompassing a fraction $p$ of the distribution can be obtained by multiplying the appropriate $t$ value (Table 3.9) by the experimental standard uncertainty. The number of degrees of freedom in case of an average is $n - 1$, where $n$ is the number of observations. For example, for three observations and a level of confidence of 95.45 %, $t = 4.53$ and the confidence interval is 4.53 times the standard uncertainty estimated by Equation (3.28) if the uncertainty refers to a single measurement or Equation (3.29) if

*Table 3.9* Values of t from the t-distribution for three confidence intervals as a function of the degrees of freedom

| Degrees of freedom | $p = 68.27$ % | $p = 95.45$ % | $p = 99.73$ % |
|---|---|---|---|
| 1 | 1.84 | 13.97 | 235.80 |
| 2 | 1.32 | 4.53 | 19.21 |
| 3 | 1.20 | 3.31 | 9.22 |
| 4 | 1.15 | 2.87 | 6.62 |
| 5 | 1.11 | 2.65 | 5.51 |
| 10 | 1.05 | 2.28 | 3.96 |
| 20 | 1.03 | 2.13 | 3.42 |
| 50 | 1.01 | 2.05 | 3.16 |
| 100 | 1.005 | 2.025 | 3.077 |
| $\infty$ | 1.000 | 2.000 | 3.000 |

See the ISO/IEC Guide 98-3 for a more complete table (ISO/IEC, 2008).

the uncertainty refers to the average. For a large number of observations the $t$ value for a confidence level of 95.45 % approaches $t = 2$.

## 3.4 Detectors

### 3.4.1 General Characteristics

#### *3.4.1.1 Photon Energy Response*

The objective in dosimetry is often to use a detector (or detectors) to estimate the absorbed dose that would be produced at a specific point in the medium of interest or phantom, typically water or tissue, in the absence of the detector, due to exposure to the same radiation field to which the detector was exposed. Ideally, only a detector made of the same material as the medium of interest would interact with the radiation field in the same way and not disturb the radiation field. Solid-state detectors such as those used in TL and OSL dosimetry do not have the same chemical composition as water or tissue, and therefore, their response depends on the photon energy.

In a first approximation, the energy dependence can be attributed to the dependence of the mass energy absorption coefficient with the atomic number $Z$ of the material (Figure 3.5a), where the largest contribution is due to the dependence of the photoelectric cross-section with $Z$. The dependence is typically of the order of $Z^m$, where $m$ is a number in between 3 and 4 (Figure 3.5b). However, this is only an approximation, since the photo-ionization cross-section has characteristic step increases whenever the photon energy is sufficient to ionize electrons from a deeper shell within the atom, as seen for Al in Figure 3.5a.

In conditions of charged-particle equilibrium and for detectors that are large in comparison with the range of secondary electrons (photon detectors), the absorbed dose in a medium is determined by the photon fluence $\Phi$, the average probability of interaction in

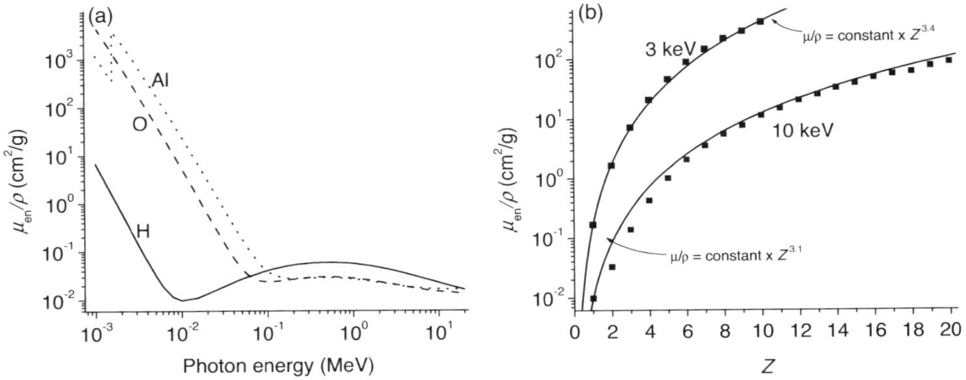


***Figure 3.5*** *Mass energy absorption coefficients $\mu_{en}/\rho$: (a) for various elements as a function of photon energy and (b) for specific photon energies as a function of the element atomic number Z. Based on data from Hubbell and Seltzer (2004).*

the material, and the photon energy $h\nu$. The dose to the medium is then given by:

$$D_m = \Phi h\nu \left(\frac{\mu_{en}}{\rho}\right)_m, \tag{3.35}$$

where the mass energy absorption coefficient $\mu_{en}/\rho$ represents the fraction of energy that is deposited locally by secondary charged particles (i.e., excluding energy losses by radiative processes such as bremsstrahlung that will not deposit energy in the volume). Applying Equation (3.35) to the material of the medium and the OSL detector, the ratio between the absorbed dose in the medium of interest and in the OSL detector when both are exposed to identical photon fluence $\Phi$ is:

$$\frac{D_{OSL}}{D_m} = \frac{(\mu_{en}/\rho)_{OSL}}{(\mu_{en}/\rho)_m}. \tag{3.36}$$

Equation (3.36) shows that the energy deposited in the OSL detector depends on the ratio between the mass energy absorption coefficients of the OSL detector and the medium of interest. Figure 3.6 shows the ratio between the mass energy absorption coefficients of different TL and OSL materials and tissue as a function of photon energy. This figure shows that the dose deposited in $Al_2O_3$ can be three to four times higher than the dose deposited in the tissue. The situation is even worse for materials such as $CaF_2$, in which the dose deposited can be more than 10 times higher when compared to tissue. The dose deposited in BeO is slightly lower than tissue for low photon energies.

Detector materials for which the values of $(\mu_{en}/\rho)_{OSL}/(\mu_{en}/\rho)_{tissue}$ are similar to 1 are called tissue-equivalent detectors. For tissue-equivalent detectors, the energy deposited in the OSL material should be identical to that in tissue when both are exposed to the same photon fluence $\Phi$. The effective atomic number, $Z_{eff}$, is often used to describe how much a material approximates or deviates from tissue in its interaction with the photon radiation

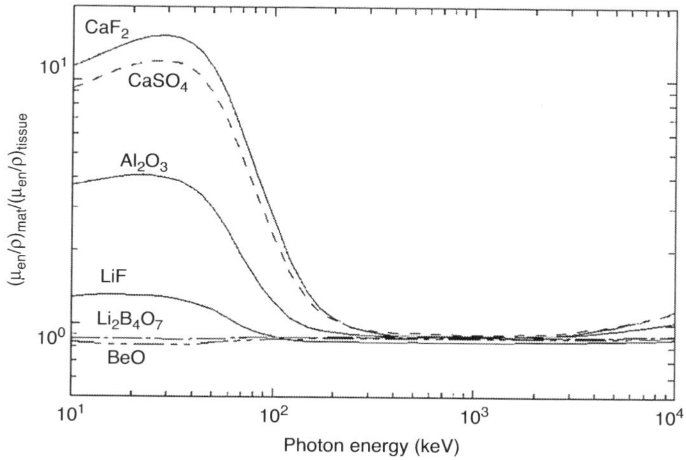


***Figure 3.6*** *Ratio between the mass energy absorption coefficients for some dosimetric materials with respect to tissue as a function of photon energy. Reprinted from Nuclear Instruments & Methods in Physics Research B, Bos, A.J.J., High sensitivity thermoluminescence dosimetry. Vol. 184, 3–28. Copyright (2001) with permission from Elsevier.*

field. The effective atomic number is defined as:

$$Z_{\text{eff}} = \left( \sum_i a_i Z_i^m \right)^{1/m} \tag{3.37}$$

where $a_i$ is the fractional electron content of element $i$ of atomic number $Z_i$ given by

$$a_i = \frac{f_i Z_i / A_i}{\sum_j f_j Z_j / A_j}, \tag{3.38}$$

and where $f_i$ and $A_i$ are the weight fraction and atomic mass of the $i$th element, and $m$ is a value in the 3–4 range, with 3.5 considered a reasonable value (Attix, 2004; Bos, 2001). The effective atomic numbers of some OSL materials were presented in Table 2.2. The effective atomic number is only used as a guide, since the actual interaction with the radiation field cannot be described by a single parameter.

Photon energy dependence may also be related to factors other than the differences in absorbed dose caused by the mass energy absorption coefficients. One should recall that Equation (3.36) is only an approximation, valid in conditions of charged particle equilibrium and for detectors large in comparison with the range of secondary electrons produced in the medium. For low-energy photons, the detector may attenuate the field more than the medium. For high-energy photons, the range of the electrons may become comparable to the size of the detectors. In the limiting case of small detectors (Bragg–Gray cavity) the detector will respond to the secondary electrons crossing the volume of the detector. Applying Equation (3.8) to the dose deposited in the detector material and in the medium when both are exposed to the same field of secondary charged particles, we obtain:

$$\frac{D_{\mathrm{OSL}}}{D_{\mathrm{m}}} = \frac{\overline{(S_{\mathrm{c}}/\rho)}_{\mathrm{OSL}}}{\overline{(S_{\mathrm{c}}/\rho)}_{\mathrm{m}}} \tag{3.39}$$

where $\overline{(S_{\mathrm{c}}/\rho)}$ is the mass collision stopping power for the specific material, averaged over the fluence spectrum of secondary particles crossing the volume. This equation shows that the energy deposited will depend on the energy spectrum of secondary electrons crossing the detector and the ratio between the mass collision stopping power in the material and in the medium (Rogers, 2009).

In an intermediate situation in which the detector is not small enough to be considered a Bragg–Gray cavity nor large enough to be considered a photon detector, Monte-Carlo calculations of the energy deposition process can be used to estimate the difference in dose deposition processes between the medium and the detector, therefore providing information on the expected energy dependence of the detector (e.g. Mobit, Agyingi and Sandison, 2006; Mobit, Mayles and Nahum, 1996). This type of energy dependence has been called “absorbed dose energy dependence” (Rogers, 2009).

However, another important cause for the energy dependence of TL and OSL detectors has to do with intrinsic processes of conversion of deposited energy into luminescence signal. We noted in Section 2.2.3 (and discuss in more detail in Chapter 4) the effect of ionization density created by charged particles in the OSL signal of the detectors. In the case of photon irradiation, the energy spectrum of the secondary electrons ultimately responsible for the energy deposition in the detector varies with the incident photon energy. This leads to variations in the ionization densities created by the secondary electrons which may affect the luminescence efficiency in a way that depends not on the matrix of the detector, but on the type and concentration of defects responsible for the TL and OSL emission. If the ionization density is high, trapping centers may become locally saturated resulting in less OSL signal. For this reason, the photon energy response may be different than the values predicted based on the dose deposition given, for example, by Equation (3.36).

A classical example of this intrinsic energy dependence is the photon energy response of LiF. Figure 3.7 shows the photon energy response of LiF:Mg,Ti and LiF:Mg,Cu,P. Although the crystal matrix is the same for both detectors, the energy response varies considerably due to intrinsic variations in the luminescence efficiency. This type of energy dependence is not taken into account when simulating the processes of energy deposition using Monte-Carlo calculations, although more sophisticated models based on track structure theory can be used to explain and predict the energy response (Olko, 2002). The influence of ionization density on the photon energy response of $Al_2O_3$:C OSL detectors will be discussed in more detail in Section 3.4.2.5.

### *3.4.1.2 Dose Response*

An ideal detector should have a signal that is linearly proportional to the absorbed dose of radiation for a given radiation field. However, deviations from linearity are predicted to occur as a result of the dynamics of the process of charge capture between different defects in the material. Although the details of this process require understanding of the mechanism prevalent in a particular type of detector, the general behavior of most TL and OSL materials is to show a linear relationship between signal and absorbed dose for low

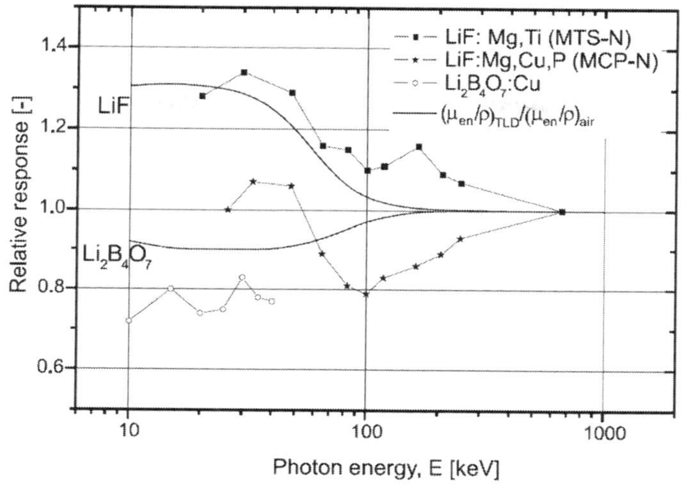


***Figure 3.7*** *Photon energy response relative to the response to $^{137}Cs$ gamma rays for some TLD materials (LiF:Mg,Ti, LiF:Mg,Cu,P and $Li_2B_4O_7$:Cu), as well as the ratio between the mass energy absorption coefficients with respect to air. Reprinted from Gasparian, P.B.R., Methodological developments for application of optically stimulated luminescence (OSL) in medical dosimetry. M. S. Thesis. Copyright (2009) with permission from Oklahoma State University.*

doses, approaching saturation for higher doses as the trapping centers become saturated. This can be approximated by a saturating exponential function of the type:

$$S(D) = a\left(1 - e^{-bD}\right) \tag{3.40}$$

where $a$ is a proportionality constant and $b$ is a free parameter. At low doses, this function approximates the linear relation $S(D) = abD$. At high doses the material response (signal per dose) is less than the response at low doses, characterizing a sub-linear dose response.

Competition between different trapping and recombination centers during either irradiation or readout can cause the material to produce more signal that would be expected based on the response at low doses. This is called supralinear dose-response. An example of a competition process that could lead to supralinearity is the presence of a competing deep trapping center. At low doses, the deep trapping center competes for the capture of charges in the delocalized band during both irradiation and OSL readout. As a result, the OSL emission related to charges trapped at the main trap (the "dosimetric trap") is lower than would have been observed in the absence of the deep trap. This effect was shown earlier in Figure 2.18. However, as the dose increases, the deep trap becomes saturated, more charges are available for trapping at the main dosimetric trap during irradiation and less charges stimulated from the main dosimetric trap during readout are captured by the deep trapping center. The final result is that more light is produced than would be expected from the response at low doses, when the deep traps are empty.

Several models have been investigated to explain various non-linear types of dose response behavior in TL materials (Chen, McKeever and Durrani, 1981; Lee and Chen, 1995;

Mische and McKeever, 1989; Sunta, Yoshimura and Okuno, 1994). These models generally rely on analytical approximations or numerical solutions for the rate equations discussed in Section 2.2.8 for a particular system. The same type of analysis has been applied to the OSL process to explain supralinearity and other effects.

The degree of supralinearity can be expressed by the dose response function or supralinearity factor *f(D)*, defined as:

$$f(D) = \frac{S/D}{S_0/D_0} \tag{3.41}$$

where $S_0$ is the signal obtained for an absorbed dose $D_0$ in the linear range of the dose response (Chen and McKeever, 1994).

The exact shape of the dose response curve or the values of the supralinearity factor as a function of dose depends on the details of the OSL mechanism in the material. However, the data can sometimes be approximated by a function derived by parameterizing the data. For example, Sawakuchi (2007) described the supralinearity factor of $Al_2O_3$:C using the following function:

$$f(D) = \frac{\left(1 - e^{-a_0 D}\right) + a_1\left(1 - e^{-a_2 D^{a_3}}\right)}{a_0 D} \tag{3.42}$$

where $a_i$ ($i = 0, \ldots, 3$) are fitted parameters. The first term on the right-hand side of the equation represents the linear-sublinear-saturation behavior, whereas the second term represents a "more-than-linear" followed by sublinear-saturation behavior. The dose response *f(D)D* and the supralinearity factors *f(D)* are presented in Figure 3.8 for one particular set of parameters.

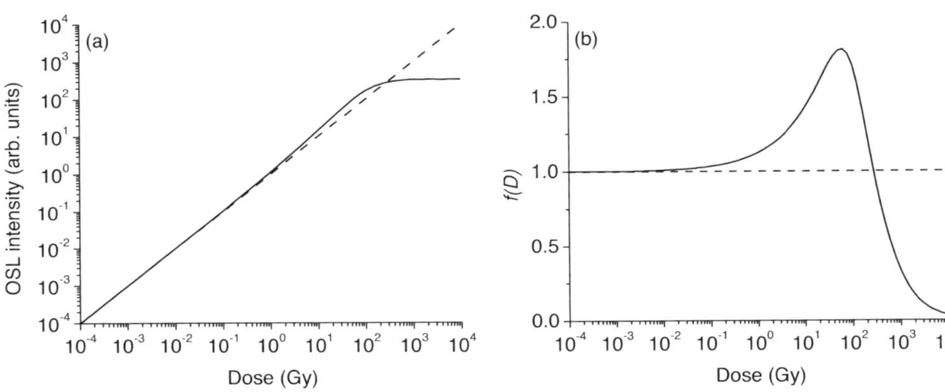


***Figure 3.8*** *(a) OSL signal and (b) supralinearity factor as a function of the absorbed dose calculated based on Equation (3.42) with parameters $a_0 = 4.53 \times 10^{-3}$ $Gy^{-1}$, $a_1 = 0.535$, $a_2 = 1.04 \times 10^{-3}$ $Gy^{-a3}$, $a_3 = 1.57$ (Sawakuchi, 2007).*

#### 3.4.1.3 Detection Limits

The definition of detection limit depends on the intended objective, for example to make a decision on whether a real signal (and, therefore, dose) was detected (qualitative analysis) or whether a satisfactory estimation of the expected value can be obtained (quantitative analysis). It also depends on the maximum permissible levels for the errors of the first kind (e.g., deciding that a dose was present when in fact it was not) and for the second kind (deciding that a dose was not present, when in fact it was). These issues are discussed in detail by Currie (1968), who also provides "working expressions" for different situations.

For the discussion in this book, it suffices to adopt a simple definition of minimum detectable level or minimum detectable dose (MDD) as that dose corresponding to $3\sigma_{BG}$, where $\sigma_{BG}$ is the experimental standard deviation of the background. The value of $\sigma_{BG}$ can be obtained using blank detectors, that is, detectors that are identical to those used for the measurements, except that they were not irradiated. For the estimation of $\sigma_{BG}$, the blank detectors should be evaluated in the same way as detectors that were irradiated for the determination of the dose response.

### 3.4.2 $Al_2O_3$:C Detectors

#### 3.4.2.1 Available Forms

The $Al_2O_3$:C currently used in personal dosimetry is grown in the form of single crystals. Single crystals of 5 mm in diameter by 0.9 mm thickness cut from boules have been widely used in TL dosimetry since the 1990s. Some were commercialized under the trade name TLD-500 (Thermo Fischer Scientific, Inc., former Harshaw Inc.). The single crystals have the advantage that they can be used almost indefinitely, since the detector can be annealed to reset its sensitivity. The disadvantage is the inherent variability in sensitivity from crystal to crystal due to inhomogeneities in the concentration of defects, especially along the length of the original crystal boule. For this reason, pre-selection of the detectors or individual detector calibration is necessary to achieve good precision.

To overcome sample-to-sample differences, the crystals produced by Landauer Inc. are crushed after growth, sieved into a powder with grain size below 105 μm and mixed to produce a powder with homogeneous sensitivity to ionizing radiation. The powder is then mixed with an organic binder and used in the manufacture of tapes consisting of polymeric material of 0.3 mm total thickness containing a layer of $Al_2O_3$:C powder. These tapes are cut into different shapes for use as detectors in the Luxel or InLight dosimetry systems (Figure 3.9). The advantage of these detectors is a more uniform sensitivity within the same batch and less variability from batch to batch, although control of the thickness of the detector layer is important for good uniformity in the sensitivity. The disadvantage is that the detectors can only be bleached, since annealing would destroy the polymeric tape.

#### 3.4.2.2 Preparation and Handling

$Al_2O_3$:C is an inert material and is therefore safe for handling. The single crystal detectors are very stable and not affected by humidity. $Al_2O_3$:C single crystals can be annealed (heated to high temperature) or bleached (illuminated with light of appropriate wavelength)

**Figure 3.9** *Available forms of $Al_2O_3$:C detectors, including round single crystals ~5 mm in diameter by 0.9 mm thickness, 0.3 mm thick tapes containing a detector layer consisting of $Al_2O_3$:C powder and organic binder, and pieces of tape of sensitive area ~1.7 cm × 2.0 cm used in the Luxel dosimetry system (Luxel detectors).*

to empty the trapping centers associated with the OSL signal and erase the OSL signal due to natural background radiation or previous irradiation.

Annealing is important to reset the crystal's sensitivity. Sensitivity changes in $Al_2O_3$:C are observed and have been attributed to a build-up of electrons and holes in deep traps and associated change in concentration of defects (see Section 3.4.2.4 below). Based on the understanding of the influence of deep traps on the sensitivity of $Al_2O_3$:C, heating the crystal to 900 °C for 15 min is recommended after each use. This temperature is much lower than the melting temperature of the crystal (2072 °C) and does not change the properties of the crystal. If the crystal has not been irradiated with a high dose, heating to a lower temperature (e.g., 400 or 500 °C) will empty the trapping centers associated with the main OSL signal, but deep traps will remain unaffected and the crystal's sensitivity may change with the accumulated dose.

Bleaching of the $Al_2O_3$:C detectors is effective in emptying the trapping centers associated with the main OSL signal, as shown in Figure 3.10. However, there are some aspects that must be carefully considered. First of all, illumination with light sources having short wavelength components ($\lambda < 450$ nm) should be avoided to prevent photo-ionization of defects in the crystal, which in turn can generate free electrons, giving rise to a light-induced OSL signal, and to prevent photo-transfer from charges in deep traps to the dosimetric traps. As a result, when using fluorescent lamps, longpass filters (e.g., Schott GG-495) or UV blocking filters should be used. Second, the spectrum of the bleaching light should be similar to the spectrum of the light used for readout. For example, if the OSL is read using blue LED stimulation and bleaching is done using green LED stimulated, the bleaching spectrum may not be able to empty effectively all the trapping centers that are stimulated with blue light (Umisedo *et al.*, 2010). If the crystal has been irradiated with high doses, bleaching may also cause photo-transfer of charge carriers from deep traps to the main dosimetric trap (Emfietzoglou and Moscovitch, 1996; Oster, Weiss and Kristianpoller, 1994; Umisedo *et al.*, 2010; Walker *et al.*, 1996). Another phenomenon that has been observed is

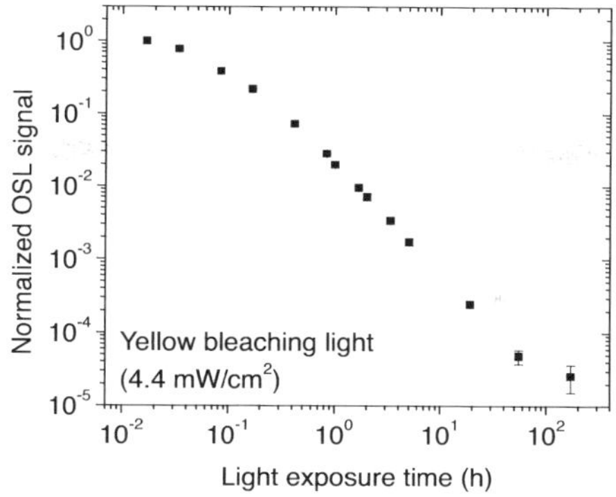


***Figure 3.10*** *OSL signal of $Al_2O_3$:C Luxel detectors irradiated with 100 mGy after different bleaching times under a yellow light (fluorescent lamp filtered by Schott GG-495 filter) with irradiance of 4.4 mW/cm$^2$ at the detector position. The data were obtained using an InLight microStar reader and normalized to the OSL intensity after irradiation. Reprinted from Gasparian, P.B.R., Methodological developments for application of optically stimulated luminescence (OSL) in medical dosimetry. M. S. Thesis. Copyright (2009) with permission from Oklahoma State University.*

an increase in the OSL signal of bleached detectors when stored in the dark at a rate that is higher than the rate expected from simple background radiation and that is proportional to the previous dose received by the detector (Jursinic, 2010).

Once the detectors have been prepared, they should be protected from light to avoid the introduction of light-induced signal before irradiation, and loss of the OSL signal after irradiation (Moscovitch, Tawil and Svinkin, 1993; West, Kearfott and Bernal, 2006). Figure 3.11

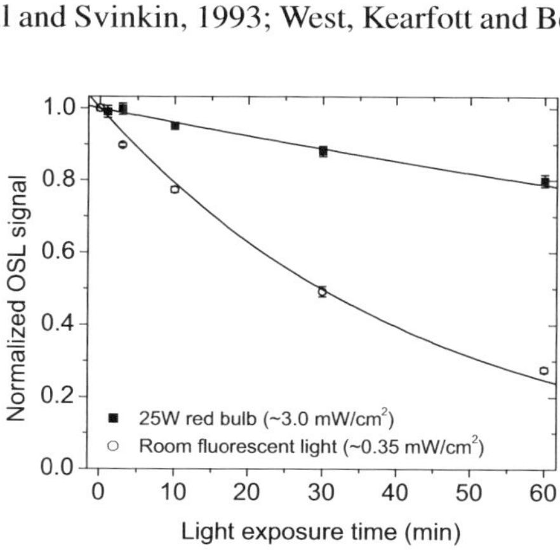


***Figure 3.11*** *OSL signal of $Al_2O_3$:C Luxel detectors irradiated with 100 mGy and after different exposure times under red light (red incandescent light bulb) or room fluorescent light. The irradiance at the detector position is shown in the figure. The data were obtained using an InLight microStar reader and normalized to the OSL intensity after irradiation. Reprinted from Gasparian, P.B.R., Methodological developments for application of optically stimulated luminescence (OSL) in medical dosimetry. M. S. Thesis. Copyright (2009) with permission from Oklahoma State University.*

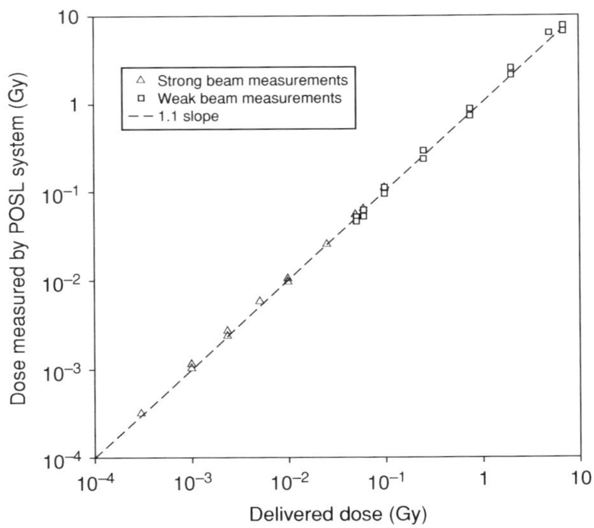


***Figure 3.12*** *Dose response of $Al_2O_3$:C powder measured using the POSL technique and two stimulation intensities, as strong laser beam (1.2 W average power) and a weak laser beam (0.01 W average power). Reprinted Radiation Protection Dosimetry, Akselrod, M.S., McKeever, S.W.S., A radiation dosimetry method using pulsed optically stimulated luminescence. Vol. 81, 167–176. Copyright (1999) with permission from Oxford University Press.*

demonstrates the decrease of OSL signal under a red bulb and in room fluorescent light as a function of time.

### *3.4.2.3 Dose Response*

The OSL response of $Al_2O_3$:C is reported to be linear over a wide range of doses, especially in the dose range of interest for personal dosimetry, as shown in Figure 3.12 for the range from $\sim 10^{-4}$ to 10 Gy.

However, at high doses the response deviates from linearity, becoming supralinear and then sub-linear. The degree of supralinearity is observed to depend on the choice of signal (initial OSL intensity or total OSL area) and on the emission band monitored (only *F*-center emission, only UV emission, or a combination of both bands). The dependence on the choice of signal is illustrated in Figure 3.13, which presents the initial OSL signal or total OSL area (both normalized to a dose of 1 Gy) as a function of dose for doses of almost 600 Gy. At low doses ($<1$ Gy), the initial OSL intensity and the total OSL area are proportional to each other. However, as the dose increases, the initial OSL intensity becomes more supralinear than the total OSL area.

The uncoupling between initial OSL intensity and total OSL area is related to a change in the shape of the OSL curve as the dose increases. This effect is illustrated in Figure 3.14. In Figure 3.14a one can see that at low doses the OSL decay curve is independent of the dose for doses less than $\sim 1$ Gy. However, as the dose increases, the OSL decay becomes more accentuated (Figure 3.14b). One should notice that the data in Figure 3.14

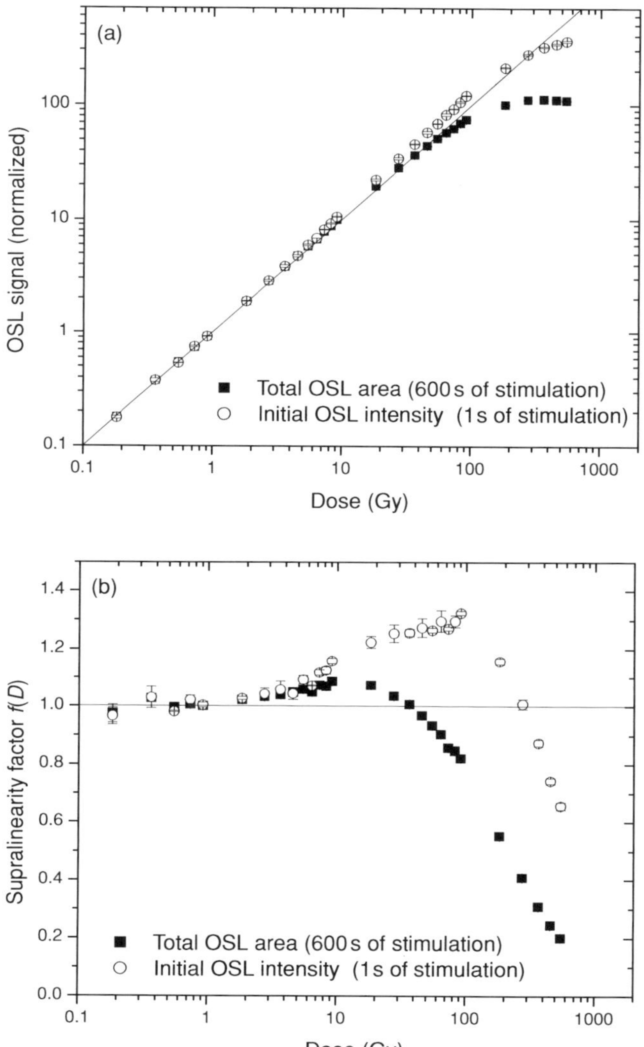


***Figure 3.13*** *(a) High-dose range dose response and (b) supralinearity factor for* $Al_2O_3$*:C Luxel detectors measured with green stimulation using a Risø reader and Hoya U-340 filters (7.5 mm thickness) plus a 2 mm thick Schott WG-360 filter in front of the PMT. The data were calculated using both the initial OSL intensity of total OSL area.*

were obtained using a filter combination that transmits only the $F$-center emission band in $Al_2O_3$:C and, therefore, the changes in OSL curve shown in this figure are not affected by the UV emission band. When both emission bands are detected, the situation is more complicated because the $F$-center and UV emission band have different characteristic OSL decay rates (Yukihara and McKeever, 2006b).

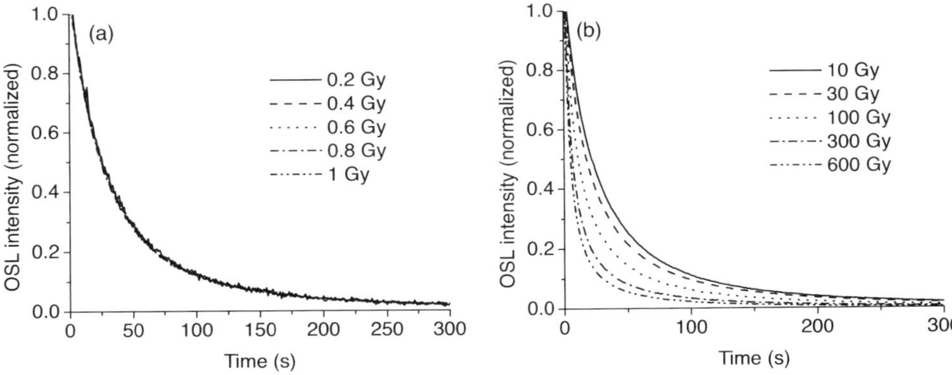


*Figure 3.14* OSL curves of $Al_2O_3$:C Luxel detectors for various doses, measured with green stimulation using a Risø reader and Hoya U-340 filters (7.5 mm thickness) plus a 2 mm thick Schott WG-360 filter in front of the PMT. Each curve corresponds to the average of three detectors.

The dependence of the dose response on the OSL emission band is illustrated in Figure 3.15. It can be seen that when the *F*-center emission band is monitored the degree of supralinearity is smaller than when the UV emission band is monitored. The data in this figure show that the contribution from the UV emission band is higher at high doses than at low doses, demonstrating a dependence of the relative intensity between these emission bands with the ionization density created in the crystal.

In general, the dose response of $Al_2O_3$:C has been explained using a model containing deep electron traps that become unstable at ~900 °C and deep hole traps that become

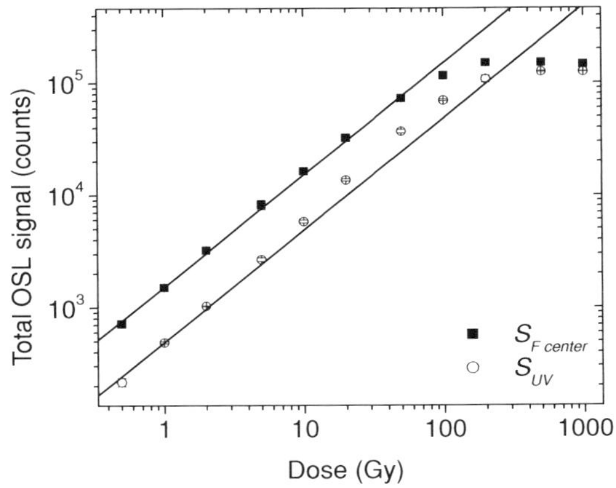


*Figure 3.15* Dose response of $Al_2O_3$:C Luxel detectors obtained using a time-resolved OSL technique to separate the F-centers and UV emission bands in $Al_2O_3$:C (Yukihara and McKeever, 2006b). The data show that the OSL from the F-centers presents less supralinearity than the OSL from the UV emission band.

unstable at ~600 °C, in addition to the main dosimetric trap, $F$-centers and $F^+$-centers (Chen, Pagonis and Lawless, 2006a; Pagonis, Chen and Lawless, 2006; Yukihara *et al.*, 2004a). During irradiation, electrons and holes are created and captured by the different defects. $F^+$-centers are converted to $F$-centers by electron capture, whereas $F$-centers are converted to $F^+$-centers by hole capture. In the absence of defects, there is an equilibrium concentration of $F$-centers and $F^+$-centers due to the following processes:

$$F\text{-center} + \mathrm{h}^+ \rightarrow F^+\text{-center} \tag{3.43}$$

and

$$F^+\text{-center} + \mathrm{e}^- \rightarrow F\text{-center} \tag{3.44}$$

However, due to capture of electrons and holes, the balance of the equations can tilt in favor of an increased concentration of one defect or the other.

In this model supralinearity results from two combined effects: (i) during irradiation electrons are captured at deep traps, resulting in an increase in the concentration of recombination centers ($F^+$-centers) and increased sensitivity and (ii) reduced competition due to filling of deep electron traps leaves more electrons available for capture by the main dosimetric trap and recombination, resulting in more OSL signal.

Moreover, a decrease in the sensitivity is observed after saturation due to two effects: (i) during irradiation holes are captured at deep hole traps, resulting in a decrease in the concentration of recombination centers ($F^+$-centers) and decreased sensitivity and (ii) the deep hole traps may act as competing recombination centers, leaving less electrons available for recombination at $F^+$-centers, resulting in $F$-center luminescence.

The model discussed above is supported by the fact that when the deep hole trap is emptied by heating, holes are released resulting in an increased concentration of $F^+$-centers and increased sensitivity. When the deep electron traps are emptied, the opposite is observed (Yukihara *et al.*, 2003, 2004a). However, it has not been demonstrated yet that other observed phenomena, such as the different dose response associated with the $F$-center emission band and the UV emission band in $Al_2O_3$:C or the changes in the shape of the OSL curve, can be explained by such model.

The data presented in this section exemplify one of the important aspects in OSL dosimetry: the OSL response is dependent on the choice of signal (integration period for calculation of the signal, emission band selected). In TL dosimetry, this situation is equivalent to using the area under different peaks of the TL curve as the signal.

#### *3.4.2.4 Sensitivity Changes*

The supra- and sub-linear behaviors of the OSL dose response are related to a change in OSL sensitivity as a function of accumulated dose in the crystal. Figure 3.16 shows how the OSL sensitivity, defined as the OSL response following bleaching and irradiation with a fixed test dose, changes as a function of the previous dose received by the dosimeters. It can be observed that the OSL sensitivity increases in the dose region in which the dosimeters exhibit supralinearity, and decreases for higher doses. However, the correlation between the supralinearity factor in Figure 3.13b and the sensitivity in Figure 3.16 is not perfect. It is particularly interesting to notice that the supralinearity factor in Figure 3.13b is higher for the initial OSL intensity, whereas the degree of sensitization in Figure 3.16 is higher

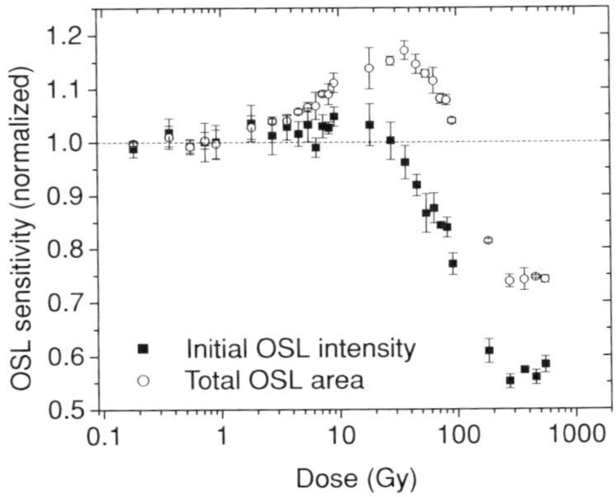


***Figure 3.16*** *OSL sensitivity of $Al_2O_3$:C Luxel detectors following bleaching and irradiation with a fixed test dose of 0.91 Gy, as a function of previous dose received by the dosimeter. The bleaching was carried out using fluorescent light for 10 days. The initial intensity corresponds to the average of the first 3 s of stimulation, and the total OSL area correspond to the total OSL detected over 600 s of stimulation. The OSL readouts were performed with green stimulation using a Risø TL/OSL reader with Hoya U-340 (7.5 mm thickness) and WG-360 filters (2 mm thickness) in front of the PMT. The data points represent the average and experimental standard deviation for three detectors.*

for the total OSL area. This effect may be related to how filling of the deep traps affect the shape of the OSL curve to a test dose, which is seen in Figure 3.17 to change considerably. It may be possible to explain such behavior with a kinetic model, as done by Yukihara *et al.* (2004a) for the change in OSL curves versus dose, but a detailed investigation of this effect has not been carried out.

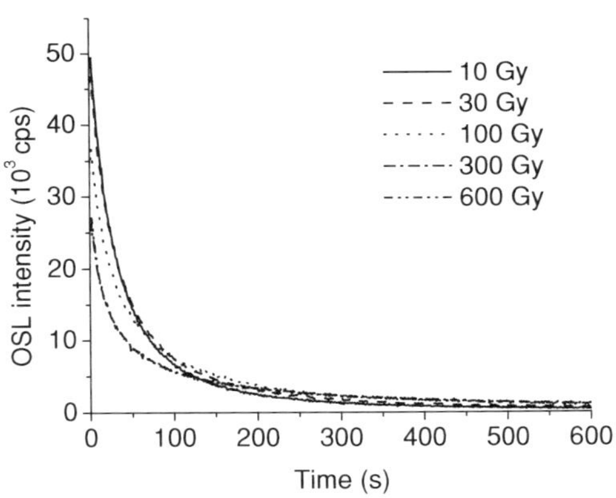


***Figure 3.17*** *OSL curves of $Al_2O_3$:C Luxel detectors following a fixed test dose (0.91 Gy) for detectors that were previously irradiated with various doses (indicated in the figure) and bleached. The OSL was measured with green stimulation using a Risø reader and Hoya U-340 (7.5 mm thickness) plus WG-360 filters (2 mm thickness) in front of the PMT. Each curve corresponds to the average of three detectors.*

#### 3.4.2.5 Energy Response

As discussed in Section 3.4.1.1, $Al_2O_3$ has a higher mass energy absorption coefficient than tissue, water or air for X-ray energies below ~400 keV, which leads to an increased response to these energies when compared to tissue-equivalent dosimeters. This section reviews the data on the energy response of $Al_2O_3$:C detectors irradiated free-in-air. This should be distinguished from the energy responses of OSL dosimeter systems, which include the influence of badge, reader and analysis, and radiation scattered by the phantom.

The energy dependence of $Al_2O_3$:C single crystals (cylindrical pieces 5 mm in diameter by 0.9 mm thickness) and $Al_2O_3$:C Luxel detectors (cut in circular pieces 7 mm in diameter by 0.3 mm thickness) was characterized in terms of air kerma by Gasparian (2009) using the ISO narrow spectrum X-ray reference fields, $^{137}$Cs and $^{60}$Co gamma-ray reference fields (ISO, 1996a). The OSL detectors were irradiated with air kerma values between 0.3 and 50 mGy, read using different readers (InLight microStar, a POSL reader and the Risø reader), and analyzed using the initial OSL intensity and total OSL area whenever data on the entire OSL curve were available (e.g., when using a POSL reader or Risø reader). A build-up layer was used for the $^{137}$Cs (2 mm PMMA) and $^{60}$Co (4 mm PMMA) gamma-ray exposures.

Table 3.10 shows the energy response of $Al_2O_3$:C Luxel detectors, normalized to the response to the $^{60}$Co gamma-ray field. Generally speaking, the OSL response follows the expected behavior discussed in connection with Figure 3.6, the response increasing with decreasing X-ray energies. However, one should notice in these data that the energy response is dependent on the reader, choice of signal (initial OSL intensity or total OSL area) and optical filters (i.e., whether the optical filter transmits only the $F$-center luminescence or a combination of $F$-center and UV emission bands). For a reader that detects only the $F$-center emission band, the response for the initial OSL intensity is typically 5–10% higher than the response obtained for the total OSL area. For a reader that detects a combination of UV and $F$-center emission bands (e.g., Risø reader with Hoya U-340 filter), this difference increases to 10–30%. As in the case of the dose response, the difference between the response obtained using the initial OSL intensity or the total OSL area is related to changes in the OSL curves, probably due to the differences in ionization density created by the radiation fields. Figure 3.18 illustrates the changes in OSL decay curves as a function of radiation field. These differences in the OSL decay curve as a function of ionization density for X-rays, beta rays and protons have been investigated by Jain, Bøtter-Jensen and Thomsen (2007).

Table 3.11 shows the energy response of $Al_2O_3$:C single crystals read using the Risø reader and two filter combinations, also normalized to the response to the $^{60}$Co gamma-ray field. In general the same observations regarding the differences due to choice of signal (initial OSL intensity and total OSL area) and due to emission band detected apply here. However, the energy response from $Al_2O_3$:C single crystals is different than the energy response from $Al_2O_3$:C Luxel detectors.

The energy response for $Al_2O_3$:C single crystals and Luxel detectors are compared in Figure 3.19 with the ratio of mass energy absorption coefficients for $Al_2O_3$ and air. The energy response follows the ratio of mass energy absorption coefficient down to ~50 keV, but diverges for lower energy. The reason for the decrease in efficiency has not been elucidated, but it may be associated with details of the energy deposition in $Al_2O_3$ due to the different dimensions and compositions of the detectors, or due to ionization density effects in $Al_2O_3$:C. A detailed investigation on the energy deposition by Monte Carlo would be

***Table 3.10*** *Energy response of $Al_2O_3$:C Luxel detectors exposed free-in-air to ISO narrow spectrum X-ray reference fields, $^{137}$Cs and $^{60}$Co gamma-ray reference fields (ISO, 1996a) for various readers and choice of signal (initial OSL intensity or total OSL area)*

| ISO reference field | $\bar{E}$ [a] (keV) | microStar[b] | POSL reader[c] Initial OSL intensity ($S_i$) | POSL reader[c] Total OSL area ($S_a$) | Risø reader[d] Initial OSL intensity ($S_i$) U-340 | Risø reader[d] Total OSL area ($S_a$) U-340 |
|---|---|---|---|---|---|---|
| N-15 | 12 | 2.13 ± 0.10 | 2.05 ± 0.11 | 1.76 ± 0.06 | 3.09 ± 0.21 | 2.25 ± 0.11 |
| N-20 | 16 | 2.89 ± 0.07 | 3.04 ± 0.08 | 2.73 ± 0.04 | 4.34 ± 0.16 | 3.38 ± 0.10 |
| N-25 | 20 | 3.16 ± 0.09 | 3.15 ± 0.08 | 2.86 ± 0.04 | 4.28 ± 0.14 | 3.44 ± 0.09 |
| N-30 | 24 | 3.29 ± 0.06 | 3.37 ± 0.08 | 3.14 ± 0.05 | 4.42 ± 0.15 | 3.73 ± 0.08 |
| N-40 | 33 | 3.24 ± 0.10 | 3.25 ± 0.06 | 3.03 ± 0.05 | 4.18 ± 0.16 | 3.52 ± 0.08 |
| N-60 | 48 | 2.96 ± 0.11 | 2.97 ± 0.06 | 2.81 ± 0.04 | 3.73 ± 0.19 | 3.20 ± 0.13 |
| N-80 | 65 | 2.07 ± 0.04 | 2.11 ± 0.04 | 1.98 ± 0.03 | 2.63 ± 0.12 | 2.26 ± 0.08 |
| N-100 | 83 | 1.47 ± 0.04 | 1.50 ± 0.04 | 1.405 ± 0.023 | – | – |
| N-120 | 100 | 1.28 ± 0.04 | 1.25 ± 0.03 | 1.168 ± 0.016 | 1.60 ± 0.06 | 1.36 ± 0.04 |
| N-150 | 118 | 1.20 ± 0.03 | 1.19 ± 0.05 | 1.12 ± 0.05 | 1.53 ± 0.05 | 1.30 ± 0.03 |
| N-200 | 164 | 1.11 ± 0.03 | 1.10 ± 0.03 | 1.041 ± 0.020 | 1.33 ± 0.06 | 1.17 ± 0.04 |
| N-250 | 208 | 1.034 ± 0.019 | 1.041 ± 0.023 | 0.997 ± 0.019 | 1.22 ± 0.05 | 1.10 ± 0.03 |
| N-300 | 250 | 1.005 ± 0.023 | 0.98 ± 0.03 | 0.942 ± 0.025 | 1.14 ± 0.04 | 1.041 ± 0.024 |
| Cs-137 | 667 | 1.000 ± 0.025 | 1.01 ± 0.04 | 1.00 ± 0.07 | 1.00 ± 0.05 | 1.00 ± 0.04 |
| Co-60 | 1225 | 1 | 1 | 1 | 1 | 1 |

The data are normalized to the response to the $^{60}$Co gamma-ray field (Gasparian, 2009).
[a] Mean energy of the radiation field.
[b] Data from the InLight microStar reader correspond primarily to the initial intensity of the *F*-center emission of $Al_2O_3$:C and, therefore, it is equivalent to the initial intensity of the POSL reader.
[c] Data based on a POSL reader with green LEDs and Corning 5-58 filters in front of the PMT. In this arrangement only the *F*-center emission band of $Al_2O_3$:C is detected.
[d] Data based on the Risø reader with CW-OSL readout and Hoya U-340 filters in front of the PMT. In this arrangement, both *F*-centers and UV emission bands of $Al_2O_3$:C are detected.

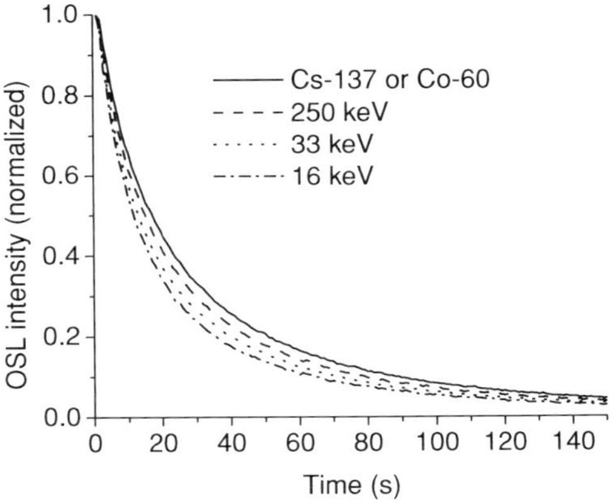


***Figure 3.18*** *OSL curves of $Al_2O_3$:C Luxel detectors following exposure in different ISO narrow-spectrum series X-ray (mean energy indicated in the figure) or gamma reference fields. The OSL was measured with green stimulation using a Risø reader and Hoya U-340 filters in front of the PMT. Each curve corresponds to the average of at least three detectors.*

***Table 3.11*** *Energy response for $Al_2O_3$:C single-crystal detectors (~5 mm diameter by 0.9 mm thickness) exposed free-in-air to ISO narrow spectrum X-ray reference fields, $^{137}$Cs and $^{60}$Co gamma-ray reference fields (ISO, 1996a) relative to the response to reference $^{60}$Co gamma-ray field for various optical filters and choice of signal (initial OSL intensity or total OSL area).*

| ISO reference field | $\bar{E}$[a] (keV) | Risø reader[b] Initial OSL intensity ($S_i$) U-340+WG-360 | Risø reader[b] Total OSL area ($S_a$) U-340+WG-360 | Risø reader[c] Initial OSL intensity ($S_i$) U-340 | Risø reader[c] Total OSL area ($S_a$) U-340 |
|---|---|---|---|---|---|
| N-15 | 12 | 0.63 ± 0.12 | 0.60 ± 0.08 | 0.91 ± 0.09 | 0.74 ± 0.07 |
| N-20 | 16 | 1.46 ± 0.03 | 1.30 ± 0.03 | 1.71 ± 0.08 | 1.426 ± 0.029 |
| N-25 | 20 | 2.13 ± 0.14 | 1.82 ± 0.07 | 2.45 ± 0.16 | 2.11 ± 0.05 |
| N-30 | 24 | 2.73 ± 0.09 | 2.59 ± 0.07 | 3.29 ± 0.28 | 2.73 ± 0.18 |
| N-40 | 33 | 2.91 ± 0.12 | 2.666.0.027 | 3.35 ± 0.24 | 2.92 ± 0.22 |
| N-60 | 48 | 2.99 ± 0.14 | 2.75 ± 0.10 | 3.22 ± 0.23 | 2.94 ± 0.10 |
| N-80 | 65 | 2.22 ± 0.14 | 2.07 ± 0.12 | 2.42 ± 0.16 | 2.16 ± 0.06 |
| N-100 | 83 | 1.50 ± 0.06 | 1.39 ± 0.03 | 1.73 ± 0.10 | 1.49 ± 0.08 |
| N-120 | 100 | 1.23 ± 0.06 | 1.161 ± 0.017 | 1.40 ± 0.12 | 1.25 ± 0.06 |
| N-150 | 118 | 1.15 ± 0.07 | 1.07 ± 0.03 | 1.38 ± 0.07 | 1.18 ± 0.05 |
| N-200 | 164 | 1.059 ± 0.028 | 0.999 ± 0.026 | 1.11 ± 0.05 | 1.00 ± 0.04 |
| N-250 | 208 | 0.98 ± 0.03 | 0.94 ± 0.03 | 1.06 ± 0.04 | 0.97 ± 0.04 |
| N-300 | 250 | 0.978 ± 0.026 | 0.933 ± 0.009 | 1.01 ± 0.03 | 0.930 ± 0.019 |
| Cs-137 | 667 | 0.970 ± 0.021 | 0.965 ± 0.015 | 1.006 ± 0.022 | 0.997 ± 0.019 |
| Co-60 | 1225 | 1 | 1 | 1 | 1 |

The data are normalized to the response to the $^{60}$Co gamma-ray field (Gasparian, 2009).
[a] Mean energy of the radiation field.
[b] Data based on the Risø reader with CW-OSL readout and Hoya U-340 and Schott WG-360 filters in front of the PMT. In this arrangement, only the *F*-center emission band of $Al_2O_3$:C is detected.
[c] Data based on the Risø reader with CW-OSL readout and Hoya U-340 filters in front of the PMT. In this arrangement, both *F*-centers and UV emission bands of $Al_2O_3$:C are detected.

helpful to understand how much this effect is associated with energy deposition and how much it is related to ionization density effects. However, the changes in the OSL decay curve as a function of the mean energy of the photon field indicates that ionization density effects are partly involved in the process.

#### *3.4.2.6 Fading*

A preliminary study of the fading of $Al_2O_3$:C detectors did not allow a definitive conclusion on the presence of long-term fading for time intervals up to 85 days (Bøtter-Jensen *et al.*, 1997). Fading is generally quoted to be 3–5% per year (Kortov, 2007; Pradhan, Lee and Kim, 2008). Short-term fading is generally not an issue for personal dosimetry, but will be important for medical applications (see Chapter 5).

#### *3.4.2.7 Temperature Dependence*

The effects of irradiation and readout temperature on $Al_2O_3$:C are generally small. The OSL signal from $Al_2O_3$:C increases with readout temperature in the 25–100 °C range, possibly

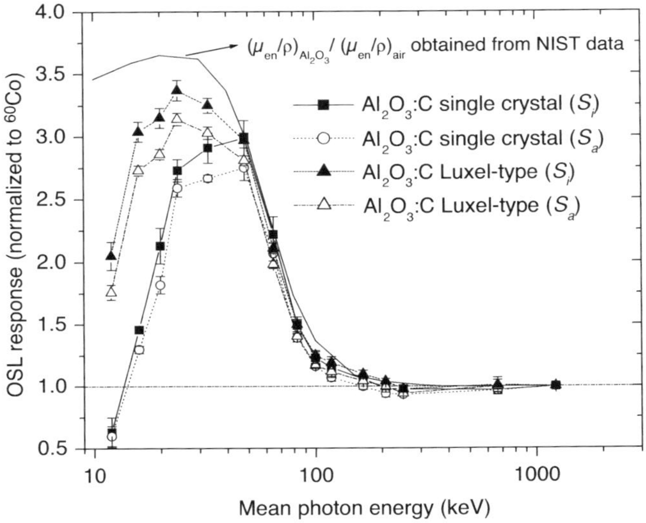


***Figure 3.19*** *Energy response of $Al_2O_3$:C single-crystal (5 mm in diameter by 0.9 mm thickness) and Luxel detectors (cut in pieces 7 mm in diameter by 0.3 mm thickness) based on the initial OSL intensity ($S_i$) and total OSL area ($S_a$). The $Al_2O_3$:C single crystals were read with green stimulation using a Risø reader with Hoya U-340 and Schott WG-360 filters in front of the PMT. The Luxel detectors were read using a portable POSL reader with Corning 5-58 filters in front of the PMT. In both cases, only the F-center luminescence is detected. Reprinted from Gasparian, P.B.R., Methodological developments for application of optically stimulated luminescence (OSL) in medical dosimetry. M. S. Thesis. Copyright (2009) with permission from Oklahoma State University.*

due to the influence of shallow traps (Markey, Colyott and McKeever, 1995). This temperature range is not sufficient to cause thermal quenching of the *F*-center luminescence, which starts to affect the luminescence efficiency above 100 °C (see Section 2.2.9). However, data on materials with a small concentration of shallow traps also show a small increase in OSL intensity with readout temperature in the temperature range from −30 °C to 100 °C (Akselrod *et al.*, 1998a), indicating the possibility of a small degree of thermal assistance in the process of optical stimulation of the trapped electrons.

The effect of irradiation temperature has been investigated by a few researchers for $Al_2O_3$:C Luxel detectors. Miller and Murphy (2007) observed a slight increase in OSL with temperature, but the OSL response was within ±1.5% from the room temperature value for a temperature range between ~5 and 37 °C. No trend was observed by Jursinic (2007) between 10 and 40 °C, the values also being within ±1.5%. Similar results were reported by other authors (Yukihara *et al.*, 2008a). In these cases, the OSL detectors were read out at room temperature after irradiation.

### 3.4.3 BeO Detectors

Ceramic beryllium oxide (BeO) is an OSL material which has been explored for TL and OSL dosimetry and is receiving renewed attention (Sommer and Henniger, 2006; Sommer,

Freudenberg and Henniger, 2007, 2008). In addition to being highly sensitive to ionizing radiation, it has an effective atomic number ($Z_{eff} = 7.2$) similar to water ($Z_{eff} \sim 7.51$). Some of the luminescence properties of BeO have been discussed in Section 2.7.1.1.

### 3.4.3.1 *Available Forms*

The BeO detectors currently used as OSL dosimeters are sintered ceramics of 99.5% beryllium oxide (99.5% Thermalox, Brush Ceramic Products, Brush Wellman, Inc.), having a nominal density of 2.85 g/cm$^3$. Sommer, Jahn and Henniger (2008) have investigated two forms of detectors, namely discs of 4 mm in diameter by 0.8 mm thickness and square chips 4.7 mm × 4.7 mm × 0.5 mm. Other sizes are also available from the manufacturer.

### 3.4.3.2 *Preparation and Handling*

Handling of the sintered ceramics is considered safe, but precautions should be taken to avoid abrasion and powder formation (Scarpa, 1970). Inhaling of beryllium oxide may cause a chronic lung disease (chronic beryllium disease) which may be fatal over time. Beryllium is also considered to be carcinogenic.

With respect to annealing procedures, Sommer and Henniger (2006) investigated the effect of two different annealings (heating up to 400 °C at 5 °C/s or heating up to 500 °C with a hold time of 30 s at that temperature) on the OSL response of BeO to a 50 mGy dose. The authors concluded that the reproducibility was independent of the type of annealing and better than 5%. The ~200 °C TL peak of BeO can be increased by almost 20% using slow cooling rates (Borio *et al.*, 1990), but the OSL signal is known to be associated with a trapping center that becomes unstable at ~340 °C, not the ~200 °C TL peak.

In principle, bleaching of the detectors with light of a spectrum similar to that used for the OSL readout should be sufficient to erase the OSL signal due to accumulated dose. However, optical stimulation can transfer charge carriers from deep traps (unstable in the 400–650 °C temperature range) to the trapping centers responsible for the TL and OSL signals. This photo-transferred signal may become a problem for detectors previously irradiated with high doses. Heating to 650 °C is sufficient to eliminate the source of photo-transferred charges (Bulur and Göksu, 1998).

Pre-heating to 50–125 °C may be necessary to remove unstable components in the OSL signal (see Section 3.4.3.5). The OSL signal did not seem to be affected by humidity (Sommer, Freudenberg and Henniger, 2007). As any OSL detector, BeO should be protected from light after exposure to radiation.

### 3.4.3.3 *Dose Response*

The OSL response of BeO is linear over six orders of magnitude, between 5 μGy and 5 Gy, starting to saturate at higher doses (Figure 3.20). Deviation from linearity is less than 5% at 30 Gy (Sommer and Henniger, 2006). Doses of the order of 1 μGy can be detected using BeO and a system in which the detector is stimulated with LEDs (broad band centered at 455 nm) from one side and the OSL signal is detected from the other side (Sommer, Freudenberg and Henniger, 2007).

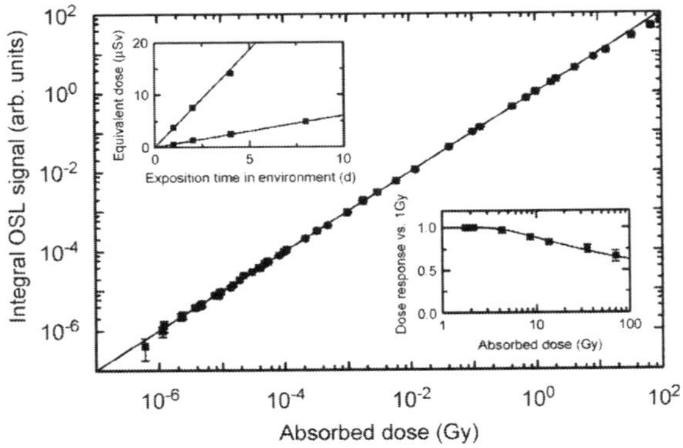


***Figure 3.20*** *Dose response of BeO OSL detectors for doses up to 100 Gy. Reprinted from Radiation Measurements, Sommer, M., Freudenberg, R., Henniger, J., New aspects of a BeO-based optically stimulated luminescence dosimeter. Vol. 42, 617–620. Copyright (2007) with permission from Elsevier.*

### *3.4.3.4 Energy Response*

One of the advantages of BeO over $Al_2O_3$:C is the low effective atomic number ($Z_{eff}$ = 7.2), which makes the material nearly tissue-equivalent (Bos, 2001). Figure 3.21 shows the energy response of BeO in comparison with calculated values. The data illustrate a small under-response to low-energy X-rays, which is consistent with the absorbed dose energy dependence (Section 3.4.1.1).

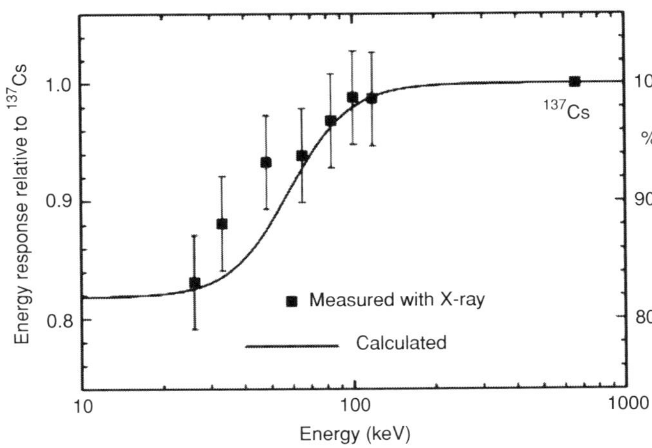


***Figure 3.21*** *Photon energy response of BeO OSL detectors relative to* $^{137}$*Cs. Reprinted from Radiation Protection Dosimetry, Sommer, M., Henniger, J., Investigation of a BeO-based optically stimulated luminescence dosimeter. Vol. 119 (1–4) 394–397. Copyright (2006) with permission from Oxford University Press.*

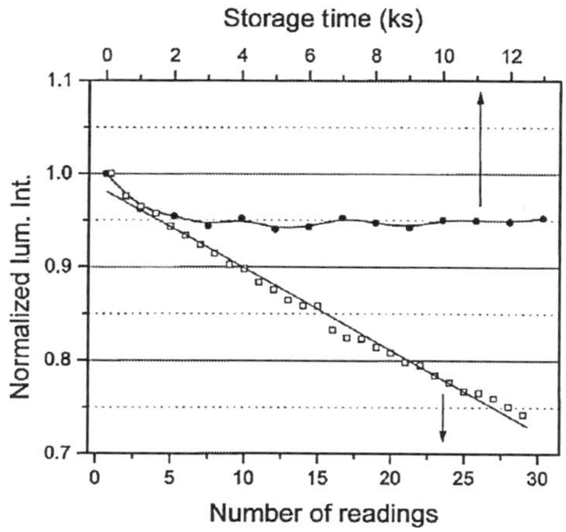


*Figure 3.22* OSL intensity from BeO as a function of number of readings and storage time. Reprinted from Radiation Measurements, Bulur, E., Göksu, H.Y., OSL from BeO ceramics: new observations from an old material. Vol. 119 (1–4), 394–397. Copyright (2006) with permission from Elsevier.

### *3.4.3.5 Fading*

The OSL signal from BeO is known to fade by ~5% in the first hour following irradiation, remaining relatively stable after that (Figure 3.22). This is confirmed by results from Sommer and Henniger (2006), which indicates a fading of ~6% in the first day when compared to 30 min after irradiation, and less than 1% afterwards up to 6 months even when stored at 50 °C. The unstable signal is related to shallow trapping centers that can be emptied by a short heating period prior to the OSL readout. Sommer and Henniger (2006) suggest a 50 °C for 5 min, whereas Bulur and Göksu (1998) used heating to 125 °C.

## 3.5 Dosimetry Systems

In this section we describe the Luxel and InLight dosimetry systems (Landauer Inc., USA), which by the time of writing were the only OSL dosimetry systems available commercially.

### 3.5.1 Luxel+ Dosimetry System

The Luxel dosimetry system (Landauer Inc., USA) was the first system employing the OSL technology for personal and environmental monitoring of X-rays, gamma rays and beta radiation to be introduced in commercial scale. The filter pack that is part of the dosimeter has been redesigned and now the system is called "Luxel+." Luxel+ dosimeters may also include neutron dosimetry capability provided by a plastic nuclear track detector (CR-39), but since this detector is not based on the OSL technology it will not be discussed any further.

#### 3.5.1.1 Detector and Badge

The Luxel+ dosimeter consists of a Luxel detector ($Al_2O_3$:C powder in a tape as described in Section 3.4.2, with dimensions of ~1.7 cm × 2.0 cm by 0.3 mm thickness) enclosed in a multi-element filter pack, packaged in a light-tight paper wrapper and sealed in a polycarbonate package (see Figure 3.23).

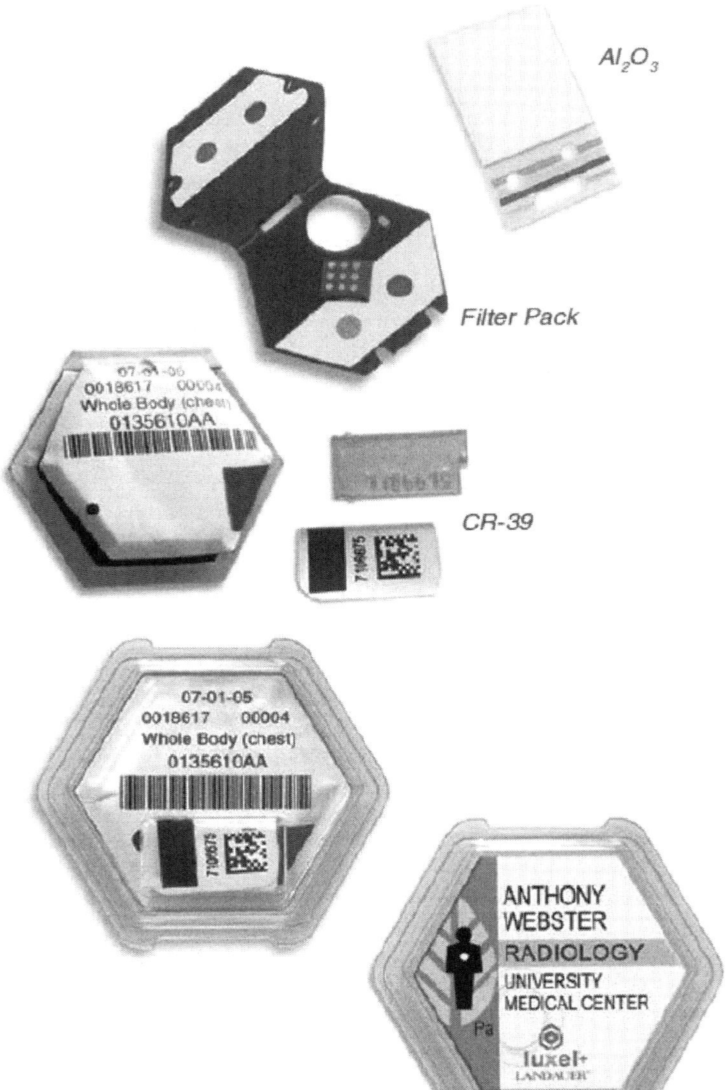


**Figure 3.23** *Luxel+ badge and its components. Reprinted with permission from Landauer Inc. (USA).*

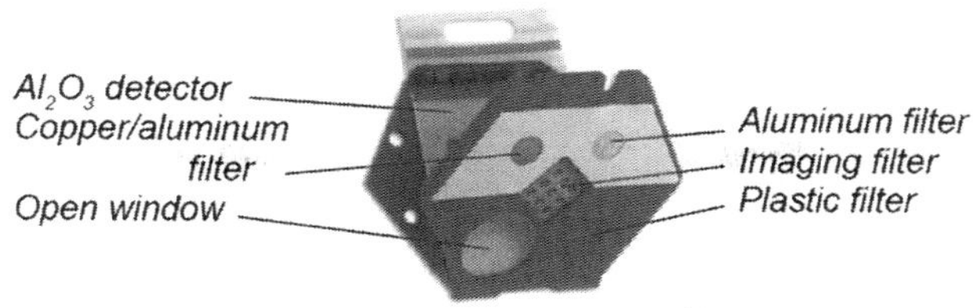


***Figure 3.24*** *$Al_2O_3$:C detector and filter pack. Reprinted with permission from Landauer Inc. (USA).*

The multi-element filter pack is designed for determination of mean photon energy and determination of the contribution from beta radiation. The readout of the detector area behind the different filters will produce different OSL signals $S_g$ that can be used with a dose algorithm to determine the deep dose, lens-of-eye dose and shallow dose. The filter combination used in the Luxel+ dosimeters is shown in Figure 3.24 and includes an aluminum filter, a copper/aluminum filter, a plastic filter, and an open window. In addition, the central area has a patterned imaging filter for distinction between static and dynamic exposure (see Section 3.5.1.3).

### *3.5.1.2 Reader and Dose Evaluation*

Luxel+ dosimeters are read at Landauer's laboratories using a system based on the same principles as the system shown in Figure 2.41. The reader is based on the POSL readout approach. A pulsed laser stimulates different regions of the Luxel detector for a short period of time, generating multiple signals $S_g$ from the detector areas behind the different filters. These signals are combined using a dose algorithm for determination of the quantities of interest. Since the stimulation is very short, most of the OSL signal is preserved and the detector can be re-stimulated to confirm the measurement. This is automatically done for doses above 5 mSv.

### *3.5.1.3 Imaging Capability*

For low-energy photons and beta doses above 5 mSv, the OSL detector area behind the imaging filter can be stimulated using an expanded pulsed laser beam and imaged by a CCD detector to obtain a pattern that may indicate static or dynamic exposure. This information can help identify abnormal exposures and provide additional information in case the exposure needs to be investigated. A typical example of abnormal exposure is the intentional or accidental irradiation of the dosimeter by itself. Such an exposure may take place rapidly and/or while the dosimeter is in a stationary position such that the subsequent image indicates a "static" exposure. In contrast, if the exposure occurred over an extended period of time while the dosimeter was being worn properly one would expect the image to be blurred, indicating a "dynamic" exposure.

The principle behind this procedure is described by Akselrod *et al.* (2000) and illustrated in Figure 3.25. The CCD image is processed using a 2-D discrete Fourier transform to remove high-frequency components associated, for example, with pixel noise. A numerical parameter can then be defined, based on the filtered imaged to determine the probability of

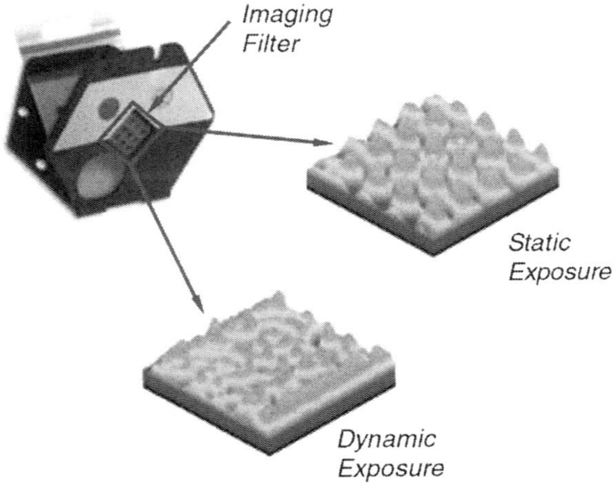


***Figure 3.25*** *Illustration of the OSL image (OSL intensity versus position) obtained from the detector area behind the imaging filter for a dynamic and static exposure. Reprinted with permission from Landauer Inc. (USA).*

static or dynamic exposure. The method has shown to be effective for low-energy photons (less than 150 keV).

#### *3.5.1.4 Other Technical Specifications*

The quoted energies X-ray and gamma-ray energies detected for the Luxel+ dosimetry system are from 5 keV to $\geq$40 MeV, with a dose range from 10 μSv to 10 Sv. For beta particles, the quoted energy range is from 150 keV to $\geq$ 10 MeV, with a dose range between 100 μSv to 10 Sv.

A recent study compared the performance of the Luxel dosimeters with film, TLD and radiophotoluminescence (RPL) dosimeters (Garcier *et al.*, 2007). Tests were performed for accuracy, energy response for photons from X-rays with mean energy of 45 keV up to $^{60}$Co gamma rays, and resulted in a suggestion to implement the OSL dosimetry system because of its overall performance.

### 3.5.2 InLight Dosimetry System

The InLight dosimetry system (Landauer Inc., USA) comprises a dosimeter badge containing four OSL detectors (elements) and different reader models, the designs of which follow that of the Panasonic TLD reader. Landauer has used the InLight dosimetry system as a service or provided users with a reader for in-house operation (Perks, Le Roy, and Prugnaud, 2007).

#### *3.5.2.1 Detector and Badge Design*

The OSL detectors used in the InLight dosimeters are identical to the Luxel detectors, but cut into round pieces of ~5 mm in diameter. Each InLight dosimeter contains a slide with

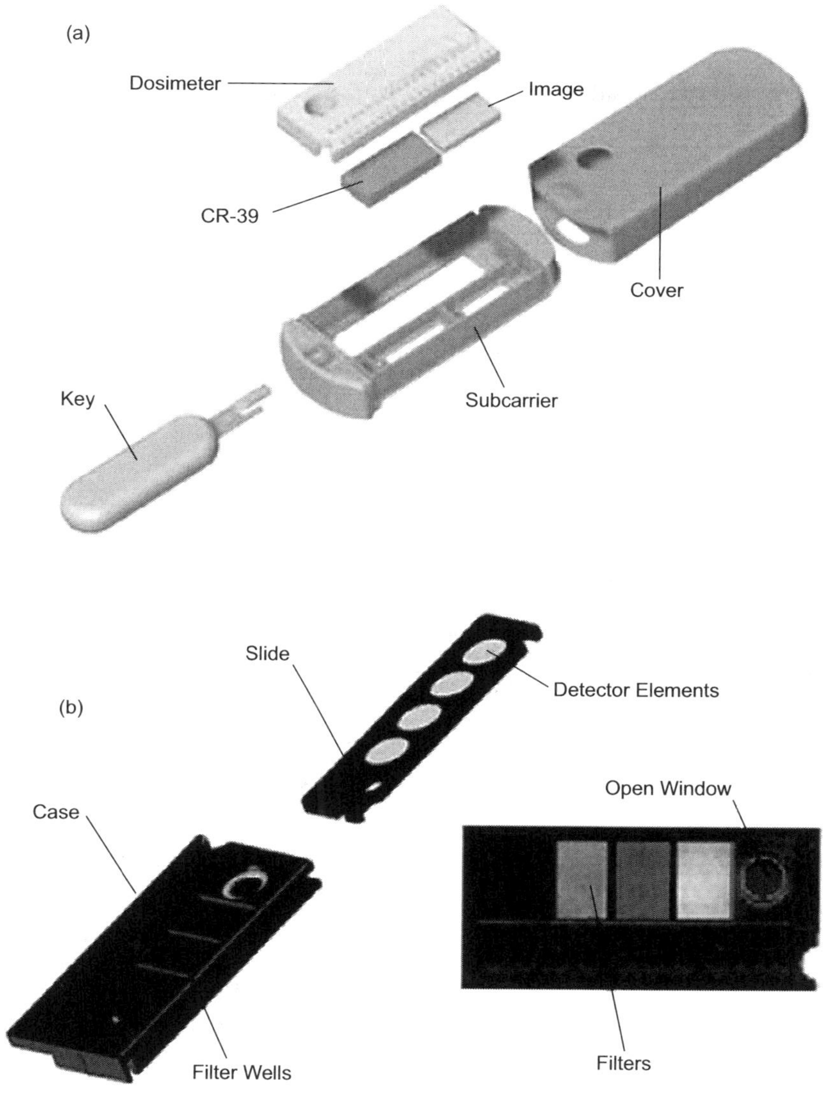


***Figure 3.26*** *(a) InLight badge with dosimeter and additional components (CR-39 and imaging detector) and (b) dosimeter case with slide containing four OSL detectors. Reprinted with permission from Landauer Inc. (USA).*

four of such OSL detectors, as shown in Figure 3.26. When the slide is inside the case, each detector is positioned behind different filters providing different radiation attenuation conditions. The signal from each OSL detector is used in conjunction with a dose algorithm to evaluate different dosimetric quantities, as discussed for the Luxel dosimetry system. Two types of filter combinations have been used, as shown in Table 3.12.

***Table 3.12*** *Plastic or filter mass thicknesses (mg/cm$^2$) for two InLight badge models (Ford, Hanify, and Perks, 2004)*

| Model 1 badge | | Model 2 badge | |
|---|---|---|---|
| Open window: | 28.9 mg/cm$^2$ front<br>134.0 mg/cm$^2$ back | Open window: | 28.9 mg/cm$^2$ front<br>134.0 mg/cm$^2$ back |
| Plastic: | 274.5 mg/cm$^2$ front<br>282.8 mg/cm$^2$ back | Plastic: | 274.5 mg/cm$^2$ front<br>282.8 mg/cm$^2$ back |
| Copper: | 544.7 mg/cm$^2$ front<br>553.0 mg/cm$^2$ back | Copper: | 544.7 mg/cm$^2$ front<br>553.0 mg/cm$^2$ back |
| Lead: | 982.2 mg/cm$^2$ front<br>990.5 mg/cm$^2$ back | Aluminum: | 374.6 mg/cm$^2$ front<br>382.9 mg/cm$^2$ back |

The slides contain a two-dimensional bar code for identification of the detectors and their sensitivities. The badge also contains a bar code for identification of the dosimeter and filter combination. The reader reads the bar codes automatically. The badges may also contain an additional CR-39 detector for neutron dosimetry and an OSL detector for distinguishing between static and dynamic exposure in case of low-energy photons and low-energy beta-particle exposures, although the capability to read the latter is not included in the InLight readers themselves.

Dosimeters for single-point measurements (InLight Dot and nanoDot dosimeters) have also been introduced aiming particularly at clinical measurements of skin entrance dose and other applications (see Figure 2.43d). These single-point dosimeters are placed in an adapter for readout in the InLight microStar reader.

### *3.5.2.2 Readers*

The InLight readers are designed for OSL readout at a rate of almost 10 s per dosimeter. There are three models available, the portable InLight microStar (32.7 cm width × 23.2 cm depth × 10.9 cm height), and two automated ones with capacity for either 200 badges (~110 cm width × 46 cm depth × 38 cm height), or 500 badges (~110 cm width × 49 cm depth × 77 cm height).

In all models the OSL readout of each detector is performed using an array of 38 green LEDs operated in CW-OSL mode during a short period of time (~1 s). The number of LEDs used for stimulation depends on the dose range: for low doses all 38 LEDs are used (strong beam), whereas only six LEDs are used for high doses (weak beam). The dose range is evaluated prior to the measurement using one LED and the cut-off point is an adjustable parameter (Perks *et al.*, 2004).

The combination of short stimulation time and low stimulation intensity results in only a small depletion of the OSL signal per readout. Figure 3.27 shows the mean OSL signal for four irradiated detectors read repeatedly using either the weak stimulation or the strong stimulation. Based on the fitted curves of the type $y = a + bx$ we estimate that, for this specific reader, the weak stimulation depletes the OSL signal at a rate of (0.07 ± 0.04)% per readout, whereas for the strong stimulation the depletion rate is (0.25 ± 0.03)% per readout. Data similar to those shown in Figure 3.27 can be obtained for each individual

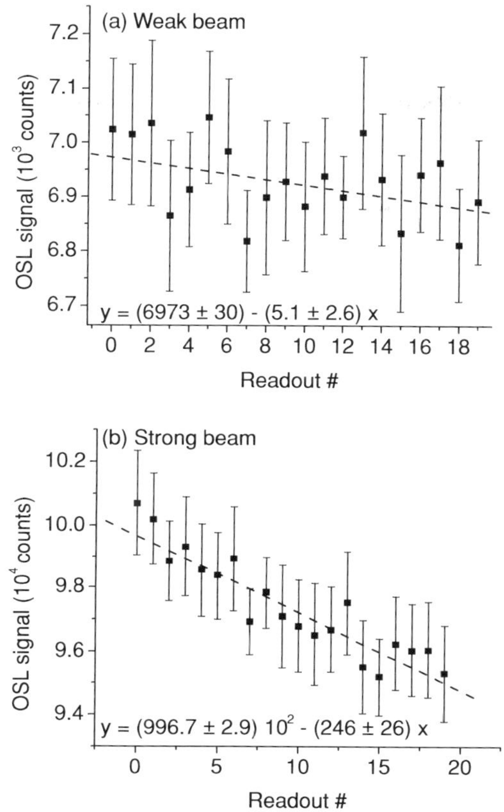


***Figure 3.27*** *Mean OSL signal and experimental standard deviation based on four OSL detectors irradiated with ~100 mGy and read using the InLight microStar reader using (a) weak stimulation and (b) strong stimulation. The data were fitted with a linear function of the type* $y = a + bx$*, where y is the OSL signal and x is the readout number, which is an approximation of the exponential decay* $y = a\exp(-bx/a)$ *for the case of small b/a values.*

reader to correct for signal depletion, if desired (see Ford, Hanify and Perks, 2004). Because the readout is almost non-destructive, the dosimeters can be used over an extended period of time, while still providing information on the accumulated dose.

### *3.5.2.3 Calibration and Dose Calculation*

Calibration of the reader can be performed using a set of calibration (pre-irradiated) dosimeters provided with the reader. As an example, in the microStar (Landauer, 2008) the expression used to determine the calibration factor in the linear model of dose response is:

$$\mathrm{CF} = \frac{1}{M}\sum_{j=1}^{M}\frac{1}{E_j S_j}\left(\frac{1}{N}\sum_{i=1}^{N} C_i - B_{\mathrm{avg}}\right), \tag{3.45}$$

where CF is the calibration factor, $B_{avg}$ is the average background from control or blank, $C_i$ is the counts for window position $i$ and $N$ is the number of read positions per dosimeter, $S_j$ is the sensitivity of dosimeter $j$, $E_j$ is the dose for dosimeter $j$ and $M$ is the number of calibration dosimeters. The average background is given by:

$$B_{avg} = \frac{1}{L}\sum_{j=1}^{L}\left(\frac{1}{N\,S_j}\sum_{i=1}^{N} C_i\right) \tag{3.46}$$

where $L$ is the number of control dosimeters. Equation (3.46) gives the background per position averaged over all positions and control (or blank) dosimeters. The control dose is given by $B_{avg}$/CF. A polynomial calibration function of the form $y = ax^2 + bx + c$ is also available to account for the non-linear response of $Al_2O_3$:C for doses above 3 Gy (Landauer, 2008). Two dose calculation algorithms have been developed for the InLight dosimeters, depending on the filter configuration in the badge.

#### 3.5.2.4 Other Technical Specifications and New Developments

The InLight dosimeters are rated for photon energies from 5 keV to 20 MeV with linear responses in the range from 10 μSv to in excess of 10 Sv. Regarding fading, dosimeters irradiated with 10 mSv indicated a dose of $(9.7 \pm 0.4)$ mSv after 168 h, therefore identical to the initial value within the experimental uncertainties (Nascimento and Hornos, 2010). (See also Chapter 5.)

## 3.6 Neutron-Sensitive OSL Detectors

$Al_2O_3$ has a small cross-section for neutron interaction, resulting in practically zero neutron sensitivity. For example, the neutron sensitivity of $Al_2O_3$:C at 14 MeV neutrons is $\sim$20% of the neutron sensitivity of $^7$LiF:Mg,Ti, which is commonly used as a neutron-insensitive TLD (Klemic, Azziz and Marino, 1996). The lack of neutron sensitivity of $Al_2O_3$:C, combined with the fact that this is the most important OSL material used in personal dosimetry, has been pointed out as one of the major disadvantages of the OSL technique when compared with the TL technique applied to personal dosimetry (McKeever and Moscovitch, 2003). Clearly, however, this is a problem of the material, not the OSL technique per se.

Several approaches have been attempted to address this problem. One is to develop a new OSL material containing a high content of hydrogen or a material that could be enriched with isotopes of high cross-section for neutron capture (e.g., $^6$Li, $^{10}$B), such as thallium-doped ammonium salts such as $NH_4Br$, $(NH_4)_2SiF_6$, $(NH_4)_2BeF_4$ (Le Masson *et al.*, 2004a, 2004b) and barium aluminoborate glasses ($20Al_2O_3{\cdot}50B_2O_3{\cdot}30BaO$) (Yoshimura and Yukihara, 2006a). Unfortunately fading of the OSL signal has been one of main problems with these new materials.

Another proposed approach is to embed $Al_2O_3$:C grains in a plastic matrix, so that neutron interaction with hydrogen in the matrix would generate recoil protons that could deposit energy in the $Al_2O_3$:C detector, giving rise to an OSL signal (Markey, Colyott and McKeever, 1995). This approach has been used in the case of TL of $Al_2O_3$, but the relative

*Table 3.13* Main isotopes used as neutron converters in luminescence detectors and corresponding natural abundance, thermal neutron cross-section and products (Knoll, 2000; van Eijk, 2004)

| Isotope | Natural abundance | $\sigma^*$ (barns) | Products |
|---|---|---|---|
| $^{6}$Li | 7.4% | 940 | $^{3}$H (2.75 MeV) + $^{4}$He (2.05 MeV) |
| $^{10}$B | 19.8% | 3840 | $^{7}$Li (1.0 MeV) + $^{4}$He (1.8 MeV) |
| | | | $^{7}$Li (0.83 MeV) + $^{4}$He (1.47 MeV) + $\gamma$ (0.48 MeV) |
| $^{157}$Gd | 15.7% | 255 000 | $^{158}$Gd + $\gamma$s + conversion $e^-$ + X-rays (29–182 keV) |
| $^{155}$Gd | 14.8% | 60 900 | $^{156}$Gd + $\gamma$s + conversion $e^-$ + X-rays (39–199 keV) |

* Cross-section for capture of thermal neutrons.

response decreases steeply with the reduced neutron energy and range of the recoil protons (Spurny *et al.*, 1976). A similar approach was also proposed by Henniger *et al.* (1982) using $CaF_2$:Mn and OSL. Obviously, the use of a plastic matrix is more convenient in conjunction with the OSL technique than with the TL technique since heating of the detectors is not necessary.

A third approach is to embed or coat $Al_2O_3$:C grains with materials containing neutron converters such as $^{6}$Li, $^{10}$B or Gd ($^{157}$Gd or $^{155}$Gd) having a high neutron-capture cross-section. The secondary radiation created by the neutron-capture reaction can then deposit energy in the $Al_2O_3$:C, giving rise to an OSL signal (see Table 3.13). The advantage of this approach is that it can lead to a composite material with essentially the same OSL properties as $Al_2O_3$:C in terms of emission and stimulation wavelength, but sensitive to neutrons.

In this section we discuss the development of neutron-sensitive OSL detectors using the latter approach.

### 3.6.1 Development of Neutron-Sensitive OSL Detectors

The practical considerations when choosing the appropriate neutron converter to increase the neutron sensitivity of OSL detectors involves not only the capture cross-section for neutron interaction, but also how the secondary radiation will interact with the OSL material, how the neutron converter material may affect the interaction of the OSL material with photons, the natural abundance of the isotope and cost considerations.

For example, from Table 3.13 one can see that $^{157}$Gd and $^{155}$Gd have a high neutron-capture cross-section and relatively high natural abundance. A neutron imaging detector has, in fact, being developed based on a mixture of $Al_2O_3$:C and Gd neutron converter (Kobayashi *et al.*, 2005). However, the secondary radiation generated is mostly low-LET radiation (gamma rays, X-rays, conversion electrons) that have a higher probability of escaping the detector and depositing energy elsewhere (particularly if the detector consists of small grains of $Al_2O_3$:C). Moreover, because of its high atomic number, Gd ($Z = 64$) has a photoelectric cross-section higher than that of $Al_2O_3$. This in turn affects the energy response of the composite material such that a mixture of $Al_2O_3$ + $Gd_2O_3$ would have a higher response to low-energy photons than $Al_2O_3$ alone. The next-best choice is $^{10}$B,

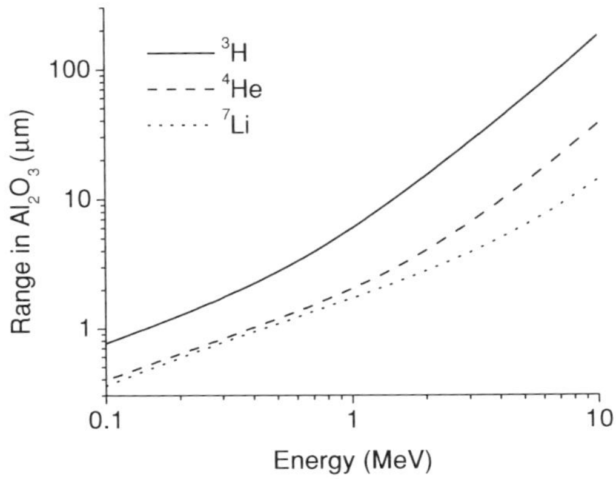


***Figure 3.28*** *Range of particles produced in neutron-capture reactions with $^{10}B$ and $^{6}Li$ in $Al_2O_3$ as a function of particle energy. Calculated using SRIM (Ziegler, Ziegler and Biersack, 2008).*

which has high neutron-capture cross-section and low atomic number ($Z = 5$). The neutron-capture reaction with $^{10}$B results in heavy charged particles ($^{7}$Li and $^{4}$He) that can deposit energy in the detector material if $^{10}$B is in close contact with the $Al_2O_3$:C grains. However, one can see in Figure 3.28 that the $^{4}$He and $^{7}$Li ions produced by the neutron-capture reaction (energies below 2 MeV) have a range in $Al_2O_3$:C less than 5 μm. If the $Al_2O_3$:C grains are considerably larger than 5 μm, the energy will be deposited in a thin layer close to the surface of the grains. The neutron-capture reaction with $^{6}$Li, on the other hand, results in 2.75 MeV $^{3}$H particles which have a range in $Al_2O_3$:C of almost 30 μm.

The consequence of the discussion above can be seen in Figure 3.29, which shows the OSL curves of $Al_2O_3$:C powder mixed with different neutron converters following exposure to neutron sources in identical conditions. Figure 3.29a demonstrates that the

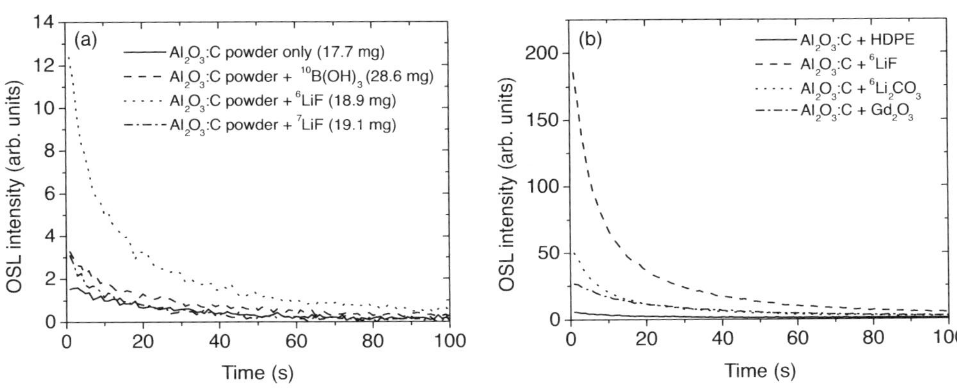


***Figure 3.29*** *OSL curves of a mixture of $Al_2O_3$:C powder and various neutron converters following exposure to (a) a PuBe neutron source and (b) a $^{252}Cf$ neutron source. The data were obtained using a Risø TL/OSL-DA-15 reader with green stimulation LEDs (Mittani et al., 2007).*

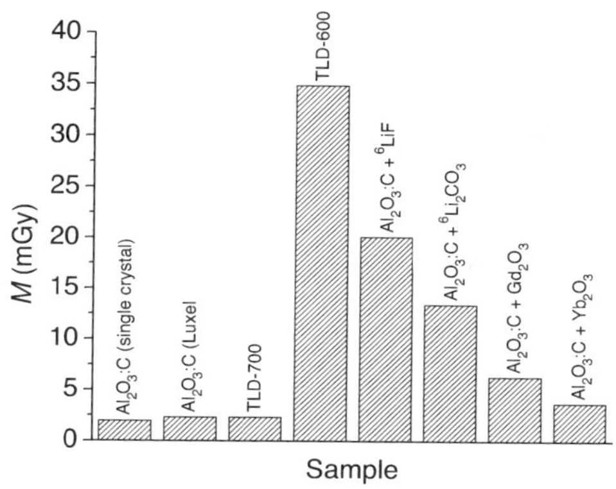


***Figure 3.30*** *OSL indicated value (in mGy) of pellets prepared using $Al_2O_3$:C powder and various neutron converters following exposure to a $^{252}$Cf neutron source in identical conditions. Data on TLD-600 and TLD-700 were also included for comparison. The $Al_2O_3$:C + $Yb_2O_3$ sample was used to demonstrate that part of the enhanced response of $Al_2O_3$:C + $Gd_2O_3$ sample is due to the high effective atomic number of Gd, since Yb has also a high atomic number, but low neutron-capture cross-section (Yukihara et al., 2008b).*

OSL intensity from $Al_2O_3$:C mixed with $^6$LiF is higher than that from $Al_2O_3$:C mixed with $^{10}B(OH)_3$ (boric acid). The latter is very similar to the OSL curves of pure $Al_2O_3$:C powder or $Al_2O_3$:C mixed with $^7$LiF, which are only due to gamma-ray "contamination" in the neutron radiation field. In a comparison, which includes $Al_2O_3$:C powder mixed with other neutron converters such as high-density polyethylene (HDPE), $Gd_2O_3$ and $^6Li_2CO_3$, the mixture containing $^6$LiF still presents the highest OSL intensity (Figure 3.29b).

Mittani *et al.* (2007) investigated the response of pellets prepared using a mixture of $Al_2O_3$:C powder and several types of neutron converters following exposure to a $^{252}$Cf neutron source. Figure 3.30 shows the indicated value $M$ for the various pellets, here defined as the $^{60}$Co gamma dose to water necessary to produce an OSL (or TL) signal of the same intensity as the one following neutron irradiation and calculated by:

$$M = \frac{S}{S_R} D_R \tag{3.47}$$

The indicated value $M$ was used instead of the OSL or TL signals because it does not depend on the sample mass or reader sensitivity. In Figure 3.30, the indicated values from $Al_2O_3$:C, TLD-700 and $Al_2O_3$:C + $Yb_2O_3$ are mostly due to the gamma-ray component of the radiation field (the latter is higher than $Al_2O_3$ or TLD-700 because of the higher effective atomic number). The data indicate that pellets of $Al_2O_3$:C mixed with $^6$LiF can achieve a neutron response comparable to TLD-600.

The ratio between the amounts of $Al_2O_3$:C and neutron converter is also an important parameter to be controlled. One would expect that the more neutron converter in the mixture, the better the neutron sensitivity. Figure 3.31a shows that this is in fact observed, with the neutron sensitivity increasing with increasing neutron converter to $Al_2O_3$:C mass ratio. In this figure the ratio between the neutron and the gamma indicated values of the samples

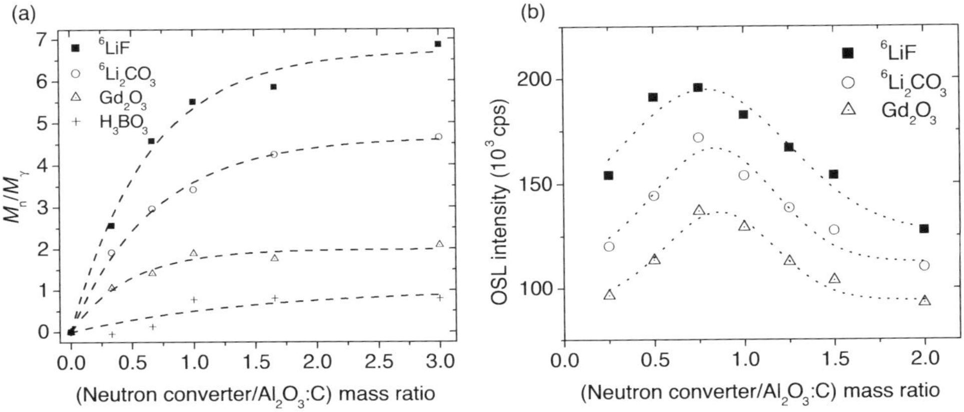


***Figure 3.31*** *(a) Ratio of the indicated value (in mGy) due to neutrons and gamma rays for samples of $Al_2O_3$:C mixed with different neutron converters (indicated in the figure) exposed to PuBe neutron source, as a function of the neutron converter to $Al_2O_3$:C mass ratio; (b) OSL signal versus neutron converter to $Al_2O_3$:C mass ratio for the same samples.*

($M_n/M_\gamma$) is used as a surrogate for the neutron sensitivity since, ideally, a good neutron detector would have a high response to neutrons and a low response to gammas.

The OSL signal also shows an increase with increasing proportion of neutron converter in the mixture (Figure 3.31b). However, as one would expect, after a certain point the OSL signal starts to decrease as the amount of $Al_2O_3$:C in the sample decreases. Therefore, Figure 3.31 shows that there is a compromise between increasing the neutron sensitivity (which is related to the amount of neutron converter) and increasing the overall signal of the detector (which is related to the amount of $Al_2O_3$:C). Based on the data shown in Figure 3.31, Mittani *et al.* (2007) concluded that a mass ratio of 1:1 between neutron converter and $Al_2O_3$:C represents the optimum compromise.

The neutron sensitivity of the $Al_2O_3$:C and neutron converter powder mixture can be improved further by decreasing the $Al_2O_3$:C grain size. As shown in Figure 3.28, the range of $^3$H produced by the neutron capture reaction with $^6$Li can penetrate almost 30 μm inside the $Al_2O_3$:C grain. This means that for grain sizes >30 μm in radius there will be a core that is not sensitive to neutrons. Therefore, by decreasing the size of the $Al_2O_3$:C grains it is possible to increase the detector volume exposed to the $^3$H, resulting in a larger neutron sensitivity relative to the gamma sensitivity. This is exemplified in Figure 3.32, where the ratio $M_n/M_\gamma$ was determined using neutron-irradiated samples prepared using $Al_2O_3$:C with grain sizes less than 105 μm or less than 38 μm. One can see that decreasing the grain size increases the neutron sensitivity for the $^6$Li neutron converters, but decreases the sensitivity for the samples mixed with $Gd_2O_3$.

### 3.6.2 Properties of OSLN Detectors

Based on the results presented in the previous section, new neutron-sensitive OSL detectors (hereafter called "OSLN" detectors) were developed by Landauer Inc. using a manufacturing process similar to that used to produce the Luxel detectors (Figure 3.33), but using

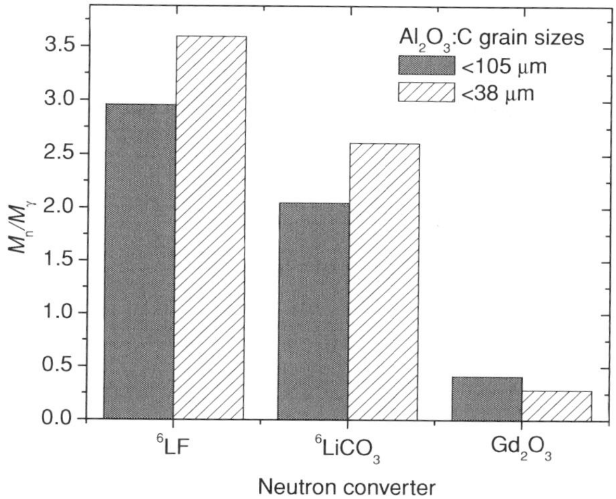


***Figure 3.32*** *Ratio of indicated value (in mGy) due to neutrons and gamma rays for samples of $Al_2O_3$:C (two different grain size distributions) mixed with different neutron converters (indicated in the figure) exposed to PuBe neutron source.*

a composite of $Al_2O_3$:C and $^6Li_2CO_3$ (Yukihara *et al.*, 2008b). The choice of $^6Li_2CO_3$ as neutron converter is based on economics aspects. Landauer is currently developing a new albedo dosimeter for the InLight dosimetry system based on this new OSLN detector (Perks *et al.*, 2008). In this section we describe some properties of this detector.

Figure 3.34 shows the indicated value $M$ of the OSLN detectors after subtraction of the indicated value from control detectors $M_0$ in comparison with regular $Al_2O_3$:C Luxel detectors, TLD-600 ($^6$LiF:Mg,Ti) and TLD-700 ($^7$LiF:Mg,Ti) as a function of the personal dose equivalent $H_p(10)$ from exposure to $^{252}$Cf. The indicated value of regular $Al_2O_3$:C Luxel detectors and TLD-700 to this radiation field is very similar and is primarily due to

***Figure 3.33*** *OSLN detector tape, produced using a mixture of $Al_2O_3$:C and $^6Li_2CO_3$ powder. The OSLN tapes are similar to the Luxel detector tapes shown in Figure 3.9.*

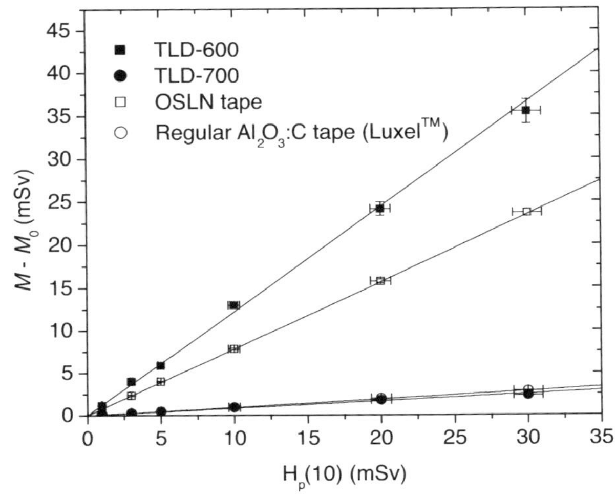


***Figure 3.34*** *Indicated value R (in mGy) minus indicated value from control detectors for TLD-600, TLD-700, OSLN tape, and regular* $Al_2O_3$*:C Luxel detectors irradiated with different values of personal dose equivalent* $H_p(10)$ *from a* $^{252}$*Cf source. The irradiations were carried out with the detectors on a phantom.*

the gamma component of the radiation field (~5–10%). The indicated value in excess of this value is due to the neutron component in the radiation field. It can be observed that the OSL from the new OSLN detectors is linear with the neutron dose and the sensitivity is ~60% of the sensitivity of TLD-600.

One disadvantage of the new OSLN tape in comparison with regular OSL detectors is the lower gamma sensitivity. Because of the lower $Al_2O_3$:C content, the gamma sensitivity of the OSLN detector is ~35% of the sensitivity of the regular Luxel detectors (Yukihara *et al.*, 2008b). The photon energy response of OSLN also seems to be affected by the $^6Li_2CO_3$ content, showing a slightly lower over-response to low-energy X-rays compared to regular OSL detectors (Perks *et al.*, 2008).

One of the limitations of this approach is the neutron energy dependence of the detectors, particularly the low response to fast neutrons. Table 3.14 shows the neutron response to various neutron radiation fields. The neutron-capture cross-section for $^6$Li decreases approximately with the inverse of the velocity of the neutrons, that is, with $E^{-0.5}$ (Figure 3.35), showing that the response to fast neutrons ($E > 10$ keV) is orders of magnitude lower than the response to thermal neutrons ($E < 0.5$ eV). At the same time, the neutron fluence to personal dose equivalent $H_p(10)$ increases by almost two orders of magnitude as the neutron energy increases from ~100 keV to 20 MeV due to a combination of increased scattering and higher values for the quality factor (ICRU, 2001).

The strong reduction in the probability for neutron capture as the neutron energy increases can be partially compensated by using the detector as an albedo dosimeter. Albedo dosimeters are designed to detect the low-energy albedo neutrons scattered by the body.

In this situation, both TL or OSL badge and body should be considered as the dosimeter and the response is higher than the response of identical detectors irradiated free-in-air.

***Table 3.14*** *Neutron response (indicated value in mGy divided by the conventional true value) for OSLN detectors irradiated in different neutron fields*

| Radiation field | Facility | Neutron response (mGy/mSv) |
|---|---|---|
| Thermal neutrons | SCK-CEN BR1 | $R_n/H^*(10) = 103.8 \pm 0.9$ |
| 43.6 keV | NPL | $R_n/H_p(10) = 4.07 \pm 0.09$ |
| $^{252}$Cf (moderated) | PNNL | $R_n/H_p(10) = 1.78 \pm 0.03$ |
| $^{252}$Cf (bare) | PNNL SCK-CEN | $R_n/H_p(10) = 0.226 \pm 0.008$ |
| | | $R_n/H_p(10) = 0.213 \pm 0.023$ |
| AmBe source | PNNL | $R_n/H_p(10) = 0.160 \pm 0.006$ |
| | Landauer Inc. | $R_n/H_p(10) = 0.164 \pm 0.006$ |

For comparison, the response for TLD-600 to the bare $^{252}$Cf is (0.32 ± 0.05) mGy/mSv. Data from Shah (2009) and Yukihara *et al.* (2008b).

Albedo dosimeters may incorporate different badge designs and detector elements to discriminate between gamma and neutrons, and between incident and albedo neutrons. Several designs have been investigated in albedo dosimetry using TL detectors (Piesch and Burgkhardt, 1985, 1988). Figure 3.36 shows one example of a design for an albedo dosimeter, and Figure 3.37 shows a typical response of an albedo detector with neutron energy. The neutron energy response of OSLN irradiated on the surface of a PMMA phantom 30 cm × 30 cm × 15 cm is shown in Figure 3.38.

### 3.6.3 Ionization Density Effects

Since the energy deposited in the $Al_2O_3$:C grains contained in the OSLN detectors is mostly caused by secondary heavy charged particles ($^3$H and $^7$Li) generated during the neutron-

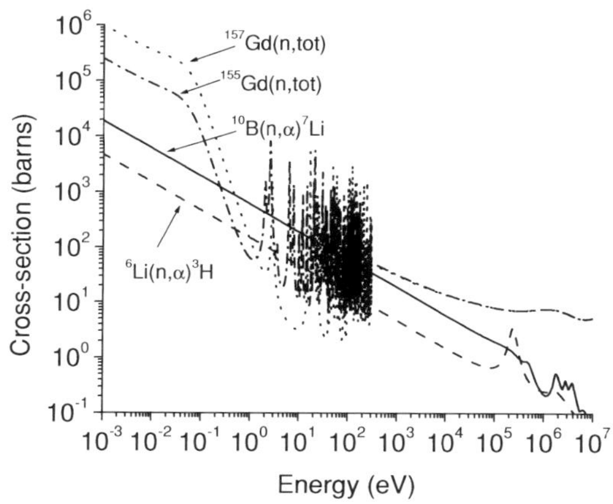


***Figure 3.35*** *Neutron capture cross-section as a function of neutron energy for common neutron capture reactions used in neutron detection. Based on Chadwick et al. (2006) (available at http://www.nndc.bnl.gov/, accessed on 14 June 2010).*

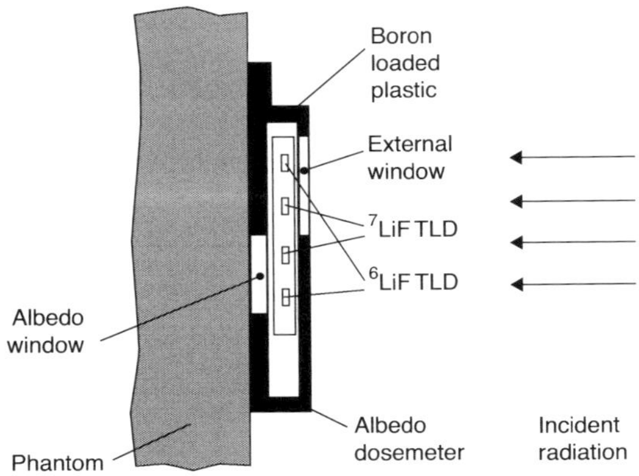


**Figure 3.36** *Example of an albedo TL badge based on $^6$LiF and $^7$LiF pair of detectors. Reprinted from Journal of the ICRU, ICRU Report 66: Determination of operational dose equivalent quantities for neutrons. Vol. 1 (3), 1–93. Copyright (2001) with permission from Oxford University Press.*

capture reaction with $^6$Li, the OSL signal from OSLN detectors irradiated with neutrons is affected by ionization density effects.

Ionization density effects are seen in Figure 3.39, which shows the OSL curves of two different type of detectors ($Al_2O_3$:C mixed with $^6$LiF and $Al_2O_3$:C mixed with $Gd_2O_3$) irradiated with gamma rays or neutrons ($^{252}$Cf source). The OSL of $Al_2O_3$:C mixed with $^6$LiF irradiated with neutrons decays considerably faster than the OSL from the same

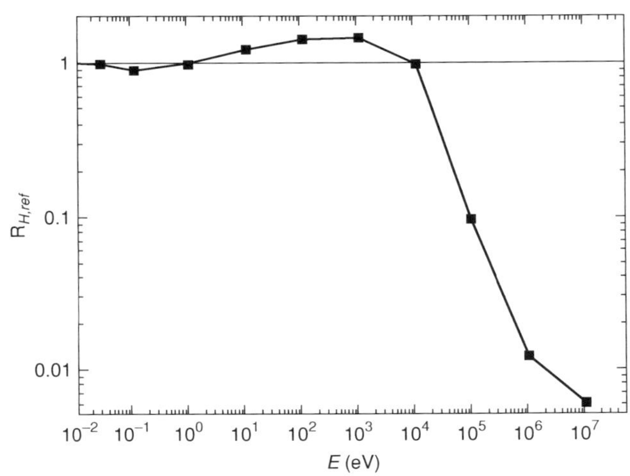


**Figure 3.37** *Relative response of a TLD albedo detector as a function of neutron energy. Reprinted from Journal of the ICRU, ICRU Report 66: Determination of operational dose equivalent quantities for neutrons. Vol. 1 (3), 1–93. Copyright (2001) with permission from Oxford University Press.*

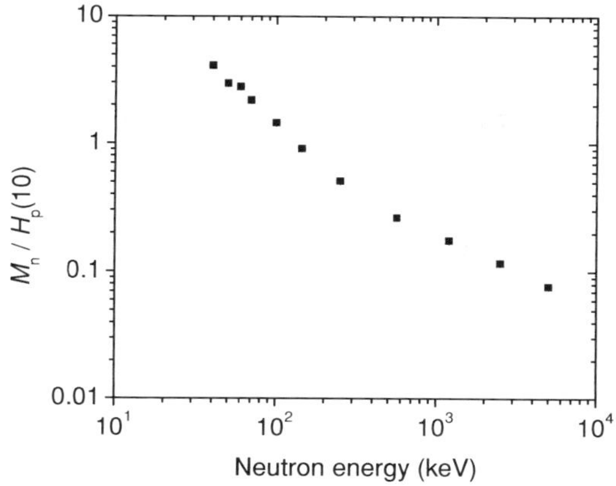


***Figure 3.38*** *Neutron response from OSLN detectors as a function of neutron energy (Shah, 2009).*

sample after gamma irradiation. The increase in the OSL decay rate is characteristic of high ionization densities created by high doses of gamma rays or low doses of high-LET radiation (see Section 2.2.10.2). This is further confirmed by the fact that the OSL decay curve from $Al_2O_3$:C mixed with $Gd_2O_3$ exposed to neutrons is similar to the OSL decay curve from $Al_2O_3$:C mixed with $^6$LiF irradiated with gamma rays, since in this case the energy is deposited mostly by low-LET radiation (gamma rays, X-rays, and conversion electrons).

Another manifestation of the high ionization density created by neutron irradiation is the increase in the intensity of the OSL emission band in the UV with respect to the *F*-center

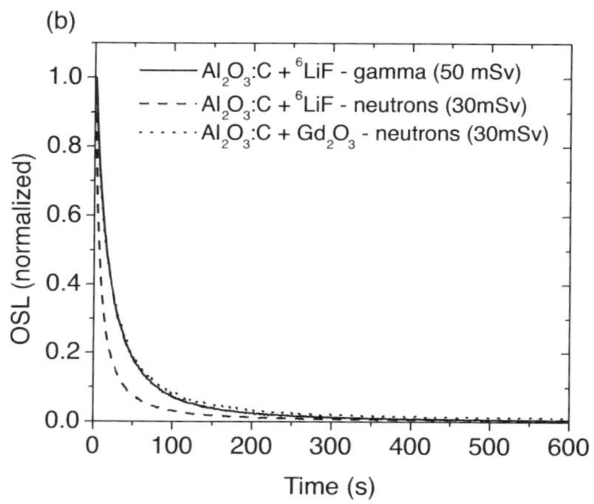


***Figure 3.39*** *OSL curves for samples consisting of a mixture of $Al_2O_3$:C and different neutron converters after irradiation with gamma rays ($^{60}$Co source) or neutrons ($^{252}$Cf source) (Mittani et al., 2007).*

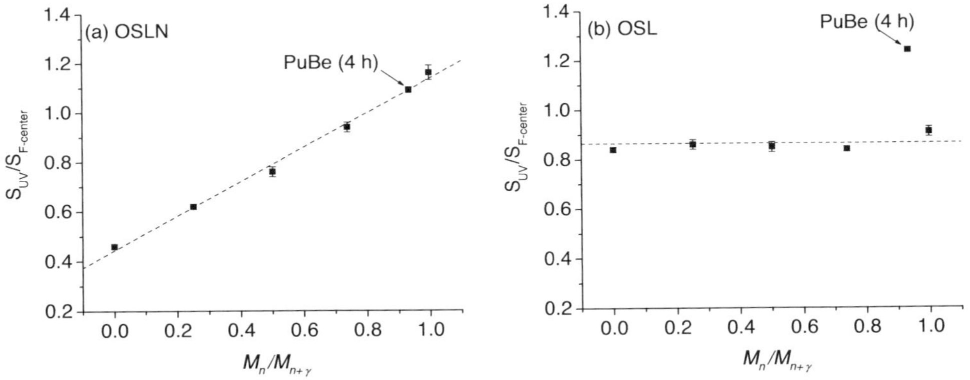


***Figure 3.40*** *Ratio between the OSL signal from the UV emission band and the F-center emission band from $Al_2O_3$:C in (a) OSLN detectors and (b) OSL detectors ($Al_2O_3$:C Luxel detectors) as a function of the OSL signal due to neutrons or due to gamma rays. The data were obtained by exposing the samples to thermal neutrons or beta rays ($^{90}Sr/^{90}Y$ source), the latter used to simulate the gamma irradiation (Shah, 2009).*

emission band, when compared to the low-LET radiation. In principle this effect can be used for discrimination between low- and high-LET irradiation (e.g., gamma rays and neutrons) in a mixed-radiation field.

This possibility is illustrated in Figure 3.40, in which the ratio between the signal from the UV emission band and from the *F*-center emission band (determined using a time discrimination to measure the two emission bands separately) is plotted as a function of the ratio between the indicated value due to neutrons and the total indicated value

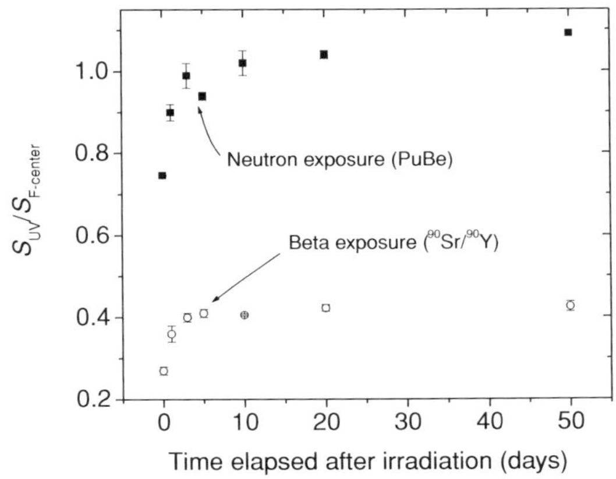


***Figure 3.41*** *Ratio between the OSL signal from the UV emission band and the F-center emission band from $Al_2O_3$:C in OSLN detectors as a function of time following neutron exposure (PuBe source) or beta rays exposure ($^{90}Sr/^{90}Y$ source) (Shah, 2009).*

(due to neutrons and gamma rays). The data points correspond to OSLN and OSL detectors exposed to different thermal neutron doses and a $^{90}Sr/^{90}Y$ beta source simulating the gamma irradiation. The point $M_n/M_{n+\gamma} = 0$ indicates a pure beta-ray irradiation, whereas the point $M_n/M_{n+\gamma} \sim 1$ indicates a pure thermal neutron irradiation. For the OSLN detectors (Figure 3.40a), the ratio $S_{UV}/S_{F\text{-center}}$ increases linearly in proportion to the signal fraction due to neutrons. In the case of the OSL detector (Figure 3.40b), the ratio remains constant since the detectors are not sensitive to neutrons and the signal is due to the beta-ray irradiation only.

One limitation of this approach is the time-dependence of the OSL signal from the UV emission band, which is known to increase with time elapsed since irradiation (Yukihara and McKeever, 2006b). Figure 3.41 shows that this leads to an increase in the ratio $S_{UV}/S_{F\text{-center}}$ in the first 10 days following either a neutron or a beta exposure.

# 4
# Space Dosimetry

## 4.1 Introduction

Sometime within the next few decades astronauts will lift off the Earth's surface at the start of a mission to place humans on the surface of the Moon. When this will occur is currently an open question, as is the nationality of the men or women who will make the journey. Global and national economies, national interests, politics and mankind's curiosity drive and restrain space exploration efforts and yet the interest by the space nations of the world in returning to our nearest-neighbor planetary[1] body is undeniable (e.g. Bell *et al.*, 2008). Although at the time of writing the timetable of such exploration is uncertain, we can be confident that manned exploration beyond Earth's orbit will occur once again.

Whatever program lifts men and women from the Earth and back to the Moon, or beyond, and whenever this occurs, the design stage of the next generation of manned space exploration vehicles will be somewhat different from past efforts. For the first time in the history of manned space flight, from the Soviet Vostok and US Mercury spacecraft to the International Space Station (ISS), space vehicles will be designed with the protection of the crew from ionizing radiation as one of the driving design factors (Badavi *et al.*, 2005). The design of the space vehicles and the concomitant mission plans will be consistent with ALARA (As Low As Reasonably Achievable) radiation protection principles. As an example, NASA has discussed limiting the tissue-averaged, effective dose[2] to any individual crewmember to a maximum of 150 mSv (Gaza *et al.*, 2008). Even for some long-term, low-Earth-orbit missions on the ISS such low limits present a significant challenge. For exploration missions to the Moon (and especially to Mars), these limits will be even more problematic.

[1] Strictly speaking the Moon is not a planet, but a moon of our planet. However, it and other major moons of the Solar System's planets have long been classed as "planetary bodies" for purposes of scientific description.

[2] The terms "effective dose" (in Sieverts, Sv) and "gray equivalent" (in gray-equivalent, Gy-Eq), used in Tables 4.1 and 4.2, are explained in Section 4.3.

*Optically Stimulated Luminescence: Fundamentals and Applications* Eduardo G. Yukihara and Stephen W. S. McKeever

***Table 4.1*** *Ten-year career limits for effective dose for astronauts based on a 3% excess lifetime risk of fatal cancer*

| | Effective Dose Limits (mSv) | |
|---|---|---|
| Age at Exposure (yr) | Female | Male |
| 25 | 400 (1000) | 700 (1500) |
| 35 | 600 (1750) | 1000 (2500) |
| 45 | 900 (2500) | 1500 (3250) |
| 55 | 1700 (3000) | 3000 (4000) |

Also shown (in parentheses) are full career exposure limits. Data from NCRP (1989, 2000).

To stay within radiation dose limits for the astronauts, mission planners and space-vehicle designers are informed by calculations that use accepted models for galactic cosmic rays and worst-case solar particle events, taking into account shielding due to the spacecraft, planetary surfaces and the astronauts' bodies. In the pre-mission, pre-construction design phases such calculations, using the latest in radiation transport and Monte-Carlo computer codes, are the only tools available to enlighten engineers with respect to radiation protection and its relationship to spacecraft design. Nevertheless, as stated by Gaza *et al.* (2008): "Radiation measurements will ultimately provide confirmation of the calculated vehicle radiation protection capabilities." One might say that *only* radiation measurements will provide this necessary confirmation. This is not solely of esoteric engineering interest. It is not merely a question of seeing how successful the engineers were in achieving the necessary radiation protection, but rather it is a question of realizing that there may be very serious health, ethical and, possibly, litigation consequences if these protection limits are exceeded. Current NASA rules, for example, require that no astronaut will receive a total radiation exposure in the course of his or her active space flight career that will increase their lifetime risk of fatal cancer by more than 3%. Ten-year career stochastic effective dose limits based on a 3% excess lifetime risk were recommended to NASA by the National Council on Radiation Protection and Measurements (NCRP) and are given in Table 4.1 (NCRP, 1989, 2000, 2002). In addition, again through direction from the NCRP, NASA assigns dose limits to minimize the probability of deterministic effects. Recommended organ dose limits to inhibit deterministic effects are given in Table 4.2 (NCRP, 1989, 2000,

***Table 4.2*** *Gray-equivalent limits for astronauts (in Gy-Eq) for the prevention of deterministic effects due to radiation exposure. Data from NCRP (2000)*

| | Gray-equivalent Dose Limits (Gy-Eq) | | | | |
|---|---|---|---|---|---|
| | Skin | Eye Lens | Heart[a] | CNS[b] | Bone Marrow |
| Career | 4.0 | 4.0 | 1.0 | 1.5 | *[c] |
| 1 year | 3.0 | 2.0 | 0.50 | 1.0 | 0.50 |
| 30 days | 1.5 | 1.0 | 0.25 | 0.25 | 0.25 |

[a] Averaged over the heart muscle.
[b] Central nervous system, calculated at the hippocampus.
[c] NCRP considers that the ten-year career limits for stochastic effects are sufficient to protect the bone marrow against deterministic effects (NCRP, 2002).

2002). The task of monitoring the radiation levels received by astronauts falls to the various space agencies from the various countries, including the NASA Space Radiation Analysis Group (http://srag-nt.jsc.nasa.gov/), the European Space Agency's Medical Operations Group (MedOps; Straube *et al.*, 2010) and similar organizations from other national space agencies.

As often stated in the literature (e.g., Benton and Benton, 2001), space radiation represents one of the greatest challenges in radiation protection and dosimetry. Lessons learned and advances made in the dosimetry of energetic charged particles from space have overlapping benefit to other areas of dosimetry, particularly in the Earth-bound field of radiation medicine since protons and carbon ions are currently used for radiation therapy. In the sections to follow we discuss the various methods that are currently in use for measuring radiation exposure to astronauts with emphasis on the importance of passive luminescence dosimeters in general, and with specific emphasis on the potential of OSL dosimeters (OSLDs). We begin, however, with descriptions of the space radiation environment to which the astronauts are exposed and estimations of the health risks associated with that exposure.

## 4.2 Space Radiation Environment

Primary sources of radiation to which astronauts are exposed are galactic cosmic rays (GCR), solar particle events (SPEs), the Earth's radiation belts (ERB), and secondary particles due to interactions of the primary particles with the spacecraft walls, equipment, the astronauts themselves and planetary surfaces. Included in the latter are secondary and albedo neutrons. The exposure within the spacecraft is affected by altitude, inclination, orientation and, of course, spacecraft design. In addition, mission profiles such as the number and duration of extravehicular activities and the length of time on planetary surfaces or in lander modules (on exploration missions) also have a significant impact on the radiation exposure for individuals. An additional influence, outside the control of mission planners and designers, is the degree of solar activity, which, when high, decreases the GCR flux, but increases the probability of energetic SPEs, and vice versa when low.

Schematic summaries of the space radiation fields relevant to low Earth orbit and exploration missions are shown in Figures 4.1 and 4.2.

### 4.2.1 Galactic Cosmic Rays (GCR)

Galactic cosmic rays are an isotropic source of energetic, charged (fully ionized) particles originating outside the Solar System. The composition is 98% baryons (of which ~87% are protons, ~12% He ions, with the remaining 1% composed of heavier charged particles) and 2% leptons (electrons and positrons) (Simpson, 1983). Energies range from tens of MeV per atomic mass unit (u) to greater than 10 GeV/u. The intensity of the GCR flux varies in anti-phase with solar activity due to modulation of the flux by the interplanetary magnetic fields from the solar wind during the 11-year solar cycle. Consequently, GCR fluxes are highest during solar minimum. Modulation by the Earth's magnetosphere reduces the flux at the Earth's surface. The strength of the decrease in the GCR flux due to interaction with magnetic fields depends on the energy of the particle, with the lower-energy particles being affected the most. For the latter particles one can expect variations in particle flux of a

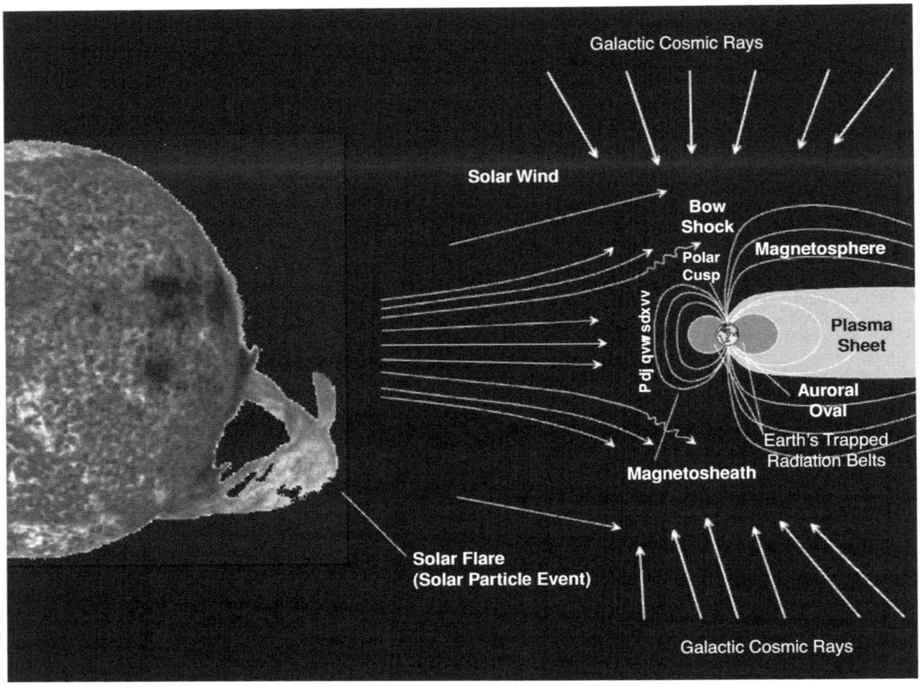


*Figure 4.1* The three principal sources of space radiation, namely galactic cosmic rays (GCR), solar particle events (SPEs) and the Earth's (trapped) radiation belts (ERB). Charged particles from all three sources are affected by the Earth's magnetosphere. Reprinted from Nuclear Instruments & Methods in Physics Research B, Benton, E. R., Benton, E. V., Space radiation dosimetry in low-Earth orbit and beyond. Vol. 184, 255–294.

factor of 10 over the solar cycle, while for particles with energies $>10$ GeV/u the variation is $<20\%$ (Robbins, 1997). Typical spectra for several species of GCR particle are shown in Figure 4.3 while the modulation of the GCR flux with solar cycle is shown in Figure 4.4.

The degree to which the GCR particles are modulated by the Earth's magnetic field is determined by the particle's momentum per unit charge (rigidity) with all particles with the same rigidity following the same curvature in a given magnetic field. Below a certain rigidity threshold (the geomagnetic cut-off) the particles are deflected away and are unable to reach the Earth's surface. At the equator, where the field lines are parallel to the surface and the rigidity threshold is high, particles are deflected away and the incident number of particles is correspondingly low. The rigidity threshold decreases towards the magnetic poles, however. Here the field lines direct the particles towards the Earth's surface. Because of the geomagnetic cut-off, only a fraction of the incident GCR particles reaches orbiting spacecraft (in low Earth orbit) and the GCR dose inside the spacecraft will depend upon the inclination of the orbit, with higher inclinations experiencing a higher GCR flux (Reitz, 2006). The Mir space station had a high inclination orbit, as does the ISS, and as a result GCR form a significant part of the astronaut doses in these space vehicles.

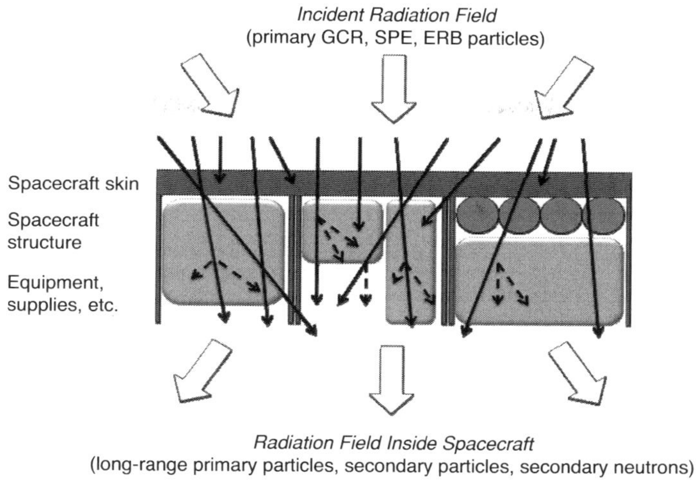


***Figure 4.2*** *Schematic representation of the radiation field to which astronauts may be exposed, whether inside or outside the spacecraft. Outside, the sources are galactic cosmic rays (GCR), solar particles events (SPEs) and (in low Earth orbit) the Earth's radiation belts (ERB). Inside, the field consists of those long-range energetic primary particles that penetrate the spacecraft structure, equipment racks and so on, plus the secondary particles created by nuclear interactions with the atoms of spacecraft structure and equipment. Reprinted from Nuclear Instruments & Methods in Physics Research B, Benton, E. R., Benton, E. V., Space radiation dosimetry in low-Earth orbit and beyond. Vol. 184, 255–294. Copyright (2001) with permission from Elsevier.*

Several authors have presented cosmic-ray abundance data since Simpson (1983), including isotopic ratios (Mewaldt, 1989). Illustrative abundance data are shown in Figure 4.5 (NRC, 2008). Similar data are given in NCRP Report 132 (NCRP, 2000). Several models for the GCR flux are available, as discussed by Benton and Benton (2001). To be noted in the models and illustrated in Figure 4.5 is the fact that the GCR abundances drop off rapidly after the low-$Z$ (nuclear charge) elements, which dominate the spectra. However, there is a clear peak in the GCR abundance at Fe ($Z = 26$). This is of significance in radiation protection since these high-$Z$ elements are characterized by a high quality factor ($Q$; see Section 3.2.3 for definitions) and, as a result, they are responsible for a significant proportion of the dose equivalent. In general, the heavy ion component of the GCR field is known as the high-charge, high-energy (HZE) component. The high quality factor for the HZE particles is a result of their high linear energy transfer (LET), which in turn creates a high-ionization density within the particle track as the particles lose energy traversing the target material.

### 4.2.2 Earth's Radiation Belts (ERB)

The Van Allen Belts (named after their discoverer, James Van Allen) are regions around the Earth consisting of electrons and protons trapped by the Earth's magnetic field. The trapped charges spiral around the geomagnetic field lines in a cyclotronic motion, converging

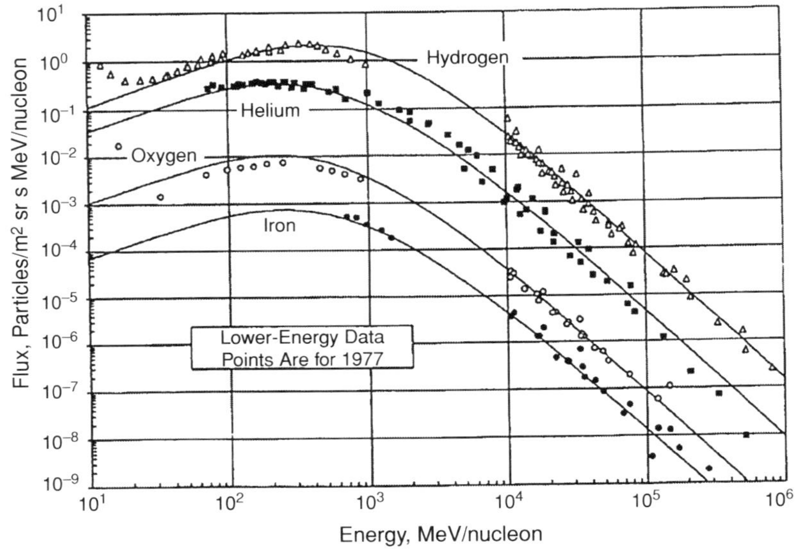


***Figure 4.3*** *Differential energy spectra for GCR protons, helium, carbon and iron ions for the 1973–1977 solar minimum at the Earth's surface. The lines are fits using the model of Badhwar and O'Neill (1996). Reprinted from Radiation Research, Badhwar, G. D., The radiation environment in low-Earth orbit. Vol. 148, S3–S10. Copyright (1997) with permission from Radiation Research Society.*

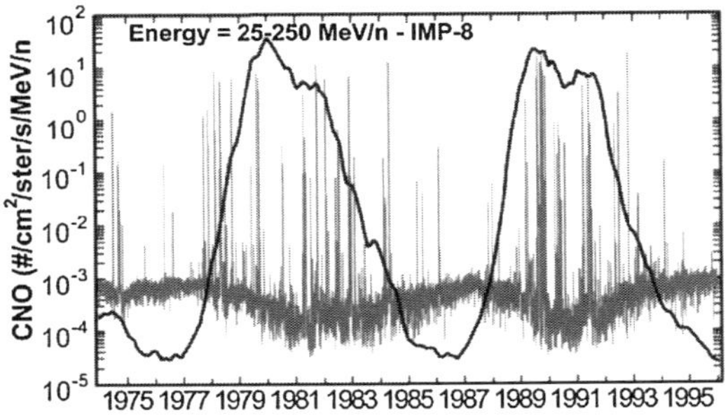


***Figure 4.4*** *Variation with time of the interplanetary GCR flux as measured at the IMP-8 satellite, from 1974 to 1996. The GCR flux (carbon-nitrogen-oxygen (CNO) flux; 25–250 MeV/u; gray line) is anticorrelated with the 11-year solar cycle, as measured by the average number of sun spots (dark line). The spikes in the CNO count are actually solar particle events (SPEs), which can be seen to be more concentrated during the periods of solar maximum. Reprinted from IEEE Transactions on Nuclear Science, Barth, J., Dyer, S.C., Sassinopulos, E.G., Space, atmospheric, and terrestrial radiation environments. Vol. 50, 466–482. Copyright (2003) with permission from IEEE.*

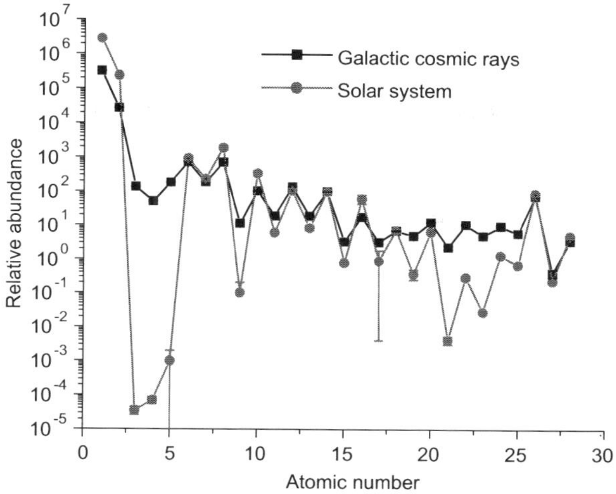


***Figure 4.5*** *Relative abundances of elements in galactic cosmic rays and in the Solar System (NRC, 2006). Reproduced from data available at http://helios.gsfc.nasa.gov/ace/abund_plot.html, accessed on 12 May 2010.*

towards the magnetic poles. Trapped protons drift westward and trapped electrons drift eastward. The trapped electrons occur in two belts – the inner and the outer. The inner belt, extending to about 2.4 Earth radii ($R_E$), consists primarily of electrons with energies up to about 5 MeV, while the outer belt extends from approximately 2.8 $R_E$ to about 12 $R_E$ and consists primarily of electrons with energies up to about 7 MeV. Trapped protons occur primarily in the inner belt with energies ranging from a few MeV to several hundred MeV (Benton and Benton, 2001; NCRP, 1989, 2000; Robbins, 1997).

Figure 4.6 shows the contour maps of trapped particles fluxes for protons of >10 MeV and electrons of >1 MeV, calculated using trapped proton and electron models (Sawyer and

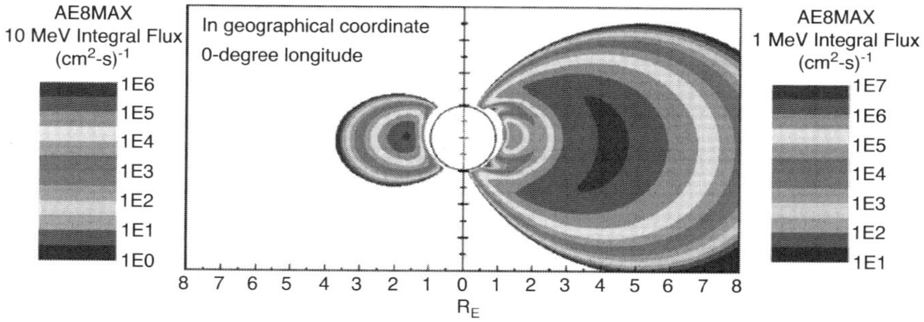


***Figure 4.6*** *Contour plot showing ≥1 MeV electron (right side) and ≥10 MeV proton (left side) integral fluxes at Earth. The models used are the AP-8 and AE-8 solar maximum models for protons and electrons, respectively. The contour plane is for the zero meridian. Reprinted from Radiation Protection Dosimetry, Barth, J., S. C., Sassinopulos, E. G., Comparison of high-energy trapped particle environments at the Earth and Jupiter. Vol. 116, 50–54. Copyright (2005) with permission from Oxford University Press.*

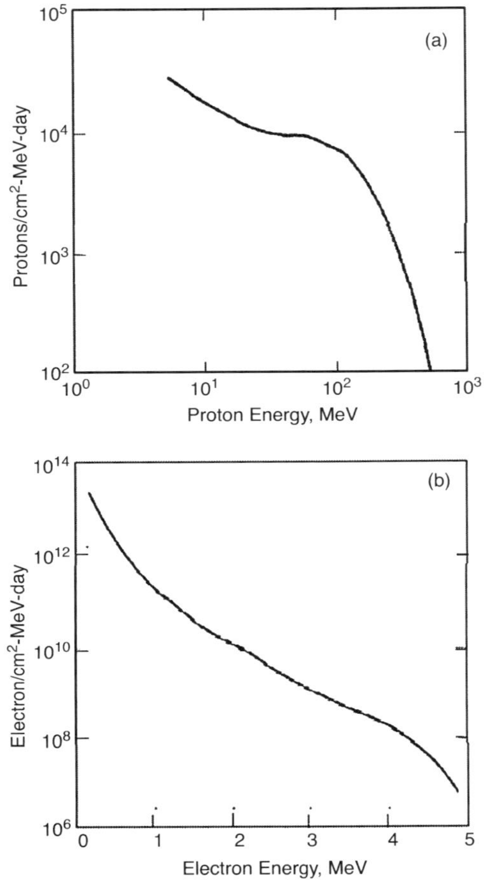


***Figure 4.7*** *Differential energy spectra integrated over one day for (a) trapped protons and (b) trapped electrons using models AP-8 (for protons) and AE-8 (for electrons). The spectrum for protons is calculated at solar minimum, 28.5° inclination and 450 km altitude. The electron spectrum is calculated at 0° inclination for geosynchronous orbit at an altitude of 35 790 km and 160° longitude. Figure (a) was reprinted from Advances in Space Research, Curtis, S.D., et al., Radiation environments and absorbed dose estimations on manned space missions. Vol. 6, 269–274. Copyright (1986) with permission from Elsevier. Figure (b) was reprinted from Advances in Space Research, Curtis, S.D., et al., Radiation environments and absorbed dose estimations on manned space missions. Vol. 6, 269–274. Copyright (1986) with permission from Elsevier.*

Vette, 1976; Vette, 1976). As discussed by Jun and Garrett (2005), the electron contours show twin maxima indicating that the electron distributions fall into inner and outer regions, whereas only one region (inner belt) is seen for protons. Figure 4.7 shows the trapped proton and trapped electron differential energy spectra, also calculated using trapped particle models AP-8 and AE-8, respectively. The trapped proton data are calculated at solar minimum and the electron data are for solar maximum.

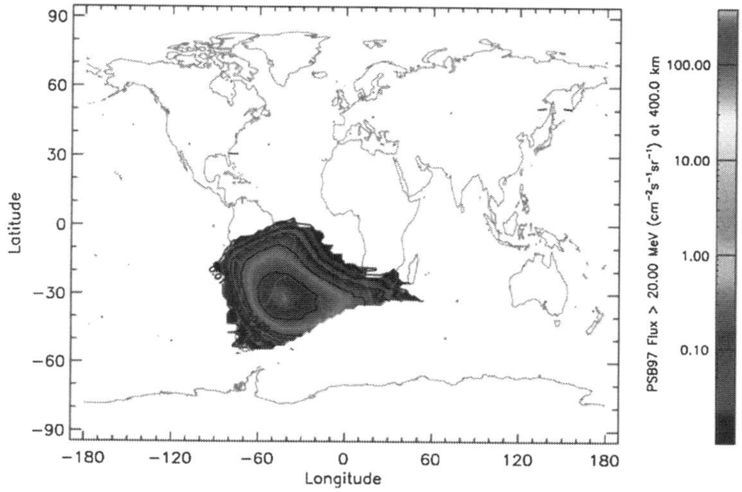


***Figure 4.8*** *The South Atlantic Anomaly revealed in a contour plot of proton (>20 MeV) fluence at an altitude of 400 km and calculated using the trapped proton model PSB97 and the SPENVIS software package (http://www.spenvis.oma.be/spenvis/). Reprinted from Journal of Atmospheric and Solar-Terrestrial Physics, Heynderickx, D., Radiation belt modelling in the framework of space weather effects and forecasting. Vol. 64, 1687–1700. Copyright (2002) with permission from Elsevier.*

The Earth's magnetic field lines dip towards the surface of the Earth in a region over the South Atlantic, off the coast of Brazil. This effect is known as the South Atlantic Anomaly. The South Atlantic Anomaly is caused by an anomaly in the geomagnetic field and the displacement of the center of the magnetic dipole from the Earth's center. Those spacecraft orbits that are at low inclination pass through the South Atlantic Anomaly and give rise to an increased radiation exposure to crews and equipment (Figure 4.8) during each passage (Heynderickx, 2002). Observation of variations in the geomagnetic field have led to predictions that the South Atlantic Anomaly will become a greater radiation hazard for space crews in the future, as the exposure and the geographical area covered by the South Atlantic Anomaly will both increase compared to the present (Heirtzler, 2002).

### 4.2.3 Solar Particle Events (SPEs)

The primary health concerns associated with SPEs are the induction of deterministic effects rather than stochastic processes. Rather than the chronic, low-to-mild exposure characteristic of GCR, SPEs are characterized by periods of high-intensity, short-duration exposures, which may cause a wide range of detrimental health issues, including debilitating prodromal syndrome (e.g., vomiting), central nervous system damage and ultimately even death. Protection against SPEs relies upon spacecraft shielding design, availability of "safe-havens," careful mission planning and accurate space weather forecasting.

The most hazardous types of SPE are those in which the energetic particles are accelerated to high velocity by the shock from a coronal mass ejection (see http://helios.gsfc.

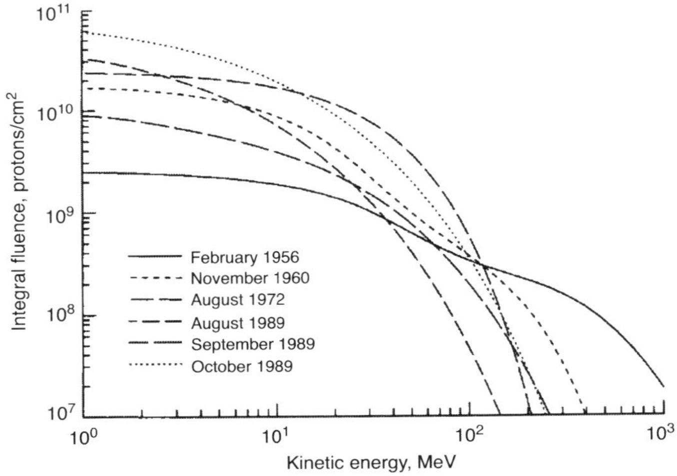


***Figure 4.9*** *Integral proton fluence spectra for large SPEs observed over several solar cycles. Reprinted from Radiation Measurements, Wilson, J. et al., Shielding from solar particle event exposures in deep space. Vol. 30, 361–383. Copyright (1999) with permission from Elsevier.*

nasa.gov/cme.html). High fluences ($>10^9$ particles per square centimeter) of high-energy ($>10$ MeV) protons, with smaller contributions of electrons and heavier ions, can be expected from this type of SPE with a gradual onset and a long duration (of the order of days). The alternative form of SPE is that associated with solar flares. These impulsive, fast-onset flares consist of smaller fluences ($10^7$–$10^8$ cm$^{-2}$) and lower energies, and are mostly electrons.

The most important health hazard for astronauts comes from the shock-accelerated SPE, with proton energies in the range 30–100 MeV. Particles with energies below this range are easily shielded by the spacecraft walls, while the fluence drops off steeply for energies $>100$ MeV. Protons contribute most to the absorbed dose, but the contribution of helium ions may have to be considered when calculating the gray-equivalent dose. The SPE frequency is difficult to predict but is generally anti-correlated with the GCR intensity with most events occurring during solar maximum (see Figure 4.4).

Typical energy spectra for several recorded major SPEs are shown in Figure 4.9 (Wilson *et al.*, 1999). Using such spectra, estimates can be made of the potential dose delivered to astronauts behind assumed shielding depths. For example, Gaza *et al.* (2008) used the August 1972 event, as parameterized by King (1974), to calculate doses during the design of the *Orion* crew exploration vehicle using a variety of shielding and crew placement scenarios. In order to have minimum impact on the mass of the crew exploration vehicle, the ability for the crew to reposition themselves in safe havens is essential during an SPE storm. This in turn requires an ability to predict or forecast such storms, and their temporal evolution. Examples of measured temporal profiles for several SPEs are shown in Figure 4.10. Multiple SPEs can often be observed, as can the peak caused by the arrival of the coronal mass ejection shockwave (e.g. October 21, 1989). Reames (1999) argues

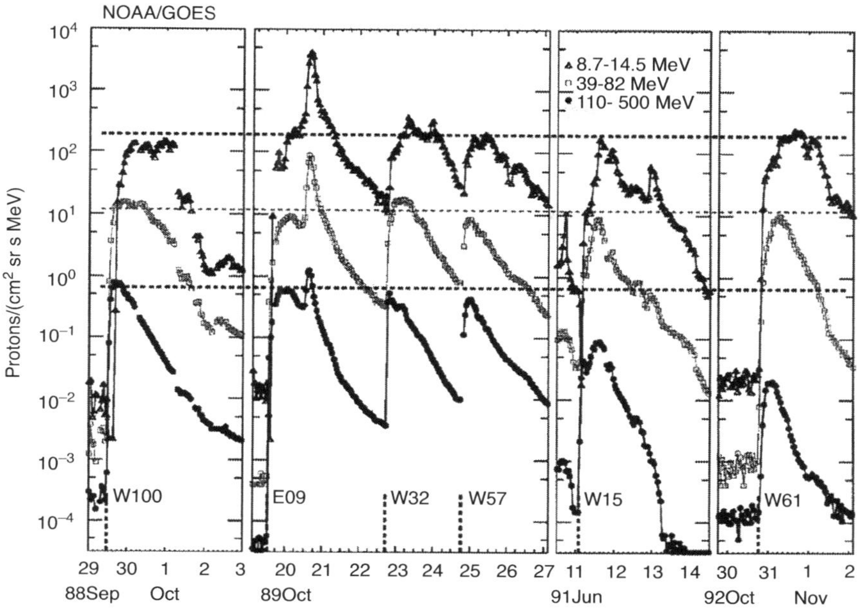


***Figure 4.10*** *Intensity–time profiles for protons for several large solar particle events (SPEs) as monitored by the NOAA Geostationary Observational Environment Satellite (GOES). Data from three proton energy channels are shown. The observed rapid onset and gradual decay is typical. In the October 1989 event the spike around October 21 is due to the shock wave from the coronal mass ejection reaching the satellite. Multiple SPEs can be observed in several of the events. Reprinted from Advances in Space Research, Reames, D. V., Shielding from solar particle event exposures in deep space. Vol. 34, 381–390. Copyright (2004) with permission from Elsevier.*

that although true forecasting of an SPE may not yet be possible, the period between initial detection of the event and the point at which the storm reaches full intensity may often be enough to allow crews to take shelter in the more shielded parts of the spacecraft. The issue is a difficult one, however, and the Space Radiation Analysis Group at NASA would ideally like up to 24 hours advance warning (NRC, 2006). Hu *et al.* (2009) modeled health effects for astronauts using various historical SPE data and concluded that SPEs similar to the August 1972 event could cause moderate acute radiation syndrome. Thus, although there is much variety in the time profiles and energy spectra of the different SPEs, the August 1972 event has been adopted by NASA as the "standard" SPE for assessing risk to astronauts during SPE exposure, with an often-used "worst case" assumed to be an SPE with an intensity of four times that of the August 1972 event.

### 4.2.4 Secondary Radiation

As illustrated in Figure 4.2, the radiation field to which an astronaut is exposed is not the same as the field incident on the spacecraft. Interaction of the primary high-energy particles with the material of the spacecraft, the astronaut's body and, if on the surface of the Moon or Mars, the planetary surface leads to the creation of secondary particles because of nuclear interactions. High- and very-high-energy particles, for which the range is greater than the mean free path for nuclear interactions, are likely to lose energy via nuclear interactions, as well as through ionization. In contrast, lower-energy particles lose energy through ionization processes only (Reitz, 2006). Nuclear interactions produce neutrons, protons and heavier charged particles, with a concomitant increase in the mean LET of the overall particle distribution. The more energetic secondary particles are the result of fragmentation of heavy primary particles. The dose of interest, therefore – that is the dose to internal body organs – is the result of irradiation by primary particles as well as a complex field of secondary particles that vary substantially with location within the spacecraft, or external to the spacecraft, and the location of the critical organ within the body. (The deeper the organ, the greater the contribution from secondary particles.) The production of secondary neutrons is particularly troublesome, both because they are difficult to measure and because the quality factor $Q$ for neutrons can be substantial (Benton and Benton, 2001; Dudkin *et al.*, 1996), resulting in a large contribution to the dose equivalent. Benton and Benton (2001) estimated the neutron contributions to the exposure of ISS crews to be between 30% and 60%. In calculations of the doses received by spacecrews operating in various locations and environments the inclusion of secondary particles in the radiation transport codes is essential.

## 4.3 Quantities of Interest

At this juncture it is necessary to define several of the dosimetric quantities of importance in space-radiation dosimetry before we can discuss how OSL can be used as a means to determine these quantities. The main framework is identical to that discussed in personal dosimetry (Section 3.2). For example, we already saw in Tables 4.1 and 4.2 that dose limits are specific in terms of effective dose ($E$) for stochastic effects, and in terms of the gray-equivalent for deterministic effects. However, the complex composition of the space radiation environment poses unique challenges when trying to apply these dosimetry concepts. This is primarily caused by the presence of heavy charged particles (HCPs), which are particles having mass much greater than the electron, and the different effects they have on biological tissue and the detector material. We begin with a discussion of absorbed dose, from which all other dosimetric quantities are derived. In the remainder of this chapter we will ignore the contribution from neutrons, because of the complexity that it would add to the discussion and the fact that OSL detectors currently in use are not sensitive to neutrons. (Neutron-sensitive OSL detectors described in Section 3.6 have not been tested in space.)

### 4.3.1 Absorbed Dose, $D$

The basic quantity to be measured in dosimetry is the absorbed dose, $D$, defined in Equation (3.1) as the energy per unit mass imparted to matter. In luminescence dosimetry, surrogate measurements are made of quantities that are indirectly related to the amount of absorbed energy. Usually, this occurs via the measurement of an entity that is directly related to the degree of ionization within the material as a result of the energy absorption. For OSL, the entity is luminescence emission following release (by light stimulation) of non-equilibrium trapped charges, which are themselves created by ionization during the original energy absorption. However, the radiation protection quantities are based on absorbed dose to specific organs or tissues, whereas the OSL is proportional to the absorbed dose to the dosimeter. Here lies the first fundamental challenge in space dosimetry, which is the problem of calibrating the dosimeters, that is, determining the relationship between the OSL emitted and the absorbed dose to tissue. The problem of determining absorbed dose (or other dosimetric quantities) from the OSL signal is not trivial and we will postpone this discussion until Section 4.5.

### 4.3.2 Dose Equivalent, *H*

Absorption of the same amount of energy, but from different radiation types does not guarantee the same biological end effect in a given tissue or organ. The biological damage produced in a given tissue for the same absorbed dose depends strongly on the type of radiation, such that the damage resulting from gamma rays and HCPs can differ enormously. To account for this the concept of dose equivalent ($H$) was introduced in Equation (3.15) as the absorbed dose weighted by the quality factor $Q$ of the radiation.

The quality factor $Q$ was expressed in Equation (3.16) in terms of $L$, the unrestricted linear energy transfer (LET) of the radiation. For a heavy charged particle, $L$ is given by the Bethe formula:

$$\frac{L}{\rho} = 4\pi \frac{e^4}{m_0 c^2} \frac{N_A Z z^2}{A\beta^2} \left[ \ln \frac{2m_0 c^2 \beta^2}{I\left(1-\beta^2\right)} - \beta^2 - \frac{C}{Z} - \frac{\delta}{2} \right] \tag{4.1}$$

where $N_A$ is Avogadro's constant, $m_0$ is the electron mass, $\beta$ is the particle velocity in terms of the speed of light ($\beta \equiv v/c$), $N_A Z/A$ is the number of electrons per gram of the stopping medium, $z$ is the particle charge, $I$ is the mean excitation potential for the particular absorber, and $C$ and $\delta$ are correction factors (Attix, 2004; ICRU, 1970). The LET increases with $(z/\beta)^2$, that is, increases with the particle charge and decreases with the particle velocity.

For irradiation with multiple particle types, as experienced in space, $H$ is given by:

$$H = \int_L Q(L) D(L) \mathrm{d}L \tag{4.2}$$

where $Q(L)$ and $D(L)$ are the spectral distributions of $Q$ and $D$ in terms of the LET (Reitz *et al.*, 2009).

### 4.3.3 Equivalent Dose, $H_T$

If the average absorbed dose over tissue T for radiation of type R is written $D_{T,R}$ then one may define the equivalent dose, $H_T$, as in Equation (3.12) based on the radiation weighting factor $w_R$ for radiation of type R listed in Table 3.3.

As noted in NCRP Report 142 (NCRP, 2002), for complex mixtures of charged particles, as experienced in space, the point quantity $H$ (dose equivalent) may be averaged over the tissue to yield $\overline{H}_T$ and this may be used as an approximation of the equivalent dose $H_T$.

Furthermore, $w_R = 20$ is a poor approximation for particles heavier than alpha particles. The biological effectiveness is therefore better expressed by the function $Q(L)$ (Reitz *et al.*, 2009). Thus, the dose equivalent given by Equation (4.2) averaged over the tissue, that is, $\overline{H}_T$, is generally used instead of Equation (3.12) in operational space dosimetry.

### 4.3.4 Effective Dose, *E*

The quantity of interest specifically for stochastic effects is the effective dose, $E$ (also in Sv), defined in Equation (3.14). Based on the discussion above, the effective dose can be approximated by:

$$E = \sum_T w_T H_T \approx \sum_T w_T \overline{H}_T \tag{4.3}$$

where $w_T$ is the tissue weighting factor for tissue T. It is a factor relating the stochastic risk to tissue T for a given radiation type compared to the stochastic risk to the whole body when uniformly irradiated by the same radiation type, independent of the radiation type (ICRP, 1991; NCRP, 2002). Values for $w_T$ are listed in Table 3.4. The career limits for stochastic effects noted in Table 4.1 are in terms of effective dose.

### 4.3.5 Gray-Equivalent, $G_T$

The NCRP recommends use of the quantity gray-equivalent, $G_T$, to determine career limits for deterministic effects (NCRP, 2000, 2002). Conceptually, the gray-equivalent is defined in similar fashion to the equivalent dose in that it is related to the absorbed dose by a dimensionless factor known as the relative biological effectiveness (RBE), $R_i$, thus:

$$G_T = R_i D_T \tag{4.4}$$

where $R_i$ is the recommended value of the relative biological effectiveness for particles of type $i$. $G_T$ is dimensionally the same as the gray, but the unit is indicated by "Gy-Eq" (NCRP, 2000). Values for $R_i$ are given in Table 4.3 (ICRP, 1991; NCRP, 2002). $R_i$ values can depend upon whether the effect being monitored is an early-stage effect or a late-stage effect, and may vary with the dose used to determine the value (NCRP, 2002). As discussed by Hu *et al.* (2009), it is also preferred that each biological endpoint for a given particle is associated with its own RBE. Endpoint-specific RBEs are generally not available, however.

**Table 4.3** *Biological effectiveness for various particle radiation types important in space radiation. Data from ICRP (1991) and NCRP (2002)*

| Particle Type | $R_i$ (range) |
|---|---|
| 1–5 MeV neutrons | 6.0 (4–8) |
| 5–50 MeV neutrons | 3.5 (2–5) |
| Heavy ions (helium and heavier) | 2.5 (1–4) |
| Protons (>2 MeV) | 1.5 (–) |

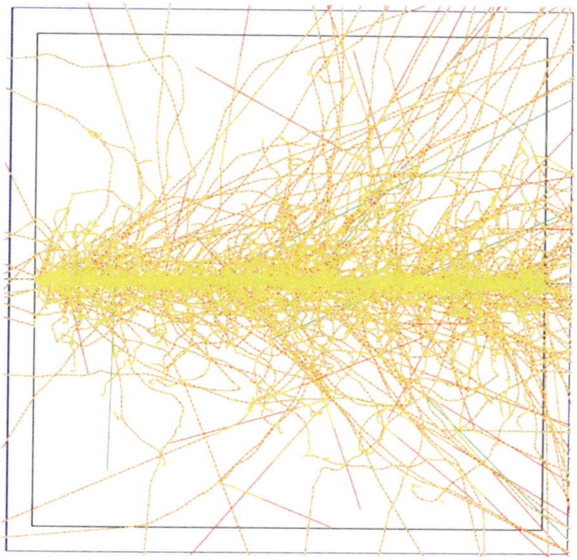

***Figure 2.4*** *Illustration of the pattern of ionizations and secondary electrons created when a He ion with energy of 200 MeV/u crosses a water volume of 100 μm × 100 μm × 100 μm horizontally from left to right, calculated by the Monte Carlo method (courtesy of Dr. Gabriel O. Sawakuchi, Carleton University, Canada). The figure shows that the ionization density is very high close to the particle track, decreasing as the distance from the track increases.*

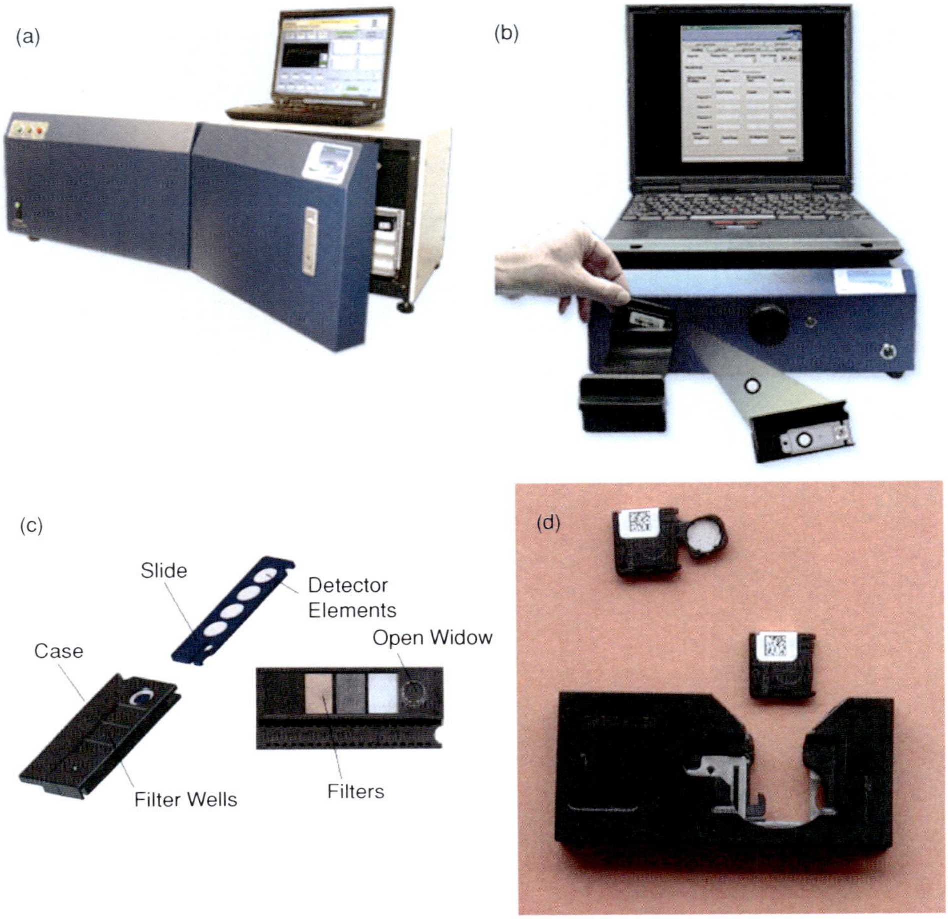


***Figure 2.43*** *InLight dosimetry system: (a) 200 unit automatic reader; (b) microStar reader; (c) InLight dosemeter; and (d) nanoDot dosimeter and adapter. Reprinted with permission from Landauer Inc. (USA).*

***Figure 3.9*** *Available forms of $Al_2O_3$:C detectors, including round single crystals ~5 mm in diameter by 0.9 mm thickness, 0.3 mm thick tapes containing a detector layer consisting of $Al_2O_3$:C powder and organic binder, and pieces of tape of sensitive area ~1.7 cm × 2.0 cm used in the Luxel dosimetry system (Luxel detectors).*

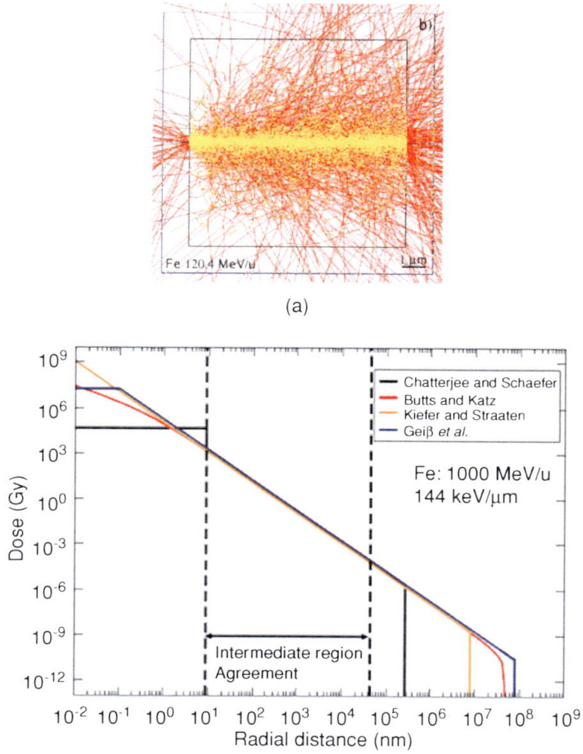

(b)

***Figure 4.11*** *(a) GEANT-4 calculations of the energy deposition from the passage of a 124 MeV/u Fe ion through a water target. The lines represent delta rays (secondary electrons). The dose deposition is clearly greater along the core of the track. This is seen in (b) in which the radial dose distributions in water around the path of a 1000 MeV/u Fe ion are shown as determined using the models of Chatterjee and Schaefer (1976); Butts and Katz (1967); Kieffer and Straaten (1986) and Geiβ, Krämer and Kraft (1998).*

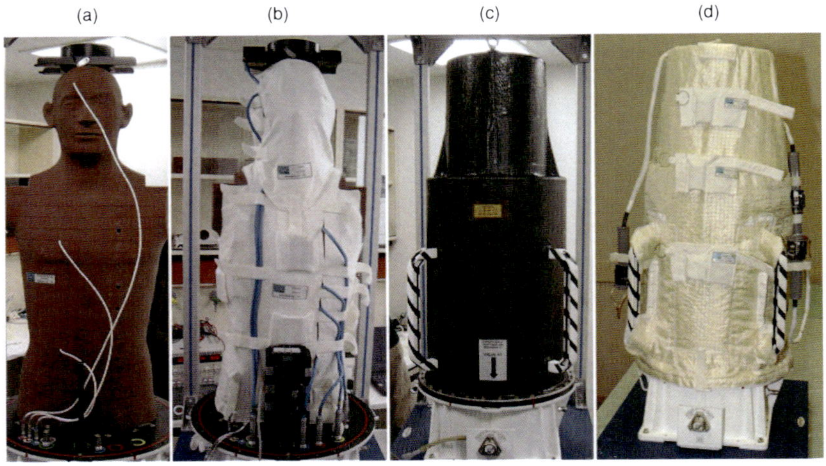


*Figure 4.29* The MATROSHKA anthropomorphic phantom. Left to right: (a) upper torso equipped with active and passive dosimeters. The torso is cut into 33 horizontal slides, each 25 mm thick. The base structure houses electronics for data acquisition (from the active dosimeters as well as temperature and pressure). Each slice is equipped with polyethylene tubes holding in total, over all slices, 4800 TL and OSL detectors; (b) torso inside the poncho and hood, equipped with passive detectors for skin dose measurements; (c) carbon fiber "suit," used to simulate an astronaut's space suit; (d) final package, ready for launch, with multilayer insulation for thermal protection. The whole package was mounted outside the Zvezda Module on the ISS for 539 days until its retrieval in August, 2005. The dosimeters were removed and returned to Earth for analysis by the collaborating laboratories. Reprinted from Reitz et al. (2009), with permission from the Radiation Research Society.

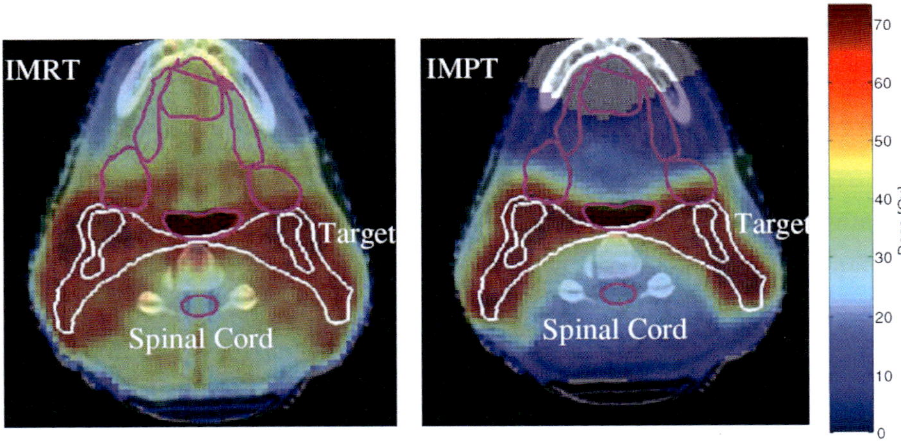


*Figure 5.1* Example of calculated dose distributions (plan) for a nasopharynx case using X-ray intensity modulated radiation therapy (IMRT) and intensity modulated proton therapy (IMPT). The target is indicated by the white contour lines, whereas critical structures such as the spinal cord are indicated by the pink contour lines. The steep dose gradients separating the high and low dose regions are particularly noticeable in the IMPT plan. Reprinted from Bortfeld (2006) with permission from IOP Publishing Ltd.

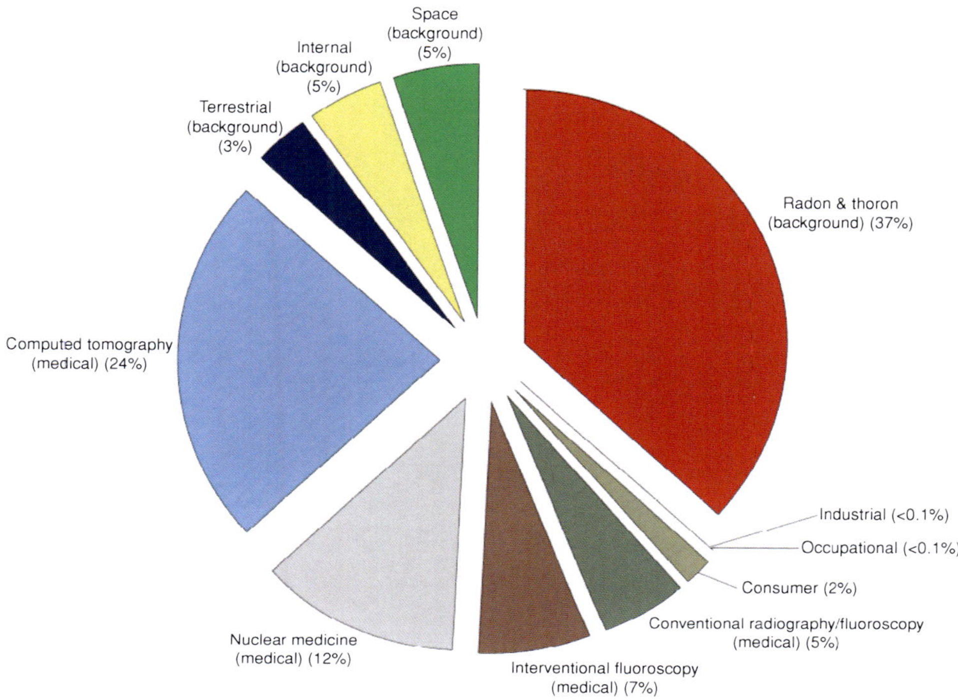


***Figure 5.2*** *Contribution of different sources of radiation exposure to the collective dose to the United States population in 2006. Reproduced from NCRP (2009) with permission from the National Council on Radiation Protection and Measurements.*

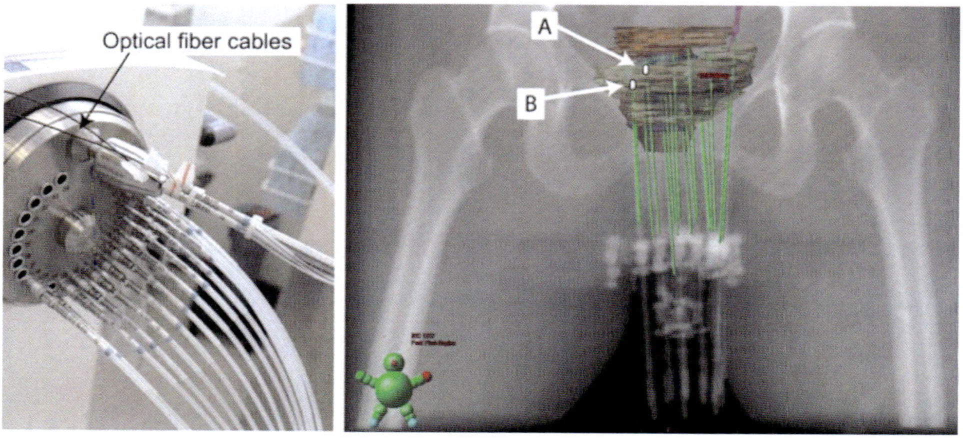


***Figure 5.37*** *Afterloader with $^{192}Ir$ source and guide tubes used in the brachytherapy treatment. The right-hand figure indicates the position of the OSL probes (A and B). Reprinted from* Andersen et al. *(2009b) with permission from Medical Physics Publishing.*

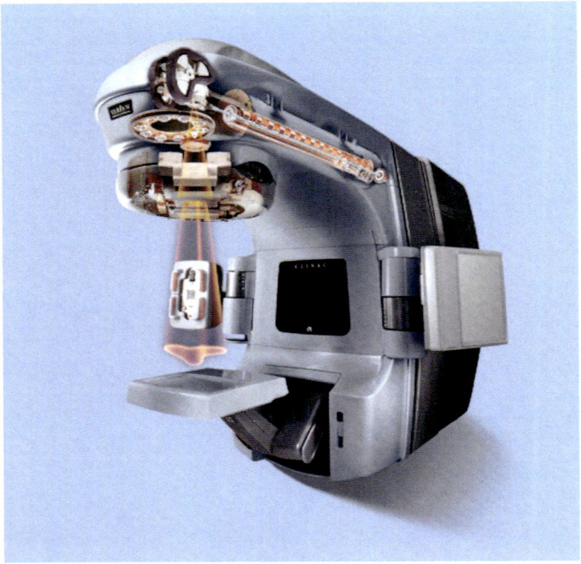

***Figure 5.4*** *Illustration of a medical linear accelerator used for radiotherapy treatment using megavoltage X-ray and electron beams. Image courtesy of Varian Medical Systems, Inc., all rights reserved.*

(a)

(b)

***Figure 5.9*** *Photos of the (a) cyclotron proton accelerator and (b) beam transport system from the ProCure Treatment Center in Oklahoma City, OK, USA. Photo courtesy of ProCure Treatment Centers, Inc.*

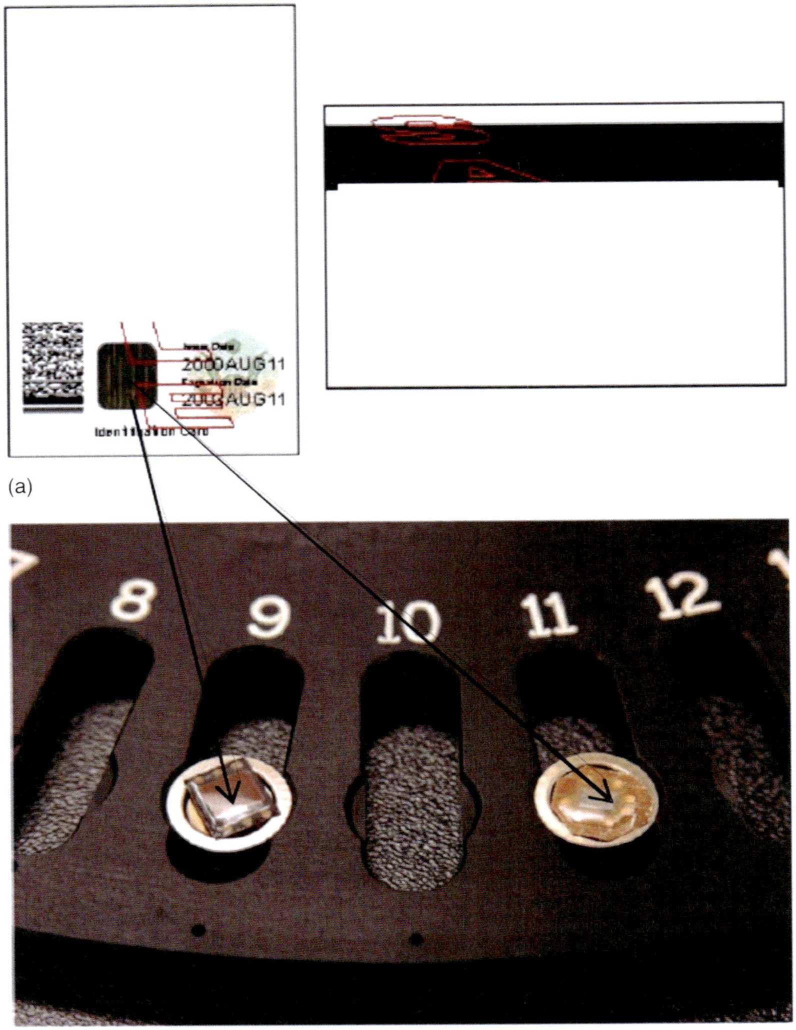


(a)

(b)

***Figure 6.8*** *(a) Photograph of ID card containing the (b) electronic chips investigated as a potential OSL detector for accident dosimetry. Reprinted from Mathur* et al. *(2007) with permission from Elsevier BV.*

## 4.4 Health Risk

Damage to living cells by radiation can be repaired by several natural chemical repair mechanisms. If, however, the rate of irradiation is sufficiently high that the damage cannot be repaired quickly enough permanent cell damage can result. For severe damage the cell may die. If sufficient cells die within a particular tissue or organ the radiation exposure can lead to organ malfunction, complete failure and even death. Such health detriments are described as deterministic effects and are the major health concerns associated with high-radiation-rate exposure by SPEs (NRC, 2006, 2008). Dose limits to maintain the probability of such processes to acceptable levels were listed in Table 4.2.

Alternatively, during the lower-intensity but longer-term exposure from GCR, the radiation effects may be such that the cell continues to live, but genetic damage acquired during the irradiation can be passed on to progeny cells during cell division (NRC, 2006, 2008). The degree of cell repair and the effectiveness of such are stochastic. Even at low doses damaged cells can lead to the growth of more damaged cells via cell division with a concomitant increased probability of cancer. The dose estimates for cancer production are difficult to judge precisely because such effects are probabilistic and occur at doses much lower than those for which deterministic effects are observed. However, calculations by Cucinotta, Kim and Ren (2005) of the probabilities for cancer induction suggest severe limitations on the mission scenarios that can currently be attempted. For example, Cucinotta, Kim and Ren (2005) calculated the probability of cancer-induced death on a 600 day Mars "swing-by" mission, assuming a mean shielding thickness of 20 g/cm$^2$. The results indicate that changes to the mission scenario or shielding design would have to be made if the astronauts' increased lifetime risk of fatal cancer is to remain lower than the mandated maximum of 3%.

The uncertainties in calculations of the type noted above arise primarily from uncertainties in the functional form of $Q(L)$ for space radiation, in spacecraft design, in space radiation and transport models, in fragmentation cross-sections, in mission scenarios, in solar activity and in many other factors (Cucinotta, Kim, and Ren, 2005; Cucinotta and Durante, 2006). Furthermore, attempts to relate specific cancers to particular radiation exposures in epidemiological studies are notoriously difficult due, among other issues, to the difficulties and uncertainties associated with retrospective assessment of the level of radiation exposure to individuals after the event (Muirhead, 2008; Simon *et al.*, 2007). Nevertheless, using current knowledge one can estimate the lifetime risk of fatal cancer and the associated confidence intervals for various missions under different assumed conditions. Example results are given in Table 4.4.

Even assuming that one can accurately model and calculate the risk one is then faced with the challenge of measuring the doses actually received by the astronauts. Dosimetry measurements on the ISS agree with model predictions only within ±10–25% (Badhwar and Cucinotta, 2000), although whether this is a weakness of the dosimetry or the calculations or both is unclear. As already noted (Benton and Benton, 2001; Yasuda, 2006), reliable dosimetry in such complex environments is a significant challenge. However, past experience with all manned space flights to date has led to the clear recommendation that it is essential for personal dosimeters to be worn by astronauts at all times and that the dosimeters should consist of low-LET detectors such as TLDs or OSLDs, and high-LET detectors

***Table 4.4*** *Calculations of absorbed dose, D, effective dose, E, and 'Radiation Exposure Induced Death', REID (including confidence intervals) for 40-year-old male and female astronauts for three different missions at solar minimum behind an assumed shielding of 5 g/cm² aluminum. Data from Cucinotta and Durante (2006)*

| | | | REID% (95% confidence)[a] | |
|---|---|---|---|---|
| | Absorbed Dose (Gy) | Effective Dose (Sv) | Men (40 years old) | Women (40 years old) |
| 180-days Lunar mission | 0.06 | 0.17 | 0.68 (0.20–2.4) | 0.82 (0.24–3.0) |
| 600-days Mars orbit | 0.37 | 1.03 | 4.0 (1.0–13.5) | 4.9 (1.4–16.2) |
| 1000-days Mars exploration | 0.42 | 1.07 | 4.2 (1.3–13.6) | 5.1 (1.6–16.4) |

[a] Radiation exposure induced death, expressed as a percentage with 95% confidence.

such as PNTDs (NCRP, 2002). The potential for and use of OSL in these applications is the topic of the following sections.

## 4.5 Evaluation of Dose in Space Radiation Fields Using OSLDs (and TLDs)

### 4.5.1 The Calibration Problem for Space Radiation Fields

OSL dosimeters can be calibrated in terms of the absorbed dose to water or an other quantity of interest using a reference calibration field, usually gamma rays from a $^{60}$Co source, as discussed in Chapter 3. However, when the dosimeter is exposed to a radiation field different than the reference calibration field, two problems arise: (i) the energy deposited in the detector material (i.e., absorbed dose) may not be the same as the energy deposited in water under the same exposure conditions; (ii) the luminescence produced per absorbed dose (luminescence efficiency) may not be the same as the luminescence efficiency after exposure to a reference radiation field. This problem is similar to the photon energy dependence observed for luminescence detectors that are tissue equivalent (see Section 3.4.1.1), but in space dosimetry the concept of tissue equivalency must also include the interaction of HCPs with the detector.

For HCPs the situation becomes even more complex. Not only does one have to account for the range of the primary particle in the detector, but also for the radial distribution of dose around the incident particle track. Figure 4.11a shows the pattern of secondary electrons (delta rays) generated along the path of a 124 MeV/u Fe ion (traversing a water target from left to right). The pattern of energy deposition is such that the dose is highest along the core of the particle track. Several estimates of the radial dose distribution around a charged particle track are shown in Figure 4.11b. The figure shows the results of using several different analytical expressions to define the radial dose distribution around a charged particle track, in this case a 1000 MeV/u Fe ion (with $L = 144$ keV/μm in water). Although there are discrepancies between the models at very small radial distances (and the definition of the core radius) and at very large radial distances, there is full agreement in the intermediate ranges where the dose approximates an $r^{-2}$ law (where $r$ = radial distance).

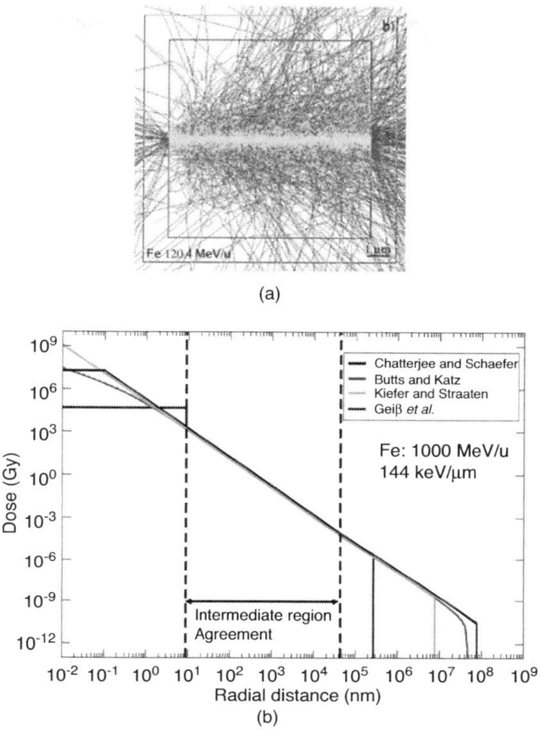


***Figure 4.11*** *(a) GEANT-4 calculations of the energy deposition from the passage of a 124 MeV/u Fe ion through a water target from left to right. The lines represent delta rays (secondary electrons). The dose deposition is greatest along the core of the track. This is seen in (b) in which the radial dose distributions in water around the path of a 1000 MeV/u Fe ion are shown, as determined using the models of Chatterjee and Schaefer (1976); Butts and Katz (1967); Kieffer and Straaten (1986) and Geiβ, Krämer and Kraft (1998). Figure (a) was reprinted from Sawakuchi, G. O., Characterization and modeling of relative luminescence efficiency of optically stimulated luminescence detectors exposed to heavy charged particles. Ph.D. Thesis. Copyright (2007) Oklahoma State University. Figure (b) was reprinted from Sawakuchi, G. O., Characterization and modeling of relative luminescence efficiency of optically stimulated luminescence detectors exposed to heavy charged particles. Ph.D. Thesis. Copyright (2007) Oklahoma State University.*

The overall dose range, from the core to the outer extremities, is huge, varying by over 20 orders of magnitude.

Figure 4.11b also shows that the dose goes to zero for radial distances larger than the maximum range of electrons $R_{max}$. Although various definitions of $R_{max}$ exist (Sawakuchi, 2007), it can be shown by conservation of mass and energy that the maximum energy that can be transferred to an electron by an HCP depends on the relativistic velocity $\beta$ of the primary particle (Rossi and Zaider, 1996). Therefore, the radial dose distributions can be different, even for particles with the same $L$, but a different combination of $Z$ and $\beta$, showing that $L$ cannot fully represent the distributions.

With the assumption that all the secondary delta rays stop inside the material, the basic relationship for determining total dose deposited by an HCP traversing a medium is:

$$D = L\frac{\Phi}{\rho} \tag{4.5}$$

where $\Phi$ is the particle fluence (number of particles per cm$^2$), $L$ is the LET in the medium and $\rho$ is the medium density. If $L$ is in keV/μm and $\rho$ is in g/cm$^3$, Equation (4.5) can be re-written as:

$$D(\text{Gy}) = 1.602 \times 10^{-9} L(\text{keV}/\mu\text{m})\frac{\Phi(\text{cm}^{-2})}{\rho(\text{g cm}^{-3})} \tag{4.6}$$

so that the absorbed dose is given in units of Gy (J kg$^{-1}$).

However, Equation (4.5) only gives the *total* absorbed dose to the medium. From Figure 4.11 it is clear that the energy deposition is spatially highly non-uniform, and around each track the doses vary over an extremely wide range. This in turn means that the OSL (or TL) is not simply a characteristic of the total absorbed dose, as expressed by Equation (4.5), but is instead a convolution of OSL/TL signals emerging from regions within the detector characterized by very different doses. Difficulties arise because the OSL and TL signals from a detector not only saturate at high doses, but they also change curve shape (decay shape for OSL, glow-curve shape for TL) as the dose increases. The net signal, therefore, is a convolution of these different OSL/TL curve shapes in response to the radial dose distribution. The overall result is that the OSL signal that one obtains from a given high-energy photon dose (e.g. from $^{60}$Co gamma rays) is not the same as the signal that one obtains from the same dose of charged particles.

To express the variation in OSL or TL signal induced by exposure to charged particles of LET $L$, as compared to a reference gamma-radiation field, one can define a relative luminescence efficiency $\eta_L = \eta(L)$ as:

$$\eta_L = \frac{S_L^{\text{OSL}}/D_L}{S_\gamma^{\text{OSL}}/D_\gamma} \tag{4.7}$$

where $S_\gamma^{\text{OSL}}$ is the OSL signal following a gamma dose $D_\gamma$, and $S_L^{\text{OSL}}$ is the OSL signal following a charged particle dose $D_L$ (assuming both are in the linear region of the dose response)[1] of LET $L$.

If $D_L$ and $D_\gamma$ in Equation (4.7) represent the absorbed dose to the detector material (i.e., $D_\gamma^{\text{det}}$ and $D_L^{\text{det}}$), $\eta_L$ should be equal to 1 for low-LET particles (e.g. high-energy protons), since the ionization densities created in the material are in principle similar to gamma radiation and not sufficient to cause local saturation of the trapping centers. However, in the literature $D_L$ and $D_\gamma$ are usually defined in water (i.e., $D_\gamma^{\text{w}}$ and $D_L^{\text{w}}$ are used instead of $D_\gamma^{\text{det}}$ and $D_L^{\text{det}}$). This is for convenience, because the quantities in absorbed dose to

[1]The difficulty with this definition, however, is that it is limited to application at low doses where the TL or OSL dose response is linear. For high-enough particle energies, sublinearity can be detected at low particle fluence (and, therefore, doses) and it becomes problematic to determine the efficiency based on this definition. As pointed out by Gaza, Yukihara and McKeever (2006), however, the OSL dose response curve for charged particles can be approximated to a saturating exponential $S(D) = a(1 - e^{-bD})$, and the signal in the linear region (low doses) can be approximated by $S(D) \cong abD$, such that $\eta$ can be calculated by curve fitting the whole of the dose response curve to determine parameters $a$ and $b$.

water are readily available from the beam-calibration and beam-monitoring instruments, whereas for the same charged particle fluence $D_\gamma^{det}$ and $D_L^{det}$ will be different for different detector materials. As discussed by Sawakuchi (2007), when a detector is exposed to a given gamma-ray fluence or charged-particle fluence, the energy deposited in the detector will be different than the energy deposited in water (if water was exposed in the same conditions and to the same fluence). Therefore, from Equation (4.7) one can see that the ratio between $\eta_L^{w}$ (where the doses are expressed in water) to $\eta_L^{det}$ (where the doses are expressed in the medium) is given by the ratio of the gamma dose in the material to the gamma dose in water, multiplied by the ratio of the charged-particle doses in the two media, thus:

$$\frac{\eta_L^{w}}{\eta_L^{det}} = \frac{D_\gamma^{w}}{D_\gamma^{det}}\frac{D_L^{det}}{D_L^{w}} \tag{4.8}$$

According to Equations (3.36) and (3.39), these in turn are given by the ratios of the mass energy absorption coefficients $(\mu/\rho)$ and of the mass stopping powers $(L/\rho)$ in the two media. That is:

$$\frac{\eta_L^{w}}{\eta_L^{det}} = \frac{(\mu_{en}/\rho)^{w}}{(\mu_{en}/\rho)^{det}}\frac{(L/\rho)^{det}}{(L/\rho)^{w}} \tag{4.9}$$

This expression is an approximation considering that the OSL detector is large compared to the range of the secondary electrons, and that $L$ is constant as the particles traverse the detector material. The value of this ratio is a function of the gamma photon energy and the LET of the charged particle. As shown by Sawakuchi *et al.* (2008c), for a $^{60}$Co reference source and a 230 MeV proton source in $Al_2O_3$:C, $\eta_L^{w}/\eta_L^{det} = 0.934$. As defined $\eta_L^{w}$ is not properly a relative luminescence efficiency, but a relative response relating the absorbed dose to water indicated by the dosimeter exposed to a non-reference radiation field (in this case, HCPs) to the absorbed dose to water from a reference radiation field (e.g. $^{60}$Co gamma rays). Keeping this subtle, but important, point in mind, from now on we will consider the doses expressed in dose to water, except when explicitly stated otherwise.

When one includes the additional complexity that space-radiation fields are not characterized by one particle type, but instead are composed of a very wide range of different particles moving in different directions, with a range of energies and LET, each with their own radial dose distribution, then one is presented with a calibration problem of significant magnitude.

If the OSL signal $S_\gamma^{OSL}$ resulting from a gamma dose $D_\gamma$ can be written as:

$$S_\gamma^{OSL} = \xi D_\gamma \tag{4.10}$$

where $\xi$ is a calibration factor (proportionality constant between the OSL signal and the absorbed dose to water from gamma rays), then the OSL signal $S_\gamma^{OSL}$ produced by irradiation with a dose $D_L$ from a single type of charged particle of LET $L$ is:

$$S_L^{OSL} = \eta_L \xi D_L. \tag{4.11}$$

It is obvious that $\xi$ is constant only in the linear range of the dosimeter dose response, so this simplified treatment is only valid in these circumstances; for doses in the non-linear region of the dosimeter response, a calibration curve needs to be determined.

For dosimetry in a mixed space-radiation field of multiple particle types, each with its own LET and each with its own imparted dose $D_L$, one then has for the net OSL signal:

$$S^{\mathrm{OSL}} = \xi \sum_L \eta_L D_L, \tag{4.12}$$

where the summation is over all particles of different values of LET, $L$. The desired quantity, however, is $D = \sum_L D_L$, which for sufficiently small incremental differences in $L$ can be written in integral form as $D = \int D(L)\mathrm{d}L$, where $D(L)$ is now the differential dose spectrum. Thus, the net OSL can be written:

$$S^{\mathrm{OSL}} = \xi \int \eta(L)D(L)\mathrm{d}L. \tag{4.13}$$

However, since $S^{\mathrm{OSL}}$ is a single number, whereas $\eta(L)$ and $D(L)$ are functions, then the problem of extracting $D$ from $S^{\mathrm{OSL}}$ is clearly ill posed.

One possible solution to this problem is to obtain more information about the radiation field using not only the signal $S$, but other information or signals that can be obtained from a single detector or from a combination of different detector materials. However, before dealing with the application of OSL to mixed fields, let us first discuss properties of TL and OSL materials that will be important to tackle this problem, particularly the changes in the shape of OSL (and TL) signals as a function of the gamma (or beta) radiation dose.

### 4.5.2 Thermoluminescence, TL

There is a wide range of potential TL materials for use in space-radiation dosimetry, but only a handful have been used in practice and only one, LiF:Mg,Ti, has been utilized to a major degree. Other TL materials, utilized to different extents, include LiF:Mg,Cu,P, $CaF_2$:Tm, $CaSO_4$:Dy, $Al_2O_3$:C and $Mg_2SiO_4$:Tb, with the latter two TL materials playing only a minor role. For the most part we concentrate the following discussion on the properties of LiF:Mg,Ti when exposed to radiation typically found in space. This discussion is not merely for completeness, due to the historical popularity of this material, but rather its purpose is to establish the main conceptual understanding of the behavior of luminescence dosimeters when exposed to space-radiation fields and to show how this understanding can be extended to OSL.

LiF:Mg,Ti was perhaps the earliest material to be utilized fully in TL dosimetry following its development at the University of Wisconsin (USA) in the 1950s. (See McKeever (1985) for a historical overview.) It is available in three isotopic forms – namely with Li in its natural isotopic abundances, $^6$Li-enriched or $^7$Li-enriched. In these forms it is popularly known as TLD-100, TLD-600 and TLD-700, respectively, following the marketing names provided to the first commercial forms of the dosimeter by what was then the Harshaw Chemical Company (US). Other similar versions of the material are known and have been extensively used in space, for example MTS-6 and MTS-7 from the Institute of Nuclear Physics in Krakow (Poland).

As described in earlier chapters, the TL or OSL signal from a luminescence detector arises from thermally or optically stimulated electron–hole recombination as trapped charge carriers are excited from their traps by the absorption of energy from heat or light. There are no known TL or OSL materials, however, for which the luminescence signal originates from one trap type only. In all cases the overall signal is a sum of emissions due to detrapping of charge from multiple traps of different types. Understanding how the overall signal responds to increasing dose, therefore, requires an understanding of how the individual trap types respond to increased dose. LiF:Mg,Ti in particular displays multiple TL peaks in the overall TL glow curve. Figure 4.12a shows glow curves (from TLD-700) following irradiation with either $^{60}$Co gamma rays or $^{56}$Fe ions, with multiple peaks caused by the presence of different trap types in this material.

To be noted in Figure 4.12 is the main TL peak – known as "peak 5" – appearing near 200 °C, and a lower-intensity complex of peaks at higher temperatures, generically described as the "high-temperature" signal, between ~250 and 350 °C. The latter region is a combination of several, overlapping TL peaks (Berger *et al.*, 2006b, 2006c; Berger and Hajek, 2008b; Horowitz, Oster and Datz, 2007; Shachar and Horowitz, 1988; Yasuda, 2001). Normally, peak 5 is the only signal used in dosimetry. The TL signal associated with this peak can be obtained using curve-fitting procedures, using the maximum intensity of the TL curve or using the integrated intensity (area under the curve) between two temperatures defining the region of interest in the TL curve. (See reviews on the use of LiF:Mg,Ti in dosimetry in Horowitz (1983), McKeever (1985), and McKeever, Moscovitch and Townsend (1995), among others.)

Figure 4.12a illustrates that the different TL peaks from LiF:Mg,Ti do not respond the same way to the different radiation types (here $^{60}$Co and $^{56}$Fe ions). In particular the ratio of the high-temperature TL peaks to the main peak 5 is found to be a function of the LET of the irradiating particle, leading to the suggestion that the ratio of these signals can be used to determine the LET of the irradiation particle. The concept of using the differential behaviors of various TL peaks to provide information about the LET of the irradiation, using LiF and $CaF_2$ TLDs in particular, has its genesis in the late 1970s (e.g., Hoffmann and Prediger, 1984; Lucas and Kaspar, 1977; see also McKeever, 1985 for additional examples). Vana and colleagues (e.g., Schöner, Vana and Fogger, 1999; see also Yasuda and Fujitaka, 2000) used the so-called "high temperature ratio" (HTR) as a means of determining the "mean" LET of the space radiation field – that is to use the TLD not just as a dosimeter but also as an "LET meter." An example of the effect for particular charged particles is shown in Figure 4.12b, in this case for TLD-600, where the relative intensity of the high-temperature region compared the height of peak 5 is seen to increase with the LET of the radiation source. The HTR is defined as the ratio of the area under the TL curve between two defined temperatures, as indicated in the figure, to the height of peak 5 for the particle of interest, divided by the same quantity for $^{60}$Co gamma photons. By definition, therefore, the HTR for $^{60}$Co is 1. An example of the HTR versus the LET of various particles is given in Figure 4.13. Moyers and Nelson (2009) show similar data for $CaF_2$:Tm.

The differential growth rates of TL peak 5 and the high-temperature region of the glow curve can be explained from the different dose response curves of each to gamma irradiation. In particular, the high-temperature region demonstrates earlier supralinearity as a function of dose than is seen for peak 5 (Berger and Hajek, 2008a, b). Examples of dose response functions $f(D)$ for TLD-100 are illustrated in Figure 4.14, where the dose response function $f(D)$ was previously defined in Equation (3.41). As defined, $f = 1$ indicates the linear region of the dose response curve while $f > 1$ indicates supralinearity and $f < 1$ indicates sublinearity. Figure 4.14 shows the $f(D)$ curves for the different peaks that make up the TL glow curve of TLD-100 and it can be seen that supralinearity ($f > 1$) starts

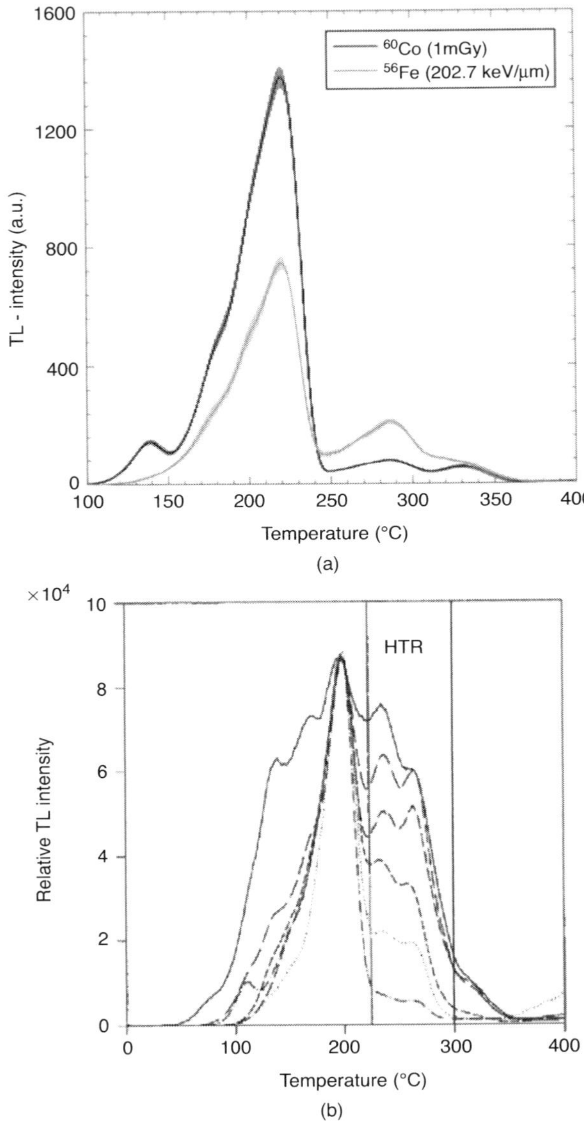


***Figure 4.12*** *(a) TL glow curve from TLD-700 after 1 mGy irradiation with $^{60}$Co gamma photons compared with that from 500 MeV/u $^{56}$Fe ions ($L$ = 202.7 keV/μm), showing the differential growth of the high-temperature region of the glow curve compared with the main peak (so-called "peak 5"). Reprinted from Berger and Hajek (2008a) with permission from Elsevier BV; (b) Glow curves of TLD-600 after irradiation with (from highest intensity to lowest in the high-temperature region): $^{241}$Am alpha particles ($L$ = 174 keV/μm); thermal neutrons ($L$ = 130 keV/μm); $^{19}$Fe ions ($L$ = 102 keV/μm); $^{19}$Fe ions ($L$ = 30 keV/μm); $^{12}$C ions ($L$ = 12.8 keV/μm); $^{60}$Co gamma photons. All curves have been normalized to the height of TL "peak 5." Figure (a) was reprinted from Radiation Measurements, Berger, T. And Hajek, M., 2008. On the linearity of the high-temperature emission from 7LiF:Mg,Ti (TLD-700). Vol. 43, 1467–1473. Copyright (2008) with permission from Elsevier. Figure (b) was reprinted from Radiation Protectin Dosimetry, Schöner, W., Vana, N., Fogger, M., The LET dependence of LiF:Mg,Ti dosemeters and its application for LET measurements in mixed radiation fields. Vol. 85, 263–266. Copyright (1999) with permission from Oxford University Press.*

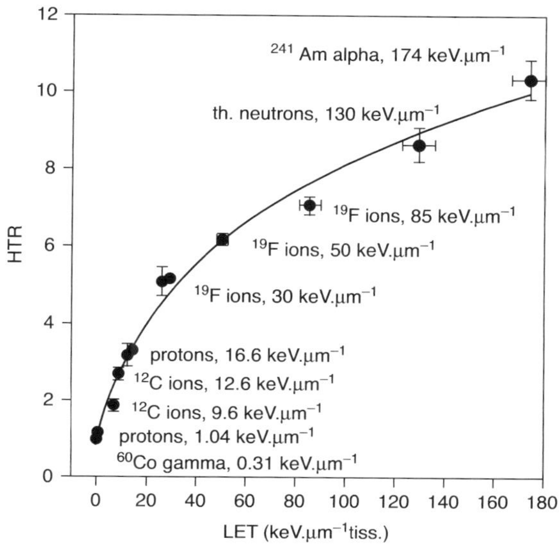


***Figure 4.13*** *The high-temperature ratio HTR as a function of the LET of the absorbed radiation in TLD-600. Reprinted from Radiation Protectin Dosimetry, Schöner, W., Vana, N., Fogger, M., The LET dependence of LiF:Mg,Ti dosemeters and its application for LET measurements in mixed radiation fields. Vol. 85, 263–266. Copyright (1999) with permission from Oxford University Press.*

at lower doses for the high-temperature peaks than for peak 5. As discussed in Section 4.5.1 and illustrated in Figure 4.11, the dose distribution surrounding the track of a charged particle ranges from very high doses for which $f < 1$, through medium doses for which $f > 1$, to low doses for which $f = 1$. Thus, the net TL curve shape is a convolution of TL glow curves from different regions surrounding the track represented by low, medium and high doses. Because of the high ionization densities, the intensity of the high temperature peaks relative to peak 5 following heavy charged particle irradiation is higher than in the case of low-LET radiation.

Moreover, since much of the signal is from regions of the track for which $f \neq 1$, the net signal following irradiation with heavy charged particles will give a different intensity than the signal obtained after $^{60}$Co gamma irradiation for the same net absorbed dose. This is expressed by the efficiency term $\eta_L$ defined in Equation (4.7). The efficiency for energetic charged particles is not expected to equal 1, even when the absorbed doses are expressed in dose to the detector material. Indeed, for heavy charged particles (e.g. low-energy He ions and heavier charged particles) it is generally found that the sub-linear ($f < 1$) regions surrounding the track dominate the signal and therefore $\eta_L < 1$. For some of the lighter and/or more energetic particles (e.g. protons and high-energy helium ions) it is found that the supralinear ($f > 1$) regions dominate and therefore $\eta_L > 1$. The net result, for particles of interest in space dosimetry, is a $\eta(L)$ curve of the type illustrated in Figure 4.15, where for low values of LET the efficiency can be greater than 1 while the general trend is one of decreasing efficiency with increasing LET.

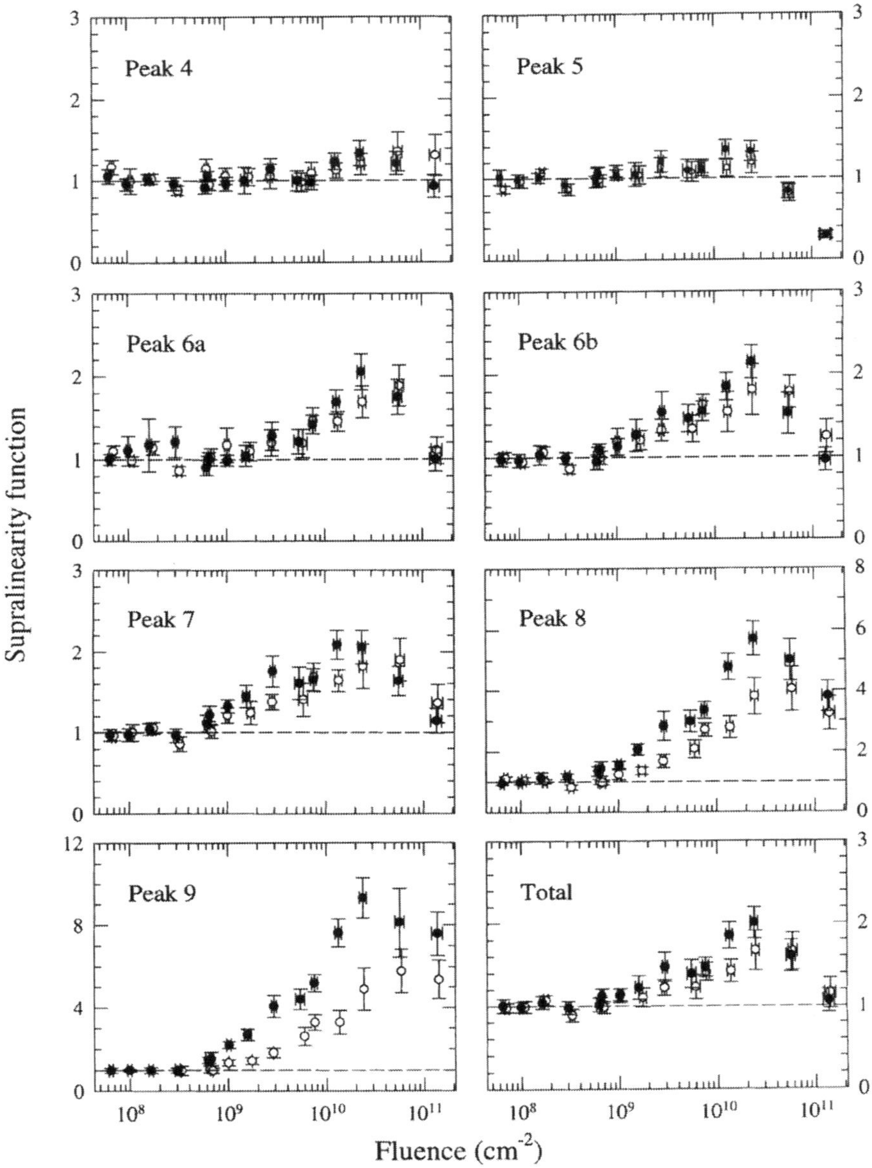


***Figure 4.14*** *Dose response functions (or supralinearity index) for individual TL peaks from TLD-100 for 3.0 MeV and 7.5 MeV He ions. Reprinted from Nuclear Instruments & Methods in Physics Research B, Rodríguez-Villafuerte, M., Study of the TL response of LiF:Mg,Ti to 3 and 7.5 MeV helium ions: Measurements and interpretation in terms of the track interaction model. Vol. 160, 377–386. Copyright (2000) with permission from Elsevier.*

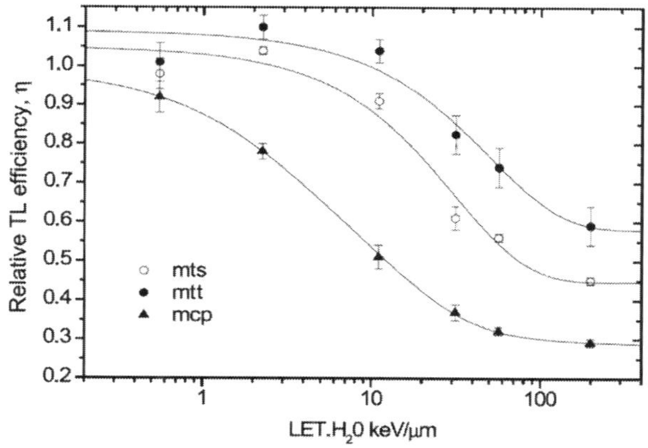


***Figure 4.15*** *Typical efficiency versus LET curves for different forms of LiF TLDs. Note that for a given TLD the efficiency $\eta_L$ may be greater than 1 or less than 1, and that the LET regions over which $\eta_L$ may be greater than or less than 1 vary with dosimeter type. In this figure MTT and MTS are two different forms of LiF:Mg,Ti, while MCP is a form of LiF doped with Mg, Cu and P. All samples were produced at the Institute of Nuclear Physics in Krakow, Poland. The data are for irradiations with protons (150 MeV), He (144 MeV/u), C (386 MeV/u), Si (490 MeV/u), Ne (368 MeV/u) and Fe (418 MeV/u) ions. The curves are fits to arbitrary polynomials. Reprinted from Radiation Protection Dosimetry, Bilski, P., Dosimetry of densely ionizing radiation with three LiF phosphors for space applications. Vol. 120, 397–400. Copyright (2006) with permission from Oxford University Press.*

Several factors are of major importance when examining, and using, $\eta(L)$ curves such as those shown in Figure 4.15. The first is obvious from the figure – namely that the shape of the curve depends upon the material used. Less obvious, but nonetheless of equal importance, is that the shape of the $\eta(L)$ curve critically depends upon the experimental details used when recording the TL glow curve. This is because the dose response functions, which ultimately dictate the shape of the $\eta(L)$ curve, depend upon whether one uses peak height or peak area. They also depend upon the heating rate used (e.g., Mische and McKeever, 1989). Thus, one can search the published literature and readily find a range of $\eta(L)$ curve shapes, even for the same material, depending on the experimental details used by each research group (e.g., Benton, Frank and Benton, 2000; Berger *et al.* 2006a; Bilski, 2006a, b; Horowitz, 1983; Weiss, Horowitz and Oster, 2009; Yasuda, 2001). Each is valid in its own right.

With curves such as those in Figure 4.13 and Figure 4.15 it is now possible to evaluate an unknown charged particle dose using a LiF:Mg,Ti TLD. For a single type of charged particle one can determine $L$ from Figure 4.13 by measuring the HTR value following the charged particle irradiation. With $L$ thus estimated, one can then use the $\eta(L)$ curve and determine the appropriate value for the efficiency, $\eta_L$, from Figure 4.15. This then allows the determination of the absorbed dose using an expression equivalent to Equation (4.11) for TL and the measured TL intensity (using peak 5).

This procedure works well for irradiation with a single charged-particle type. For a mixed particle field, however, the procedure is less clear since there is no single value for $L$ and, therefore, for $\eta$. Our investigation of this issue is delayed until Section 4.5.4. First, however, we turn to an examination of the properties of OSL detectors in charged-particle fields.

### 4.5.3 Optically Stimulated Luminescence, OSL

#### *4.5.3.1 OSL Efficiency*

With the information learned from the TL of LiF:Mg,Ti we now extend the discussion to investigate the characteristics of OSL after exposure to space radiation fields. As with TL, the key is to find two parts of the radiation-induced OSL signal that respond differently to particles of different LET. To find such signals we begin with observations of the dose response of OSL from the most popular OSL material, $Al_2O_3$:C. At the time of writing this material remains the only commercial OSL detector to have been flown in space.

As already discussed in Chapter 3, the shape of the OSL decay curve from $Al_2O_3$:C changes with dose. Figure 3.14 illustrates several OSL decay curves, each normalized to the initial intensity, following beta irradiation of $Al_2O_3$:C OSL detectors. Normalization of the curves in the manner shown in the figure emphasizes that the curve shape does not remain the same as the dose increases. For doses less than the minimum shown, all curves are the same shape, but as the doses increases beyond this value the decay of the OSL with stimulation time becomes faster. Jain *et al.* (2007) quantified the change in the OSL decay rate with dose by defining a decay constant for the initial part of the decay (which is approximated by these authors to an exponential). It is observed that the decay constant (i.e. decay rate) is invariant with absorbed dose at low doses ($<\sim$1 Gy), then increases as the dose increases (over the range from $\sim$1 to $\sim$100 Gy) and finally becomes invariant again at high doses ($>\sim$100 Gy). Jain and colleagues also observe that the decay constant starts at a higher value as the ionization density of the irradiation increases, even for photon irradiation. Thus, the initial decay rate after $^{90}Sr/^{90}Y$ beta irradiation is slower than that observed after irradiation with 50 kV X-rays, while the decay rate is faster still after 10 kV X-ray irradiation (Jain *et al.*, 2007).

The OSL decay curve is made up of several individual decay curves as multiple traps empty during optical stimulation (Akselrod and Akselrod, 2002; Jain *et al.*, 2007; Yukihara and McKeever, 2006a). Thus, a possible cause of the increase in decay rate with dose for the net OSL curve may be due to the differential growth of the different components, such that the slower components saturate first as a function of dose and leaving the faster components to dominate at higher doses (Yukihara *et al.*, 2004a; Yukihara and McKeever, 2006a). Alternatively, Yukihara *et al.* (2004a) showed that the kinetics of the OSL processes may also give rise to different OSL decay curve shapes (see Figure 2.16 and associated discussion in Section 2.2.8). Although such arguments do not provide an definite explanation for changes in the OSL curve of $Al_2O_3$:C, the relationship between ionization density and changes in the shape of the OSL curves is well documented (Jain *et al.*, 2007; Yukihara and McKeever, 2006a).

As one might expect, the dose response also depends upon whether one defines the OSL signal as the initial OSL intensity or the area under the OSL decay curve. This can be seen in Figure 3.13 in which the dose response for the OSL from the same material is illustrated, both for the initial intensity and the area as measured with CW-OSL. The dose response for the initial intensity exhibits significantly more supralinearity ($f > 1$) than its counterparts

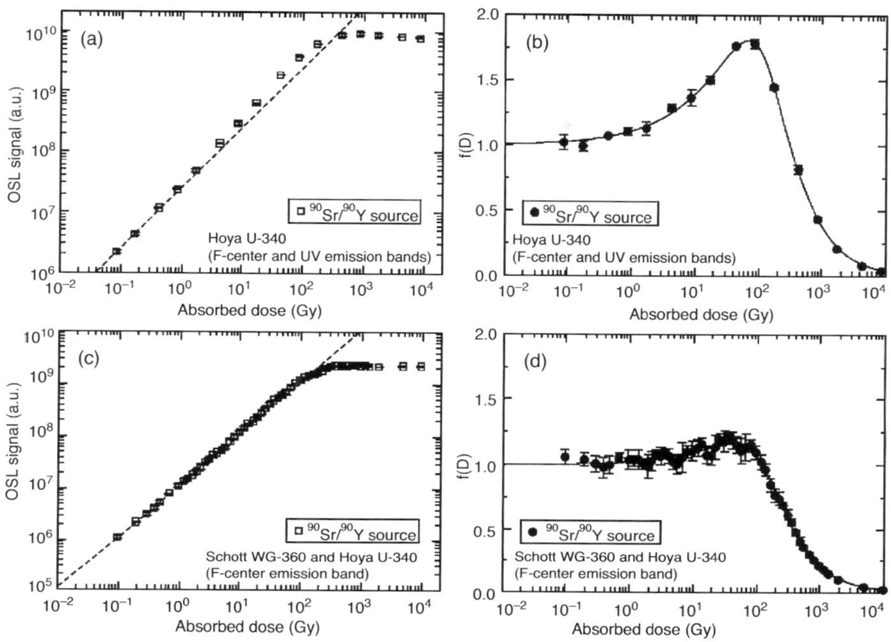


***Figure 4.16*** *Dose response curves for OSL from $Al_2O_3$:C Luxel detectors. The OSL is detected in either both the 420 nm and 330 nm regions ((a) and (b)), or in the 420 nm region only ((c) and (d)). The dose responses, illustrated in (b) and (d) as f(D) curves, is different in each case. Reprinted from Sawakuchi, G. O., Characterization and modeling of relative luminescence efficiency of optically stimulated luminescence detectors exposed to heavy charged particles. Ph.D. Thesis. Copyright (2007) Oklahoma State University.*

for the area measurement, whereas the curve for the total OSL area exhibits sublinearity ($f < 1$) at a much lower dose than that seen for the initial OSL intensity. Using what was learned from the study of TL from LiF:Mg,Ti, one may expect that the efficiency curves $\eta(L)$ for the different signal types will be quite different. Before examining this point, however, we first look at the other parameters that affect the shape of the dose response curve.

The $f(D)$ curves for $Al_2O_3$:C Luxel detectors are shown in Figure 4.16 where we see that not only does the dose response curve depend on how the signal is defined (in this case, total area), but it also depends on the detection filters used to monitor the OSL emission. The OSL from $Al_2O_3$:C consists of a strong emission centered near 420 nm, due to relaxation of excited $F$-centers, and emission in the UV, centered at 330 nm, possibly due to excited $F^+$ centers (Section 2.7.1.1). When detected in both wavelength regions the dose response is different (curves (a) and (b)) than when detected in the 420 nm region only (curves (c) and (d)). Comparison of curves (b) and (d) in particular show that when the UV signal is included, the $f > 1$ region is accentuated, whereas for only $F$-center emission the linear ($f = 1$) region extends to higher doses. Thus, we can also expect different $\eta(L)$ efficiency curves, depending upon the chosen wavelength range for monitoring the OSL emission.

Similarly, differences in $f(D)$ are observed, depending upon whether the OSL detector is an $Al_2O_3$:C single crystal or Luxel detector (Yukihara *et al.*, 2004a) and, likewise, the $\eta(L)$ efficiency curves also vary with the form of the detector (Gaza, Yukihara and McKeever, 2004; Yukihara *et al.*, 2004b).

What, therefore, might we expect for the OSL response to a flux of heavy charged particles? As noted previously, the wide range of doses within the track of a charged particle means that the net OSL signal after charged-particle irradiation will be made up of signals from regions of the track with small volumes and very high doses ($f < 1$; close to the core), larger volumes of intermediate doses ($f > 1$; moving away from the core) and yet larger volumes of smaller doses (for which $f = 1$; away from the core). The situation is illustrated in Figure 4.17. Here, the radial dose gradient within the track passing through

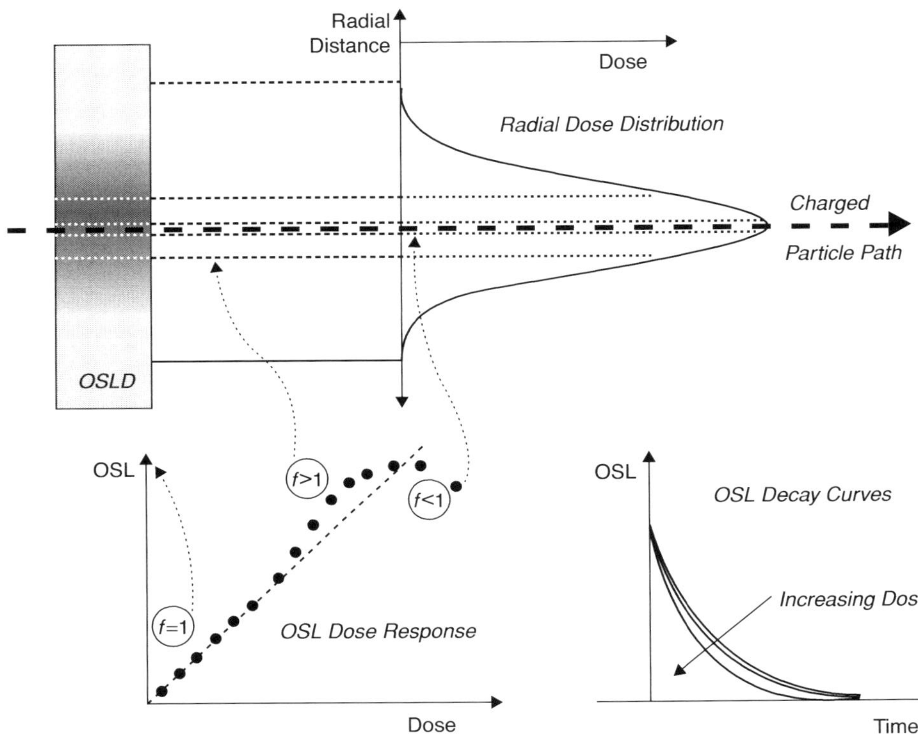


***Figure 4.17*** *Schematic diagram illustrating the radial dose distribution surrounding the trajectory of a charged particle through an OSL material. The dark regions in the OSLD represent regions of high dose, centered on the path of the particle. The corresponding radial dose distribution consists of large volume regions of low dose, smaller volumes of intermediate dose and small volumes of very high dose. The OSL signal will therefore be the sum of all those signals coming from the irradiated region, some of which correspond to the linear part of the dose response (f = 1), some from the supralinear part of the dose response (f > 1) and some from the sub-linear part of the response (f < 1). The net OSL will be a convolution of those OSL decay curves corresponding to each of these regions.*

the OSL detector is illustrated by the shaded region, with the darkest region along the path of the particle's trajectory representing the highest dose. (Refer also to Figure 4.11.) The radial dose distribution shows that the regions of lower dose ($f = 1$) extend over much larger volumes than do those of very high dose ($f < 1$) along the track core. Thus, the net OSL signal is a convolution of OSL signals for which $f = 1, f > 1$ and $f < 1$. Whether this composite signal yields an efficiency value of $\eta_L$ equal to 1, greater than 1 or less than 1 depends on both the shape of the dose response curve and the shape of the radial dose distribution. The latter, in turn, depends upon the energy and mass of the particle – and thus, indirectly, upon $L$. Furthermore, as already noted, the shape of the dose response curve, in its turn, depends upon the conditions of measurement of the OSL signal and the physical form and type of the dosimeter. Armed with this understanding and these observations, we can now describe the OSL properties of $Al_2O_3$:C after irradiation with energetic heavy charged particles.

Yasuda, Kobayashi and colleagues were the first to observe that the shape of the OSL decay curve from $Al_2O_3$:C irradiated with heavy charged particles varied with the LET of the particle (Yasuda and Kobayashi, 2001; Yasuda, Kobayashi and Morishima, 2002). Figure 4.18 shows a typical set of normalized OSL decay curves for a series of charged particles, compared with that for $^{90}Sr/^{90}Y$ beta particles. In these experiments (from Yukihara 2004b) the net dose delivered to the sample was the same in each case (100 mGy), but the shape of the curves changed dependent upon the charged particle used. In particular, the higher the value of $L$ the faster the OSL decay. This is reminiscent of the change in the shape of the OSL decay as a function of beta dose, seen in Figure 3.14, and is more pronounced when UV emission is detected than when not. Such effects are a result of the different radial dose distributions around the charged-particle paths, illustrated schematically in Figure 4.17. An appreciation of the different shapes of the dose distributions for different charged particles

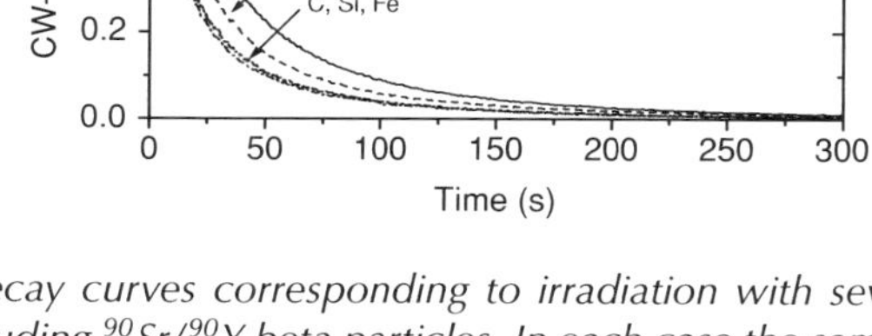


***Figure 4.18*** *OSL decay curves corresponding to irradiation with several different types of charged particle, including* $^{90}Sr/^{90}Y$ *beta particles. In each case the same dose (100 mGy) has been delivered. The OSL decay curves after exposure to high-LET radiation (He and heavier) are faster than those after exposure to low-LET radiation (gamma, beta and high-energy protons) as a consequence of the high doses within the charged-particle tracks created by high-LET radiations. Reprinted from Radiation Measurements, Yukihara, E.G. et al., Optically stimulated luminescence and thermoluminescence efficiencies for heavy charged particle irradiation in* $Al_2O_3$*:C. Vol. 38, 59–70. Copyright (2004) with permission from Elsevier.*

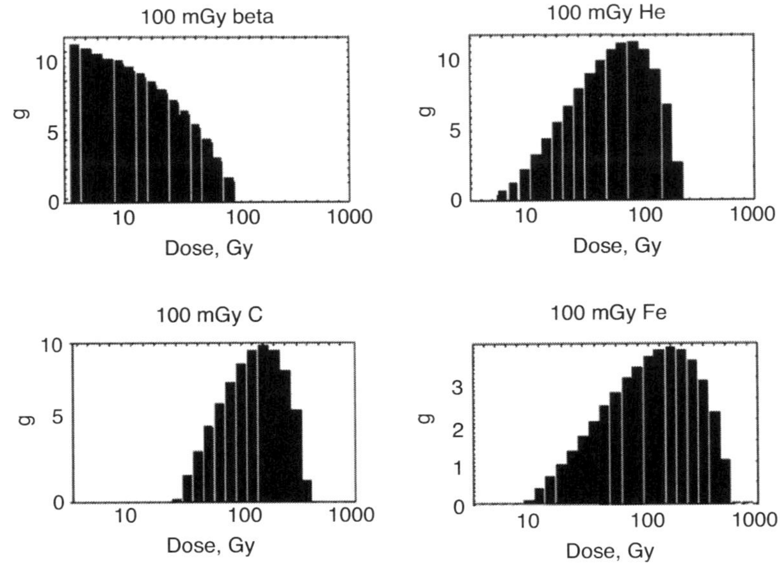


***Figure 4.19*** *Estimated dose distributions by deconvolving data similar to those shown in Figure 4.18 using curves such as those shown in Figure 3.14. Parameter g is a weighting factor expressing the relative strength of the OSL signal emerging from regions of the track that received the dose indicated. Dose distributions correspond to* $^{90}Sr/^{90}Y$ *beta particles, 144.2 MeV/u He, 385.5 MeV/u C and 122.9 MeV/u Fe ions.*

may be obtained by assuming that the net OSL after a given charged-particle irradiation is the sum of a set of OSL decay curves obtained for different beta doses. Thus, approximations to the dose distributions can be obtained by deconvolving the observed OSL decay curve for a given charged particle (Figure 4.18) using an observed set of OSL decay curve shapes, such as those seen in Figure 3.14b. The result is a set of dose distributions, as illustrated in Figure 4.19. Although the distributions shown are simply crude approximations to the actual distribution, one can see several important points. Firstly, the peak of the dose distribution shifts to higher doses as the LET of the particle increases – that is, the OSL signal becomes dominated by those regions closer to the high-dose core of the track rather than the low-dose outer regions. Secondly, although the net dose delivered to the whole sample was the same in each case (100 mGy) the doses within the tracks are orders of magnitude higher than this. Since these doses were deposited as each particle traverses the sample, the corresponding local dose rates within the track volumes are also very high. It should be pointed out that the data in Figure 4.19 represents only a first attempt to obtain these distributions; this procedure has not been fully developed or tested.

Knowing the radial dose distribution (for example, Figure 4.11) and the dose response function (for example, Figure 4.16) one can combine the two and determine the net OSL signal, per incident particle, for a wide variety of particle types and energies. In this way it is then possible to estimate the dependence of the OSL efficiency $\eta$ on the LET of the charged particle. This is the basis for the analysis of the TL (and OSL) response to

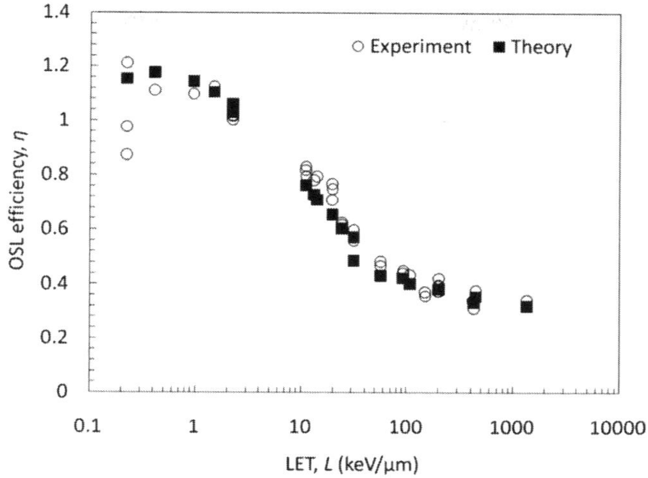


***Figure 4.20*** *Experimental OSL efficiency ($\eta$) versus LET for $Al_2O_3$:C Luxel detectors where the emission is detected in the 420 nm and UV bands (open circles). Data from Sawakuchi et al. (2008c). The experimental data are compared to the efficiencies calculated using a version of the Butt and Katz model, as modified by Sawakuchi (2007), and the dose response data of Figure 4.16b (filled circles). The LET is $L_{\infty}^{H_2O}$. Theory data from Sawakuchi (2007).*

charged particles using track structure theory; for a recent example see Weiss, Horowitz and Oster (2009). Sawakuchi (2007) applied track structure theory for OSL from $Al_2O_3$:C Luxel detectors using the models for the radial dose distribution shown in Figure 4.11. Example results, using Sawakuchi's modified version of the Katz model, are plotted in Figure 4.20 where they are compared to experimental data from Sawakuchi *et al.* (2008c). Of note are the efficiency values $<1$ for all LET values greater than $\sim$2 keV/μm, while for values of $L$ smaller than this the efficiency is $>1$. The dose response function used in these calculations is shown in Figure 4.16a and Figure 4.16b.[1] Here the supralinear ($f > 1$) region is dominant over the dose range from $\sim$1 to $\sim$100 Gy. It is clear, therefore, that the dose distribution in the tracks of the low-LET particles used in these experiments (all high-energy protons) is dominated by doses in this dose range (Sawakuchi *et al.*, 2008a). Conversely, the high-LET particles ($L > 2$ keV/μm; He and heavier ions) have efficiencies $< 1$ since the dose distributions for these particles are dominated by much higher doses ($> \sim$100 Gy) for which $f < 1$. These conclusions are consistent with the evaluated dose distributions displayed in Figure 4.19.

The efficiency function, however, does not depend exclusively on the value of the LET. Horowitz and colleagues pointed out the lack of uniqueness between the efficiency and the LET in the TL response of LiF:Mg,Ti (Horowitz and Kalef-Ezra, 1981; Horowitz, 1983) in

[1]Horowitz and colleagues (Horowitz, 1983, 2001; Horowitz and Olko, 2004) have shown that in order not to over-estimate the efficiency of TL (and, by inference, OSL) production by charged particles by using beta and/or gamma dose response functions one should take care to use a reference source for which the secondary (delta) electron energy spectrum is similar to that from the charged particles. Thus, in what these authors term modified track structure theory, they show that for low-energy particles one should use low-energy photon sources for the reference irradiation rather than high-energy photon or beta sources.

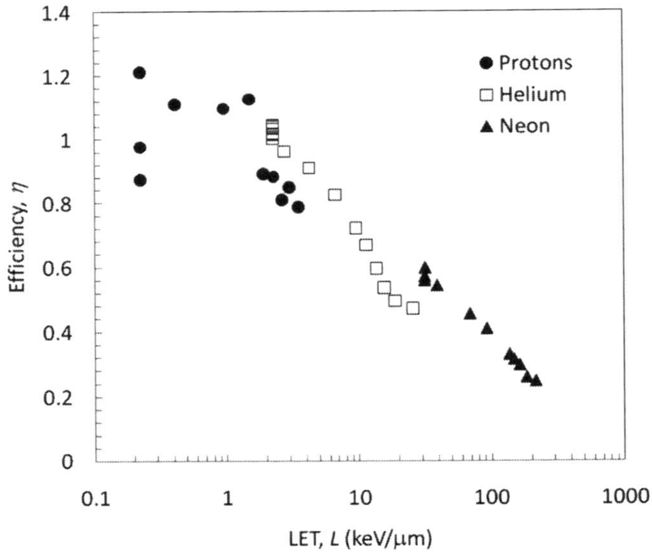


***Figure 4.21*** *Experimental OSL efficiency ($\eta$) versus LET for $Al_2O_3$:C Luxel detectors where from the emission is detected in the 420 nm and UV bands. Replotted using data from Sawakuchi et al. (2008c). Three data sets are shown, corresponding to protons (40 MeV), helium ions (150 MeV/u) and neon ions (400 MeV/u) with different binary filters placed in front of the beam to degrade the beam energy in order to increase the LET of the particles. Each particle type has its own unique $\eta(L)$ curve.*

that a given particle species (of a certain $E$ and $Z$), but with the same $L$ as another particle species (but of different $E$ and different $Z$) gave different values of the efficiency, that is $\eta_L$ was not uniquely dependent on the value of $L$. Berger and Hajek (2008b) found the same for high-$E$, high-$Z$ particles found in space-radiation fields. Not surprisingly, this same result is found for OSL, as illustrated in Figure 4.21, using data from Sawakuchi *et al.* (2008c).

In view of these data one might wonder how a single $\eta(L)$ curve, such as shown in Figure 4.20, can be used to calibrate the OSL efficiencies for the space particles for a given dosimeter type and a given dosimetry method, given that each particle has its own unique $\eta(L)$ curve. The answer is the energy spectrum of the various particles to which the astronauts are exposed is dominated by energies in the range from 10 to 1000 MeV (e.g. Figure 4.3). Thus, for these energies, the spectrum of particles is dominated by low-LET values and the single $\eta(L)$ curve (Figure 4.20) is therefore justified since it can be regarded as the envelope of the low-LET efficiency values for each particle type. (This can easily be seen by comparing the theory curve of Figure 4.20 with the single-particle data of Figure 4.21. Those parts of the single-particle $\eta(L)$ curves that overlap with the theory curve from Figure 4.20 are those that correspond to those particle energies that dominate the space radiation spectrum.)

A comprehensive list of measured $\eta_L$ values for the OSL from $Al_2O_3$:C is given in Sawakuchi *et al.* (2008c).

#### 4.5.3.2 Fluence Response

We have seen how the shapes of OSL decay curves (and, by similar argument, the shapes of TL curves) are governed by the radial dose distribution within the tracks and the beta (or gamma) dose response. Imagine, therefore, a detector with just one charged-particle track traversing the material. The characteristics of the OSL signal are defined by the principles that have been discussed herein. A detector with two particle tracks gives an OSL signal of identical shape, but of twice the intensity; three particle tracks would yield a same-shaped signal that is three times as intense, and so on. In other words, despite the fact that the doses within the tracks are in the non-linear range of the beta (or gamma) dose response, the OSL dependence on the *net* dose of charged particles to the whole detector is simply linear – that is, the OSL dependence on charged-particle fluence is linear. This can be seen in Figure 4.22, where we show the fluence dependence of OSL from $Al_2O_3$:C for several charged particles.

The dose (i.e. the net charged-particle dose) is proportional to the particle fluence through Equation (4.5). The linear dependence will remain as long as each track is independent of its neighbor and no track interaction effects occur. We might ask, therefore, what changes can we expect to see if interaction effects do occur? As the fluence increases, the mean distance between the tracks' centers decreases. "Overlap" will occur when the wings of the radial dose distributions start to occupy the same volume of the dosimeter. For low-energy particles with narrow radial dose distributions such effects will occur for high fluences, whereas for high-energy particles with broader radial dose distributions overlap will occur at lower fluences. As we have discussed, the properties of the measured OSL signal depend upon the balance between the central core region, of small volume and high dose, and the outer regions of much smaller dose, but correspondingly much larger volume. Since the overlap effects will occur in the wings of the distributions one might expect that, for a given particle type, the OSL curve shape changes and non-linear effects will be seen first for those particles for which the radial dose distribution is broad.

The above effects were demonstrated by Gaza *et al.* (2006b) for OSL in $Al_2O_3$:C irradiated with (relatively) low-energy protons (1–4 MeV). For these particles supralinearity is not observed, but sub-linearity was seen to occur at higher fluence. The smaller the energy of the protons the higher the fluence at which such effect are observed, in accordance with the above expectations. Sawakuchi *et al.* (2008a) examined similar effects, but for particle types and energies more relevant to the space environment. These authors observed that weak supralinearity ($f > 1$) is seen for protons for fluence values above $\sim 5.6 \times 10^9$ cm$^{-2}$, whereas for heavier particles no supralinearity is observed. However, these authors also show that sub-linearity ($f < 1$) occurs for all particles at the approximately the same dose, $\sim$50 Gy. This dose also corresponds to that at which shape changes in the OSL decay curves also start to occur.

Estimating the average distance $d$ between the centers of two adjacent tracks for a particle fluence $\Phi$ using the relation $\pi(d/2)^2 = \Phi^{-1}$, Sawakuchi *et al.* (2008a) showed that for Fe particles (of $E = 120.4$ MeV/u and $L = 431.8$ keV/μm) 50 Gy corresponds to a fluence of $\sim 7 \times 10^7$ cm$^{-2}$ and $d = 1326$ nm, whereas for Ne particles (of $E = 371.9$ MeV/u and $L = 31.5$ keV/μm) the corresponding fluence is $\sim 9 \times 10^8$ cm$^{-2}$ and $d = 358$ nm. Thus, it is observed that the high-energy Fe ions ($E = 120.4$ MeV/u) interact at greater distances than the lower-energy Ne particles ($E = 371.9$ MeV/u).

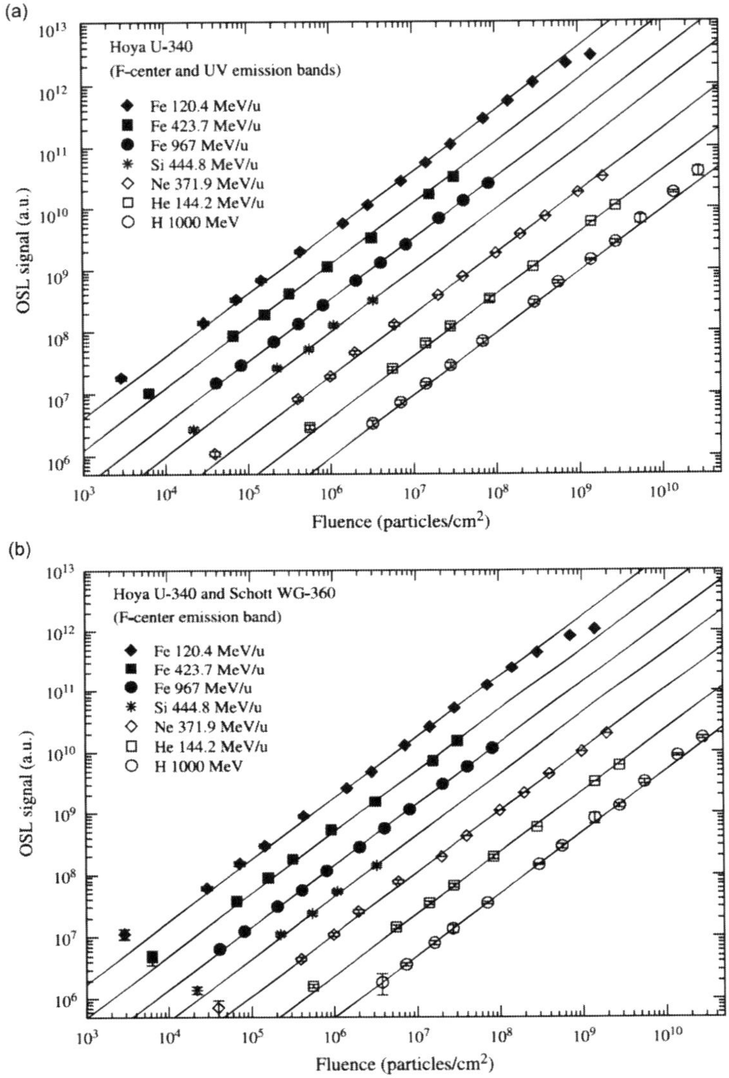


***Figure 4.22*** *OSL signal versus fluence from $Al_2O_3$:C Luxel detectors exposed to H, He, Ne, Si and Fe beams (energies indicated in the data legend): (a) The F-center and UV emission bands were detected with Hoya U-340 filters; (b) the F-center emission band was detected with combination of Hoya U-340 and Schott WG-360 filters. Reprinted from Radiation Measurement, Sawakuchi, G. O., et al. Optically stimulated luminescence fluence response of $Al_2O_3$:C dosimeters exposed to different types of radiation. Vol. 43, 450–454. Copyright (2008) with permission from Elsevier.*

### 4.5.4 OSL Response in Mixed Fields

Determining the absorbed dose $D$ and the dose equivalent $H$ when a dosimeter is exposed to a single particle type of LET $L$ is in principle a relatively straightforward process. First of all, one can use the LET dependence of the observed OSL (or TL) curve shape or other property, as exemplified in TL by Figure 4.13, to determine $L$. The OSL (or TL) efficiency $\eta_L$ can then be found using the appropriate curve $\eta(L)$ (e.g., Figure 4.15 or Figure 4.20). Finally, one can then use Equation (4.11) to obtain $D_L^{\text{OSL}}$ as

$$D_L^{\text{OSL}} = \frac{S_L^{\text{OSL}}}{\xi \eta_L}, \tag{4.14}$$

and the dose equivalent by

$$H_L^{\text{OSL}} = Q_L D_L^{\text{OSL}} = Q_L \frac{S_L^{\text{OSL}}}{\xi \eta_L}. \tag{4.15}$$

For irradiation with a space-radiation mixed field of different particles, however, the situation is less clear. The net dose equivalent for irradiation with multiple particle types was given in Equation (4.2) (namely, $\int_L Q(L)D(L)\mathrm{d}L$). The difficulty when attempting to evaluate $H$ with integrating dosimeters becomes immediately clear, namely that such dosimeters return a single number (e.g. OSL intensity or TL intensity) representing the net effect of the entire spectrum of particles types and energies incident on the dosimeter. Doses due to individual particles, $D(L)$, are not determined. As a result it becomes pertinent to inquire if a mean or effective LET value, $L_{\text{eff}}$, can be found, from which a single value quality factor $Q_{\text{eff}} = Q(L_{\text{eff}})$ and, similarly, a single value efficiency $\eta_{\text{eff}} = \eta(L_{\text{eff}})$ can each be determined. If so, then Equations (4.14) and (4.15) can be written as:

$$D^{\text{OSL}} = \frac{S^{\text{OSL}}}{\xi \eta_{\text{eff}}} \tag{4.16}$$

and

$$H^{\text{OSL}} = Q_{\text{eff}} D^{\text{OSL}} = \frac{Q_{\text{eff}} S^{\text{OSL}}}{\xi \eta_{\text{eff}}}, \tag{4.17}$$

where $\xi$ is the previously defined gamma-ray calibration factor for the OSL dosimeter.

This approach requires a method to determine the "effective LET." The favored method is to use a characteristic property from the measured OSL (or TL) curves that varies with LET and to compare the measured OSL (or TL) value for that property with the known dependence of this parameter on $L$. The value of $L$ so obtained is then defined as the "effective LET," $L_{\text{eff}}$. Such an approach was first proposed for TL from LiF:Mg,Ti by Vana and colleagues (Berger *et al.*, 2006a, b, c; Schöner, Vana and Fogger, 1999) who used the variation in the high temperature TL peak ratio (HTR) with the LET, measured the HTR for TLDs exposed to a space-radiation field, and proceeded to evaluate $H$ according to the procedure in Equation (4.17).

A similar approach can be devised for OSL. Here one can use the variation with $L$ of (a) the ratio of the initial CW-OSL peak intensity to the area under the whole OSL curve (Yukihara *et al.*, 2004b), (b) the ratio of time constants obtained by fitting the OSL decay curve to multiple exponentials (Gaza, Yukihara and McKeever, 2006), or (c) the ratio of the OSL signal emitted at UV wavelengths to that emitted at $F$-center wavelengths (determined from POSL measurements) (Yukihara *et al.*, 2006). An example of the latter result is shown in Figure 4.23a where it is compared with similar data (Figure 4.23b) for the HTR from LiF:Mg,Ti from Berger *et al.* (2006c).

An additional example for OSL curves is shown in Figure 4.24a from an analysis of CW-OSL curves from $Al_2O_3$:C single crystals in which the ratio $R$ of the initial CW-OSL peak intensity to the area under the whole OSL curve is calculated, as a function of the LET of the irradiating particle. The similarity in the shapes between Figure 4.23a, Figure 4.23b and Figure 4.24a is striking, with each data set showing an increase followed by a decrease; the peak value for the POSL emission data (Figure 4.23a) is reached at approximately 20 keV/μm, whereas the TL data (Figure 4.23b) peaks at about 80 keV/μm. The CW-OSL ratio (Figure 4.24a) peaks at about 30 keV/μm. Indeed, this general shape, as a function of the LET of the incident particles, is observed for different luminescence measurements from a variety of materials (Berger *et al.*, 2006c; Bilski, 2006b; Gaza, Yukihara and McKeever, 2004; Yasuda, 2001, 2002, 2006; Yukihara *et al.*, 2004b). Yasuda (2001, 2002, 2006) points out the similarity of this shape with that of the RBE or $Q$-factor as functions of $L$, for example, as given by the ICRP (Equation (3.16)).

Yasuda (2001, 2002, 2006) proposes combining different luminescence characteristics in order to mimic the shape of the RBE or $Q$ functions. In general, if one measures several different luminescence parameters, denoted here as $\varepsilon_i$, as a function of $L$, then one might try to approximate $Q(L)$ by:

$$Q(L) \approx \sum_i k_i \varepsilon_i(L), \tag{4.18}$$

where $k_i$ is the weighting factor of each function $\varepsilon_i(L)$. Since $H = \int_L Q(L)D(L)\mathrm{d}L$ we then have:

$$H \approx \int_L \left[\sum_i k_i \varepsilon_i(L)\right] D(L)\mathrm{d}L. \tag{4.19}$$

Yasuda (2001) illustrates the potential usefulness of this approach by choosing the efficiency, that is $\varepsilon_i(L) \equiv \eta_i(L)$, of two ($i = 1, 2$) different components of the TL glow curve from TLD-700. Thus, Equation (4.19) becomes:

$$\begin{aligned} H &= \int_L Q(L)D(L)\mathrm{d}L = \int_L [k_1\eta_1(L) + k_2\eta_2(L)]\, D(L)\mathrm{d}L \\ &= k_1 \int_L \eta_1(L)D(L)\mathrm{d}L + k_2 \int_L \eta_2(L)D(L)\mathrm{d}L \\ &= k_1 \frac{S_1}{\xi_1} + k_2 \frac{S_2}{\xi_2} \end{aligned} \tag{4.20}$$

where $S_1$ and $S_2$ are the two TL signals and $\xi_1$ and $\xi_2$ are the corresponding proportionality constants between the TL signals and the reference gamma dose. As Yasuda notes with respect to TL, selecting more than two ($i = 1, 2, 3\ldots$) signals may result in a greater accuracy through a better approximation to the $Q(L)$ function. Thus, Yasuda (2006) used a combination of the efficiency functions of TL from LiF and BeO and the efficiency of phosphate glass photoluminescence detectors to simulate the $Q(L)$ function. The method has been used for TL only so far and has not yet been demonstrated for OSL.

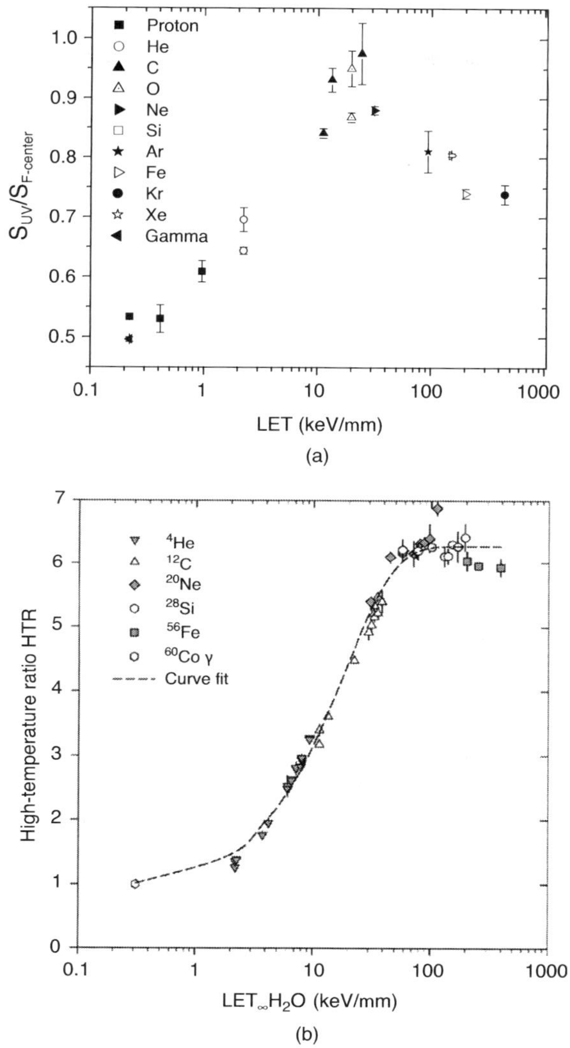


***Figure 4.23*** *(a) Ratio between the UV and F-center emission components of the OSL from $Al_2O_3$:C Luxel detectors for various particle/energy combinations as a function of the LET of the particle. The data were measured using pulsed OSL (POSL) as described in Yukihara and McKeever (2006a). (b) The HTR from TLD-700 (LiF:Mg,Ti) as a function of the LET of the irradiating particle. Figure (a) was reprinted from Radiation Measurement, Yukihara, E. G., Application of the optically stimulated luminescence (OSL) technique in space dosimetry. Vol. 41, 1126–1135. Copyright (2006) with permission from Elsevier, and Figure (b) was reprinted from Advances in Space Research, Berger, T. et al., Comparison of various techniques for the exact determination of absorbed dose in heavy ion fields using passive detectors. Vol. 37, 1716–1721. Copyright (2006) with permission from Elsevier.*

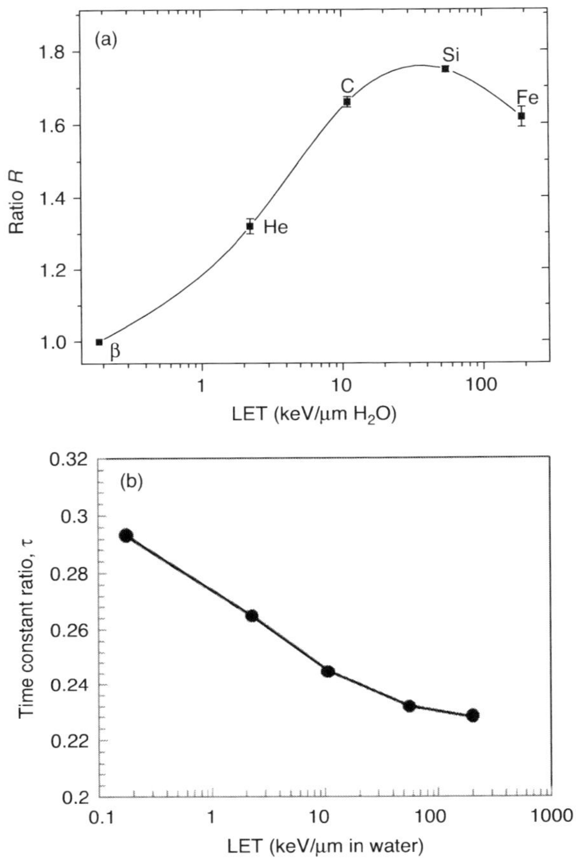


***Figure 4.24*** *(a) Ratio, R, of the initial intensity of a CW-OSL signal to the area under the whole CW-OSL curve, for $Al_2O_3$:C single crystals irradiated with different charged particles of various LET values. Reprinted from Yukihara et al. (2004b) with permission from Elsevier BV; (b) Ratio, τ, of the time constants extracted from a fit of the CW-OSL curve from $Al_2O_3$:C Luxel detectors to multiple exponentials. Figure (a) was reprinted from Radiation Measurements, Yukihara, E.G et al, Optically stimulated luminescence and thermoluminescence efficiencies for heavy charged particle irradiation in $Al_2O_3$:C. Vol. 38, 59–70. Copyright (2004) with permission from Elsevier.*

Despite the possibilities with the Yasuda method it has not been widely used. An alternative approach is to simply use Equations (4.16) and (4.17) and this leads to the question of how effective this approach is for the determination of $D$ and $H$, that is, how valid is the assumption $D \cong D^{\mathrm{OSL}}$ and $H \cong H^{\mathrm{OSL}}$. This question was addressed within the framework of the ICCHIBAN[1] project, which is an international collaboration to

[1] InterComparison of Cosmic Rays with Heavy Ion Beams at NIRS (National Institute of Radiological Sciences, Chiba, Japan).

compare results and performance of various forms of solid-state passive dosimeters and active detectors for potential applications in space dosimetry. The first experiment for passive detectors was ICCHIBAN-2 (May 2002) performed at the Heavy Ion Medical Accelarator in Chiba (HIMAC) at NIRS, in Chiba, Japan, followed later by ICCHIBAN-4 (May 2003), ICCHIBAN-6 (June, 2004) and ICCHIBAN-8 (September, 2005). The experiments are summarized by Yasuda *et al.* (2006) and by Uchihori and Benton (2004, 2008). Similar inter-laboratory comparison experiments have been performed at the NASA Space Radiation Laboratory (NSRL), New York, USA (NRSL-ICCHIBAN), the Loma Linda, CA, USA synchrotron (proton-ICCHIBAN), CERN-EU High Energy Reference Field, Switzerland (CERF-ICCHIBAN) and experiments onboard the International Space Station (Space-ICCHIBAN) (Uchihori *et al.*, 2006). ICCHIBAN-1, 3, -5 and -7 were reserved for active detectors (Uchihori *et al.*, 2002). The overall objectives of the ICCHIBAN experiments were to determine the response of space radiation dosimeters to heavy ions of charge and energy similar to those found in GCR, SPE and ERB fields, to compare the response and sensitivity of various space-radiation monitoring instruments to aid in reconciling differences in measurements made by various radiation instruments during space flight, and to establish and characterize a heavy ion "reference standard" against which space radiation instruments can be calibrated (Uchihori *et al.*, 2006).

An outcome of this extensive series of experiments has been the ability to determine the efficacy and limitations of the above-described method (viz. Equations (4.16) and (4.17)) to determine $D$ and $H$ for mixed charged-particle fields. To demonstrate this we describe here the results of a set of ICCHIBAN experiments and analyses based on OSL, with additional data for TL given in published reports (Uchihori and Benton, 2004, 2008). For the sake of discussion, the combined results of ICCHIBAN-2 and -4 are discussed here for the OSL detectors used in these experiments.

A summary of the exposures at the HIMAC facility in these two ICCHIBAN experiments is given in Table 4.5. Several combinations of radiation types were used in "blind" exposures, of unknown identity and dose. OSL detectors consisted of $Al_2O_3$:C single crystals (viz. TLD-500) and $Al_2O_3$:C Luxel detectors. The procedure adopted to determine the dose follows from the discussion of this chapter and is:

***Step 1:*** Measure the OSL following unknown exposure; determine the OSL signal $S^{OSL}$ (either from the area under the CW-OSL curve, from the initial intensity of the CW-OSL signal or from the POSL signal).

***Step 2:*** Irradiate with a reference gamma dose to water, $D_R$, and measure the corresponding OSL signal, $S_R^{OSL}$, and calculate the calibration factor $\xi$ using Equation (4.10), that is, $\xi = S_R^{OSL}/D_R$.

***Step 3:*** Determine the "effective LET" from a measurement of the chosen OSL characteristic. For example, if using POSL this might be the ratio between the UV and $F$-center emission bands in $Al_2O_3$:C, as shown in Figure 4.23a. If using CW-OSL, this might be the initial-intensity/total-area ratio $R$, as shown in Figure 4.24a. A difficulty with both of these measurements, however, is that the mean LET so obtained is not a unique value over some ranges. For example, using Figure 4.24a, an $R$-value of $\sim$1.75 could correspond to a particle with an LET of approximately 10 keV/μm or approximately 100 keV/μm. For this reason, following a similar suggestion by Yasuda (2001), Gaza, Yukihara

***Table 4.5*** *Delivered doses and fluences as part of the ICCHIBAN-2 and -4 experiments. Data from Uchihori and Benton (2004, 2008)*

| | | Particles and nominal energies[a] | | | | | | | |
|---|---|---|---|---|---|---|---|---|---|
| | Exposure | $^{60}$Co | $^{137}$Cs | $^{4}$He 150 MeV/u | $^{12}$C 400 MeV/u | $^{20}$Ne 400 MeV/u | $^{28}$Si 490 MeV/u | $^{56}$Fe 500 MeV/u | *D* |
| ICCHIBAN-2 | 1 | | | 100.0 | 1.0 | | 1.0 | 1.0 | 103.0 |
| | 2 | | | 2.0 | 2.0 | | 2.0 | 2.0 | 8.0 |
| | 3 | | | 5.0 | 5.0 | | 5.0 | 5.0 | 20.0 |
| | 4 | | | *1000* | *1000* | | *1000* | *1000* | |
| | 5 | | | 1.0 | 3.0 | | 7.0 | 12.0 | 23.0 |
| | 6 | | | | | | 5.0[b] | 1.0[c] | 6.0 |
| | 7[d] | | | | 2.0 | | | | 2.0 |
| | 8[e] | | | | | | | *5000* | |
| ICCHIBAN-4 | 9 | 27.8 | | | | | | | 27.8 |
| | 10 | | 27.8 | | | | | | 27.8 |
| | 11 | | | 25.0 | | | | | 25 |
| | 12 | 11.1 | | 1.0 | *1000* | *1000* | | *1000* | 12.6 |
| | 13 | 2.0 | | 2.0 | 2.0 | 2.0 | | 2.0 | 10.2 |
| | 14 | | | | *1000* | *1000* | | *1000* | |
| | 15 | | | | | | | 1.0 | 1.0 |
| | 16 | | | | 25.0 | | | | 25.0 |

[a] All doses are in mGy; all fluences (in italics) are in $cm^{-2}$.
[b] Behind 10 g/cm$^2$ Al
[c] Behind 5.4 g/cm$^2$ Al
[d] Behind 10 g/cm$^2$ Al and 5 g/cm$^2$ PMMA
[e] Spread-Out Bragg Peak (SOBP)

and McKeever (2004, 2006) fitted the OSL decay curve to a sum of exponentials, thus:

$$I_{\text{CW-OSL}} = \sum_k A_k e^{-t/t_k} \tag{4.21}$$

and defined a parameter $\tau = t_1/t_2$ from a two or three exponential fit. The parameter $\tau$ was found to be monotonically dependent on $L$ over the LET range used in the experiments, as illustrated in Figure 4.24b. In either case, one would use either Figure 4.23a, Figure 4.23b or Figure 4.24b, depending on what OSL parameter is being measured and, from the OSL data for the unknown exposure, determined the "effective LET," $L_{\text{eff}}$.

***Step 4:*** With $L_{\text{eff}}$ so determined, one would then use data such as those shown in Figure 4.20 to determine the corresponding value of the efficiency $\eta(L_{\text{eff}})$. The exact shape of the $\eta(L)$ curve depends upon exactly what part of the OSL data is used as the signal, $S$, and the exact form of the detector. Example shapes are shown in Figure 4.25. Clearly the efficiency can be >1 or <1 depending not only on LET, but also on the type of measurement (TL, CW-OSL, POSL) and the actual signal measured (initial intensity, area under the curve,

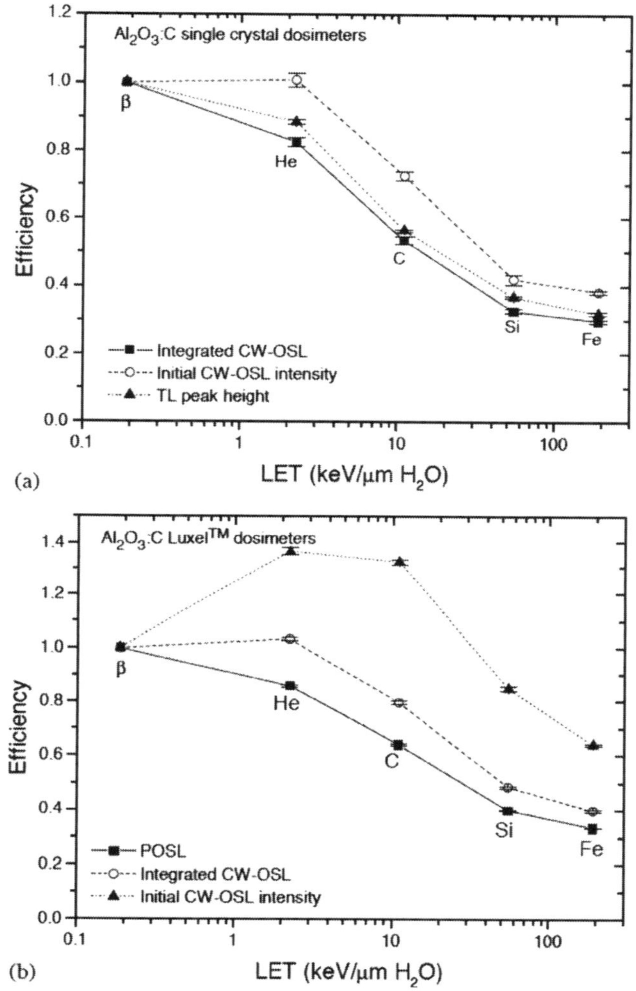


***Figure 4.25*** *Efficiencies of (a) $Al_2O_3$:C single crystal and (b) $Al_2O_3$:C Luxel detectors to high-energy heavy charged particles as a function of LET. The data show the variation in the shape of the $\eta(L)$ curve depending upon the type of measurement (TL, CW-OSL or POSL) and the signal (initial intensity, area under the curve). Reprinted from Radiation Measurements, Gaza, R, et al. The response of thermally and optically stimulated luminescence from $Al_2O_3$:C to high energy heavy charged particles. Vol 38, 417–420. Copyright (2004) with permission from Elsevier.*

POSL). It is important, therefore, to be consistent with the measurement type throughout the experiment.

***Step 5:*** With $L_{\mathrm{eff}}$ and $\eta_{\mathrm{eff}} = \eta(L_{\mathrm{eff}})$ so determined, the absorbed dose $D^{\mathrm{OSL}}$ is then estimated using Equation (4.16), and the dose equivalent $H^{\mathrm{OSL}}$ using Equation (4.17), where $Q_{\mathrm{eff}} = Q(L_{eff})$ is determined as discussed in the beginning of this section.

***Table 4.6*** *Comparison between determined absorbed doses and dose equivalents and the actual dose and dose equivalents using the procedure described in the text. Data from McKeever, Gaza and Yukihara (2004) and Gaza, Yukihara and McKeever (2006)*

| | Absorbed Dose (mGy) | | Dose Equivalent (mSv) | | Actual Doses | |
|---|---|---|---|---|---|---|
| Exposure[a] | *R*-method[b] | $\tau$-method[c] | *R*-method[b] | $\tau$-method[c] | *D* (mGy) | *H* (mSv) |
| 1 | $99.7 \pm 11$ | — | 99.7 | — | 103.0 | 138.8 |
| 5 | $19.57 \pm 14$ | — | 19.57 or 415.1[d] | — | 23.0 | 372.7 |
| 9 | — | $28.5 \pm 0.3$ | — | $28.5 \pm 0.3$ | 27.8 | 27.8 |
| 11 | — | $25.3 \pm 0.4$ | — | $25.3 \pm 0.4$ | 25.0 | 25.0 |
| 12 | — | $12.5 \pm 0.2$ | — | $12.5 \pm 0.3$ | 12.6 | 20.4 |
| 13 | — | $11.2 \pm 1.0$ | — | $35.7 \pm 1.0$ | 10.2 | 64.9 |
| 16 | — | $25.4 \pm 1.3$ | — | $25.4 \pm 1.3$ | 25.0 | 33.0 |

[a] Described in Table 4.5.
[b] CW-OSL; $Al_2O_3$:C Luxel detectors; initial intensity; *R*-method (viz. Figure 4.24a).
[c] CW-OSL; $Al_2O_3$:C Luxel detectors; initial intensity; $\tau$-method (viz. Figure 4.24b).
[d] Using the *R*-method gives an *R*-value of ~1.7. This yields a value for $L_{mean}$ of either ~7 keV/μm or ~200 keV/μm using data such as those shown in Figure 4.24a. These in turn give values for $Q_{mean} = Q(L_{mean})$ of either 1.0 or 21.2. Thus, $H = Q_{mean}D$ has estimated values of either 19.57 mSv or 415.1 mSv.

How effective is this procedure? In Table 4.6 we summarize some example results extracted from the exposures described in Table 4.5. The exposure package is shown in Figure 4.26 and consists of an array of $Al_2O_3$:C OSL single-crystal detectors, TLD chips (typically LiF: Mg,Ti) and $Al_2O_3$:C Luxel detectors. The whole is encapsulated in a black cover to keep it light tight.

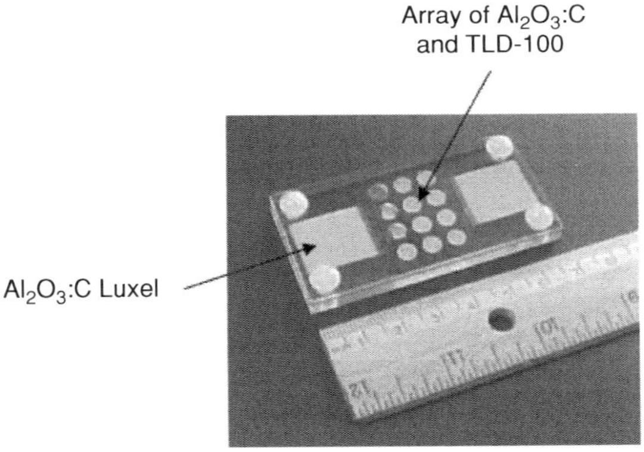


***Figure 4.26*** *Typical OSL detector configuration for ICCHIBAN exposures. In this case, 14 TL/OSL detectors are arranged in a polycarbonate holder of dimensions 7 cm × 3.5 cm area by 0.7 cm thickness. The ion beams traverse 0.2 cm of polycarbonate before reaching the detectors. In the picture shown there are nine unpolished $Al_2O_3$:C single crystals, three TLD-100 (LiF: Mg,Ti) chips, and two 1 cm × 1 cm $Al_2O_3$:C Luxel detectors. The whole is wrapped in black plastic to ensure that the detectors are kept in the dark during transportation and exposure.*

The results are revealing and are summarized thus:

- When exposed with low-LET single particles (exposure #11) the procedure described above is able to yield acceptable values for both $D$ and $H$, to within approximately $\pm 5\%$.
- When exposed to single particles with high-LET (exposure #16) the results are acceptable for $D$ ($\pm\sim 5\%$), but highly inaccurate for $H$ (factor of 1.5 in error) due to the uncertainty in deriving the value of $L_{\text{eff}}$ and the sensitivity of the $Q(L)$ function.
- When exposed to a mixed radiation field, dominated by low-LET particles (exposure #1 and #12), again the estimation for $D$ is acceptable ($\pm\sim 5\%$), but the estimate for $H$ induces significant errors (between $\sim 40\%$ and $\sim 100\%$).
- When exposed to a mixed radiation field dominated by high-LET particles (exposure #5) the estimate for $D$ is more uncertain ($\pm\sim 15\%$), but the estimate for $H$ now becomes very problematic. This is especially so with the $R$-method for which two values for $L_{\text{eff}}$ may be determined, only one of which yields a result anywhere close to the actual value.
- When exposed to a mixed radiation field with equal doses for each particle (exposure #13), the value for $D$ is less accurate ($\pm\sim 10\%$), and the value for $H$ is totally unacceptable (factor of $\sim 2$ in error).

Clearly, the method described is acceptable for $D$ in most cases (certainly to within $\sim 10\%$). This is illustrated in a different way in Figure 4.27, in which we show the values for $D$ obtained by several different OSL detectors, both without correction for the efficiency, that is using Equation (4.16) with $\eta_{\text{eff}} = 1$, and with the appropriate value for $\eta_{\text{eff}}$ for three different exposures. In each case the dose with correction is consistent among the detector types and closer to the actual delivered dose.

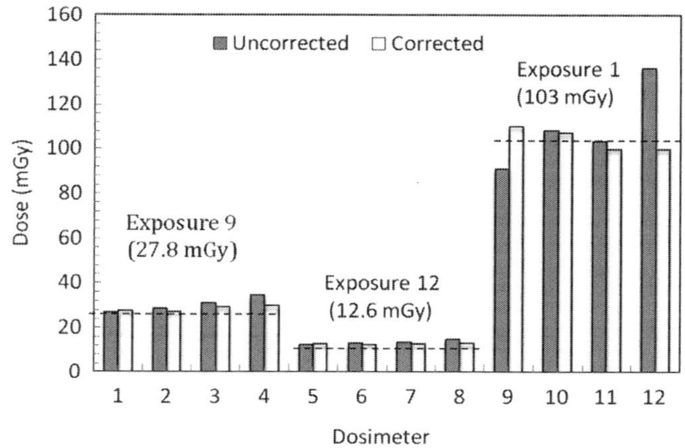


***Figure 4.27*** *Evaluated absorbed doses, corrected for the determined efficiency and uncorrected, for three exposures (with known doses indicated on the figure; refer to Table 4.5 for details). The data show that in all cases the efficiency correction yields improved dose estimates. The dosimeter types and readout methods are: (1,5,9) CW-OSL area,* $Al_2O_3$*:C single crystals; (2,6,10) CW-OSL initial intensity,* $Al_2O_3$*:C single crystals; (3,7,11) CW-OSL area,* $Al_2O_3$*:C Luxel detectors; (4,8,12) CW-OSL initial intensity,* $Al_2O_3$*:C Luxel detectors. Data from McKeever, Gaza and Yukihara (2004) and Gaza, Yukihara and McKeever (2006, 2008).*

However, the method is unacceptable for $H$ when the detectors are exposed to mixed fields due to a combination of the contributions to the dose equivalent from the high-LET particles, the decrease in efficiency of the OSL (and TL) detectors at high LET, the uncertainties with respect to the calculated value of $L_{\mathrm{eff}}$ and the sensitivity of the $Q(L)$ function to the value of $L$, for $L > 1$. What is needed to deal with this problem is a combination of passive detectors that will retain the sensitivity, accuracy and ease of use of OSL (and TL) detectors at low LET, but gain sensitivity for high-LET particles. To achieve this goal a combination of OSLDs and plastic nuclear track detectors (PNTDs) is required, as described in the next section.

## 4.6 Applications

### 4.6.1 Use of OSLDs (and TLDs) in Space-Radiation Fields

As we have observed, OSLDs alone are unable to determine the dose equivalent in cases where there are mixed fields with significant high-LET components. To do so requires the addition of a dosimeter that is sensitive to the high-LET contributions and can be used to determine the dose due to each component.

Over the period of the space program various combinations of passive dosimeters have been used for crew personal dosimeters (CPDs) and radiation area monitors (RAMs), as listed in Table 4.7. The favored type of personal dosimeter that meets the required criteria for current space applications is a plastic nuclear track detector (PNTD) in combination with TLDs and OSLDs, as recommended by the National Council on Radiation Protection and Measurements (NCRP, 2002) and illustrated in Figure 4.28.

PNTDs are plastics capable of registering the passage of a charged particle through the material via the processes of ionization and radiation damage in the path of the particle (Fleischer, Price and Walker, 1975; Knoll, 2000). Upon etching the particle tracks can be made visible using optical microscopes, or very lightly etched tracks can be observed using atomic force microscopes. The size, shape and depth of the resulting etch pits can be analyzed to yield information about the LET, energy and fluence of the various incident particles. Application of PNTDs in space dosimetry has been popular since the beginning of the space program and in modern times the most popular material is a polycarbonate known as CR-39, used to measure the LET spectrum above ~5 keV/μm (Benton and Benton, 2001).

Clearly, when using two different types of dosimeter with different registration efficiencies over different LET ranges it is necessary to devise a method of integrating the results from the two devices in such a way as to avoid double counting in the region of overlapping efficiencies. Thus, as we have seen, OSLDs have good efficiency for $L$ less than approximately 10 keV/μm with decreasing, albeit non-zero, efficiencies for larger values of $L$. As noted, however, CR-39 PNTDs lose registration efficiency below approximately 5 keV/μm, but retain good efficiency for $L$ values greater than this (e.g., Yasuda, 2009). Thus, one cannot simply add the results from the OSLD to those of the PNTD to obtain the absorbed dose and it becomes necessary to account for the region of $L$ in which the efficiencies overlap.

A proposal to achieve this was described by Doke *et al.* (1995), but currently the most commonly used procedure is that initially developed by the group at the University of

*Table 4.7* Dosimeters used in different spacecraft along with the measured quantities. Modified from Reitz (2006)

| Type | Program | Measured Quantity |
|---|---|---|
| *Crew Personal Dosimeter* | | |
| OSLD | STS, ISS | Absorbed dose |
| TLD-100 | All programs | Absorbed dose |
| TLD-300, -600, -700 | STS and ISS | Absorbed dose, neutron dose, dose equivalent |
| PNTD (Lexan, CR-39) | Apollo, Skylab, STS, Mir, ISS | LET-spectra, neutron dose |
| Fission Foils | Apollo, Skylab, STS, Mir | Neutron dose |
| *Area Dosimeters* | | |
| OSLD | STS, ISS | Absorbed dose |
| TLD-100 | STS, Mir, ISS | Absorbed dose |
| TLD-300, -600, -700 | Apollo, Skylab, STS, Mir | Dose equivalent |
| PNTD (Lexan, CR-39) | Apollo, STS | LET-, $Z$- and energy-spectra, neutron dose |
| Bubble detectors | Mir | Neutron dose |
| Fission foils | Apollo, Skylab, STS, Mir | Neutron dose |
| Nuclear emulsions | Apollo, Mir, STS | Neutrons, nuclear disintegrations |
| Silicon detectors | Mir, STS, ISS | LET, $Z$ and energy spectra |

*Glossary*

STS – Space transportation system (Shuttle)
ISS – International Space Station
PNTD – Plastic nuclear track detector
TLD – Thermoluminescence dosimeter
OSLD – Optically stimulated luminescence dosimeter
Mir – Mir Space Station
LET – Linear energy transfer
$Z$ – Electronic charge

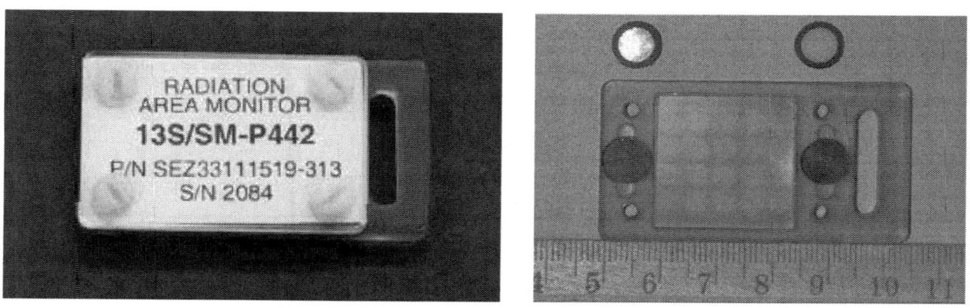


(a) (b)

***Figure 4.28*** *(a) US passive radiation area monitor (RAM) package; (b) the area dosimeter and the crew personal dosimeter (CPD) packages for US astronauts consist of an array of 20 TL detectors (TLD-100), two OSL detectors ($Al_2O_3$:C, in circular, black Teflon holders) and a CR-39 plastic PNTD film. Courtesy of NASA, used by permission.*

San Francisco (Benton, Benton and Frank, 2002a, b) and proceeds as follows. Firstly, one selects a demarcation value for $L$, below which the dose will be evaluated using the OSLDs (or TLDs), and above which the dose from the PNTD will be used. $L = 10$ keV/μm is a convenient value since $Q(L < 10\ \text{keV}/\mu\text{m}) = 1$ and thus numerically $D_{L<10\text{keV}/\mu\text{m}} = H_{L<10\text{keV}/\mu\text{m}}$ (with $D$ in Gy and $H$ in Sv). Thus:

$$D_{\text{Total}} = D^{\text{OSLD}}_{L<10\text{keV}/\mu\text{m}} + D^{\text{PNTD}}_{L\geq 10\text{keV}/\mu\text{m}} \tag{4.22}$$

and

$$H_{\text{Total}} = H^{\text{OSLD}}_{L<10\text{keV}/\mu\text{m}} + H^{\text{PNTD}}_{L\geq 10\text{keV}/\mu\text{m}}, \tag{4.23}$$

where the superscripts denote doses measured using the OSLD or PNTD, respectively.

Defining $D^{\text{OSL}}_{\text{raw}}$ as the dose indicated by the dosimeter without correction for the efficiency, that is, using Equation (4.16) with $\eta_{\text{eff}} = 1$,[1] then:

$$D^{\text{OSLD}}_{L<10\text{keV}/\mu\text{m}} = D^{\text{OSL}}_{\text{raw}} - \sum_{L_i=10\text{keV}/\mu\text{m}}^{1500\text{keV}/\mu\text{m}} \eta_{\text{i}} D^{\text{PNTD}}_i, \tag{4.24}$$

where $D^{\text{PNTD}}_i$ is the dose due to particles of type $i$, as determined with the CR-39 PNTD and the summation is over the LET range from 10 keV/μm to a selected upper limit of 1500 keV/μm. As previously defined, $\eta_i$ is the OSL efficiency for the type-$i$ particles (Figure 4.20). As already noted, the "effective LET" ($L_{\text{eff}}$) is a convolution of the radiation field properties and the OSLD properties. Since the efficiency of the OSLD decreases with increasing LET, the "effective LET" values determined by the methods described so far are biased towards low LET. This being the case, it is reasonable to correct the low-LET dose value, that is $D^{\text{OSLD}}_{L<10\text{keV}/\mu\text{m}}$, using the corresponding effective efficiency $\eta_{\text{eff}} = \eta(L_{\text{eff}})$, thus

$$D^{\text{corr}}_{L<10\text{keV}/\mu\text{m}} = \frac{D^{\text{OSLD}}_{L<10\text{keV}/\mu\text{m}}}{\eta_{\text{eff}}}. \tag{4.25}$$

To obtain the total absorbed dose, one then adds back to this quantity the dose due to the larger LET ($L \geq 10$ keV/μm) components, as measured by the PNTD, thus

$$D_{\text{Total}} = D^{\text{corr}}_{L<10\text{keV}/\mu\text{m}} + \sum_{L_i=10\text{keV}/\mu\text{m}}^{1500\text{keV}/\mu\text{m}} D^{\text{PNTD}}_i. \tag{4.26}$$

Finally, to calculate the dose equivalent one then uses the appropriate $Q_i$ value for each particle, and proceeds as follows:

$$H_{\text{Total}} = H_{L<10\text{keV}/\mu\text{m}} + \sum_{L_i=10\text{keV}/\mu\text{m}}^{1500\text{keV}/\mu\text{m}} Q_i D^{\text{PNTD}}_i, \tag{4.27}$$

with $H_{L<10\text{keV}/\mu\text{m}}$ in Sv numerically equal to $D^{\text{corr}}_{L<10\text{keV}/\mu\text{m}}$ in Gy.

---

[1] $D^{\text{OSL}}_{\text{raw}}$ is the reference gamma dose that would produce an OSL signal of the same magnitude as the exposure to HCPs. In the space dosimetry literature, this is sometimes referred to as "equivalent gamma-ray dose" (NCRP, 2002). However, here we avoid using this term to prevent confusion with other radiation protection quantities (e.g., equivalent dose $H_{\text{T}}$ and dose equivalent $H$). A better term is simply "indicated value" ($M$), as adopted in international standards and discussed in Chapter 3.

The choice of $L = 10$ keV/μm as the demarcation, or threshold, point was influenced by the fact that $Q\,(L < 10\ \text{keV}/\mu\text{m}) = 1$. Others have used different $L$ values and, since CR-39 has sensitivity down to ~5 keV/μm, any value of $L$ between 5 keV/μm and 10 keV/μm would be suitable.

From Equations (4.26) and (4.27), the average Quality Factor, $\overline{Q}$ is defined as:

$$\overline{Q} = \frac{H_{\text{Total}}}{D_{\text{Total}}}. \tag{4.28}$$

Similarly, mean dose rate and mean dose equivalent rate may be determined by dividing $D_{\text{Total}}$ and $H_{\text{Total}}$ by the exposure period, respectively.

An earlier suggested method by Doke *et al.* (1995) to combine TLDs and PNTDs differed from the above by defining a constant $\kappa$ such that Equation (4.24) is replaced by:

$$D^{\text{OSLD}}_{L<10\text{keV}/\mu\text{m}} = D^{\text{OSL}}_{\text{raw}} - \kappa \sum_{L_i=10\text{keV}/\mu\text{m}}^{1500\text{keV}/\mu\text{m}} D_i^{\text{PNTD}} \tag{4.29}$$

and thus

$$D_{\text{Total}} = D^{\text{OSL}}_{\text{raw}} + (1-\kappa) \sum_{L_i=10\text{keV}/\mu\text{m}}^{1500\text{keV}/\mu\text{m}} D_i^{\text{PNTD}}. \tag{4.30}$$

Similarly, for the dose equivalent:

$$H_{\text{Total}} = D^{\text{OSL}}_{\text{raw}} - \kappa \sum_{L_i=10\text{keV}/\mu\text{m}}^{1500\text{keV}/\mu\text{m}} D_i^{\text{PNTD}} + \sum_{L_i=10\text{keV}/\mu\text{m}}^{1500\text{keV}/\mu\text{m}} Q_i D_i^{\text{PNTD}}, \tag{4.31}$$

where $D^{\text{OSL}}_{\text{raw}} - \kappa \sum_{L_i=10\text{keV}/\mu\text{m}}^{1500\text{keV}/\mu\text{m}} D_i^{\text{PNTD}}$ in (Gy) is numerically equal to the dose equivalent (in Sv) for the dose range determined by the OSLD.

The validity of approximating $\sum_{L_i=10\text{keV}/\mu\text{m}}^{1500\text{keV}/\mu\text{m}} \eta_i D_i^{\text{PNTD}} \simeq \kappa \sum_{L_i=10\text{keV}/\mu\text{m}}^{1500\text{keV}/\mu\text{m}} D_i^{\text{PNTD}}$ has not been tested. Doke *et al.* (1995) used a demarcation value for $L$ of 3.5 keV/μm rather than 10 keV/μm and dealt only with TLDs rather than OSLDs.

### 4.6.2 Example Applications

As an example of the effectiveness of the above-described method, exposure #12 from ICCHIBAN-4 used CR-39 PNTDs to determine the dose from the high-$L$ components.[1] Whereas in Table 4.6 the dose equivalent for exposure #12 was determined to be (12.5 ± 0.03) Sv when using Equations (4.16) and (4.17), the value determined after application of Equations (4.22)–(4.23) is (18.3 ± 1.9) Sv, demonstrating a considerable improvement compared to the nominal actual value of $H_{\text{Total}} = 20.4$ Sv. However, apart from this one (unpublished) example, testing of the integration method for PNTDs and OSLDs (or TLDs) has not been thoroughly examined in ground-based experiments and room exists to test the method more thoroughly to determine uncertainties and efficacies.

[1] Courtesy Dr. Eric Benton, Oklahoma State University.

Despite the scarcity of ground-based testing the technique has, nevertheless, been applied on multiple occasions in space experiments and indeed for astronauts. Including the use of TLDs as well as OSLDs, examples of the use of the integration method can be found for experiments and personal dosimetry applications on Mir, the ISS and STS. Experiments in which the method has been used include MATROSHKA, DOBIES, MESSAGE and BRADOS, among others. Here we describe a few selected results both for astronaut dosimetry and for some scientific experiments.

#### 4.6.2.1 Personal and Area Dosimetry

LiF TLDs were used in conjunction with CR-39 PNTDs by Benton, Benton and Frank (2002a,b) for measuring absorbed dose and dose equivalent both inside and outside the Mir orbiting space station. Although TLDs were used in these measurements it is instructive to examine the results since they give a clear understanding of the distribution of dose and dose equivalent both inside and outside the spacecraft. We summarize the data in Table 4.8, where we see the change in the radiation environment as one goes from outside the spacecraft (with minimal shielding) to inside (with shielding in excess of 40 g/cm$^2$). Outside the spacecraft (dosimeters 1–16) the calculations of the dose and dose equivalent for the LET regions $<10$ and $\geq 10$ keV/μm show that the high-LET components contribute a significantly higher proportion of the dose equivalent than they do for the absorbed dose. Furthermore, the contribution for the high-LET components becomes greater the higher the degree of shielding. Inside the spacecraft the percentage of dose equivalent from the high-LET components ranges from 35% to over 60%, whereas for the dose the percentage contribution varies from only 5 to 11%.

The increase in the contribution from the high-LET components as a function of shielding is also illustrated in the value of the average Quality Factor, $\overline{Q}$. Using Equation (4.28) the average quality factor can be seen to vary from 1.00 to 1.43 as the shielding increases from 0.0052 g/cm$^2$ to 1.475 g/cm$^2$ (Al equivalent depth) for the exterior dosimeters. Inside the spacecraft, for shielding depths $>40$ g/cm$^2$, the average quality factor is approximately 2 or above.

Likewise, Doke *et al.* (1995) used their slightly different method (Equations (4.30) and (4.31)) to combine the readings from $Mg_2SiO_4$:Tb TL detectors and CR-39 PNTDs to assess the internal doses inside the Space Shuttle (on mission STS-47). Using a value of $L = 3.5$ keV/μm as the demarcation point between the low-LET dose and the high-LET dose, Doke *et al.* determined $\overline{Q}$ values varying from 1.23 to 1.49, with an average value for the dose rate of 153 μGy/d and an average dose equivalent rate of 207 μSv/d.

The NASA Johnson Space Center, Space Radiation Analysis Group has made similar measurements on several Shuttle flights and ISS missions. In these cases, however, both OSLDs and TLDs have been used to determine the low-LET dose and dose equivalent components. Example measurements and results are described by Zhou *et al.* (2007, 2007, 2009a, 2009b) and illustrative results are listed in Table 4.9. From these data the Space Radiation Analysis Group group calculated that the average contribution to the total dose equivalent is 68% from high-LET particles while 32% comes from low-LET components, with $\overline{Q}$ values ranging from 2.53 to 2.68, depending upon location. Doses determined by the OSLDs and TLDs are in close agreement.

An important note, however, is that the procedure adopted by NASA Space Radiation Analysis Group group does not include the efficiency correction step, given by Equation

**Table 4.8** *Results of integration of data from LiF TLDs and CR-39 PNTDs inside and outside Mir*

| Dosimeter | Location or shielding ($g/cm^2$ Al) | Total Dose Rate (μGy/d) | High-LET Dose Rate (μGy/d) | High-LET Contribution to Dose (%) | Total Dose Equivalent Rate (μSv/d) | High-LET Dose Equivalent Rate (μSv/d) | High-LET Contribution to Dose Equivalent (%) | Average Quality Factor |
|---|---|---|---|---|---|---|---|---|
| ***Interior*** | | | | | | | | |
| APD-1 | Core, door to engineer's cabin | 335 ± 16 | 29.5 ± 1.1 | 9 | 706 ± 23 | 400 ± 16 | 57 | 2.11 |
| APD-2 | Core, ceiling panel | 285 ± 14 | 30.9 ± 1.3 | 11 | 644 ± 21 | 390 ± 16 | 61 | 2.26 |
| APD-3 | Core, beneath command console | 367 ± 16 | 19.2 ± 0.3 | 5 | 537 ± 16 | 188 ± 4 | 35 | 1.47 |
| APD-4 | Adaptor, near window #14 | 311 ± 13 | 27.7 ± 0.3 | 9 | 661 ± 13 | 377 ± 4 | 57 | 2.12 |
| ***Exterior*** | | | | | | | | |
| 1 | 0.0052 | $(6.2 \pm 0.7) \times 10^6$ | $(6.25 \pm 0.25) \times 10^5$ | 10 | $(6.2 \pm 0.7) \times 10^6$ | $(6.01 \pm 0.24) \times 10^5$ | 9 | 1.00 |
| 2 | 0.1032 | $(2.07 \pm 0.21) \times 10^4$ | 270 ± 3 | 0.1 | $(2.12 \pm 0.21) \times 10^4$ | 717 ± 10 | 3 | 1.02 |
| 4 | 0.299 | $(4.1 \pm 0.4) \times 10^3$ | 148 ± 2 | 3.5 | $(4.44 \pm 0.46) \times 10^3$ | 457 ± 9 | 10 | 1.07 |
| 8 | 0.691 | $(1.67 \pm 0.17) \times 10^3$ | 63 ± 1 | 4 | $(1.80 \pm 0.19) \times 10^3$ | 201 ± 7 | 11 | 1.08 |
| 12 | 1.083 | $(1.22 \pm 0.13) \times 10^3$ | 65 ± 1 | 5 | $(1.38 \pm 0.15) \times 10^3$ | 225 ± 7 | 16 | 1.13 |
| 16 | 1.475 | $(1.07 \pm 0.11) \times 10^3$ | 155 ± 2 | 14 | $(1.54 \pm 0.16) \times 10^3$ | 621 ± 13 | 40 | 1.43 |

The low-LET component is defined as that below $L = 10$ keV/μm and the high-LET component is that for $L \geq 10$ keV/μm. The data were obtained during a 130.1-day period from 12th January to 22nd May 1997. Data from Benton, Benton and Frank (2002a,b).

***Table 4.9*** *Results of integration of data from LiF:Mg,Ti TLDs, $CaF_2$:Tm TLDs and $Al_2O_3$:C OSLDs with that from CR-39 PNTDs, from STS-121 between 4th and 17th July, 2006*

| Passive Radiation Detector (PRD) | Dosimeter Type | Low-LET Dose (mGy) | High-LET Dose (mGy) | Total Dose (mGy) | Total Dose Equivalent (mSv) | Average Quality Factor |
|---|---|---|---|---|---|---|
| PRD1 | TLD-100[a] | 2.18 | $0.36 \pm 0.04$ | $2.54 \pm 0.05$ | $6.44 \pm 0.47$ | $2.73 \pm 0.19$ |
| | TLD-300[b] | 2.10 | | $2.47 \pm 0.06$ | $6.37 \pm 0.47$ | $2.61 \pm 0.19$ |
| | OSLD-300s[c] | 2.08 | | $2.45 \pm 0.06$ | $6.35 \pm 0.47$ | $2.55 \pm 0.19$ |
| | OSLD-3s[d] | 2.24 | | $2.61 \pm 0.08$ | $6.50 \pm 0.48$ | $2.68 \pm 0.18$ |
| PRD2 | TLD-100 | 2.79 | $0.49 \pm 0.06$ | $3.28 \pm 0.07$ | $8.53 \pm 0.68$ | $2.61 \pm 0.21$ |
| | TLD-300 | 2.64 | | $3.13 \pm 0.08$ | $8.30 \pm 0.68$ | $2.50 \pm 0.22$ |
| | OSLD-300s | 2.70 | | $3.19 \pm 0.08$ | $8.38 \pm 0.68$ | $2.65 \pm 0.22$ |
| | OSLD-3s | 2.94 | | $3.43 \pm 0.11$ | $8.68 \pm 0.69$ | $2.53 \pm 0.20$ |
| PRD3 | TLD-100 | 2.79 | $0.48 \pm 0.04$ | $3.27 \pm 0.06$ | $8.38 \pm 0.49$ | $2.56 \pm 0.15$ |
| | TLD-300 | 2.58 | | $3.06 \pm 0.10$ | $8.17 \pm 0.49$ | $2.67 \pm 0.16$ |
| | OSLD-300s | 2.49 | | $2.97 \pm 0.07$ | $8.08 \pm 0.49$ | $2.72 \pm 0.16$ |
| | OSLD-3s | 2.70 | | $3.18 \pm 0.11$ | $8.29 \pm 0.49$ | $2.61 \pm 0.15$ |
| PRD4 | TLD-100 | 2.07 | $0.38 \pm 0.04$ | $2.45 \pm 0.05$ | $6.48 \pm 0.44$ | $2.65 \pm 0.18$ |
| | TLD-300 | 1.93 | | $2.31 \pm 0.06$ | $6.34 \pm 0.44$ | $2.75 \pm 0.19$ |
| | OSLD-300s | 2.09 | | $2.46 \pm 0.06$ | $6.49 \pm 0.44$ | $2.64 \pm 0.18$ |
| | OSLD-3s | 2.25 | | $2.63 \pm 0.11$ | $6.66 \pm 0.44$ | $2.53 \pm 0.17$ |
| PRD5 | TLD-100 | 2.64 | $0.47 \pm 0.04$ | $3.10 \pm 0.06$ | $8.10 \pm 0.47$ | $2.61 \pm 0.15$ |
| | TLD-300 | 2.66 | | $3.13 \pm 0.07$ | $8.12 \pm 0.47$ | $2.20 \pm 0.15$ |
| | OSLD-300s | 2.42 | | $2.89 \pm 0.07$ | $7.88 \pm 0.47$ | $2.73 \pm 0.16$ |
| | OSLD-3s | 2.63 | | $3.09 \pm 0.10$ | $8.09 \pm 0.48$ | $2.61 \pm 0.16$ |
| PRD6 | TLD-100 | 2.39 | $0.41 \pm 0.05$ | $2.80 \pm 0.07$ | $7.12 \pm 0.64$ | $2.55 \pm 0.23$ |
| | TLD-300 | 2.17 | | $2.58 \pm 0.10$ | $6.91 \pm 0.64$ | $2.68 \pm 0.25$ |
| | OSLD-300s | 2.28 | | $2.69 \pm 0.07$ | $7.02 \pm 0.64$ | $2.61 \pm 0.24$ |
| | OSLD-3s | 2.49 | | $2.89 \pm 0.10$ | $7.22 \pm 0.64$ | $2.50 \pm 0.22$ |

The low-LET component is defined as that below $L = 10$ keV/μm and the high-LET component is that for $L \geq 10$ keV/μm. Data from Zhou *et al.* (2009b).

[a] LiF:Mg,Ti

[b] $CaF_2$:Tm

[c] $Al_2O_3$:C; CW-OSL, first 300 s integration

[d] $Al_2O_3$:C; CW-OSL, first 3 s integration

(4.25). Thus, the expression used by them for the total absorbed dose is Equation (4.26), but with $D^{\mathrm{OSLD}}_{L<10\mathrm{keV}/\mu\mathrm{m}}$ instead of $D^{\mathrm{corr}}_{L<10\mathrm{keV}/\mu\mathrm{m}}$. Tests to evaluate the effectiveness of this approach have not been performed. This last issue brings up an important point regarding the integration method as applied to dose measurements in space, namely that the dose is not independently known and there are no standard dosimeters with which comparisons can be made. A true understanding of the efficacy of the method has therefore yet to be demonstrated.

#### 4.6.2.2 *Anthropomorphic Phantom Experiments*

Since early in the space program internal-organ dose estimates have been made using anthropomorphic phantoms equipped with various types of passive dosimeter. An example

(a) (b) (c) (d)

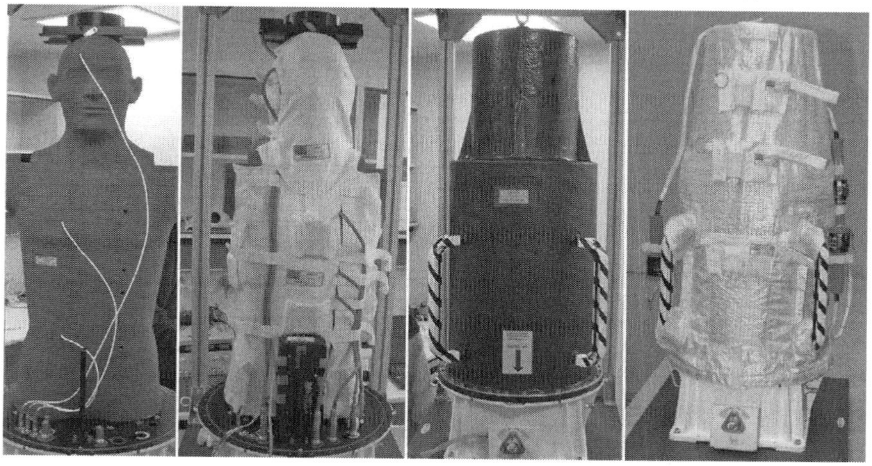

***Figure 4.29*** *The MATROSHKA anthropomorphic phantom. Left to right: (a) upper torso equipped with active and passive dosimeters. The torso is cut into 33 horizontal slices, each 25 mm thick. The base structure houses electronics for data acquisition (from the active dosimeters as well as temperature and pressure). Each slice is equipped with polyethylene tubes holding in total, over all slices, 4800 TL and OSL detectors; (b) torso inside the poncho and hood, equipped with passive detectors for skin dose measurements; (c) carbon fiber "suit," used to simulate an astronaut's space suit; (d) final package, ready for launch, with multilayer insulation for thermal protection. The whole package was mounted outside the Zvezda Module on the ISS for 539 days until its retrieval in August, 2005. The dosimeters were removed and returned to Earth for analysis by the collaborating laboratories. Reprinted from Radiation Research, Reitz, G., et al. Astronaut's organ doses inferred from measurements in a human phantom outside the International Space Station. Vol. 171, 225–235. Copyright (2009) with permission from Radiation Research Society.*

that used combined TLD and PNTD results was described by Yasuda *et al.* (2000) to measure absorbed dose and dose equivalent at various locations inside a phantom (and thereby estimate these quantities for human organs). Similar results have been described more recently for the so-called MATROSHKA phantom (Reitz *et al.*, 2009; Zhou *et al.*, 2010). In the latter experiment surface and internal (organ) doses were determined using a combination of various TLDs, OSLDs and PNTDs. Figure 4.29 shows the MATROSHKA phantom in different stages of preparation, as described in the figure caption. Although OSLDs were included in these experiments, thus far only combinations of TLD and PNTD data have been published. For completeness, some of these results are listed in Table 4.10. Previously unpublished OSLD results are listed in Table 4.11. As noted in an earlier section, the efficiency of OSL to heavy charged particles is dependent, among other factors, on the emission wavelengths monitored when measuring the OSL signal. Table 4.11 shows the values for absorbed dose from the OSLDs, obtained using the simple correction method of Equation (4.16), and using two combinations of detection filters in front of the PMT, Hoya U-340 or Hoya U-340 + Schott WG-360 (Yukihara and McKeever, 2006b). Using this approach, both filter combinations give the same absorbed dose values, within uncertainties.

***Table 4.10*** *Selected integration results for the MATROSHKA anthropomorphic phantom using LiF TLDs and CR-39 PNTDs*

| Tissue/Location | Low-LET Dose Rate[a] (mGy/d) | High-LET Dose Rate (mGy/d) | Total Dose Rate (mGy/d) | High-LET Dose Equivalent Rate (mSv/d) | Total Dose Equivalent Rate (mSv/d) | Average Quality Factor |
|---|---|---|---|---|---|---|
| Skin | 0.879 | 0.065 | 0.944 | 0.763 | 1.642 | 1.74 |
| Eye | 0.505 | 0.036 | 0.541 | 0.461 | 0.966 | 1.78 |
| Stomach | 0.211 | 0.031 | 0.242 | 0.407 | 0.618 | 2.55 |
| Poncho[b] | 0.536 | 0.065 | 0.600 | 0.762 | 1.298 | 2.16 |

The phantom was fitted with a Nomex travel suit (or "poncho") and inserted in a carbon fiber jacket (0.5 g/cm$^2$). The exposure was for 539 days on the outside of the ISS. Data from Reitz *et al.* (2009).
[a] The low-LET dose equivalent rate is numerically equal to the low-LET dose rate, but in units of Sv/d.
[b] External Nomex travel jacket.

When combined with PNTDs (Table 4.12), however, using the method outlined in Equations (4.22)–(4.28), the values obtained for the total dose are different from those obtained with the simple correction method illustrating the limitations of the simple procedure. In Table 4.12 we see that the mean quality factor values (cf. Equation (4.28)) range from 1.46 to 1.64, whereas the values for the quality factor in the high-LET (>5 keV/μm) region ranged from 5.69 to 6.17. The OSL efficiency terms (referenced to $^{60}$Co dose to water) ranged from 0.917 to 1.035.

***Table 4.11*** *Unpublished results (courtesy of Gabriel O. Sawakuchi, Carleton University, Canada) from OSL measurements from the MATROSHKA-1 experiment, for reference location inside the ISS and for different locations on the phantom*

| | Dose[a] | |
|---|---|---|
| Organ | U-340 + WG-360 (mGy) | U-340 (mGy) |
| ISS Reference 1[b] | 100.1 ± 0.8 | 98 ± 3 |
| ISS Reference 2[b] | 100 ± 2 | 98 ± 3 |
| Top of Head | 311 ± 25 | 308 ± 22 |
| Left Eye | 164 ± 7 | 156 ± 9 |
| Left Lung | 129.7 ± 1.3 | 127 ± 5 |
| Stomach | 136 ± 3 | 123 ± 3 |
| Left Kidney | 117 ± 2 | 118 ± 2 |
| Intestine | 122 ± 2 | 116 ± 3 |

The OSL data were obtained using Luxel detectors, readout with green stimulation using a Risø TL/OSL reader, and two different filter combinations in front of the PMT, Hoya U-340 or Hoya U-340 + Schott WG-360 (Yukihara and McKeever, 2006b). The doses listed are those obtained using the simple correction procedure of Equation (4.16).
[a] Corrected OSL doses using Equation (4.16).
[b] Reference dosimeters inside the ISS.

***Table 4.12*** *Unpublished results from OSL and PNTD measurements from the MATROSHKA-1 experiment, for reference locations in the ISS and for the Left Eye from the phantom*

| | Dose and Dose Equivalent[a] | | | | | | |
|---|---|---|---|---|---|---|---|
| Location (OSL filter) | Dose <5 keV/μm (mGy)[b] | Dose >5 keV/μm (mGy) | Total Dose (mGy) | Dose Equiv. <5 keV/μm (mSv) | Dose Equiv. >5 keV/μm (mSv) | Total Dose Equiv. (mSv) | Mean Quality Factor[c] |
| ISS Ref. 1 (U-340 + WG-360) | 91.9 ± 2.1 | 11.1 ± 0.3 | 103.0 ± 2.2 | 91.9 ± 2.1 | 69 ± 2 | 161 ± 2 | 1.56 |
| ISS Ref. 1 (U-340 blue + UV) | 91.8 ± 2.9 | 11.1 ± 0.3 | 102.9 ± 2.9 | 91.8 ± 2.9 | 69 ± 2 | 161 ± 3 | 1.56 |
| ISS Ref. 2 (U-340 + WG-360 blue) | 99.4 ± 1.6 | 11.1 ± 0.3 | 110.5 ± 1.6 | 99.4 ± 1.6 | 69 ± 2 | 168 ± 2 | 1.46 |
| ISS Ref. 2 (U-340) | 89.3 ± 4.4 | 11.1 ± 0.3 | 100.4 ± 4.4 | 89.3 ± 4.4 | 69 ± 2 | 158 ± 4 | 1.58 |
| Left Eye (U-340 + WG-360) | 134 ± 8 | 21.1 ± 0.7 | 155 ± 8 | 134 ± 8 | 121 ± 4 | 254 ± 8 | 1.64 |
| Left Eye (U-340) | 131 ± 9 | 21.1 ± 0.7 | 152 ± 9 | 131 ± 9 | 121 ± 4 | 252 ± 10 | 1.64 |

The OSL data were obtained using Luxel detector, readout with green stimulation using a Risø TL/OSL reader, and two different emission filter combinations in front of the PMT, Hoya U-340 or Hoya U-340 + Schott WG-360 (Yukihara and McKeever, 2006b). The dose and dose equivalent values are those obtained using the correction procedures described by Equations (4.22)–(4.28). Repeated values of dose and dose equivalent >5 keV/μm are based on the same PNTD data. (The OSL data are listed courtesy of G. O. Sawakuchi and the PNTD data are listed courtesy of E.R. Benton.) The data are to be compared with the data in Table 4.10 for different organs in the same phantom, obtained using TLDs and PNTDs.
[a] Calculated using the procedure described by Equations (4.22)–(4.27), with a threshold LET of 5 keV/μm.
[b] Equation (4.25), but with a 5 keV/μm threshold LET rather than 10 keV/μm.
[c] Total Dose Equivalent divided by Total Dose, that is Equation (4.28).

#### *4.6.2.3 Other Experiments*

The Belgian Nuclear Research Center, SCK-CEN, has flown a series of biological experiments on the ISS over a number of years. The experiments were known as MESSAGE 1 and 2, MOBILIZATION and DOBIES. As part of each experiment, each of which was designed to study the effects of space radiation on microbial organisms, the absorbed dose and dose equivalent were determined using TLDs, OSLDs and PNTDs. The effort was a collaboration between several European and US laboratories (Goossens *et al.*, 2006; Vanhavere *et al.*, 2008). As expected, the TLDs and OSLDs gave different results if uncorrected for the different efficiencies of the detectors to heavy charged particles. When corrected for the different efficiencies and combined with the PNTD data using Equations (4.22)–(4.27), with a threshold LET value of 5 keV/μm, the low-LET dose rate was determined to be 157 μGy/d and the high-LET dose rate was 23 μGy/d, for a total of 180 μGy/d.

The Institute of Biomedical Physics of the Russian Academy of Sciences in collaboration with European and US laboratories conducted an experiment known as BRADOS in the

Russian segment of the ISS (Benton *et al.*, 2005; Hajek *et al.*, 2006). OSLDs, PNTDs and TLDs were located in a stack at different shielding depths. As discussed in earlier sections of this chapter the TL peak-height ratio for LiF:Mg,Ti TLDs (the HTR method) and the CW-OSL curves shape both change as a function of the LET. Since the TLDs and OSLDs are most sensitive to the low-LET particles, the data obtained for the OSL and TL dosimeters showed that the TL and OSL curve shapes were invariant with depth, implying that the LET spectrum varied little as a function of depth. Conversely, the PNTD data indicated that there was a significant alteration in the high-LET component, with the proportion of very-high-LET particles ($>100$ keV/μm) increasing with shielding depth due to the creation of heavy secondary ions.

Combining the TLD/OSLD data with the PNTD data leads to an estimation of the total dose and dose equivalent. The total dose rate in the integrated OSLD/PNTD data decreased by 13.3% from the least-shielded to the most-shielded position, with a 14.9% decrease in the OSLD dose rate and a 12.7% decrease in the PNTD dose rate. This is in keeping with the expected attenuation of the charged particle fluence with shielding depth along with the production of high-LET secondary particles.

## 4.7 Future Directions

One of the clear messages that emerges from the above summary of ground-based and space-based experiments using OSLDs is that more ground-based studies are needed in order to properly assess the efficacy of the combined OSLD/PNTD approach as described herein. There have actually been more space-based applications of the technique (with both TLDs and OSLDs) than ground-based tests. Although this approach appears to be the best one so far suggested to evaluate the dose and dose equivalents of the low-LET and high-LET components, in accordance with the NCRP recommendations (NCRP, 2002), the method has not yet received the thorough evaluation in known radiation fields it requires and deserves, apart from the limited experiments already described in this chapter.

Another message that emerges is that for long-duration flights an on-board estimation of the crew exposures is a necessity. It is not possible when operating such long flights (long-term ISS expeditions and manned flights to the Moon and Mars) to wait until the crews return to Earth before their personal doses are evaluated. Mission managers need to know if or when exposures reach maximum permissible limits so that alternate crew duties can be assigned. To achieve this, on-board readers are needed and it is here that OSL appears to have a strong future. Although on-board readout of passive dosimeters has been performed using specially designed TLDs, using the Hungarian Pille readers (Apáthy *et al.*, 2002), it is OSL that would appears to have a technological advantage because it is an all-optical technique not requiring heating of the sample and the associated equipment complexity.

The requirements of an on-board OSL reader and dosimetry system are that it needs to have a sensitivity to dose down to $\sim$0.01 mGy, be able to accumulate dose over periods of several years without loss of performance while itself being exposed to the radiation environment, be characterized by a method of fast and simple readout, have an associated dosimeter that is easy to wear, handle and read (it must be light, small and unobtrusive), and the reader must be lightweight and consume little electrical power during use, with preferably zero power drainage during quiescent periods. In principle, OSL can meet all of these requirements. However, although some advances have been made, there are no such developments specifically for this purpose that are taking place at the time of writing.

Previous developments include those of the group at the University of Montpellier who developed a prototype integrated OSL dosimetry system for potential in-flight operation (Dusseau *et al.*, 2000; Plattard *et al.*, 2002). Miniaturized systems based on OSL from rare-earth-doped sulfide materials (CaS, SrS, MgS and BaS) were developed by the Montpellier group, using an LED as a stimulation source and aGaAsP photodiode as a light detector. The dosimeter and the readout system were integrated such that they would have to be worn by the astronauts as a single unit in the dosimeter badge. Sensitivity ranges were from 10 μGy–1 mGy depending on detector material and system design. Linearity was reported to be over 4 to 7 decades of dose. A similar concept is described by Chen, Poochinda and Stoebe (2006) who discuss an integrated dosimeter based on $Al_2O_3$:C with a GaN p-i-n diode as a light detector. To date, however, no such device has been manufactured.

In a related effort, miniaturized, robotic OSL systems are being developed for space applications, but in an entirely different area. The need for detailed chronology of the martian surface has driven the development of prototype OSL systems for dating sediments on the surface of Mars (Bøtter-Jensen, Thomsen and Jain, 2010; McKeever *et al.*, 2003). Such systems need to be small, compact, lightweight and semi-autonomous and it is likely that adaptation of such systems could easily fill the need for on-board OSL readers on manned spacecraft for reading crew personal dosimeters during long duration space flight.

# 5
# Medical Dosimetry

## 5.1 Introduction

Since the discovery of X-rays, radioactivity and sub-atomic particles in the late nineteenth and early twentieth centuries, the use of radiation in medicine for diagnosis or therapy has evolved dramatically and achieved a level of significant sophistication. In diagnostic radiology, new technologies and the rapid evolution in computer power led to the development of X-ray computed tomography (CT), which allows the internal imaging of a patient's body, slice-by-slice, using a collimated X-ray beam that rotates around the patient. In radiation therapy, patients are now routinely treated with megavoltage X-ray and electron beams. Proton beams with energies up to 250 MeV are used in many treatment centers worldwide, and a few centers are using carbon beams with therapeutical energies up to 430 MeV per atomic mass unit (u) (MeV/u) (Kitagawa *et al.*, 2010). By the end of 2009, more than 75 000 patients had already been treated with protons or carbon ions (PTCOG, 2010). Integrating the technical advances in both radiation imaging and therapy, radiation treatment planning systems allow clinicians to make extensive use of computers to manipulate a three-dimensional representation of the patient and develop an optimized radiation therapy treatment plan based on three-dimensional dose distribution calculations.

In radiation therapy, the combined advances in medical imaging, radiation delivery techniques and treatment planning have provided us with an increasing ability to characterize tumors and conform the dose distributions to complex shapes, allowing an escalation of the dose to the target while sparing healthy tissues and critical organs or structures. In intensity-modulated radiation therapy (IMRT), for example, the patient is irradiated from multiple directions with megavoltage X-ray beams of non-uniform intensities determined by a three-dimensional treatment planning algorithm (Intensity Modulated Radiation Therapy Collaborative Working Group, 2001; Williams, 2003). A similar approach for proton therapy, called intensity modulated proton therapy (IMPT), has also been proposed and

*Optically Stimulated Luminescence: Fundamentals and Applications* Eduardo G. Yukihara and Stephen W. S. McKeever

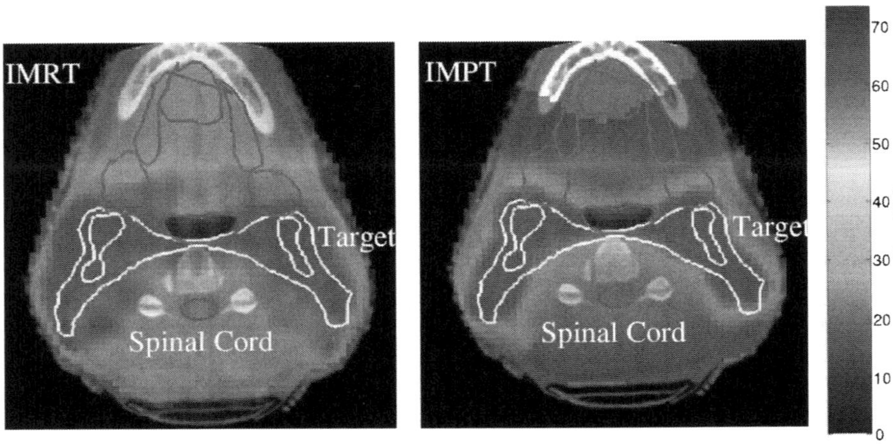


***Figure 5.1*** *Example of calculated dose distributions (plan) for a nasopharynx case using X-ray intensity modulated radiation therapy (IMRT) and intensity modulated proton therapy (IMPT). The target is indicated by the white contour lines, whereas critical structures such as the spinal cord are indicated by the grey contour lines. The steep dose gradients separating the high and low dose regions are particularly noticeable in the IMPT plan. Reprinted from Physics in Medicine and Biology, Bortfeld, IMRT: a review and preview. R363–R379. Copyright (2006) with permission from IOP Publishing Ltd.*

investigated (Bortfeld, 2006; Flynn *et al.*, 2007, 2008; Lomax, 1999). In both cases, the resultant dose distributions are characterized by steep dose gradients separating high and low dose regions (see Figure 5.1).

The precision and accuracy of modern radiation therapy techniques relies on robust and comprehensive quality assurance (QA) and quality control (QC).[1] Sophisticated dosimetry tools are required for beam calibration and characterization, verification of dose distributions delivered to patients, identification of equipment malfunctions or human error, and so on. The International Atomic Energy Agency (IAEA) states in its Code of Practice on determination of absorbed dose in external beam radiotherapy: "Modern radiotherapy has confirmed . . . the need for high accuracy in dose delivery if new techniques, including dose escalation in 3-D conformal radiotherapy, are to be applied. Emerging technologies in radiotherapy, for example modern diagnostic tools for the determination of the target volume, 3-D commercial treatment planning systems and advanced accelerators for irradiation, can only be fully utilized if there is high accuracy in dose determination and delivery" (IAEA, 2000).

[1] *Quality assurance* (QA) and *quality control* (QC) are specific terms used in quality management. As summarized by the ICRU Report 76 on Measurement Quality Assurance for Ionizing Radiation Dosimetry (ICRU, 2006), *quality assurance* is defined as "those planned and systematic actions necessary to provide adequate confidence that a structure, system, procedure, or component will perform satisfactorily and comply with agreed-upon standards." *Quality control* is defined as "the set of operations intended to maintain or improve quality." As discussed by Thwaites, Mijnheer and Mills (2005), "quality control is one part of overall quality assurance. . .concerned with operational techniques and activities used to check that quality requirements are met and to adjust and correct performance if the requirements are found not to have been met." For example, ongoing, periodic evaluation of radiation diagnostic and therapy equipment is quality control (AAPM, 2002).

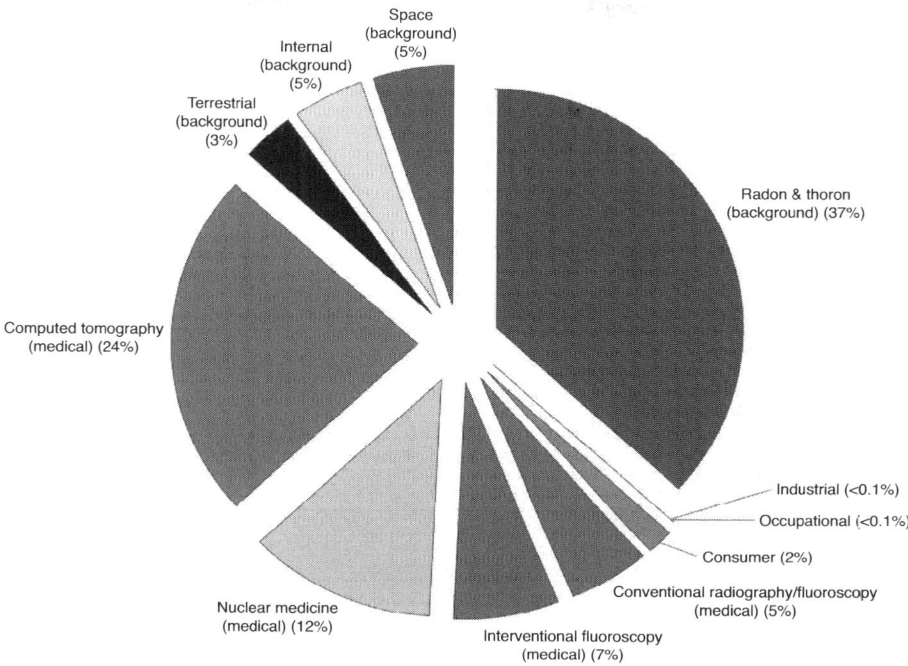


***Figure 5.2*** *Contribution of different sources of radiation exposure to the collective dose to the United States population in 2006. Reprinted from NCRP Report 160; Ionizing Radiation Exposure of the Population of the United States. Copyright (2009) with permission from National Council on Radiation Protection and Measurements, http://NCRPonline.org.*

In diagnostic radiology, appropriate dosimetry tools are also important to compare, optimize and guarantee the quality of procedures while minimizing the dose to the patient, because of the significant and increasing contribution from medical practices to the overall radiation exposure to the population. Until 2000, diagnostic medical X-ray examinations were the second largest contribution to individual exposure worldwide (~14% of the total dose), with an increasing trend due to the growth in importance of medical procedures, including X-ray CT and fluoroscopy (United Nations, 2000). However, a new report from the United States National Council on Radiation Protection and Measurements (NCRP) identified medical exposure as contributing to nearly half of the total radiation exposure to the US population in 2006 (see Figure 5.2) (NCRP, 2009).

Monitoring these doses is also becoming more difficult. In X-ray CT, for example, the introduction of multi-slice scanners with increasing numbers of detector rows, variable design and different approaches to exposure control, creates a complicated dependence between patient dose and scan parameters. To a certain extent, this increasing complexity compromises the ability of conventional dosimeters and dose descriptors to provide an understanding of the dose distributions and their relationship with scan parameters and technique (Dixon, 2006; Dixon and Ballard, 2007). Convenient techniques are sometimes required to measure not only the integrated dose but also the dose profile along the scanning axis.

**Table 5.1** *Examples of studies of application of OSL detectors to medical dosimetry*

| Application | Material | References |
|---|---|---|
| Radiation therapy | $Al_2O_3$:C | (Aznar *et al.*, 2004; Chen *et al.*, 2009; Jursinic, 2007, 2010; Meeks *et al.*, 2002; Miller and Murphy, 2007; Mobit, Agyingi and Sandison, 2006; Reft, 2009; Schembri and Heijmen, 2007; Viamonte *et al.*, 2008; Yukihara *et al.*, 2005, 2008a) |
| Proton and carbon therapy | $Al_2O_3$:C | (Andersen *et al.*, 2007; Edmund *et al.*, 2007; Reft, 2009; Sawakuchi *et al.*, 2010; Yukihara *et al.*, 2010) |
| IMRT | SrS:Ce,Sm | (Idri *et al.*, 2004) |
| Brachytherapy | $Al_2O_3$:C | (Andersen *et al.*, 2009a, 2009b) |
| Mammography | $Al_2O_3$:C | (Aznar *et al.*, 2005) |
| X-ray CT | $Al_2O_3$:C | (Bauhs *et al.*, 2007; Klein, 2008; Peakheart, 2006; Ruan *et al.*, 2010; Yukihara *et al.*, 2009) |
| Radiosurgery (Gamma Knife™) | $Al_2O_3$:C | (Gasparian *et al.*, 2010) |
| Fluoroscopy | $Al_2O_3$:C | (Gasparian *et al.*, 2010) |
| General (review papers) | Various | (Akselrod, Bøtter-Jensen and McKeever, 2007; Cygler and Yukihara, 2009; Pradhan, Lee and Kim, 2008; Yukihara and McKeever, 2008) |

OSL dosimeters (OSLDs) have only recently been investigated for medical dosimetry applications. Studies are available on OSL characteristics for applications in external beam radiation therapy, brachytherapy, mammography and X-ray CT (see Table 5.1). As discussed in previous chapters, OSL and TL detectors share many features of interest for medical applications, such as small physical size, ability to record the integrated dose and no need for cables or high-voltage (Kron, 1999). Therefore, they can also be used in applications that have traditionally used TLDs. Examples include postal dose audit in radiation therapy (e.g., Aguirre *et al.*, 2002; Han *et al.*, 2008; Ibbott, Molineu and Followill, 2006; Izewska *et al.*, 2007; Izewska, Hultqvist and Bera, 2008; Kirby, Hanson and Johnston, 1992; Tailor, Tolani and Ibbott, 2008), dose verification in radiation therapy (e.g., Low *et al.*, 1998; Nisbet *et al.*, 2004; Verellen *et al.*, 1997), in-vivo dosimetry in radiation therapy and diagnostic radiology (Essers and Mijnheer, 1999; IAEA, 2007; ICRU, 2005), determination of doses outside of target volumes (e.g., Koshy *et al.*, 2004; Mutic and Low, 1998; Pawlicki *et al.*, 2004), inter-laboratory dosimetric comparisons (Kron *et al.*, 2002) and validation of Monte-Carlo dose calculations (e.g., Das *et al.*, 1997; Kirov *et al.*, 1995; Rodrigues *et al.*, 2004).

Several aspects of the OSL technology have already been discussed in previous chapters. Optical readout allows well-controlled stimulation with the possibility of fast readout ($\sim$1 s) and re-estimation of the absorbed dose. The fact that the detector is not heated during readout avoids sensitivity changes typically observed with TL detectors due to changes in the defect concentration and defect structure (McKeever, 1985; McKeever and Moscovitch, 2003). The high sensitivity enables the use of very small detectors, minimizing perturbation of the radiation field and avoiding volume averaging effects. In principle the OSL signal can be erased with a simple exposure to visible light. However, light sensitivity requires special handling or the use of specially designed, light-tight detector holders.

The optical readout makes OSL suitable for one- and two-dimensional dose mapping and real-time, optical-fiber dosimetry. One-dimensional dose profiles can be measured to characterize beam profiles in radiation therapy and to determine dose profiles in X-ray CT. Real-time, optical-fiber dosimeters can be used, among other things, for dose verification in radiotherapy, determination of entrance and exit doses in mammography and estimations of skin dose in interventional radiology.

Significant advances are still required before the OSL technique becomes widely adopted in medical dosimetry, particularly regarding the characterization of OSL materials and dosimeters and the development of readout and calibration protocols. However, the increasing number of publications (Table 5.1) has contributed to bridging the knowledge gap between the OSL and TL techniques. Moreover, at the time of writing, a task group of the American Association of Physicist in Medicine (TG-191) is in the process of developing a set of recommendations on the clinical use of luminescent detectors.

This chapter discusses practical aspects of OSL applied to medical dosimetry. We begin with a brief overview of radiation fields found in different areas in medical dosimetry, quantities of interest, typical dose levels and requirements in terms of precision and accuracy. The use of OSL for occupational exposures of medical staff will not be addressed here; instead, we refer to Chapter 3.

## 5.2 Radiation Fields in Medical Dosimetry

The type of radiation fields, dose levels encountered, requirements in terms of precision and accuracy, and quantities of interest vary considerably across the range of procedures and modalities in diagnostic radiology and radiation therapy. Some of these aspects are summarized in Table 5.2 and discussed in more detail below. The discussion in this section is intended only as an introductory overview, not as a complete discussion. For more information, we refer to Bushberg *et al.* (1994), Handee, Ibbott and Handee (2005), Podgorsak (2005b), Rogers and Cygler (2009), and other references cited herein.

### 5.2.1 Diagnostic Radiology

Dosimetry in diagnostic radiology plays important roles in QA and QC, the establishment of relationships between image quality and dose to the patient, procedure optimization and patient dose management (AAPM, 2002; ICRP, 2000b, 2004). As stated in the IAEA Technical Report Series No. 456 (IAEA, 2007), "the main aim of patient dosimetry with respect to X rays used in medical imaging is to determine dosimetric quantities for the establishment and use of guidance levels (diagnostic reference levels) and for comparative risk assessment." The report also states that "an additional objective of dosimetry in diagnostic and interventional radiology is the assessment of equipment performance as a part of the quality assurance process" (IAEA, 2007). Dosimetry is also important for patient reassurance and to estimate risks for pregnant or pediatric patients.

Diagnostic radiology uses X-ray tubes operating with a maximum potential difference between anode and cathode between 20 and 150 kVp and currents in the range from 1–5 mA

***Table 5.2*** *Dose levels for some therapeutic and diagnostic medical procedures, adapted from Xu, Bednarz and Paganetti (2008) with additional data from the United Nations (2000), Handee, Ibbott and Handee (2005) and other sources (indicated in the table)*

| Modality | Radiation type or isotope (isotope half-life) | Energy | Typical doses |
|---|---|---|---|
| External beam radiation therapy | $^{60}$Co (5.26 y) | 1.32 and 1.17 MeV (γs) | ∼60 Gy to primary target |
| | X-rays | 4–20 MV | |
| | Electrons | 6–23 MeV | |
| | Protons | Up to ∼250 MeV | |
| | Carbon ions | Up to ∼430 MeV/u | |
| Brachytherapy | $^{192}$Ir (74.2 d) | 0.24–0.67 MeV (βs)<br>0.136–1.062 MeV (γs) | ∼60 Gy to primary target |
| | $^{125}$I (59.6 d) | 27.4 keV, 31.4 keV and 35.5 keV (γs) | |
| | $^{103}$Pd (17 d) | 20.1 and 23.0 keV (γs) | |
| | $^{198}$Au (2.7 d) | 0.96 MeV (βs)<br>0.4–1.1 MeV (γs) | |
| | others | | |
| Radioimmunotherapy | $^{131}$I (8.06 d) | 0.25–0.61 MeV (βs)<br>80–637 keV (γs) | (Prescriptions typically specified in activity.) |
| | $^{32}$P (14.3 d) | 1.71 MeV (βs) | |
| | others | | |
| Chest/abdomen radiography | X-rays | <150 keV | 0.14–2.8 mGy entrance air kerma (McCollough and Schueler, 2000)<br>0.02–0.6 mSv effective dose (McCollough and Schueler, 2000) |
| X-ray CT | X-rays | <140 keV | 30–60 mGy (Bushberg *et al.*, 1994)<br>∼30 mGy CTDI free-in-air at isocenter (McCollough and Schueler, 2000)<br>∼1–15 mSv effective dose per scan (AAPM, 2008; McCollough and Schueler, 2000) |
| Fluoroscopy | X-rays | <140 keV | ∼0.2 Sv effective dose per scan |
| Mammography | X-rays | <30 keV | 0.3–0.6 mSv effective dose per mammogram (AAPM, 2008) |

The data are not exhaustive and is only intended for comparison of order of magnitude in dose levels, since the actual doses vary considerably even within each modality. Doses in radiation therapy are generally delivered in multiple fractions.

for continuous fluoroscopy, and up to 100–1000 mA for projection radiography (Bushberg *et al.*, 1994). The X-ray spectrum consists of a filtered bremsstrahlung spectrum, with maximum photon energy determined by the peak voltage and peak intensity around one-third to one-half of maximum photon energy, in addition to the target's characteristic X-ray lines. The X-ray beam can be described in terms of its *quality* and *quantity* (Bushberg *et al.*, 1994). The quality is associated with the penetrating power of the radiation and is usually expressed in diagnostic radiology in terms of the half-value layer (HVL) in aluminum. The quantity can be described using different terms, such as photon fluence (number of photons crossing a unit area), photon flux or fluence rate (number of photons crossing a unit area per unit time), energy fluence (energy crossing a unit area) and energy flux or fluence rate (energy crossing a unit area per unit time). The term "intensity" is usually used to refer to the energy flux.

The X-ray intensity is determined by several factors, but for a specific X-ray unit and filtration the intensity is directly proportional to the tube current (mA). The total X-ray exposure and dose to the patient is proportional to the product of the tube current (mA) and the time of exposure (s), usually referred to as "mAs" ("milliampère-seconds"). The X-ray intensity also increases with the applied X-ray tube potential difference (kVp), being approximately proportional to $(\text{kVp})^2$. The quality of the X-ray beam increases with kVp as the average energy of the spectrum increases. Added filtration increases the quality of the beam by preferentially absorbing the low-energy components of the spectrum, thus decreasing its intensity.

In projection radiography, a two-dimensional static image is formed on a radiographic film or a photostimulable phosphor screen (image plate). Projection radiography systems operate at a high current (100–1000 mA) and short exposure time ($\sim$0.1 s). Mammography is a projection radiography modality in which the equipment is optimized to increase soft-tissue contrast and minimize dose to the breast. The operating electrical potential difference is below 30 keV and the X-ray spectrum is dominated by characteristic X-ray lines of targets such as molybdenum and rhodium, in the 17–23 keV range.

Fluoroscopy is a real-time X-ray imaging modality used, for example, in a variety of angiographic examinations and interventional procedures. The X-ray beam intensity and exposure rates are 200 times lower than in radiography, but the patient is exposed for periods that can add to more than one hour of exposure time (Bushberg *et al.*, 1994). The maximum skin doses at the entrance of the beam can be of the order of a few grays, that is, sufficiently high to cause deterministic effects (ICRP, 2001; Miller *et al.*, 2003a, 2003b), which must be followed and often clinically managed. Another problem in fluoroscopy is the dose to medical staff due to their proximity to the patient. Radiation scattered in the patient can lead to doses to the lens of the eye of the radiologists that exceed the threshold for deterministic effects (caractogenesis) if protective measures are not used (Vano *et al.*, 2008). Although the energy of the scattered photons is reduced by Compton scattering, this effect is partly compensated by absorption of low-energy photons within the scattered volume. The scattered X-ray spectrum has a HVL that is typically 15–40% higher than that of the X-ray beam incident on the patient, depending on the geometry (Marshall, Faulkner and Warren, 1996).

In X-ray CT, tomographic images of the patient are obtained by acquiring a large number of projections using a rotating X-ray beam and an array of detectors (rotating or fixed) positioned opposite to the X-ray tube. Each tomographic image (slice) is reconstructed

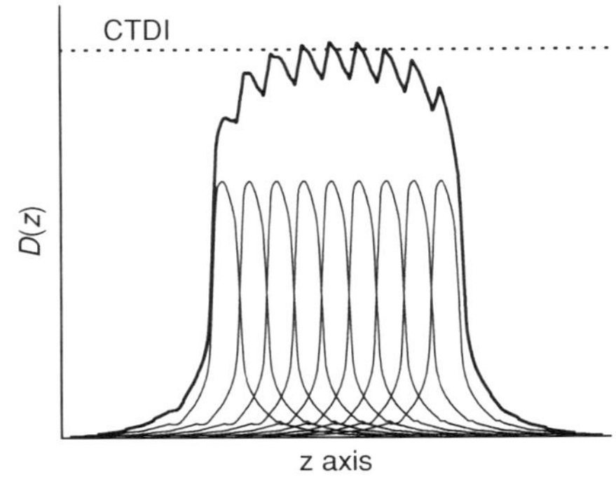


***Figure 5.3*** *Multiple-scan dose profile generated for illustrative purposes using the dose profile for a single scan measured using OSL detectors. In this figure, nine single-scan dose profiles were added to show the dose that would be obtained due to multiple contiguous scans. The dotted line represents the value of the CT dose index (CTDI) calculated from a single-scan dose profile. The CTDI estimates the equilibrium dose resulting from multiple scans, called the multiple scan average dose or MSAD (McCollough, 2008).*

by computational methods using the projections obtained after a full 360° rotation of the X-ray beam around the patient with the patient table stationary. To image the next slice, the patient table is moved in the axial direction ($z$-axis) and a new set of projections (360° rotation of the beam) is obtained. Modern X-ray CT scanners have multiple detector rows that can image simultaneously several slices with a single rotation of the X-ray tube. In the helical X-ray CT scanner the patient table is moved continuously while the projections are obtained. The dose distribution (dose as a function of the axial position) for a single scan is illustrated in Figure 5.3, showing the main peak within the primary irradiated volume and the tails due to scattered beam and penumbra effects (see McCollough, 2008). In an X-ray CT exam, various single-scan dose profiles overlap resulting in a dose distribution similar to the one illustrated in Figure 5.3.

With respect to quantities of interest used in different diagnostic radiology procedures, there are many ambiguities due to different quantities reported (e.g., kerma versus absorbed dose), material specified (e.g., air versus water), and whether or not backscattered radiation is included or not in the measurements (e.g., free-in-air or with backscatter). For this reason, the ICRU states that "the present situation in patient dosimetry for medical X-ray imaging clearly indicates the need for dose quantities recommended for the different applications and the need for using the same, self-consistent, system for names, symbols and units" (ICRU, 2005).

The ICRU recommends using air kerma quantities, except when the quantity is measured or calculated inside a phantom or patient, and using the terms *incident* (and subscript $i$) to indicate no backscatter and *entrance surface* (and subscript $e$) to indicate the inclusion of backscatter. For example, the incident air kerma (no backscatter) $K_{a,i}$ is defined as the "air kerma from the incident beam on the central X-ray beam axis at the focal spot-to-surface distance, $d_{FSD}$, that is, at the skin-entrance plane. Only the primary radiation incident on the patient or phantom and not the backscattered radiation, is included" (ICRU, 2005). Similarly, the entrance surface air kerma (with backscatter) $K_{a,e}$ is defined as "the air kerma on the central x-ray beam axis at the point where the x-ray beam enters the patient or phantom. The contribution of backscattered radiation is included" (ICRU, 2005). For deterministic effects (e.g., in fluoroscopy), "absorbed dose to the more heavily irradiated regions of tissue at the surface of the body is the radiation quantity of interest" (ICRU, 2005). These and other application-specific quantities are discussed in more detail in the ICRU Report 74 (ICRU, 2005). This system of quantities is also adopted by the IAEA international code of practice for dosimetry in diagnostic radiology (IAEA, 2007).

In X-ray CT, the primary dose descriptor is the CT dose index (CTDI), defined as the integral of the single scan dose profile divided by the nominal beam width,

$$\mathrm{CTDI} = \frac{1}{NT}\int_{-\infty}^{\infty} D(z)\mathrm{d}z \tag{5.1}$$

where $N$ is the number of tomographic sections in a single axial scan and $T$ is the thickness of each section (AAPM, 2008; ICRP, 2000b). Although this quantity is related to the total dose from a single scan distributed over the length of the patient or phantom (normalized by the slice thickness), it is equivalent to the total dose to a single slice of thickness $NT$ due to multiple contiguous scans taken along the $z$ axis, provided the scan length is long enough for the central dose to approach an asymptotic limit (Shope, Gagne and Johnson, 1981). Therefore, the CTDI estimates the dose to a central region of the scan due to multiple scans used in an examination, the multiple-scan average dose (MSAD) (see Figure 5.3). In practice, the CTDI is measured over a limited integration range using a 100 mm long, 6 $\mathrm{cm}^3$ "pencil" ionization chamber placed at the center or at the periphery of 150 mm long cylindrical acrylic phantoms of two different diameters, 32 cm (body) and 16 cm (body) (AAPM, 1990, 1993, 2008) with the patient table stationary; in this case the index is denominated $\mathrm{CTDI}_{100}$. Other estimators such as the $\mathrm{CTDI}_{\mathrm{w}}$ and the $\mathrm{CTDI}_{\mathrm{vol}}$ are also defined to take into account the difference in dose at the central or peripheral position of the phantom, and effects due to the gaps or overlaps from consecutive scans (AAPM, 2008).

In terms of the dosimetry requirements in diagnostic radiology, the IAEA and the American Association of Physicists in Medicine (AAPM) divide the measurement activities into three areas: risk assessment, quality assurance and equipment testing, and radiation surveys (AAPM, 1991; IAEA, 2007). The IAEA considers that uncertainties of 7% (expressed at a 95% confidence level, i.e., $\sim 2\sigma$) is appropriate for most cases, including risk assessment of pediatric examinations, quality assurance and equipment testing. For risk assessment of adults and radiation surveys of workplace areas and areas occupied by the general public, it is considered that an uncertainty of 20% ($2\sigma$) should be sufficient (IAEA, 2007).

### 5.2.2 Radiation Therapy and Radiosurgery

Radiation therapy may be delivered using external beams, sealed radioactive sources placed inside or adjacent to tumors (brachytherapy), or unsealed radiopharmaceuticals (Handee, Ibbott and Handee, 2005). The treatment is typically delivered in multiple fractions, except for permanent implanted brachytherapy sources and radiosurgery or palliative radiotherapy of benign conditions. Radiosurgery refers to the use of intense radiation beams to treat lesions with a single treatment, although multiple fractionation is also used (AAPM, 1995).

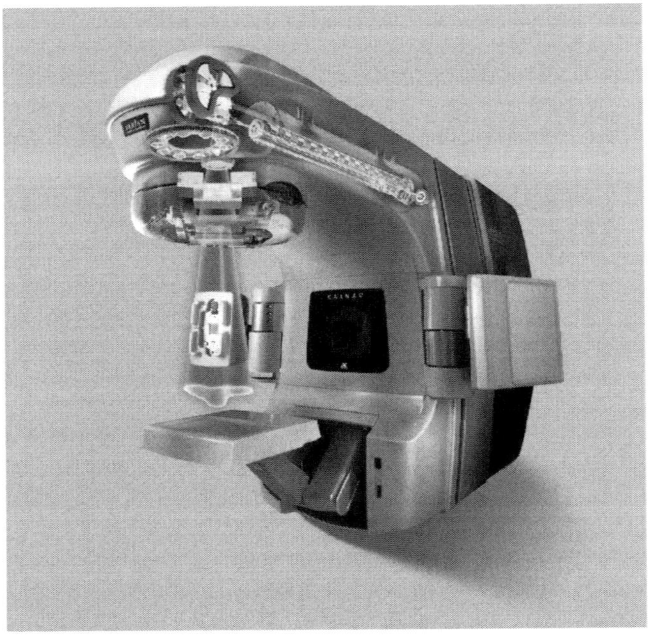

***Figure 5.4*** *Illustration of a medical linear accelerator used for radiotherapy treatment using megavoltage X-ray and electron beams. Image courtesy of Varian Medical Systems Inc. All rights reserved.*

Modern external beam radiation therapy uses mostly megavoltage X-ray (photons) and electron beams produced by clinical linear accelerators (linacs) (Handee, Ibbott and Handee, 2005). Electrons are accelerated by microwave-induced electric fields along the accelerator structure in the linac gantry (Karzmark and Morton, 1998). Figure 5.4 shows an example of a medical linear accelerator. For X-ray beam therapy, a metal target is introduced into the electron-beam trajectory and a flattening filter is used to obtain an X-ray beam with uniform intensity in the treatment field. For electron-beam therapy, the X-ray metal target is removed and beam uniformity is typically achieved using scattering foils (Grusell *et al.*, 1994). Collimators are used to define the treatment field at depth. Because electrons are easily scattered by air, an electron-beam applicator or cone is used to shape the electron beam closer to the patient's skin. X-ray beam energies are specified in MV to indicate that the photon energy spectrum consists of a distribution of energies up to the maximum energy of the accelerated electrons hitting the target, whereas electron beam energies are specified in MeV (Karzmark and Morton, 1998). An overview of treatment machines for external beam radiotherapy can be found in Podgorsak (2005a).

Linear accelerators used for megavoltage X-ray and electron-beam therapy produce pulses of radiation of typical duration ~5 μs. The frequency of the pulses is of the order of 200 Hz (Karzmark and Morton, 1998), but can be varied according to the desired dose rate. Each radiation burst has a fine pulse structure associated with the microwave frequency. The quantity of radiation delivered is controlled using monitor ionization chambers

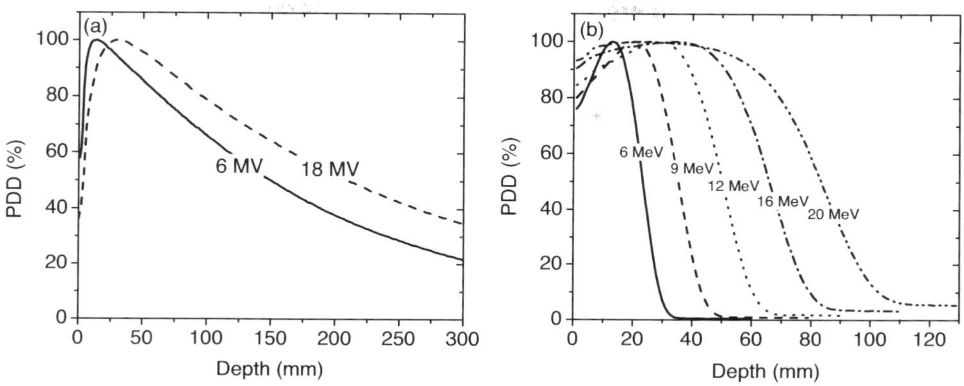


***Figure 5.5*** *Typical central-axis percentage depth-dose (PDD) curves for (a) photons and (b) electrons inside a water phantom irradiated using a linear accelerator with 10 cm × 10 cm field size, source-to-surface distance (SSD) = 100 cm. Data courtesy of Dr. Salahuddin Ahmad, University of Oklahoma Health Sciences Center.*

mounted after the target or scattering foil, indicated in the control panel as "monitor units" (MUs). The relationship between MU and dose is established by calibration during commissioning of the linear accelerator and monitored regularly as part of the quality assurance program.

The energy of the external radiation beam is usually described in terms of the central-axis depth-dose profile in the medium, usually expressed normalized to the maximum dose (percentage depth dose, PDD). Figure 5.5a shows the depth-dose profile for two photon beams and various electron beam energies from a clinical linear accelerator. The photon depth-dose profiles reflect the build-up of secondary charged particles, reduction in the photon flux due to the inverse square law, attenuation of the primary photon beam, and the presence of photons scattered in the medium (Handee, Ibbott and Handee, 2005). The electron depth-dose profiles (Figure 5.5b) are characterized by a sharp decrease with depth. Electron beams of energy up to 20 MeV are used for treating tissues within 6 cm of the body surface while sparing underlying tissue (Hogstrom and Almond, 2006). Electrons with energy above 20 MeV are not commonly used, because the depth-dose profile starts to resemble that of a photon beam due to bremsstrahlung production, and the beam broadens due to Coulomb scattering (Hogstrom and Almond, 2006). The depth-dose profiles for photon and electron beams are affected by parameters such as energy, collimation and source-to-surface distance (SSD) (Handee, Ibbott, Handee, 2005; Hogstrom and Almond, 2006). The energy spectra, dose profiles and other characteristics of 6 and 18 MV photon beams have been investigated using Monte-Carlo techniques. Figure 5.6 shows an example of the spectrum and composition of a 6 MV photon beam.

Irradiation with small fields is used in stereotactic radiosurgery and radiotherapy to treat a variety of intracranial and extracranial lesions, including benign and malignant tumors, functional disorders, vascular lesions and metastic tumors, among other conditions (Podgorsak and Podgorsak, 2005). The radiation units are based either on $^{60}$Co gamma-ray sources or linear accelerators. The small field sizes and high dose gradients (see Figure 5.7)

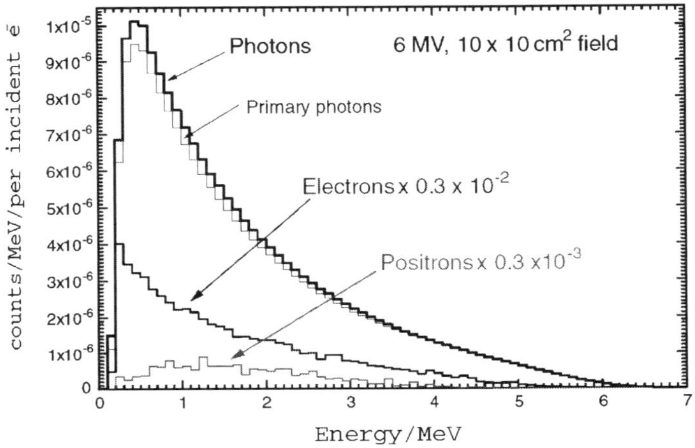


***Figure 5.6*** *Energy spectra of photons, primary photons, electrons and positrons of a 6 MV beam at a water phantom surface for a 10 cm × 10 cm field size. Reprinted from Physics in Medicine and Biology, Ding, X., Energy spectra, angular spread, fluence profiels and dose distributions of 6 and 18 MV photon beams: results of Monte Carlo simulations for a Varian 2100EX accelerator. Vol. 47, 1025–1046. Copyright (2002) with permission from IOP Publishing Ltd.*

require detectors with high spatial resolution and dimensions <2 mm. The determination of output factors for different beam collimations is required by the time of commissioning of the equipment, but for the smallest field sizes (e.g., 4 mm in Gamma Knife™) the choice of detector is limited (Kurjewicz and Berndt, 2007).

In brachytherapy, the dose is delivered by radioactive "seeds" or "wires" placed into the tumor or adjacent to the tumor, or more recently using miniature X-ray systems (Park *et al.*, 2010). High-dose-rate techniques (1.6–5.0 Gy/min) have been replacing low-dose-rate procedures (<2.0 Gy/h), allowing treatment times of the order of 1–12 min. The sources are typically $^{192}$Ir, $^{125}$I and $^{103}$Pd, producing gamma-rays during de-excitation of the daughter products or characteristic X-rays due to electron capture decay (see Table 5.2). The source dimensions vary with manufacturer, but seeds are ~3 mm long by ~0.5 mm in diameter and wires have diameters of ~0.3–0.5 mm (Nath *et al.*, 1995). The use of remote-controlled afterloaders decreases the risk of radiation exposure to the staff, and allows precise positioning of the sources in different positions for different dwell times, allowing better optimization of the dose distribution by treatment planning systems. However, the high dose rates, short dwell times and increased complexity of the procedures pose severe risks of over-exposure due to factors ranging from human error to equipment malfunction (e.g., failure in retracting the source due to a kink in the applicator needle) (ICRP, 2005; Nath *et al.*, 1997).

Though in-vivo dosimetry can assist in estimating the doses delivered to the target volume or critical structures and help identify treatment errors, dosimetry in brachytherapy is challenging. The dose rate decreases rapidly with the distance from the source, therefore requiring detectors with small volume to avoid volume averaging effects in these high-dose gradients. Other complicating factors for in-vivo dosimetry are the temperature and

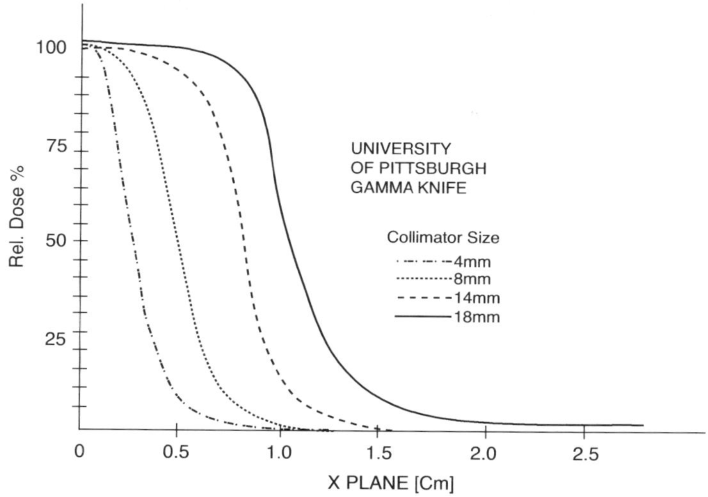


***Figure 5.7*** *Beam profiles for four beam diameters in a Gamma Knife™ unit. The figure illustrates the high dose gradients that can be obtained in radiosurgery. Reprinted from AAPM, AAPM Report 42: Stereotactic radiosurgery. Copyright (1995) with permission from American Association of Physicists in Medicine.*

directional dependence of some detectors, and changes in energy spectrum with distance from the source (Lambert *et al.*, 2007).

The requirements in terms of precision and accuracy are most strict in radiation therapy because of the need to guarantee the dose delivered to the target volume while minimizing normal tissue toxicity. It is recognized that "the available evidence for certain types of tumor points to the need of an accuracy of $\pm 5\%$ in the delivery of an absorbed dose to a target volume if the eradication of the primary tumour is sought" (ICRU, 1976). Analysis of the many sources of uncertainties (ICRP, 2000a) shows that instruments of better precision and accuracy are required if an overall uncertainty of $\pm 5\%$ is to be achieved. Ionization chambers can achieve an estimated combined standard uncertainty of the order of 0.9% for $^{60}$Co gamma-ray beams and $\sim$1.5% for megavoltage X-ray and electron beams using appropriate protocols (IAEA, 2000). In regions of high dose gradients, the precision of a detector is affected by errors in positioning and by volume averaging effects. Therefore, in these regions the concept of "distance-to-agreement" is more appropriate than a comparison between calculated and measured doses. The "distance-to-agreement" is the distance between the detector position and the nearest position in the calculated distribution with the same value of dose (Intensity Modulated Radiation Therapy Collaborative Working Group, 2001). Maximum distance to agreement values between 2 and 4 mm have been used as acceptability criteria for high-dose gradient regions in quality assurance of external photon and electron beams (Ibbott, Molineu and Followill, 2006; Van Dyk *et al.*, 1993).

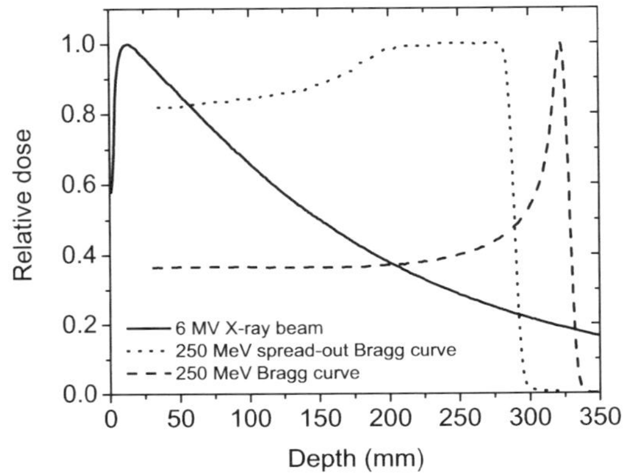


***Figure 5.8*** *Comparison between the typical depth-dose profiles for a 6 MV photon beam, a 250 MeV proton beam and a spread-out Bragg curve obtained using a 250 MeV proton beam and absorbers with different thicknesses. Data courtesy of Dr. Gabriel O. Sawakuchi, Carleton University, and Dr. Salahuddin Ahmad, University of Oklahoma Health Sciences Center.*

### 5.2.3 Proton and Heavy-ion Therapy

Protons can be used to obtain superior dose distributions when compared to photons. The energy loss per length traveled in the material (i.e., the stopping power), increases as the proton penetrates in the medium and slows down. As a result, the dose deposited by the proton is the highest near the end of the proton range, rapidly decreasing beyond that distance (the Bragg peak). As opposed to a photon beam, in which the maximum dose is delivered at a depth a few centimeters inside the medium, the dose delivered by a monoenergetic proton beam is maximum at the Bragg peak, close to the end of the proton range. Figure 5.8 shows the Bragg peak for a 250 MeV proton beam in comparison to the depth-dose profile for a 6 MV photon beam. The dose of a "pristine" Bragg peak is three to four times the entrance dose, allowing greater dose to the tumor while sparing the healthy tissue in the proximal region and sparing critical structures in the distal region of the Bragg peak (ICRU, 1998c). However, clinical beams usually employ a distribution of proton energies.

Since the Bragg peak depth varies with proton energy, it is possible to use a distribution of proton energies to create a superposition of Bragg peaks to better conform the dose to the tumor volume. This so-called spread-out Bragg peak (SOBP) is illustrated in Figure 5.8. One should notice that the ratio between the maximum dose and the entrance dose is not as large as in the case of the pristine Bragg peak. The dose can also be delivered using multiple irradiation directions and beams of variable intensity and energy (Flynn *et al.*, 2007, 2008; Lomax, 1999).

Most proton therapy facilities use cyclotrons or synchrotrons. In a cyclotron, the protons are kept in a circular trajectory by a constant magnetic field. The trajectory radius increases every time the proton is accelerated in the region between the bending sections. Cyclotron accelerators are compact, simple and inexpensive, but they produce protons of

(a)

(b)

***Figure 5.9*** *Photos of the (a) cyclotron proton accelerator and (b) beam transport system from the ProCure Treatment Center in Oklahoma City, OK, USA. Image courtesy of ProCure Treatment Centers, Inc.*

a fixed energy. Other energies may then be obtained by the use of absorbers. A cyclotron-based proton facility is shown in Figure 5.9. A list of proton- and ion-therapy facilities in operation and proposed can be found on the Particle Therapy Co-Operative Group web site (http://ptcog.web.psi.ch/, accessed on 18 May 2010).

In a synchrotron, the protons are accelerated in straight sections between bending magnets and maintained in a circular orbit of fixed radius using magnetic fields of increasing intensity. Proton beams can be extracted at different energies, making it feasible to vary the energy of the beam without the use of absorbers to control the depth of penetration of the proton beam (ICRU, 1998b). Proton beams with uniform intensity over the treatment field can be obtained using a passive scattering technique (scattering foils), or wobbling and beam-scanning techniques in which protons are deflected using time-varying magnetic fields. In the spot scanning technique, the target volume is "painted" by a pencil beam that is deflected magnetically to conform the dose to the target volume in the lateral dimension, while the beam energy is modulated to conform the dose to the distal region of the target volume.

In addition to the high dose gradients and high dose rates, particularly those found in spot scanning, a complicating factor in the dosimetry of proton beams is the relatively higher linear energy transfer (LET) of the radiation and the ionization density created in the detectors. For proton beams, the LET is ~0.4 keV/μm for the typical maximum proton energies used (250 MeV/u, range of ~37.5 cm in water), increasing as the proton slows down. A radiation weighting factor of 1.1 is typically applied to account for the increased relative biological effectiveness (RBE) (ICRU, 1998c). Carbon beams deposit energy similarly to proton beams, but the LET in water for the maximum energies typically used in radiation therapy (430 MeV/u, range 31.3 cm in water) is ~10 keV/μm, also increasing as the particle slows down. The higher LET purportedly provides an additional biological advantage due to a higher RBE (Schardt and Heavy-Ion Therapy Collaboration, 2007; Tsujii *et al.*, 2007; see also Wilkens and Oelfke, 2008 for an opposing point of view). As already discussed in Chapter 4, at these LET values the ionization density within the particle's track is very high. In the case of OSL and TL detectors, these high ionization densities correspond to local doses inside the particle track well above the saturation dose of the TL or OSL signal, leading to a reduction in the efficiency of the luminescence process. Therefore, the relative luminescence efficiency and its dependence on particle energy and charge need to be characterized.

## 5.3 Practical OSL Aspects Applied to Medical Dosimetry

### 5.3.1 A Proposed Formalism

International efforts to improve the accuracy and consistency of medical dosimetry have resulted in protocols, codes of practice and guidelines for detectors such as ionization chambers (Almond *et al.*, 1999; IAEA, 2000), diodes (AAPM, 2005), radiographic films (Pai *et al.*, 2007) and radiochromic films (Niroomand-Rad *et al.*, 1998). These efforts are driven by the goal of minimizing the uncertainties associated with dose delivered in radiotherapy and the need for consistent dosimetry in clinical trials and epidemiological studies (Almond, 2009).

In the absence of widely accepted guidelines for OSL dosimetry, we propose adopting a formalism similar to those used for other types of detectors in protocols for external beam radiation therapy (Almond *et al.*, 1999; IAEA, 2000) and dosimetry in diagnostic radiology (IAEA, 2007) and consistent as much as possible with international standards (Table 3.2). Such a formalism also follows other proposals to establish a common terminology to describe the characteristics of radiation dosimeters in clinical dosimetry for radiation therapy (Rogers, 2009).

Here we present the formalism written in standards of absorbed dose to water (indicated by the subscript "w" in the relevant quantities), but it can also be written in absorbed dose to the detector material replacing the subscript w with the appropriate one. The superscript

"OSL" was added to distinguish the terms used here from the terms used in other protocols, but it can be omitted when there is no risk of confusion. For applications in diagnostic radiology, the formalism can be re-written in standards of air kerma $K$ (see IAEA, 2007).

Although such formalism may help bridge the gap between researchers, clinical professionals, educators and the dosimetry community in general, we should recognized limitations from the outset. The main one is that the formalism assumes the correction factors to be independent from each other, which may not always be the case. For example, the dose response for $Al_2O_3$:C starts to become supralinear at doses above ~1 Gy for low LET radiation including protons, but it does not show supralinearity for particles heavier than He (Sawakuchi *et al.*, 2008c). Therefore, the correction factor for non-linearity will depend on the beam quality. These and other issues require further investigations on the properties of OSL materials that are still going on.

#### 5.3.1.1 The OSL Signal S

The dose information stored in an OSLD can be read by appropriate optical stimulation, resulting in the OSL signal $S$. If the light transducer is a photomultiplier tube (PMT) operating in photon counting mode, the OSL signal $S$ is the number of PMT pulses (counts) over a certain stimulation interval (calculated by the user or automatically). If the light transducer is a PMT operating in DC current mode, the OSL signal $S$ is the current integrated over a certain interval of stimulation (in coulombs). The OSL signal $S$ is therefore analogous to the charge collected in an air ionization chamber.

The signal $S$ obtained from the reader should also be corrected for any instrumental background, such as the signal due to PMT dark counts or current, stimulation light scattered to the PMT and light leakage in the instrument. This can be done by:

$$S = S_{\text{raw}} - S_{\text{bg}} \tag{5.2}$$

where $S_{\text{raw}}$ is the uncorrected OSL signal, and $S_{\text{bg}}$ is the estimated instrumental background. $S_{\text{bg}}$ can be evaluated, for example, using blank (unirradiated) detectors.

#### 5.3.1.2 The OSL Reading $M_{\text{raw}}^{\text{OSL}}$

The OSL reading $M_{\text{raw}}^{\text{OSL}}$ can be taken to be the OSL signal $S$ defined by the user, expressed, for example, in counts or coulombs. The situation is equivalent to using the charge collected in an ionization chamber (in coulomb) for $M_{\text{raw}}$. Alternatively, $M_{\text{raw}}^{\text{OSL}}$ can be the result of an algorithm that takes into account the signal $S$ and other factors, including calibration and detector sensitivity factors to provide the user with an indicated value $M_{\text{raw}}^{\text{OSL}}$ as a dosimetric quantity (e.g. absorbed dose or personal dose equivalent), as discussed in Chapter 3. In this case the indicated value $M_{\text{raw}}^{\text{OSL}}$ will depend on the particular calibration conditions and applied factors and will be meaningful only if the detector was used in similar conditions. $M_{\text{raw}}^{\text{OSL}}$ can also be taken to be the indicated value of the instrument in whatever units are on the scale. This is usually referred to as "rdg" (Almond *et al.*, 1999; IAEA, 2000).

Control detectors can be used to evaluate the OSL reading $M_{\text{raw},0}^{\text{OSL}}$ due to the absorbed dose received during storage and transportation, if necessary. Control detectors are detectors prepared together with the dosimeters to be used, but not exposed to the radiation field to be measured.

### 5.3.1.3 The Corrected OSL Reading $M^{\mathrm{OSL}}$

The measurement result, which consists of the measured value $M^{\mathrm{OSL}}$ and its associated uncertainty, is obtained after correction of $M_{\mathrm{raw}}^{\mathrm{OSL}}$ for all influence quantities intrinsic to the detector material, such as linearity and fading. An influence quantity is defined as a "quantity that, in a direct measurement, does not affect the quantity that is actually measured, but affects the relation between the indication and the measurement result" (ISO/IEC GUIDE 99, 2007). In the case of OSL, the influence quantities may arise from environmental factors (temperature of irradiation or readout), from the detector (fading, linearity), or from the radiation field (beam quality, dose rate, field size, depth in phantom, etc.). Correction factors related to beam quality will be treated separately in the Section 5.3.1.1.

$M^{\mathrm{OSL}}$ can be calculated from $M_{\mathrm{raw}}^{\mathrm{OSL}}$ by adopting a model function for the measurement:

$$M^{\mathrm{OSL}} = \left(M_{\mathrm{raw}}^{\mathrm{OSL}} - M_{\mathrm{raw},0}^{\mathrm{OSL}}\right)\prod_i k_i, \tag{5.3}$$

where $k_i$ is the correction factor for the $i$th influence quantity of type "F" and $M_{\mathrm{raw},0}^{\mathrm{OSL}}$ is the OSL reading of the control dosimeters (background dose), or other influence quantities of type "S" (see definitions in Section 3.3.1). The expression above assumes that the influence quantities of type "F" are independent from each other and can be corrected by the multiplicative factors $k_i$. The correction factors $k_i$ are the reciprocal of the relative response values $r$ (see Section 3.3.1 and Table 3.6). Material properties are often discussed in terms of relative responses in the literature.

Below we define a few correction factors related to environmental conditions and the detector.

*Correction factor for fading* $k_t$. The correction factor $k_t = k(t)$ can be obtained by comparing the ratio between the OSL reading at a reference time $t_0$ and at a different time $t$ using detectors irradiated with identical or similar doses $D$ and in conditions in which the other correction factors are unity:

$$k_t = \frac{\left(M^{\mathrm{OSL}}/D\right)_{t_0}}{\left(M^{\mathrm{OSL}}/D\right)_t}. \tag{5.4}$$

Fading is expected to be temperature dependent if caused by thermal stimulation of charges from the trapping centers.

*Correction factor for temperature of irradiation* $k_T$. In a similar way, the correction factor $k_T = k(T)$ can be obtained by comparing the OSL reading of dosimeters irradiated at a reference temperature $T_0$ and at a different temperature $T$ with identical or similar doses $D$ and in conditions in which the other correction factors are unity:

$$k_T = \frac{\left(M^{\mathrm{OSL}}/D\right)_{T_0}}{\left(M^{\mathrm{OSL}}/D\right)_T}. \tag{5.5}$$

It should be noticed that if the temperature of irradiation is sufficiently high to affect the trapping center responsible for the OSL signal, the correction factor will depend on other parameters such as irradiation time and dose rate. In these situations, a more complicated correction will be necessary. In the case of ionization chambers, the temperature and pressure are taken into account together in the correction factor $k_{TP}$. However, solid-state detectors used in TL and OSL dosimetry are not considered to be affected by ambient pressure.

*Correction factor for non-linearity $k_D$.* If the non-linearity in the OSL signal is small (a few percent), the correction can be represented by a factor $k_D = k(D)$. This effect can be determined by the ratio $M^{\mathrm{OSL}}/D_{\mathrm{w},Q}$ calculated at a reference dose $D_{\mathrm{w},Q} = D_0$ (typically in a range in which the detector's response is linear) divided by the same quantity evaluated at an arbitrary dose $D_{\mathrm{w},Q} = D$:

$$k_D = \frac{\left(M^{\mathrm{OSL}}/D_{\mathrm{w},Q}\right)_{D_{\mathrm{w},Q}=D_0}}{\left(M^{\mathrm{OSL}}/D_{\mathrm{w},Q}\right)_{D_{\mathrm{w},Q}=D}}. \tag{5.6}$$

This correction factor requires a priori knowledge or a first estimation of the absorbed dose and it is more appropriate to situations in which deviations from linearity are small or when the material's dose response has already been taken into account in the algorithm used by the dosimetry system when calculating from the OSL signal or signals $S_g$. In the latter case, $k_D$ corrects for any "residual" non-linearity not automatically corrected by the dosimetry system.

In the more general case, the dose response of the OSL material for a specific readout system and analysis procedure should be determined as discussed in Section 3.4.1.2. The dose response can then be used as an experimental calibration curve in the algorithm used to calculate $M_{\mathrm{raw}}^{\mathrm{OSL}}$ from the signal $S$.

#### 5.3.1.4 The Calibration Factor $N_{D,\mathrm{w},Q_0}$

Designating $M_{Q_0}^{\mathrm{OSL}}$ as the detector reading under the reference conditions used in the standards laboratory for a beam quality $Q_0$, the absorbed dose to water $D_{\mathrm{w},Q_0}$ is:

$$D_{\mathrm{w},Q_0} = M_{Q_0}^{\mathrm{OSL}} N_{D,\mathrm{w},Q_0}^{\mathrm{OSL}} \tag{5.7}$$

where $N_{D,\mathrm{w},Q_0}^{\mathrm{OSL}}$ is the calibration factor in terms of absorbed dose to water obtained by irradiation in a standards laboratory. Examples of reference conditions are the geometrical arrangement, field size, material and dimension of the irradiated phantom, and temperature of irradiation (IAEA, 2000). If $M_{Q_0}^{\mathrm{OSL}}$ has the same units as $D_{\mathrm{w},Q_0}$, the calibration factor is dimensionless. If $M_{Q_0}^{\mathrm{OSL}}$ is in arbitrary units or other unit different than $D_{\mathrm{w},Q_0}$ (rdg), then $N_{D,\mathrm{w},Q_0}^{\mathrm{OSL}}$ is, in fact, a calibration coefficient with units Gy/rdg.

#### 5.3.1.5 Beam Quality Correction Factor $k_{Q,Q_0}$

In conditions different from the reference conditions used by the standards laboratory, the absorbed dose to water is given by:

$$D_{\mathrm{w},Q} = M_{Q}^{\mathrm{OSL}} N_{D,\mathrm{w},Q_0}^{\mathrm{OSL}} k_{Q,Q_0} \tag{5.8}$$

where $D_{\mathrm{w},Q}$ is the absorbed dose to water for a beam quality $Q$, $M_{Q}^{\mathrm{OSL}}$ is the fully corrected OSL reading at beam quality $Q$, and $k_{Q,Q_0}$ is the beam quality correction factor

(IAEA, 2000):[1]

$$k_{Q,Q_0} = \frac{N_{D,\mathrm{w},Q}}{N_{D,\mathrm{w},Q_0}} = \frac{M_{Q_0}/D_{\mathrm{w},Q_0}}{M_Q/D_{\mathrm{w},Q}} \tag{5.9}$$

Ideally, the beam quality correction factor should be determined in a standards laboratory with beam quality $Q$ and $Q_0$, but often this is not possible or practical. In practice, calibration factors are often obtained against a calibrated instrument (ionization chamber). In principle, $k_{Q,Q_0}$ could also be estimated using assumptions from cavity theory or Monte-Carlo calculations of the detector properties (e.g. Chen *et al.*, 2009; Mobit, Agyingi and Sandison, 2006). However, these calculations only account for the "absorbed dose energy dependence," which is related to difference in absorbed dose in the detector and in the medium, but not for intrinsic energy dependence resultant from ionization density effects (see discussion in Section 3.4.1).

#### 5.3.1.6 *Other Influence Quantities*

In addition to the correction factor defined above, others may also be necessary.

If the material shows a dependence on the dose rate, a correction factor may be obtained by comparing the detectors indicated value to a dose rate $\dot{D}$ with that for a reference dose rate $\dot{D}_0$. However, this is difficult to apply in practice, since the dose rate may be variable in space (e.g., in case of proton and carbon beams) and in time (e.g., in case of pulsed radiation sources). Directional dependence can also be estimated by comparing the detector's response at an angle $\theta$ with that at a reference angle (e.g., assumed to be 0°). This directional dependence will depend on the radiation quality and the measurements also need to take into account possible dose gradients in the radiation field, depending on the detector's size.

The correction factors discussed above should not be considered exhaustive. A major task with any new technology is identifying the appropriate influence factors that should be taken into account.

#### 5.3.1.7 *Summary of Formalism*

The steps involved in obtaining the dose to water from the dose information stored in the OSLD are illustrated in Figure 5.10. The formalism proposed here should be considered only as a starting point. Given the dependence of the correction factors on instrument, readout approach, material and so on, the correction factors should be determined for each protocol and study, at least until a protocol in OSL dosimetry becomes widely accepted.

### 5.3.2 Calibration and Readout Protocols

The determination of the calibration factors $N^{\mathrm{OSL}}_{D,\mathrm{w},Q_0}$ for a protocol based on absorbed dose to water standards (or $N^{\mathrm{OSL}}_{K,Q_0}$ for a the protocol based on air kerma standards) consists in irradiating the dosimeters in reference conditions in a Primary Standards Dosimetry

---

[1] Here we used the symbols used in the IAEA protocol (IAEA, 2000) for the most part. The TG-51 protocol (Almond *et al.*, 1999) uses different terms, e.g., $N^{^{60}\mathrm{Co}}_{D,\mathrm{w}}$ instead of $N_{D,\mathrm{w},Q_0}$ and $k_Q$ instead of $k_{Q,Q_0}$.

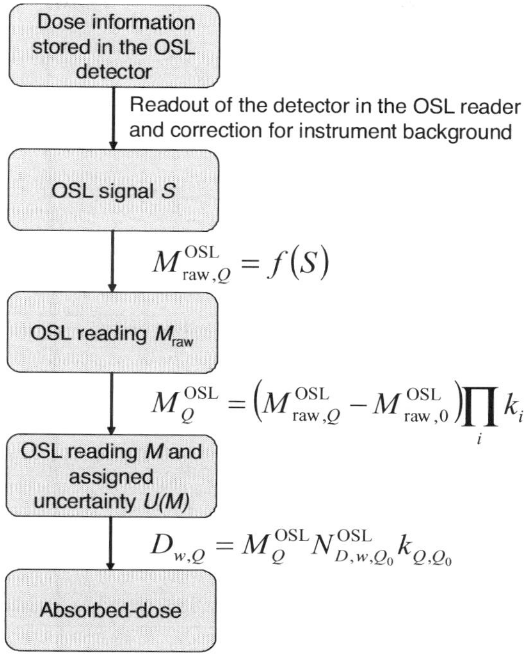


***Figure 5.10*** *Main steps in the evaluation of the dose information stored in the OSL detector. See text for more details.*

Laboratory (PSDL) or in one of the Secondary Standards Dosimetry Laboratories (SSDL) and comparing the given dose $D_{w,Q_0}$ to the OSLD indication $M^{OSL}_{Q_0}$, which has been corrected for all influence quantities.

$$N^{OSL}_{D,w,Q_0} = \frac{D_{w,Q_0}}{M^{OSL}_{Q_0}}. \tag{5.10}$$

However, for clinical applications the OSLDs will more likely be cross-calibrated against a reference instrument (typically an ionization chamber) previously calibrated in a standards laboratory.

The calibration protocol will depend on the application. For in-vivo dosimetry in external radiotherapy beams, the calibration protocols established for diodes and TLDs can be used (AAPM, 2005; Van Dam and Marinello, 2006). Figure 5.11 illustrates the setup recommended for the calibration of diodes and TLDs for determination of entrance dose, which can be applied for OSLDs. The entrance dose $D_{entrance}$ is the dose at a reference point at a depth corresponding to the depth of maximum build-up $d_{max}$ for a specific beam. The calibration factor $F_{entrance}$ is given by the ratio between the dose $D_{entrance}$ to water at the reference point and the detector's reading $M^{OSL}$ (Van Dam and Marinello, 2006):

$$F_{entrance} = \frac{D_{entrance}}{M^{OSL}}. \tag{5.11}$$

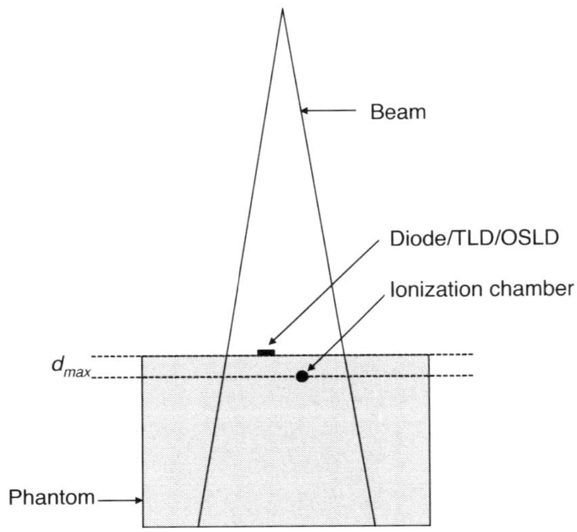


***Figure 5.11*** *Illustration of experimental arrangement for determination of entrance dose calibration factors for diodes and TLDs, based on discussion by Van Dam and Marinello (2006) and the AAPM Report 87 (AAPM, 2005).*

To determine the calibration factor $F_{\text{entrance}}$ the OSLD is placed at the upstream surface of a plastic phantom of 25 cm × 25 cm area by 10 cm thickness, whereas the dose at the reference point $D_{\text{entrance}}$ at a depth $d_{\text{max}}$ is measured with an ionization chamber. The OSLD and the ionization chamber are slightly displaced laterally to avoid the influence of one on the other. Build-up caps should be used on the OSL detectors to avoid the steep dose gradient and reduce the influence of electrons scattered in the accelerator head (electron contamination). The choice of thickness of the build-up cap is discussed by Van Dam and Marinello (2006).

For in-vivo dosimetry in diagnostic radiology, OSLDs can be calibrated in terms of the quantity of interest (e.g., the entrance dose air kerma for patient measurements) using the same protocols for TLDs described in the IAEA Technical Report Series No. 457 (IAEA, 2007).

One of the most important points in TL dosimetry which also applies to OSL dosimetry is that the dosimetry system as a whole (i.e., dosimeter, reader, and all associated electronics and procedures) has to be considered when calibrating the dosimeter (IAEA, 2007). Any variation in influence factors between the calibration and clinical conditions may affect the accuracy of the measurements. The adoption of OSLDs eliminates the sensitivity changes introduced by variations in the annealing conditions. Nevertheless, there are other aspects that need to be carefully considered. We now discuss possible calibration approaches, but one should keep in mind that, as yet, no protocol has been widely accepted.

#### *5.3.2.1 Batch Calibration*

Batch calibration is certainly the most convenient approach if the precision requirement can be satisfied, given the variability in response within the batch. This approach consists of determining a single calibration factor $N^{\mathrm{OSL}}_{D,\mathrm{w},Q_0}$ for a sub-set of dosimeters within the batch and applying the same factor to the remaining dosimeters in the batch. Investigations using commercial OSLDs (InLight 'Dot' dosimeters and microStar reader) indicates a batch uniformity of ~4%, based on an experimental standard deviation estimated from the readout of 165 detectors irradiated in the same conditions (Viamonte *et al.*, 2008).

Landauer Inc. provides a set of calibration dosimeters irradiated in reference conditions for the InLight dosimetry system. It should be noticed, however, that because of depletion of the OSL signal during readout this calibration set will drift after repeated readouts at a rate that depends on the stimulation intensity. For dosimeters irradiated with low doses, the InLight reader uses higher stimulation intensity, therefore causing a larger fractional depletion of the OSL signal (see Chapter 3). For dosimeters irradiated with high doses, the stimulation intensity is lower and the calibration dosimeters will continue to be valid even after a larger number of readouts.

#### 5.3.2.2 *Individual Calibration*

Individual calibration relies on the determination of an element correction coefficient for each detector, as discussed in Section 3.3.2. One can obtain a calibration factor $N^{\mathrm{OSL}}_{D,\mathrm{w},Q_0}$ for the batch of detectors and element correction factors relating the response of each detector to the mean response of the batch. To obtain the element correction coefficients, each single detector needs to be irradiated with a known or fixed dose prior to or after use.

The problem with individual calibration of $Al_2O_3$:C detectors is the sensitivity change exhibited by the material as a function of the accumulated dose. To illustrate this point, Figure 5.12 shows the total OSL signal $S_\mathrm{R}$ observed for a *fixed* reference dose of 0.96 Gy as a function of the total accumulated dose previously received by the detector (the detectors were bleached between irradiations). If the dose is much smaller than the reference dose (in this case 0.96 Gy), the OSL signal due to the reference dose is relatively unaffected,

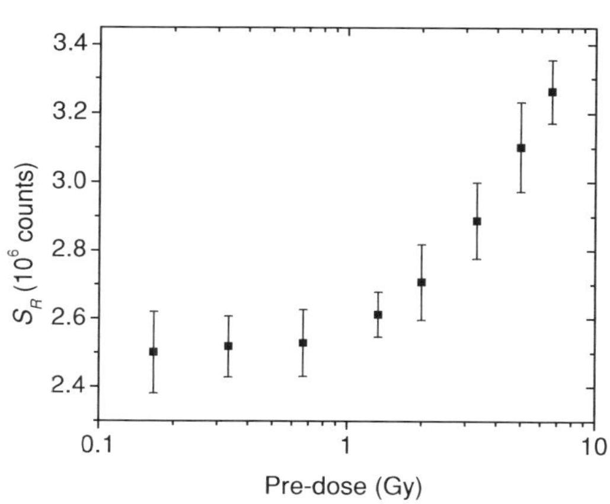


***Figure 5.12*** *Sensitization of $Al_2O_3$:C Luxel detectors as a function of pre-dose. The graph shows the total OSL signal (600 s stimulation, green LEDs at ~10 mW/cm²) measured after the detectors were irradiated with a pre-dose (indicated in the figure), bleached for 600 s (green LEDs at ~10 mW/cm²), and irradiated with a reference dose of 0.96 Gy.*

indicating a constant sensitivity. However, for pre-doses similar or higher than the reference dose, the signal due to the reference dose shows an increase that is clear evidence of the sensitization of the OSL material.

The sensitivity changes experienced by the detectors are related to the complex mechanism of charge trapping in $Al_2O_3$:C, in which many different trapping levels compete for the capture of charge during both irradiation and stimulation of the crystal (Jursinic, 2010; Yukihara *et al.*, 2004a). When the detector is irradiated, the different trapping levels are filled. Bleaching of the detector (exposure to light) can empty the main dosimetric trap, but deeper trapping levels may remain full, therefore affecting the competition mechanism and causing sensitivity changes in the crystal. A more detailed discussion on the sensitivity changes experienced by $Al_2O_3$:C during irradiation and annealing (heating) procedures can be found in Section 3.4.2.3. However, for our discussion it is sufficient to understand that the sensitivity changes are associated with the fact that bleaching is insufficient to restore the material to its original state before irradiation. This would require heating to 900 °C, which is not possible in the case of detectors containing plastic or simply not practical for other detector types.

Jursinic (2010) investigated the possibility of applying a high "conditioning" dose to fill the deep traps in the material and eliminate or reduce the sensitivity changes introduced by these deep traps. Although the results demonstrate that the supralinearity can be practically eliminated by a "conditioning" dose of 1 kGy, followed by bleaching of the detectors, it was observed that the OSL signal increased with time, in the dark, at a rate larger than the rate expected from background radiation and proportional to the previous dose received by the detector. Jursinic recommends the individual characterization of each detector if a precision of ±0.5% is to be reached.

#### 5.3.2.3 *A High-Precision Readout and Analysis Protocol for OSL Detectors*

A protocol that takes into account sensitivity changes experienced by $Al_2O_3$:C was specifically developed for OSL detectors and demonstrated to achieve a precision of ~0.7% (Yukihara *et al.*, 2005, 2008a). The main element of this protocol is an algorithm to convert the signal $S$ to the raw reading $M_{\mathrm{raw}}^{\mathrm{OSL}}$.

The algorithm requires a "calibration curve" that can be obtained as follows. A set of detectors is first irradiated with known doses $D_i$ in the dose range of interest. This set of detectors is read using the step-by-step process outlined in Table 5.3, which provides the signal $S$ due to the known dose of radiation, the signal $S_{\mathrm{R}}$ due to the fixed reference dose and the ratio $S/S_{\mathrm{R}}$. The ratio $S/S_{\mathrm{R}}$ is important because it is independent of the mass and

***Table 5.3*** *Procedure used for the determination of the algorithm calibration curve and readout of OSL detectors to achieve high precision in radiotherapy (see text for more details)*

| Step | Procedure | Result |
|---|---|---|
| 1 | Readout of OSL detector and calculation of total OSL signal $S$ | $S$ |
| 2 | Irradiation of detector with reference dose $D_{\mathrm{R}}$ | |
| 3 | Readout of OSL detector and calculation of total reference OSL signal $S_{\mathrm{R}}$ | $S_{\mathrm{R}}$ |
| 4 | Calculation of ratio between $S$ and $S_{\mathrm{R}}$ | $S/S_{\mathrm{R}}$ |

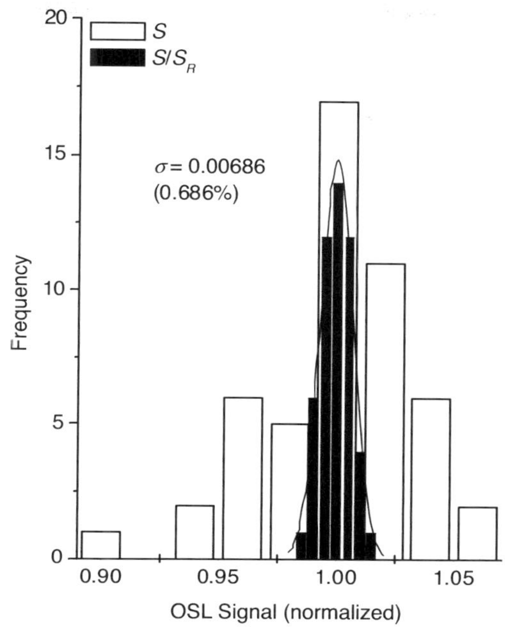


***Figure 5.13*** *Histogram of the total OSL signal S and ratio $S/S_R$ of 50 detectors irradiated with a known dose and read using the procedure outlined in Table 5.3. The signal S includes variations in the detector's mass and sensitivity.*

sensitivity of the detector. If the detector's mass or sensitivity is slightly higher, this would affect both signals $S$ and $S_R$ by the same factor. Figure 5.13 shows a histogram for both $S$ and $S_R$ for various OSL detectors with slightly different masses irradiated with the same dose. The histogram shows that the signal $S$ may vary by more than $\pm 5\%$, whereas the ratio $S/S_R$ is characterized by an experimental standard deviation of 0.7%.

The algorithm calibration curve is defined by the values of $S/S_R$ versus the known doses $D_i$, as shown in Figure 5.14. The precision in the $S/S_R$ values is demonstrated by the small error bars shown in Figure 5.14 indicating the experimental standard deviation of the data. Moreover, sensitivity changes affecting the detector are included in the calibration curve, which takes the shape of a saturating exponential function of the type:

$$\frac{S}{S_R} = \alpha \left(1 - e^{-\beta D}\right), \tag{5.12}$$

where $\alpha$ and $\beta$ are adjustable parameters, even though the dose response ($S$ versus dose) is linear in the dose range shown in Figure 5.14 (see Section 3.4.2.3). The inset in Figure 5.14 shows the relative difference between the fitted function and the experimental data, demonstrating that the agreement is within $\pm 1\%$ for doses in this interval.

Once the calibration curve is determined, OSL detectors exposed to the radiation field of interest are read using the same protocol (Table 5.3). The irradiation with reference dose $D_R$ should be identical to the one used in the determination of the calibration curve. For

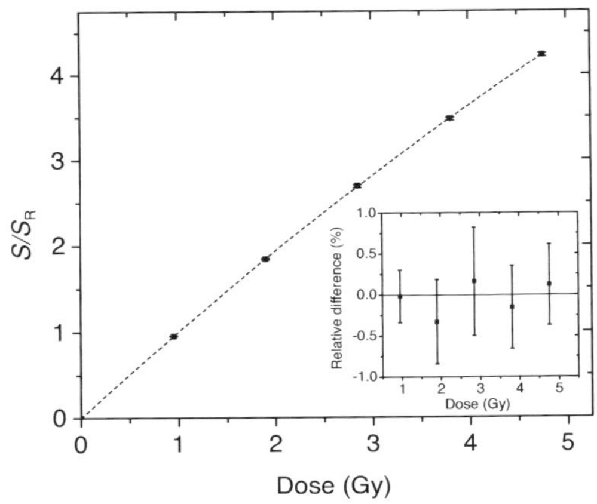


***Figure 5.14*** *Calibration curve for the high precision algorithm described in this section and relative difference between the experimental data and the fitted saturating exponential curve (inset). Each datum point corresponds to the mean and experimental standard deviation of the data based on readings from five OSL detectors. Typical experimental standard deviation values were of the order of 0.5%. This value is the experimental deviation of the data, therefore representing the spread observed in the readout of the five OSL detectors (ISO/IEC, 2008).*

each detector, the OSL reading $M_{\text{raw}}^{\text{OSL}}$ is then given by the inverse function of Equation (5.12):

$$M_{\text{raw}}^{\text{OSL}} = -\frac{1}{\beta} \ln\left(1 - \frac{S/S_{\text{R}}}{\alpha}\right). \tag{5.13}$$

If the OSL detectors were irradiated in conditions identical to those used to obtain the algorithm calibration curve, $M_{\text{raw}}^{\text{OSL}}$ is simply the dose $D$. Otherwise, $M_{\text{raw}}^{\text{OSL}}$ can be treated as an arbitrary indication (rdg) and the formalism described in Section 5.3.1 is applied to obtain the calibration and appropriate correction factors.

The protocol requires delivery of a fixed reference dose $D_{\text{R}}$ to each OSL detector and, therefore, an automated reader with built-in irradiation source (see Chapter 2) is more convenient and guarantees reproducibility in the irradiation and readout by minimizing handling of the detector. The reproducibility in the delivery of the reference dose is critical to obtain good results with this protocol.

The calibration curve of Figure 5.14 was obtained using a Risø TL/OSL reader and a set of detectors irradiated with known doses in the $\sim$1–5 Gy range, delivered by the built-in $^{90}$Sr/$^{90}$Y source from the Risø reader calibrated against a NIST standard in $^{60}$Co gamma dose to water (unpublished results). The reference dose was $\sim$0.95 Gy, delivered also using the built-in $^{90}$Sr/$^{90}$Y beta source. The OSL readouts in step 1 and 3 (Table 5.3) were carried out for a period long enough (600 s) to record the entire OSL decay curve in the current conditions (green LEDs with $\sim$10 mW/cm$^2$). The objective of the long stimulation period was to avoid the influence of fluctuations of the stimulation power in the OSL signal, since the total OSL area does not depend on the stimulation power, and to empty the main dosimetric traps in step 1 before the irradiation in step 2. The obtained calibration curve

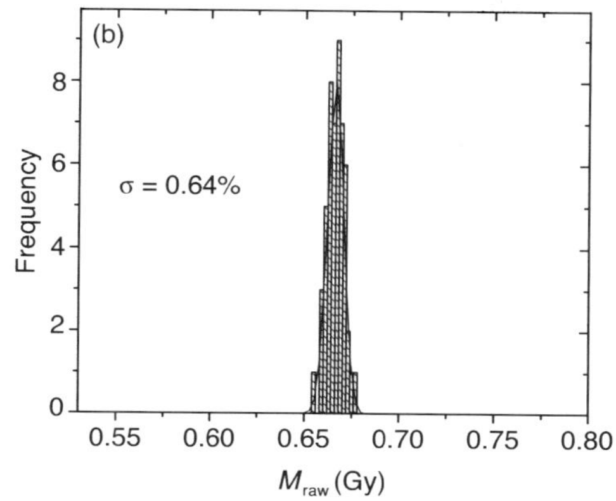


***Figure 5.15*** *Histogram of the OSL reading $M_{raw}^{OSL}$ for 50 OSL detectors irradiated in reference conditions using a 6 MV X-ray beam from a clinical linear accelerator. The graph is based on data obtained by Yukihara et al. (2005).*

(Figure 5.14) demonstrates that the uncertainties associated with the OSL readout are of the order of 0.5%, and that the saturating exponential describes very well the experimental data.

The protocol was tested in clinical conditions, using irradiations with a 6 MV X-ray beam to obtain the calibration curve (Yukihara *et al.*, 2005, 2008a). Test of the technique's precision was performed by irradiating 50 detectors in identical conditions using the 6 MV X-ray beam and obtaining the OSL readings $M_{raw}^{OSL}$. Figure 5.15 shows a histogram of the $M_{raw}^{OSL}$ values, again demonstrating that the technique provides high precision, the distribution being characterized by an experimental standard deviation of 0.64% of the mean value.

There are several benefits due to the fact that the analysis is based on the $S/S_R$ values and the reference doses are delivered by a highly reproducible radioisotope irradiator:

- The precision in $M_{raw}^{OSL}$ is very high, <0.7% for a single OSL detector (experimental standard deviation), therefore allowing the precise determination of other quantities such as calibration and correction factors.
- Dependency on individual detector sensitivity (due to variations in dosimeter mass or intrinsic sensitivity) is canceled out.
- If the measurements of $S$ and $S_R$ are made within a short period, dependency on the reader sensitivity is canceled out.
- Non-linearity of the responses are automatically taken into account.

The latter point is not obvious from the previous discussion, but it can be demonstrated when the linearity correction factor $k_D$ is calculated using the raw indication $M_{raw}^{OSL}$ taken to be the signal $S$ or the result of Equation (5.13), that is, using the calibration curve. Figure 5.16 shows $k_D$ calculated based on these two approaches. Because the OSL signal becomes supralinear after a dose of a few grays, the correction factor $k_D$ decreases as the dose increases when the OSL signal $S$ is taken as the OSL indication $M_{raw}^{OSL}$. However, when the

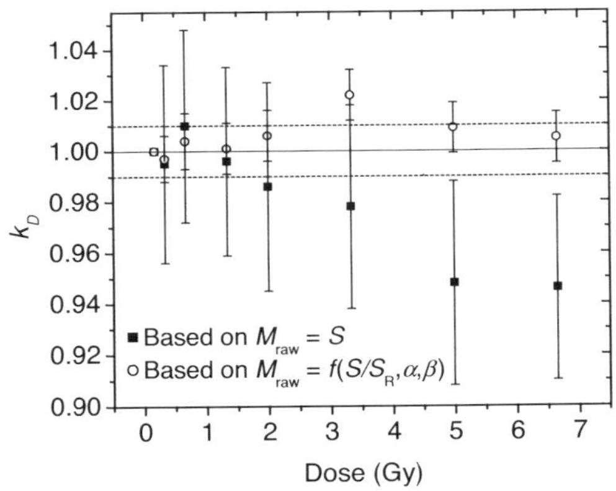


***Figure 5.16*** *Linearity correction factor $k_D$ for the OSL indicated value $M_{raw}$, when $M_{raw}$ is based on the OSL signal (total OSL area) or the result of Equation (5.13).*

OSL indication $M_{\text{raw}}^{\text{OSL}}$ is taken to be the result of Equation (5.13), the correction factor $k_D$ remains essentially constant, within the uncertainties of the experiment.

In summary, based on our experience developed during the investigations of this protocol in a clinical environment (Yukihara *et al.*, 2005, 2008a), we recommend the following implementation:

(i) Determine the calibration curve using an irradiator with high reproducibility.
(ii) Plot the calibration curve as in Figure 5.14 and fit the experimental data with the saturating exponential function shown in Equation (5.12).
(iii) Demonstrate that the calibration data are precise and well represented by the fitted function; if that is true, the value of $M_{\text{raw}}^{\text{OSL}}$ given by Equation (5.13) should be very precise.
(iv) Determine the calibration factor using $N_{D,\text{w},Q_0}^{\text{OSL}} = D_{\text{w},Q_0}/M_{Q_0}^{\text{OSL}}$ using standard calibration procedures (Almond *et al.*, 1999; IAEA, 2000).

The OSL detectors can then be applied in studies for the determination of the other correction factors or, if these correction factors are known, for dosimetry of non-reference radiation fields. The calibration curve should be acquired in the range of doses that one is expected to expose the OSL detectors.

Limitations of this readout protocol include the need for a built-in source for delivering the test dose and the long readout times required to read the entire OSL curve. Tests have not been performed using only the initial intensity as the OSL signal.

### 5.3.3 A Checklist for Reporting OSL Results

The results in OSL dosimetry depend on several experimental parameters that are often not fully specified in the literature. The situation is similar in TL dosimetry and, in fact, a checklist for the reporting measurements in TL dosimetry has been proposed to address this issue (Kron *et al.*, 1999). Although this is an important aspect for any application, it is even more relevant in medical dosimetry due to the importance of precision and accuracy,

*Table 5.4* Recommended information (not necessarily exhaustive) to be included in the methodology section when presenting results in OSL dosimetry

| Item | Information to include |
|---|---|
| Detector | • Material (single crystal, or include composition of additional materials)<br>• Detector form (shape, size and mass if appropriate)<br>• Characteristics of detector holder (if applicable) |
| Preparation | • Annealing: maximum temperature and temperature profile (e.g., heating rate, anneal time, cooling rate), crucible material<br>• Bleaching: light source, duration, filters, irradiance at detector position<br>• Packaging: composition and mass thickness of packing layers, possible air gaps, transparent or opaque (light tight?) |
| Readout | • Reader manufacturer and model<br>• Stimulation power at the dosimeter (irradiance), stimulation mode (e.g. low dose or high dose in microStar reader)<br>• Type of photodetector (e.g. PMT photocathode) and filters used in front of the PMT<br>• Light collection system (type and material for lenses, optical windows, mirrors, etc.)<br>• Readout temperature<br>• Readout approach (e.g., CW-OSL, POSL, LM-OSL)<br>• Calibration factors or conversion coefficients used by the reader<br>• Definition of OSL signal or indicated value: for example counts or mSv, period of integration<br>• Instrumental background subtraction<br>• Reader algorithm or information on type of indicated result (e.g., counts or mSv)<br>• Mode of calibration and whether or not element correction factors were automatically used<br>• Time elapsed since irradiation |
| Analysis | • Additional correction factors applied<br>• Additional calibration factors used |
| General | • Estimate of the reproducibility (e.g., relative standard uncertainty)<br>• Estimate of uncertainties |

All of the listed items are not necessarily relevant for all studies.

and the fact that the protocols and procedure are not yet well established for OSL. Thus, in Table 5.4 we offer a list of parameters that should be mentioned when reporting results from OSL measurements, adapted from the recommendations for TL dosimetry by Kron *et al.* (1999).

## 5.4 Optical-Fiber OSL Systems for Real-time Dosimetry

### 5.4.1 Basic Concept

Optical-fiber OSL systems can provide real-time assessment of the dose and dose rate at a patient's skin or inside cavities. This is useful for applications such as in-vivo dose

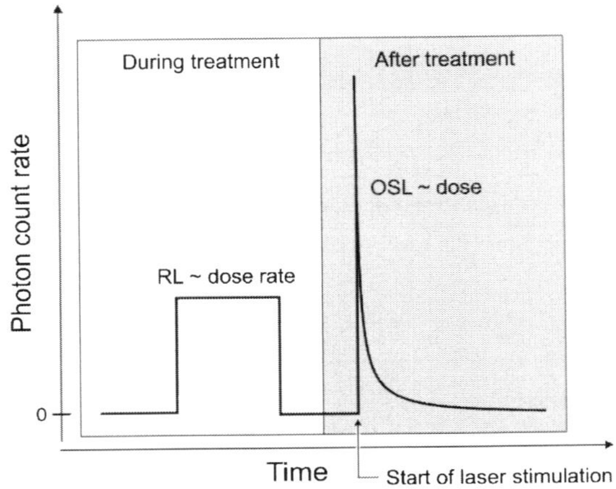


***Figure 5.17*** *Signal from an OSL probe during irradiation and upon optical stimulation after the irradiation has ended. Reprinted from International Symposium on Standards and Codes of Practice in Medical Radiation Dosimetry, Vienna. Andersen, C. E., et al. Development of optical fibre luminescence techniques for real time in vivo dosimetry in radiotherapy. 353–360. Copyright (2003) with permission from IAEA.*

verification in radiotherapy and the monitoring of patient's skin doses during fluoroscopy procedures. An optical-fiber system consists of an OSL detector optically connected to a remote reader through an optical fiber (see Figure 2.44). The optical fiber is responsible for carrying the stimulation light from the reader to the OSL detector, and the luminescence signal from the detector back to the reader.

There are two types of luminescence processes occurring in the OSL detector that can be exploited: radioluminescence (RL) occurring during irradiation, and OSL occurring during optical stimulation either during or after irradiation. RL is the spontaneous luminescence observed during irradiation and is related to the prompt recombination of radiation-induced electron–hole pairs within the detector. As in scintillation-based optical-fiber systems (Beddar, Mackie and Attix, 1992a, c; Beddar, 2006; Jordan, 1996; Suchowerska *et al.*, 2007), the RL signal provides information on the dose rate of the radiation field (see Figure 5.17). As discussed in Chapter 2, the OSL signal is related to the stimulated recombination of electron–hole pairs created in the detector during irradiation and trapped at defects in the crystal lattice (trapping centers). If the OSL detector is stimulated after the irradiation is over, the OSL signal is proportional to the net dose of radiation absorbed by the probe (see Figure 5.17). If the detector is stimulated during irradiation, RL and OSL will be observed simultaneously and an algorithm is necessary to separate these two contributions (see Section 5.4.3).

There are several aspects that make optical-fiber OSL systems attractive for medical dosimetry:

(i) In principle they can provide both real-time assessment of the dose rate (using the RL) and an independent assessment of the absorbed dose (using the OSL).
(ii) Owing to the high sensitivity of the OSL materials, the detectors can be made sufficiently small (~0.5 mm) to fit inside medical needles or catheters.

(iii) The OSL material with the most promising properties ($Al_2O_3$:C) is inert, hard, chemically stable and has a relatively low effective atomic number (11.3) (Bos, 2001).
(iv) The readout process is completely optical and, therefore, immune to electromagnetic interference.

One of the complications in real-time, optical-fiber dosimetry is the influence of spontaneous luminescence (RL) and Cherenkov radiation generated in the optical fiber itself, often called the "stem effect" (Arnfield *et al.*, 1996; Beddar, Mackie and Attix, 1992b; Deboer, Beddar and Rawlinson, 1993). Cherenkov radiation is the light produced by charged particles traveling in a medium of index of refraction $n$ with velocity higher than the phase velocity of the electromagnetic wave in that medium, that is, $v > c/n$ (Beddar, Mackie and Attix, 1992b; Jackson, 1999). In both cases, the signal varies with the length of optical fiber inside the radiation field, therefore leading to an over-estimation of the dose at the detector's position.

Several solutions have been proposed to overcome the stem effect in scintillator-based optical-fiber systems. One solution is to use two interlaced optical fibers (Beddar, 2006), one connected to the probe and the other used only to evaluate the stem effect. This approach has some technical disadvantages, such as the need for two detection systems, one for each fiber, associated electronics and compensation for differences in efficiency in the two detection systems (Clift, Johnston and Webb, 2002). Moreover, in high dose gradients the radiation field experienced by each optical fiber may be different. Another approach is the use of additional optical filtration to block the light generated in the optical fiber, provided this light and the probe's luminescence are separated in wavelength (Arnfield *et al.*, 1996; Clift, Sutton and Webb, 2000). This approach is particularly useful for continuous radiation sources, such as radioisotopes used in brachytherapy. For pulsed radiation sources, such as electron and X-ray beam from clinical linear accelerators, time-delayed gated detection of the luminescence can effectively eliminate the stem effect if the luminescence lifetime of the excited luminescence centers in the probe's material is longer than the typical radiation pulse width (Clift, Johnston and Webb, 2002; Jordan, 1996; Justus *et al.*, 2004).

More sophisticated approaches to reduce the contribution from Cherenkov radiation include the use of an air core light guide (Lambert *et al.*, 2008) and comparative analysis between different wavelength regions of the detected light spectrum (Archambault *et al.*, 2006; Fontbonne *et al.*, 2002; Frelin *et al.*, 2005).

With OSL detectors the stem effect is avoided entirely by performing the OSL readout after the irradiation is over, as illustrated in Figure 5.17. If real-time information on the radiation field is required, the techniques discussed above for scintillator-based optical-fibers systems (e.g., gated detection) can be applied to the RL signal of the detector. Another approach is to pulse the OSL stimulation during irradiation. During stimulation pulses the OSL, RL and stem-effect signals are measured, whereas between the pulses only the RL and stem effect are detected. Thus, by separating the OSL signal from the other luminescence signals, the stem effect can be eliminated or reduced. These readout approaches will be discussed in Section 5.4.3.

### 5.4.2 Optical-Fiber OSL System Designs and Materials

Different optical-fiber OSL systems have been investigated. The design varies slightly according to the detector material, necessary optical filters and technical approaches to separate the stimulation light from the detector's RL and OSL signal.

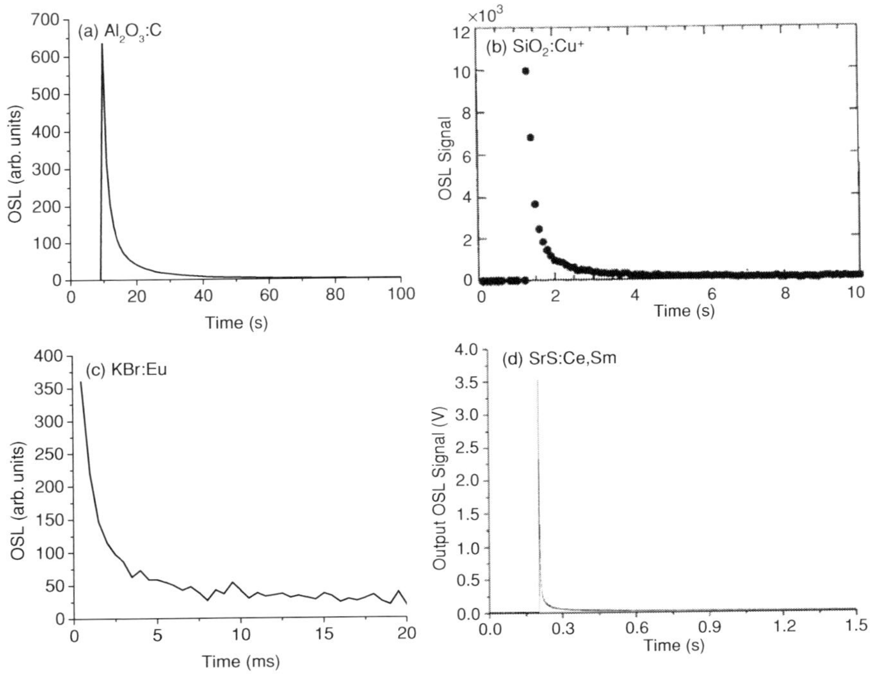


*Figure 5.18* Example of OSL curves obtained using various optical-fiber systems: (a) $Al_2O_3$:C; (b) $SiO_2:Cu^+$; (c) KBr:Eu and (d) SrS:Ce,Sm. The data for Figure (a) were provided by Dr. Claus Andersen, Risø National Laboratory; Figure (b) was from Radiation Protection Dosimetry, Justus, B. L., et al., Optically stimulated luminescence dosimetry using doped fused quartz. Vol. 84, 189–192. Copyright (1999) with permission from Oxford University Press; data for Figure (c) were provided by Dr. David Klein, University of Texas M. D. Anderson Cancer Center; Figure (d) was reprinted from IEEE Transactions on Nuclear Science, Benoit, D., et al. Real-time fibered optically stimulated luminescence dosimeter based on SrS:Ce,Sm phosphor. Vol. 55, 2154–2160. Copyright (2008) with permission from IEEE.

At least four OSL materials have been used in optical-fiber systems ($Al_2O_3$:C, MgS:Sm, $SiO_2:Cu^+$ and KBr:Eu), three of which were compared by Ranchoux *et al.* (2002). These materials differ in sensitivity, effective atomic number, characteristic decay time of the OSL curves, and fading characteristics. Figure 5.18 shows some examples of OSL curves obtained using different optical-fiber systems. The time scale of the OSL decay depends not only on the material, but also on the stimulation power; nevertheless, these curves provide examples of the time scales observed. The slowest decay is observed for $Al_2O_3$:C and the fastest for KBr:Eu. Some properties of these materials are listed in Table 2.2.

Polf *et al.* (2002) and Ranchoux *et al.* (2002) used an $Al_2O_3$:C detector connected to a bifurcated optical fiber to separate the stimulation light and the detector's RL and OSL signal. The stimulation light travels from the laser to the detector through one of the optical fibers, whereas the RL and OSL signal from the detector travels back to a PMT through another optical fiber (Figure 5.19). The $Al_2O_3$:C-based system described by Polf *et al.* uses

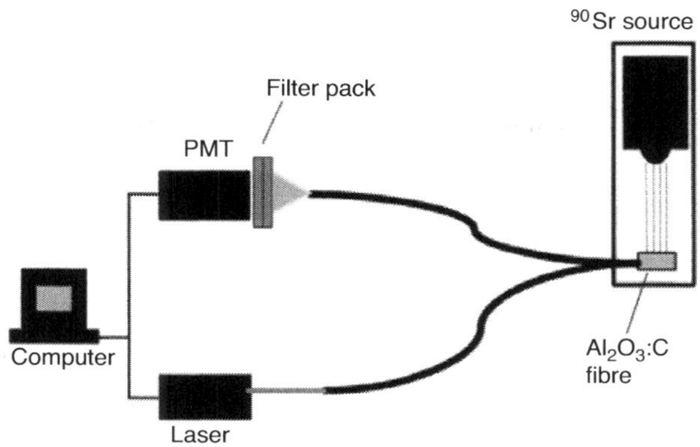


*Figure 5.19* *Diagram of optical-fiber OSL system using a bifurcated optical fiber to separate the stimulation light from the RL/OSL signal from the probe. Reprinted from Radiation Protection Dosimetry, Polf, J.C., et al. A real-time, fibre optic dosimetry system using $Al_2O_3$ fibres. Vol 100, 301–304. Copyright (2002) with permission from Oxford University Press.*

a frequency-doubled 40 mW Nd:YAG laser (532 nm) to stimulate the $Al_2O_3$:C probe and optical filters (420 nm interference filters) in front of the PMT to block the stimulation light and transmit the main luminescence from the $Al_2O_3$:C detector.

Researchers from the Risø National Laboratory (Denmark) developed and tested an optical-fiber system based on $Al_2O_3$:C (see Figure 5.20) for dosimetry of external radiation therapy beams (megavoltage X-rays and protons), mammography and brachytherapy (Andersen *et al.*, 2003, 2006, 2007, 2009a; Aznar *et al.*, 2004, 2005; Edmund *et al.*, 2007). Instead of bifurcated fibers, the system uses a dichroic mirror positioned at 45° that transmits the stimulation light from the laser and reflects the blue RL/OSL signal from the $Al_2O_3$:C probe to the PMT (see Figure 2.44). The system's repeatability was shown to be 0.2% (experimental standard deviation) and the dose response was linear over the range of doses investigated (0–3 Gy) (Andersen *et al.*, 2003).

A system based on single crystals of $Al_2O_3$:C was also described by Magne *et al.* (2008). Their reader design has an optical switch that allows the RL measurement of multiple probes during irradiation. A dichroic filter is used to separate the optical path of the stimulation light (CW diode pumped solid state laser, 532 nm, 200 mW) from the probe's RL and OSL signal. As discussed earlier, the authors observed a change in the RL and OSL sensitivity with the dose accumulated in the crystal, making it necessary to pre-irradiate the probes to "condition" (stabilize) the detectors prior to use. Repeatability between 0.6 and 1.3% was demonstrated for probes that were pre-irradiated with doses between 250 Gy and 1 kGy.

An optical-fiber system based on $Cu^+$-doped fused quartz probes has been described (Huston *et al.*, 2002; Justus *et al.*, 1997). The OSL of $Cu^+$-doped fused quartz can be excited with 790 nm or shorter wavelength, producing OSL luminescence in the blue-green region (~500 nm) (see Section 2.7.1.4). The system uses bifurcated optical fibers in which one of the fibers is optically coupled to a 790 nm solid-state diode laser and the other fiber is coupled to a PMT, eliminating the need for mirrors. Both scintillation and OSL signals were shown to be linear from 0.01 Gy to ~10 Gy. The response was tested for megavoltage

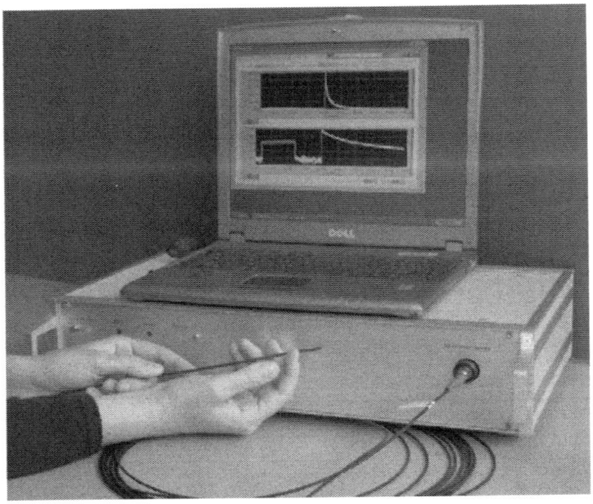

(a)

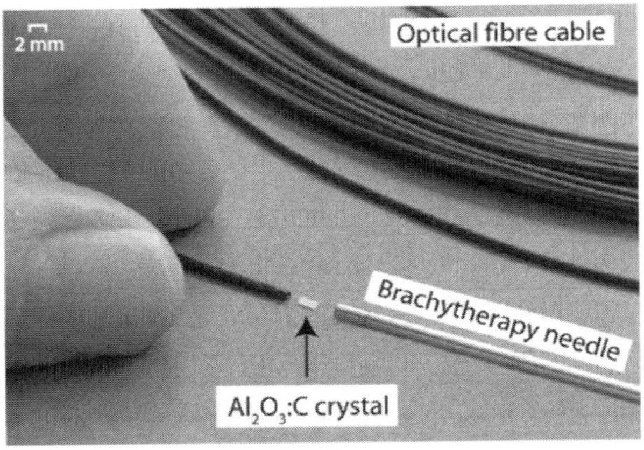


(b)

***Figure 5.20*** *Optical-fiber OSL system developed by Risø National Laboratory (Andersen et al., 2009a; Aznar et al., 2004). Figure (a) reprinted from Physics in Medicine and Biology, Aznar, M.C., et al. Real-time optical-fibre luminescence dosimetry for radiotherapy: physical characteristics and applications in photon beams. Vol. 49, 1655–1669. Copyright (2004) with permission from IOP; Figure (b) reprinted from Radiation Protection Dosimetry, Jursinic, P. A., Characterization of a fiber-coupled $Al_2O_3$:C luminescence dosimetry system for online in vivo dose verification during $^{192}$Ir brachytherapy. Vol. 36, 708–718. Copyright (2009) with permission from American Association of Physicists in Medicine.*

electrons and X-ray beams and found to be independent of energy and field size within the uncertainties of the experiment. A four-channel OSL system based on $Cu^+$-doped fused quartz is shown in Figure 5.21.

An optical-fiber system based on SrS:Ce,Sm phosphor was described by Benoit *et al.* (2008b) and tested using a $^{60}$Co gamma-ray source and a 6 MV X-ray beam. The OSL signal

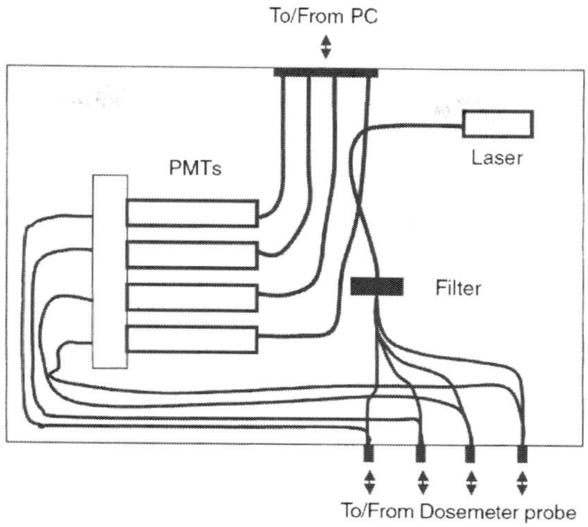


***Figure 5.21*** *Diagram of a four channel optical-fiber OSL system based on $Cu^+$-doped fused quartz and bifurcated fibers. Reprinted from Radiation Protection Dosimetry, Huston, A.L., et al. Optically stimulated luminescence glass optical fibre dosemeters. Vol. 101, 23–26. Copyright (2002) with permission from Oxford University Press.*

stimulated with a 980 nm laser decays very fast, reaching background levels in ~1.5 s. The reported repeatability of the OSL was 0.8% (one experimental standard deviation) with errors in the dose measurement within 4.9%.

Gaza and McKeever (2006) described an OSL system based on KBr:Eu for real-time dosimetry in radiotherapy. As shown in Figure 5.18, the advantage of the KBr:Eu system is the fast OSL decay when compared to other OSL materials used in optical-fiber dosimetry. One of the disadvantages is the relatively high effective atomic number ($Z_{eff} = 31.5$), which should result in strong photon energy dependence. This energy dependence was in fact verified by Klein (2008). Another disadvantage is the fact that the OSL signal is related to trapping centers unstable at room temperature, associated with TL peaks at 85 and 115 °C (Klein and McKeever, 2008). However, since the readout is so rapid and the KBr:Eu system so sensitive, the OSL signal can be read very rapidly (e.g. at rates of several hertz) such that the instability of the signal is not an issue. This type of system has been investigated for applications in fluoroscopy, brachytherapy and proton radiation therapy (Klein, 2008).

### 5.4.3 Readout Approaches

The following readout approaches have been discussed in the literature and investigated for optical-fiber OSL systems.

#### *5.4.3.1 Post-Irradiation OSL Readout*

In the post-irradiation OSL readout the OSL signal is recorded after the irradiation has ended, as illustrated in Figure 5.17, therefore avoiding the stem effect. The total OSL signal is proportional to the integrated dose absorbed by the probe. The post-irradiation

OSL signal does not provide any real-time information on the dose rate during irradiation. However, when used in conjunction with the detection of the RL signal, as illustrated in Figure 5.17 and described in the next section, the post-irradiation OSL readout provides a second, independent estimation of the absorbed dose.

#### 5.4.3.2 *Time-Delayed Gated Detection of RL*

As illustrated in Figure 5.17, the RL signal from the OSL probe during irradiation can provide information on the dose rate. However, direct measurement of the RL signal can be affected by the stem effect. A direct influence of the stem effect is the dependence of the total signal on the field size, as this affects the length of optical fiber exposed to the radiation field (Justus *et al.*, 2004).

For pulsed radiation fields such as electron and X-ray beams from clinical linear accelerators, a possible solution is to detect the RL in between the pulses of radiation using gated detection that is delayed with respect to the radiation pulses (Clift, Johnston and Webb, 2002; Jordan, 1996; Justus *et al.*, 2004). This approach requires OSL materials in which the luminescence lifetime is longer than the typical radiation pulse width. Typical pulse widths from clinical linear accelerators are of the order of 5 μs at time intervals of 5 ms (repetition frequency ~200 Hz), depending on the selected dose rate. Therefore, the time-delayed gated detection of the RL signal is suitable for OSL materials such as $Al_2O_3$:C and $Cu^+$-doped fused quartz, which have luminescence centers characterized by lifetimes of the order of 35 ms ($Al_2O_3$:C) and 50–100 μs ($Cu^+$-doped fused quartz) (see Chapter 2).

A complicating issue is the fact that the RL intensity is not simply proportional to the dose rate, but the RL sensitivity increases as the accumulated dose in the crystal increases (see Figure 5.22a). This is a phenomenon directly linked to the dynamic changes in the probabilities of trapping and recombination during irradiation, caused by filling of traps and recombination centers and consequent change in the concentration of the respective centers (see Chapter 2 for more discussion). However, these changes were observed to be reproducible as a function of the dose and, therefore, algorithms can be devised to take this effect into account (Andersen *et al.*, 2006). The algorithm requires determining the dose dependence of the RL sensitivity. The data obtained can then be used to correct for the RL sensitivity changes in real time by tracking the accumulated dose in the crystal at any instant $t$ and applying the appropriate sensitivity correction. Figure 5.22b shows an example of the data obtained when the RL signal from a probe irradiated with a constant dose rate (Figure 5.22a) is corrected. The calculated accumulated dose to the OSL detector is shown in Figure 5.22c. An improved version of the algorithm is described by Damkjær, Andersen and Aznar (2008).

The RL response of $Al_2O_3$:C can also be improved if the influence of shallow traps is taken into account. The time resolution of the system is ideally of the order of 0.1 s, corresponding to three times the lifetime of the main luminescence centers in $Al_2O_3$:C (35 ms), but shallow traps degrade this time resolution to a few seconds (Damkjær, Andersen and Aznar, 2008). Shallow traps compete with recombination centers for the capture of charges during irradiation. Therefore, if the dose rate changes instantly from zero to a specific value, as in a step function, the RL signal will increase slowly reaching an equilibrium value only when the rate of charges captured by the shallow traps is balanced by the rate of charges thermally stimulated out of the shallow traps. Likewise, if the dose rate is

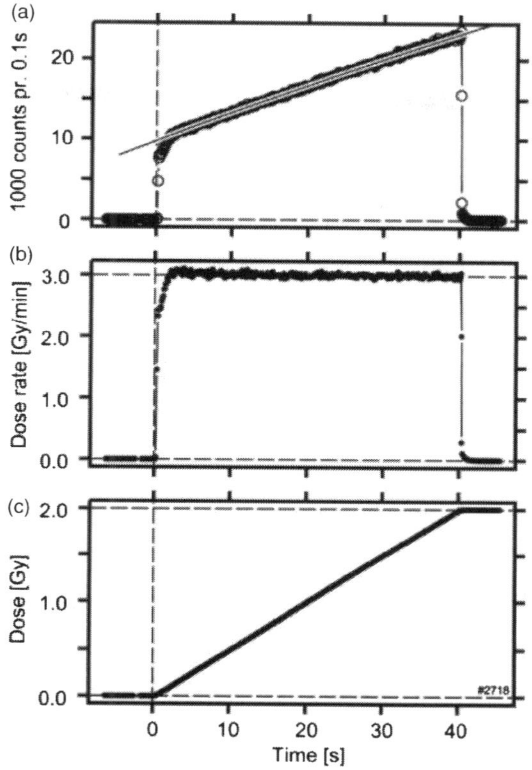


***Figure 5.22*** *(a) RL signal from an $Al_2O_3$:C probe during irradiation at a constant dose rate; (b) calculated dose rate corrected for the change in the RL signal with dose and (c) calculated dose in the crystal by integration of the dose rate. Reprinted from Radiation Protection Dosimetry, Andersen, C. E., et al. An algorithm for real-time dosimetry in intensity-modulated radiation therapy using the radioluminescence signal from $Al_2O_3$:C. Vol. 120, 7–13. Copyright (2006) with permission from Oxford University Press.*

suddenly decreased to zero, the signal from the detector will decay slowly because of the phosphorescence associated with the shallow traps. Therefore, the shallow traps slow the response of the $Al_2O_3$:C detector during irradiation. This effect can be seen in Figure 5.23, in which the raw RL signal shown in Figure 5.23a is corrected for the sensitivity resulting in the data in Figure 5.23b. In this figure, the slow decay associated with the shallow traps can be seen at the end of the irradiation.

A model including charge capture and first-order thermally stimulated release of charges from the shallow traps was used by Damkjær, Andersen and Aznar (2008) to correct for the influence of shallow traps. The parameters of the model were obtained comparing the RL signal with the ionization chamber signal shown in Figure 5.23b. The correction consists in determining for each instant $t$ the amount of decrease in the RL signal $d_1$ in an interval $\Delta t$ due to the competition by the shallow traps, and the amount of increase in the RL signal $d_2$ in the same interval due to the release of electrons from the shallow traps. The RL signal is then corrected by adding $d_1$ and subtracting $d_2$ for each time interval. The result of such

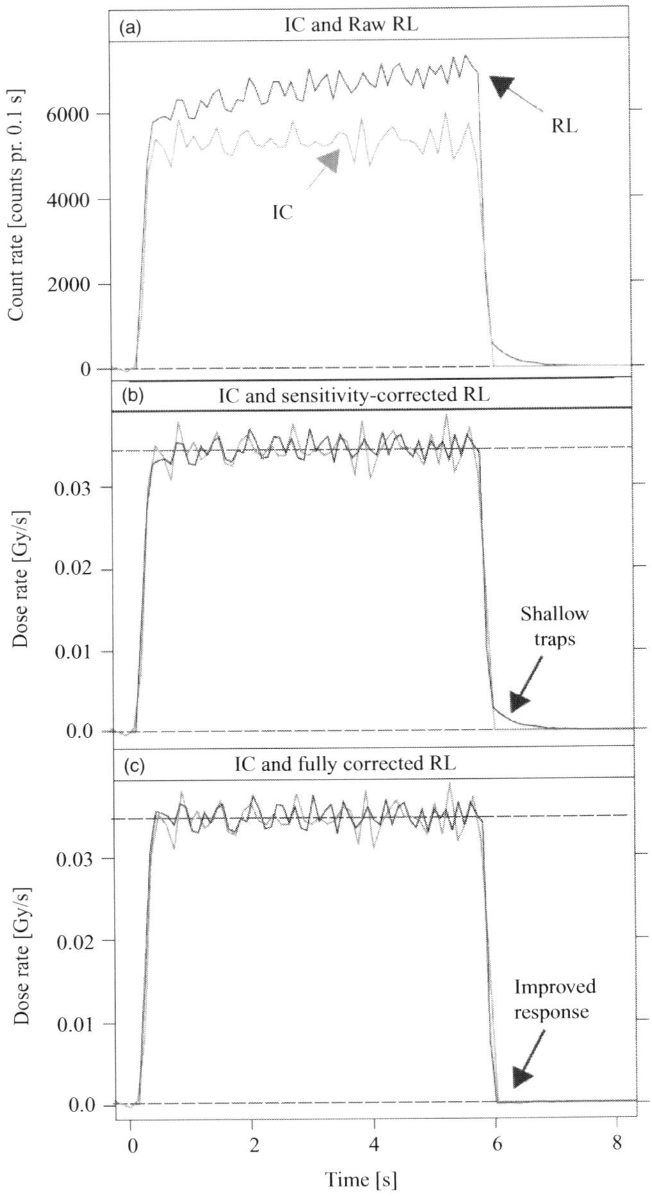


***Figure 5.23*** *(a) Comparison between the raw RL signal from the $Al_2O_3$:C detector with an ionization chamber measurement; (b) RL signal corrected for the dose-dependent sensitivity and (c) RL signal corrected for the dose-dependent sensitivity and influence of shallow traps. Reprinted from Radiation Measurements, Damkjaer, S.M.S., et al. Improved real-time dosimetry using the radioluminescence signal from $Al_2O_3$:C. Vol. 43, 893–897. Copyright (2008) with permission from Elsevier.*

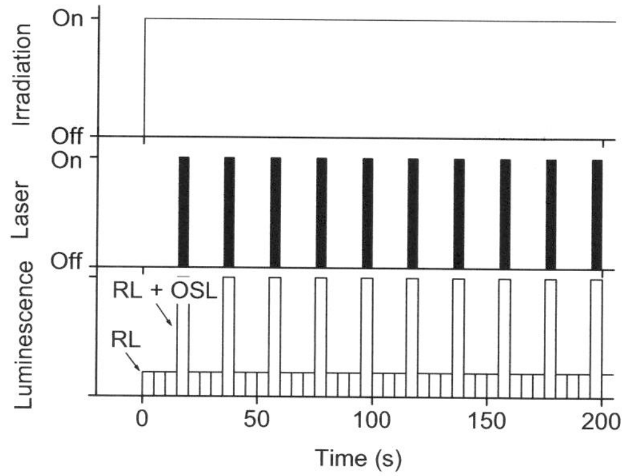


***Figure 5.24*** *Illustration of the RL/OSL discrimination by periodic stimulation of the OSL detector during irradiation. Reprinted from Radiation Measurements, Polf, J. C., et al., Real-time luminescence from $Al_2O_3$ fiber dosimeters. Vol. 38, 227–240. Copyright (2004) with permission from Elsevier.*

correction is shown in Figure 5.23c, where it can be seen that after correction the signal from the fiber system responds as fast as the ionization chamber to changes in the dose rate. Current limitations of the model are the fact that it does not include explicitly the temperature dependence of $d_1$ and possible dependence of the model parameters on the absorbed dose. For example, it has also been observed that the correction becomes less effective for doses above 3 Gy (Damkjær, Andersen and Aznar, 2008).

Optical-fiber systems based on separate RL and OSL measurements were demonstrated to provide high-precision measurements, $\sim 1\%$ experimental standard deviation (Andersen *et al.*, 2008). However, the change in RL sensitivity with dose is still considered one of the major drawbacks of this technology compared to scintillators (Beierholm *et al.*, 2008).

### *5.4.3.3 Real-time OSL Detection*

A more complex approach to eliminate or reduce the stem effect taking advantage of the OSL signal is to periodically stimulate the crystal during irradiation. In this approach, short pulses of light are delivered to the OSL probe during irradiation to stimulate the OSL signal (Figure 5.24). In between the light pulses, the signal consists mostly of RL and stem effect, whereas during the light pulses the signal is a sum of the RL, stem effect and OSL (Gaza *et al.*, 2004; Polf *et al.*, 2004). The "real-time OSL signal" can then be obtained by subtracting the PMT signal obtained in between stimulation from the PMT signal obtained during stimulation. The advantage of this readout approach is that it can be used for continuous radiation fields.

To understand the meaning of the real-time OSL signal, we should consider two extreme situations. If the trapping centers responsible for the OSL signal are completely depleted during each stimulation pulse, the real-time OSL signal is proportional to the absorbed dose accumulated since the last stimulation period. On the other hand, if the stimulation intensity is weak such that it does not significantly affect the trapping center population, the real-time OSL signal is proportional to the total dose accumulated in the crystal.

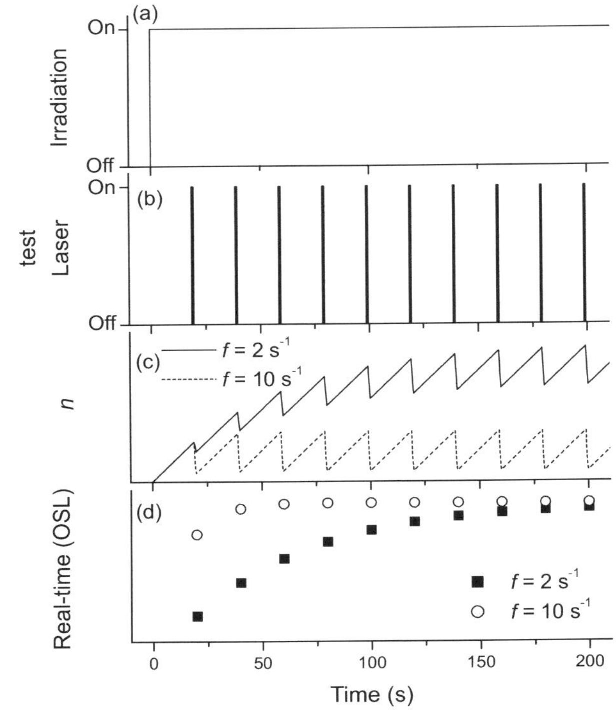


***Figure 5.25*** *Computer simulation of the real-time measurement for two optical stimulation rates: (a) dose rate, (b) laser power, (c) concentration of trapped charges and (d) real-time OSL signal. The data in Figure (c) and (d) were calculated for two OSL stimulation intensities, represented by the parameter* $f = \sigma\phi$*, where* $\sigma$ *is the defect photoionization cross-section and* $\phi$ *is the photon flux. The equilibrium level of the real-time OSL signal is the same in both cases, but equilibrium is reached faster in case of the stronger stimulation intensity (*$f = 10\ s^{-1}$*). Reprinted from Radiation Measurements, Polf, J. C., et al., Real-time luminescence from* $Al_2O_3$ *fiber dosimeters. Vol. 38, 227–240. Copyright (2004) with permission from Elsevier.*

In the case of $Al_2O_3$:C, the problem is that neither case is entirely satisfied. This is exemplified by the simulation results shown in Figure 5.25, where it can be seen that the trapping center population $n$ increases due to irradiation, but a small decrease is observed every time the detector is stimulated (Figure 5.25c). As a result, the real-time OSL signal reaches a steady-state level (Figure 5.25d) corresponding to the situation in which the number of electrons captured by the trapping centers in between stimulation pulses equals the number of electrons ejected from the trapping centers during a stimulation pulse. This behavior is confirmed by experimental data obtained using $Al_2O_3$:C detectors, shown in Figure 5.26. It can be observed that the real-time OSL signal reaches a steady-state level at a constant dose rate irradiation, instead of continuously increasing with the dose. Therefore, an algorithm is necessary to evaluate the instantaneous dose rate and the accumulated dose.

Algorithms to extract the dose rate and absorbed dose from the real-time OSL signal were investigated by Gaza, McKeever and Akselrod (2005). One of the proposed algorithms

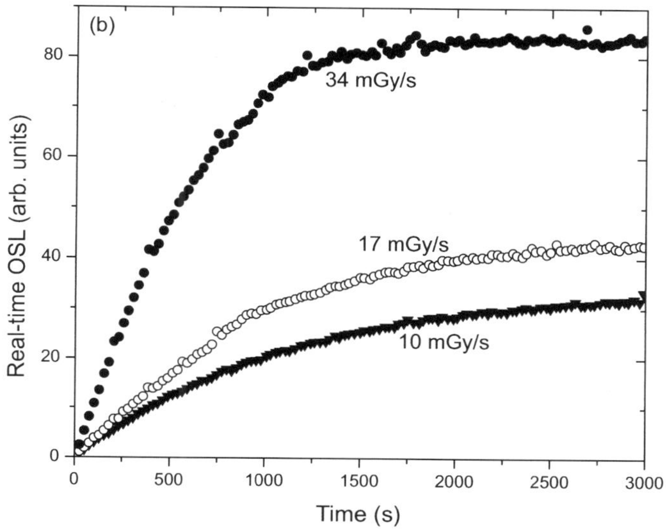


***Figure 5.26*** *Example of real-time OSL signal for an $Al_2O_3$:C detector irradiated with three different dose-rate levels. Reprinted from Radiation Measurements, Polf, J. C., et al., Real-time luminescence from $Al_2O_3$ fiber dosimeters. Vol. 38, 227–240. Copyright (2004) with permission from Elsevier.*

considers the OSL signal $I_2$ obtained during each stimulation period as composed of two components, a residual OSL signal $I_1'$ that was not read during the previous stimulation and an increment $\Delta_{OSL}$ due to the absorbed dose since the last stimulation period:

$$I_2 = I_1' + \Delta_{OSL} = I_1' + \alpha \times D. \tag{5.14}$$

Here $\alpha$ is the detector's sensitivity and $D$ is the dose absorbed since the last stimulation pulse. An approximation was used to estimate $I_1'$ based on the data from previous stimulation pulse, therefore allowing the increment dose $D$ to be calculated.

The algorithm was tested using an $Al_2O_3$:C detector irradiated with different dose-rate levels. The results shown in Figure 5.27 illustrate the performance of this algorithm. It can be observed that it takes some time for the OSL signal to achieve equilibrium after a change in dose rate. However, the dose rate calculated by the algorithm rapidly approaches a constant level. Uncertainty values as high as 8% were observed using this algorithm, which highlights the need for further research and development if the algorithm is to be used in radiotherapy applications.

The approach discussed here has three disadvantages. First of all, the precision of the algorithm relies on the proper estimation of the residual OSL signal for each stimulation pulse; small errors in each step propagate and accumulate to the next steps. A second disadvantage applies to pulsed radiation sources: the OSL and stem-effect signals during stimulation pulses and in-between the stimulation pulses are not necessarily identical. In these cases, it may be necessary to synchronize the stimulation with the radiation source. Thirdly, because the probe is being stimulated during irradiation, the net OSL signal is not

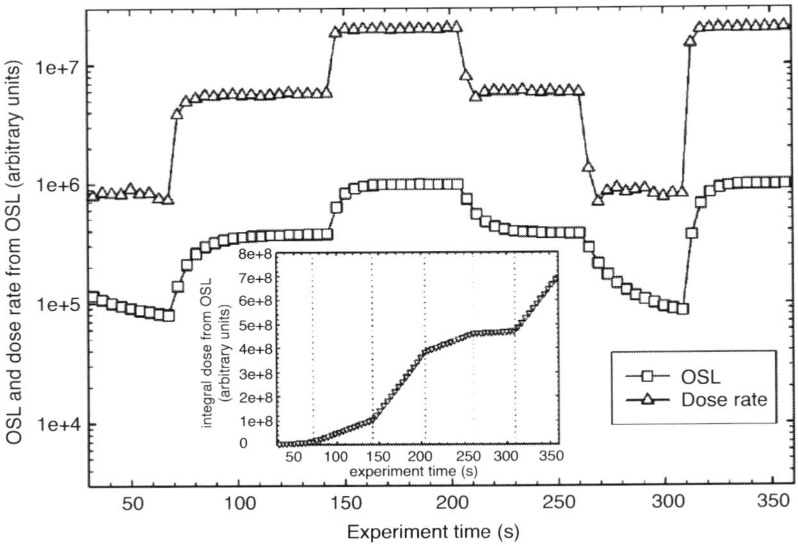


***Figure 5.27*** *Real-time OSL signal and calculated dose and dose rate for an $Al_2O_3$:C detector irradiated in six stages with three different dose rate levels (7.97 mGy/s, 55.4 mGy/s, 177 mGy/s) using an algorithm proposed by Gaza, McKeever and Akselrod (2005). Reprinted from Medical Physics, Gaza, R., McKeever, S.W.S. and Akselrod, M.S., Near-real-time radiotherapy dosimetry using optically stimulated luminescence of $Al_2O_3$:C: Mathematical models and preliminary results. Vol. 32, 1094–1102. Copyright (2005) with permission from American Association of Physicists in Medicine.*

preserved to estimate the absorbed dose after the irradiation is over. So, this is a one-time estimation only, as opposed to the previous approaches which allow the dose to be estimated independently using the RL and OSL signals from the $Al_2O_3$:C detector.

In the case of OSL materials with trapping centers characterized by high photoionization cross-sections, however, it is feasible to read an entire OSL curve in between the radiation pulses, thereby removing the need to correct for partial emptying during each stimulation pulse. This is the case of KBr:Eu OSL systems, in which the OSL curve decays to background levels in ~20 ms (see Figure 5.18c). Furthermore, in cases of pulsed irradiation the OSL signal can be read between the irradiation pulses by synchronizing the OSL stimulation pulses with those from the radiation source, as illustrated in Figure 5.28 (Klein, 2008).

## 5.5 Properties of $Al_2O_3$:C OSL Detectors for Medical Applications

Factors including precision, linearity, temperature dependence and dose rate dependence, are very dependent on the type of reader, readout approach, choice of filters in front of the PMT, OSL material and detector packaging, among others. Therefore, it is not possible to apply calibration and correction factors obtained with one set of experimental conditions to another set without verification. Until a particular OSL material, reader

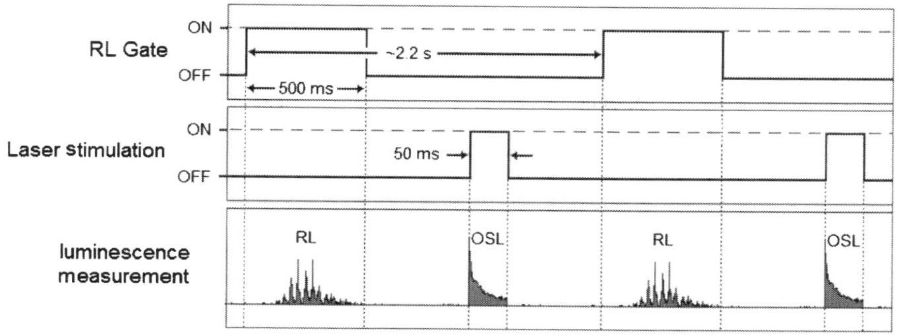


***Figure 5.28*** *Time scheme used in a KBr:Eu OSL optical-fiber system in which the OSL measurements are synchronized with irradiation pulses (proton spills) indicated by the RL intensity. In this scheme, the OSL stimulation is carried out 500 ms after the proton spill. Reprinted from Klein, D.M., 2008. Development and characterization of remote radiation dosimetry systems using optically stimulated luminescence of $Al_2O_3$:C and KBr:Eu. Ph. D. Dissertation, Oklahoma State University.*

and readout protocol become more established, each user should check whether the system performance is adequate for the intended application. Nevertheless, it is possible to learn from the experience reported in the literature, as reviewed in this section. We shall focus on $Al_2O_3$:C, which is the only OSL detector for which dedicated commercial readers are available and which has been investigated to some extent for medical dosimetry applications.

### 5.5.1 Influence Factors and Correction Factors

#### *5.5.1.1 Non-Linearity*

Based on the $Al_2O_3$:C dose response (Section 3.4.2.3), the correction factor for non-linearity $k_D$ can be considered unitary for doses in diagnostic radiology, but it must be accurately determined for high-precision measurements in radiotherapy. $Al_2O_3$:C dose response is linear below ~1 Gy and becomes supralinear for absorbed doses above that (i.e., $k_D < 1$), but the degree of supralinearity depends on the particular OSL reader, the choice of OSL signal and so on. Depending on the experimental conditions it can increase up to 20% in the range from 1 to 10 Gy.

The departure from linearity for doses above a few grays has been confirmed in a variety of studies (Jursinic, 2007; Miller and Murphy, 2007; Reft, 2009; Schembri and Heijmen, 2007). Some studies did not reported supralinearity due to the limited range of doses investigated (Viamonte *et al.*, 2008). The readout protocol described in Section 5.3.2.3 has demonstrated the need to compensate for the non-linearity of the OSL signal (see Figure 5.16). If this protocol is used to obtain $M_{\mathrm{raw}}^{\mathrm{OSL}}$, the correction factor for non-linearity can also be considered unity within $\pm 1\%$.

As pointed out earlier, supralinearity was not observed in $Al_2O_3$:C for charged particles with $Z = 2$ or higher (Sawakuchi *et al.*, 2008c).

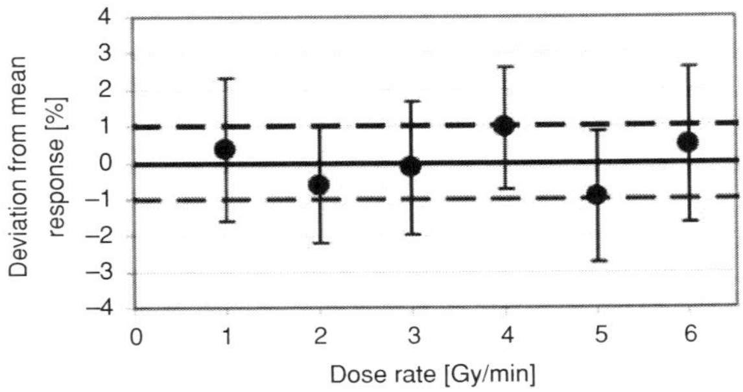


***Figure 5.29*** *OSL dependence on linear accelerator dose rate for $Al_2O_3$:C Luxel detectors (irradiation beam not specified). The error bars indicate an interval of ~2 standard deviations ($2\sigma$), and the dashed lines indicate the ±1% interval. Reprinted from Medical Physics, Schembri, V. And Heijmen, B.J.M., Optically stimulated luminescence (OSL) of carbon-doped aluminum oxide ($Al_2O_3$:C) for film dosimetry in radiotherapy. Vol. 34, 2113–2118. Copyright (2007) with permission from American Association of Physicists in Medicine.*

#### *5.5.1.2 Dose Rate*

Schembri and Heijmen (2007) investigated the effect of dose rate from 1 to 6 Gy $min^{-1}$ using a 6 MV X-ray beam from a clinical linear accelerator. The values obtained were within ±1% of the mean value (Figure 5.29), a result that was confirmed by other investigators using a 6 MV X-ray beam and a 9 MeV electron beam (Viamonte *et al.*, 2008; Yukihara *et al.*, 2008a).

The clinical linear accelerators used in the studies above are pulsed radiation sources, where changes in the mean dose rate are effectively changes in the pulse frequency, the dose per pulse being kept approximately constant. However, similar results were also obtained by Jursinic (2007) varying the dose per pulse using attenuators or changing the distance to the detector (Figure 5.30). Again, the detector signal was within ±1% regardless of the dose per pulse, with no clear trend of increase or decrease in the OSL signal with dose rate.

The overall conclusion is that the detectors are independent on the dose rate within the experimental uncertainties, at least in the conditions investigated.

#### *5.5.1.3 Fading*

There is clear evidence of a short-term decrease in the OSL signal until ~5 min to 10 min following irradiation (Jursinic, 2007; Reft, 2009). The short-term fading is likely associated with the influence of shallow traps. Phosphorescence produced by charges thermally stimulated from the shallow traps at room temperature may contribute to the detected signal, the contribution decreasing as the shallow traps are emptied. An alternative explanation is competition introduced by shallow traps. Following irradiation, the shallow traps are partially filled and, therefore, they compete less with the recombination centers for the capture of charges released by optical stimulation from the main dosimetric trap. As a result, the OSL signal is more intense following irradiation than at a later time, when the shallow traps are empty and can again compete for the capture of charges released by optical stimulation. Of course, both phosphorescence and competition processes may be simultaneously

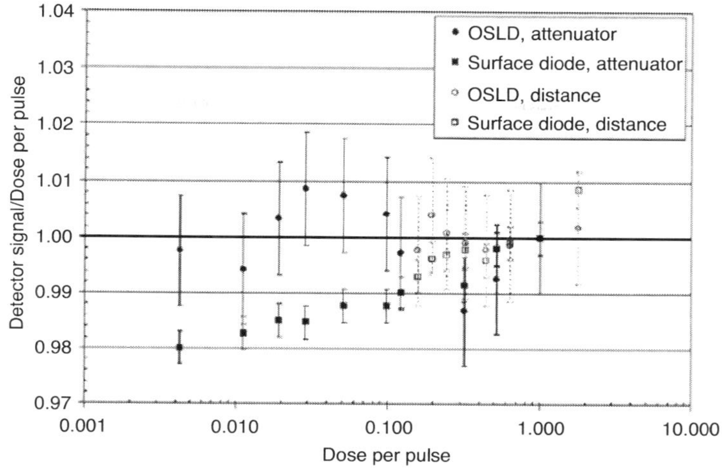


***Figure 5.30*** *OSL dependence on linear accelerator dose rate for* $Al_2O_3$*:C InLight OSL Dot dosimeters for a 6 MV X-ray beam, compared with surface diodes. The dose per pulse was varied using attenuators or by varying the distance to the OSLDs. The error bars indicate an interval of one standard deviation. Reprinted from Medical Physics, Jursinic, P.A., Characterization of optically stimulated luminescence dosimeters, OSLDs, for clinical dosimetric measurements. Vol. 34, 4594–4604. Copyright (2007) with permission from American Association of Physicists in Medicine.*

contributing to the apparent short-term fading in the OSL signal. Regardless of the explanation, a wait time of ~8 min following irradiation is recommended to avoid this short-term fading (Jursinic, 2007).

Long-term fading studies are limited. Schembri and Heijmen (2007) observed a fading of ~2% between 15 and 27 days following irradiation. A study was recently performed on the fading of the OSL signal of $Al_2O_3$:C Luxel detectors read using an InLight microStar reader (Gasparian, 2009; Yukihara *et al.*, 2010). OSL detectors were freshly irradiated with ~1 Gy and readouts of the same detectors were carried out at different interval of time. OSL detectors that had been irradiated with ~5 Gy more than one year previously to the experiment were also read to monitor the stability of the OSL reader and account for possible depletion of the OSL signal due to subsequent readouts. The readouts were carried out using weak stimulation to minimize depletion of the signal. The results shown in Figure 5.31 indicate that an initial fading of ~7% is observed immediately following irradiation until ~2 h, with a slow decrease observed for longer periods. The signal from OSL detectors irradiated more than one year before remained relatively stable, except for some depletion of the OSL signal due to multiple readouts, giving an idea of the stability of the reader over the course of the experiment. The data in Figure 5.31 show that, depending on the required precision, a wait period longer than 8 min may be necessary. The signal decayed ~2% in the 1–24 h interval after irradiation, and ~3% afterwards up to two months.

These results have not been confirmed by independent investigations and, moreover, they do not agree with a previous study that did not detect significant changes in the main OSL signal from $Al_2O_3$:C following irradiation over a period of two weeks (Yukihara and

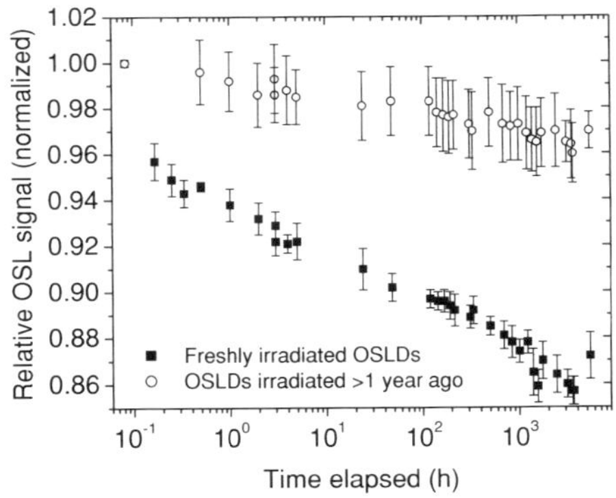


***Figure 5.31*** *Fading of $Al_2O_3$:C Luxel detectors immediately following irradiation and read using an InLight microStar reader, estimated based on detectors freshly irradiated with ~1 Gy. Detectors irradiated with ~5 Gy more than one year before the experiment were used to monitor the system's stability and depletion of the OSL signal due to repeated readouts; the results shown in the graph were corrected for these two parameters. Reprinted from Radiation Measurements, Yukihara, E. G., et al., Medical applications of optically stimulated luminescence dosimeters (OSLDs). Vol. 45, 658–662. Copyright (2010) with permission from Elsevier.*

McKeever, 2006b). The conditions of the latter experiment were considerably different, raising the possibility that fading may depend on the experimental condition, such as whether the OSL signal is the total or initial OSL intensity. In any case, there is strong evidence of a short-term fading on the OSL signal and some evidence of a potential long-term fading, depending on the experimental conditions, highlighting the need to determine the correction factors for any given protocol.

In the case of optical-fiber systems, Magne *et al.* (2008) investigated the fading for periods between irradiation and OSL readout differing by more than three orders of magnitude. The authors observed fading of the order of 1% per decade in time, but there was no evidence of the quickly-decaying components discussed earlier.

#### *5.5.1.4 Irradiation Temperature*

The effect of irradiation temperature has been investigated by a few researchers for $Al_2O_3$:C Luxel detectors. Miller and Murphy observed a slight increase in OSL with temperature, but the OSL response was within ±1.5% from the room-temperature value for a temperature range between ~5 and 37 °C (Miller and Murphy, 2007). No trend was observed by Jursinic (2007) between 10 and 40 °C, the values also being within ±1.5% (see Figure 5.32). Similar results were reported by other authors (Yukihara *et al.*, 2008a). In these cases, the OSL detectors were read at room temperature after irradiation.

For optical-fiber systems, the detector's response was observed to vary with irradiation and readout temperature, but the results are contradictory. According to Edmund and Andersen (2007), the RL signal decreases with an increase in irradiation temperature, whereas the opposite was observed by Magne *et al.* (2008). The OSL signal seems to depend on both irradiation and readout temperature, increasing with the irradiation temperature,

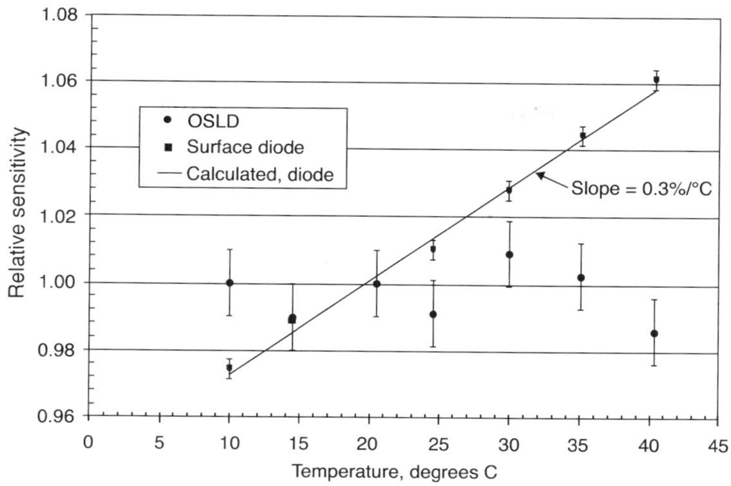


***Figure 5.32*** *Temperature dependence of $Al_2O_3$:C OSLD (InLight OSL Dot) measured using an InLight microStar reader. Reprinted from Medical Physics, Jursinic, P.A., Characterization of optically stimulated luminescence dosimeters, OSLDs, for clinical dosimetric measurements. Vol. 34, 4594–4604. Copyright (2007) with permission from American Association of Physicists in Medicine.*

but decreasing with an increase in readout temperature (Figure 5.33). The OSL sensitivity changes are of the order of 0.2–0.3%/K, but also depend on the period of OSL stimulation. It has been shown that with the appropriate choice of stimulation period, the temperature sensitivity can be canceled (Andersen *et al.*, 2008; Edmund and Andersen, 2007). The authors recommended that calibrations and readouts are carried out at the same irradiation temperature, that is, body temperature, in the case of in-vivo measurements.

#### *5.5.1.5 Directional Dependence*

The effect of the angle of irradiation on the OSL signal will depend on the beam quality and particular packing of the OSL detectors. Directional dependence was investigated for megavoltage X-rays with InLight OSL Dot dosimeters irradiated inside a cylindrical phantom, the results showing no dependence on angle of irradiation within the experimental uncertainties (Jursinic, 2007).

### **5.5.2 Correction Factors for Beam Quality**

#### *5.5.2.1 Diagnostic X-rays*

$Al_2O_3$:C has a well-known over-response to kilovoltage X-rays due to its high effective atomic number and the resultant increase in probability for the photoelectric effect (see Chapter 3). Bos (2001) has shown that the ratio between the mass energy absorption coefficients of $Al_2O_3$:C and water can be of the order of 3–4, which is approximately of the same order of magnitude as the over-response observed earlier for the TL of $Al_2O_3$:C for

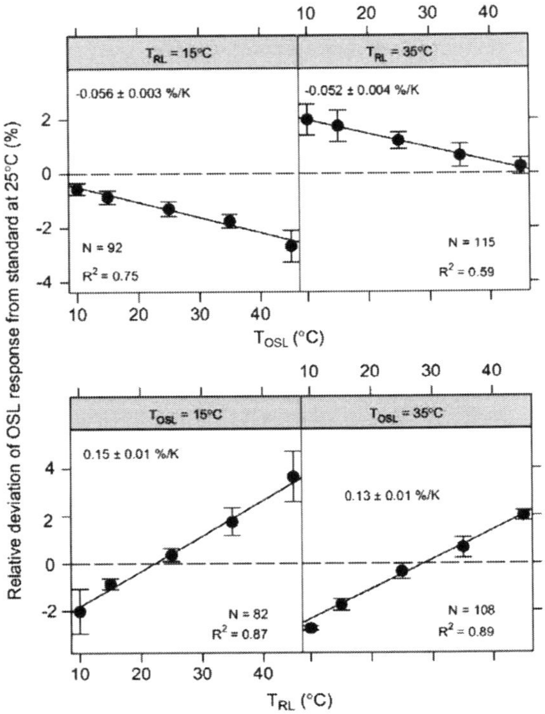


***Figure 5.33*** *Temperature dependence of the OSL signal relative to the response at 25 °C for $Al_2O_3$:C optical-fiber systems as a function of OSL readout temperature ($T_{OSL}$) and irradiation temperature ($T_{RL}$). Reprinted from Radiation Measurements, Andersen, C. E., et al. Temperature coefficients for in vivo RL and OSL dosimetry using $Al_2O_3$:C. Vol. 43, 948–953. Copyright (2008) with permission from Elsevier.*

kilovoltage X-rays relative to $^{60}$Co gamma rays (Akselrod *et al.*, 1990). This implies that correction factors substantially lower than 1 (e.g., $k \sim 0.25$–0.3) are required in diagnostic radiology. Similar values are reported by Reft (2009) and Mobit, Agyingi and Sandison (2006), the latter obtained using Monte-Carlo calculations for $Al_2O_3$ discs 2.85 mm diameter by 1 mm thickness irradiated with various X-ray energies (see Table 5.5). The discussion from Chapter 3 indicates that the degree of over-response can be dependent on experimental parameters such as the choice of OSL signal or OSL reader.

### 5.5.2.2 *High-Energy Photon and Electron Beams*

For megavoltage X-ray beams, the Monte-Carlo calculations by Mobit, Agyingi and Sandison (2006) shown in Table 5.5 predicted a variation of 2.4% for 25 MV X-ray beams relative to $^{60}$Co gamma rays, and a variation of only 1.1% between 6 MV and 25 MV X-ray beams. Various investigators also concluded that the energy dependence for megavoltage X-ray beams is small (±1%) (Jursinic, 2007; Reft, 2009; Viamonte *et al.*,

***Table 5.5*** *Energy-response factor of $Al_2O_3$:C TLD, modeled as discs of diameter 2.85 mm and thickness 1 mm, in photon beams calculated by Mobit, Agyingi and Sandison (2006). The energy response factor ${}_{\mathrm{det}}F^{Q}_{\mathrm{Co}}$ is defined as ${}_{\mathrm{det}}F^{Q}_{\mathrm{Co}} = \frac{(D_w/D_{\mathrm{det}})_{\mathrm{Co}}}{(D_w/D_{\mathrm{det}})_Q}$, where $D_w$ is the dose in water and $D_{\mathrm{det}}$ is the dose to the material of the detector.*

| Energy | Mean Energy (keV) | ${}_{Al_2O_3}F^{Q}_{\mathrm{Co}}$ | ${}_{\mathrm{LiF}}F^{Q}_{\mathrm{Co}}$ | Ratio |
|---|---|---|---|---|
| $^{60}$Co gamma rays | 1250 | 1.000 | 1.000 | 1.00 |
| 50 kV X-rays | 29 | 3.219 ± 0.3% | 1.463 | 2.20 |
| 100 kV X-rays | 60 | 2.861 ± 0.3% | 1.376 | 2.08 |
| 150 kV X-rays | 105 | 1.607 ± 0.3% | 1.245 | 1.29 |
| 250 kV X-rays | 170 | 1.449 ± 0.3% | 1.192 | 1.19 |
| 6 MV X-rays | 2020 | 0.990 ± 0.3% | 0.987 | ≈1.00 |
| 10 MV X-rays | 3050 | 0.983 ± 0.3% | 0.976 | ≈1.00 |
| 15 MV X-rays | 4180 | 0.980 ± 0.3% | 0.976 | ≈1.00 |
| 25 MV X-rays | 6600 | 0.973 ± 0.3% | 0.976 | ≈1.00 |

Data reproduced from Mobit, Agyingi and Sandison (2006) with permission from Oxford University Press.

2008; Yukihara *et al.*, 2008a), although larger variations have been observed (Schembri and Heijmen, 2007). The results from Jursinic (2007) are presented in Figure 5.34.

In the case of megavoltage electron beams, experimental data have also shown a small energy dependence, typically between ±1% (Jursinic, 2007; Reft, 2009; Yukihara *et al.*, 2008a). Larger variations were observed for electron beams (Schembri and Heijmen, 2007),

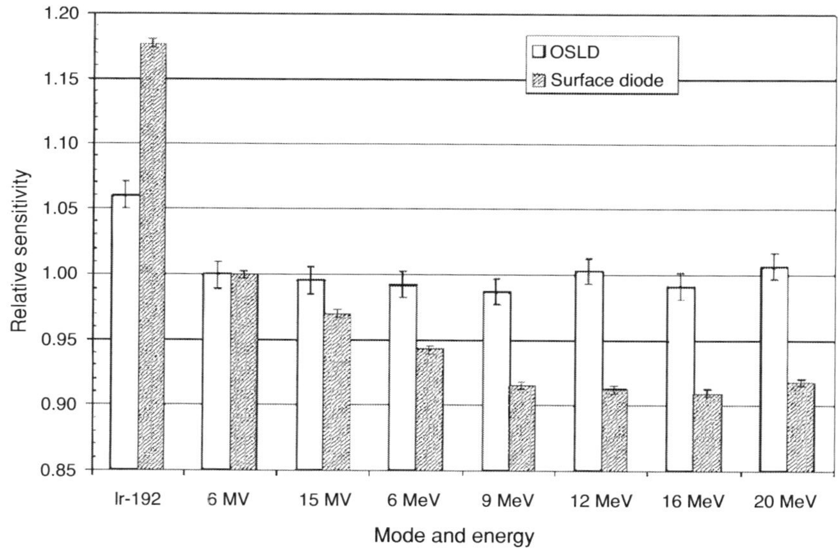


***Figure 5.34*** *OSL relative response for various radiation sources and energies, compared to a surface diode. Reprinted from Medical Physics, Jursinic, P.A., Characterization of optically stimulated luminescence dosimeters, OSLDs, for clinical dosimetric measurements. Vol. 34, 4594–4604. Copyright (2007) with permission from American Association of Physicists in Medicine.*

but with higher associated uncertainties. Optical-fiber probes based on $Al_2O_3$:C were also shown to be energy independent for electrons, the observed differences being of the order of ±0.25% (Magne *et al.*, 2008). Reft (2009) noticed a difference in the electron response from OSLDs irradiated in water or in solid water, which the author attributes to differences in the water equivalency of different phantom materials (Tello, Tailor and Hanson, 1995). However, under the same irradiation conditions a change in the electron energy did not affect the response.

Measurements of depth-dose profiles for high-energy photon and electron beams using $Al_2O_3$:C detectors and optical-fiber probes showed good agreement between OSL detectors and ionization chamber data (Aznar *et al.*, 2004; Yukihara *et al.*, 2008a). Chen *et al.* (2009) used Monte-Carlo calculations to show that beyond the build-up region the ratio between absorbed doses to the $Al_2O_3$ detector and water should remain constant. Magne *et al.* (2008) also measured depth-dose profiles for 9 MeV, 12 and 18 MeV electron beams using $Al_2O_3$:C optical-fiber probes and obtained an agreement of ±0.9% with ionization chamber measurements.

The influence of field size from a 6 MV photon beam (from 4 cm × 4 cm up to 30 cm × 30 cm) on $Al_2O_3$:C OSL detectors was investigated by Schembri and Heijmen (2007), who observed variations in the response up to 2.4%. Other investigators compared OSL detectors with ionization chambers in various field sizes and obtained an agreement of ±1% (Yukihara *et al.*, 2008a). In the conditions investigated, the response from $Al_2O_3$:C OSL detectors was independent of field size within ±1%.

#### 5.5.2.3 *Protons and Heavy Ions*

For protons and other charged particles with $Z > 2$ the relative response (i.e., the "relative luminescence efficiency") decreases with the radiation LET, as discussed in Chapter 4. The decrease in luminescence is related to the high ionization densities in regions close to the trajectory of the heavy charged particle, which cause local saturation of the OSL signal. An extensive dataset on luminescence efficiency for $Al_2O_3$:C for ions from protons to xenon with energies up to 1 GeV/u can be found in Sawakuchi *et al.* (2008c).

The relative responses depend on experimental conditions and type of detectors, but for conditions similar to some of the studies in radiotherapy (Yukihara *et al.*, 2005, 2008a) the experimental values of the relative response for low-LET radiation (LET in water $<1$ keV/μm), corresponding to the LET of protons with energy $>70$ MeV, are of the order of 0.96. For carbon ions with energies of the order of 400 MeV/u (LET $\sim$ 11 keV/μm), the relative response is of the order of 0.7 for similar experimental conditions. Reft (2009) obtained relative response values around 0.5–0.55 for a 290 MeV/u carbon beam (LET in water $\sim$40 keV/μm). The relative response decreased to $\sim$0.4 as the carbon beam was degraded using acrylic absorbers and the LET in water increased to $\sim$78 keV/μm. It is clear from these investigations that the relative response of $Al_2O_3$:C decreases with increasing $Z$ and decreasing particle energy, that is, with increasing LET of the heavy charged particle. For low-energy particles, the signal may also be reduced due to the small particle range and partial volume irradiation of the detector. As previously noted, however, the precise values of the relevant efficiency factors depend on the OSL readout conditions and not just the characteristics of the particle beam.

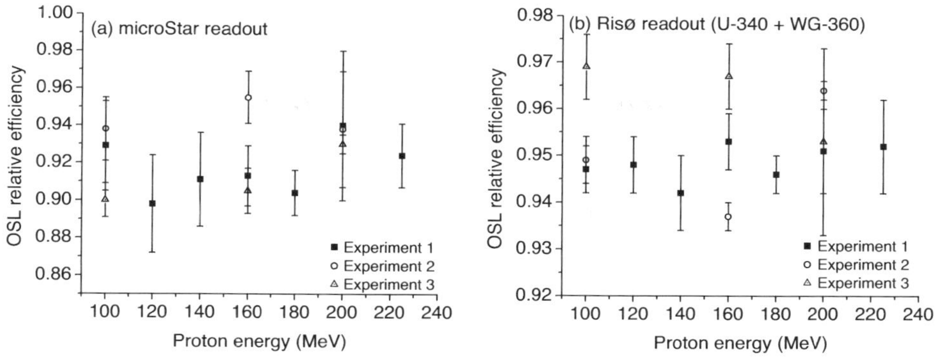


***Figure 5.35*** *OSL relative luminescence efficiency of $Al_2O_3$:C Luxel detectors irradiated with protons with energy from 100 MeV to 250 MeV. Readout was carried out with green stimulation using (a) a microStar reader or (b) a Risø TL/OSL reader (Hoya U-340 and WG-360 filters in front of the PMT) using the total OSL area for the analysis. The data represent the mean and experimental standard deviation values for three separate experiments. Data provided by Dr. Gabriel O. Sawakuchi, Carleton University.*

The relative response of $Al_2O_3$:C Luxel detectors for protons with energy between 100 and 250 MeV was investigated by Sawakuchi and colleagues and the results (unpublished) are shown in Figure 5.35 for two readout conditions. Although there is some variation from experiment to experiment, the results indicate an absence of energy dependence for protons with energy above 100 MeV. This result is supported by data from Reft (2009).

However, the relative luminescence efficiency decreases with decreasing proton energy (i.e., increasing LET). This is observed, for example, when measurements are performed at the end of the proton range. Figure 5.36 shows the depth-dose profiles (dose versus equivalent depth in water, relative to entrance dose) for $Al_2O_3$:C Luxel detectors irradiated with proton beams with nominal energies of 100 MeV, 160 MeV, 200 MeV and 250 MeV. The OSL values are lower than the ionization chamber data close to the end of the proton range, particularly as the nominal energy of the proton beam decreases. Comparison between Figure 5.36a and Figure 5.36b also shows that the results depend on the particular experimental conditions. Partial volume irradiation and uncertainties in the detector position can also be contributing to the reduced OSL values.

## 5.6 Clinical Applications

Most of the investigations on OSL applied to medicine have been focused on the preliminary characterization of the OSL properties. In this section we discuss potential applications of OSL in various areas of medical dosimetry, based on the limited number of reported studies.

### 5.6.1 Quality Assurance in External Beam Radiation Therapy

OSLDs can be used as passive detectors for in-vivo dosimetry in the same way as TLDs, keeping in mind the difference in properties as discussed in Section 5.5, with reported

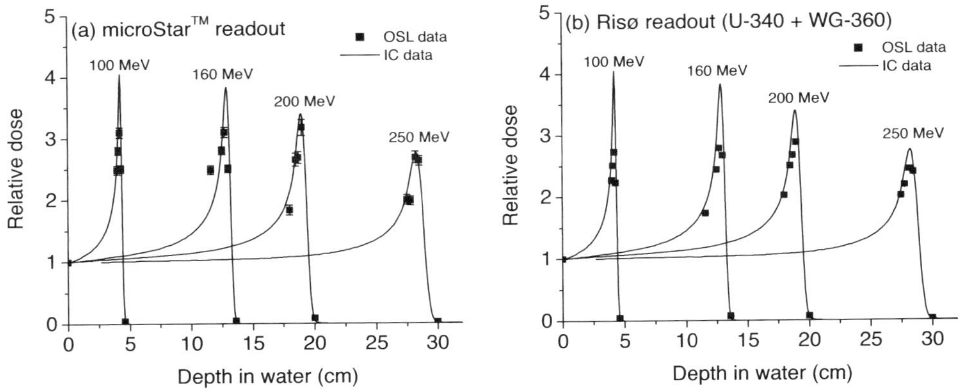


***Figure 5.36*** *Depth-dose profile (dose versus water equivalent depth), normalized to entrance dose, for $Al_2O_3$:C Luxel detectors irradiated with proton beams of nominal energies of 100 MeV, 160 MeV, 200 MeV and 250 MeV behind different thicknesses of solid water. Readout was carried out with green stimulation using (a) a microStar reader or (b) a Risø TL/OSL reader (Hoya U-340 and WG-360 filters in front of the PMT) using the total OSL area for the analysis. The data represent the mean and experimental standard deviation values for three separate experiments. The full lines represent measurements with a parallel-plate ionization chamber obtained for reference. Data provided by Dr. Gabriel O. Sawakuchi, Carleton University.*

satisfactory results in the case of $Al_2O_3$:C (Danzer *et al.*, 2007). In-vivo dosimetry for patients undergoing external-beam radiation therapy is considered the "ultimate check of the actual dose delivered to an individual patient" (Essers and Mijnheer, 1999) and its routine use is recommended as part of the quality assurance program to prevent treatment errors (ICRP, 2000a). It consists in measuring the dose using detectors placed on the patient's skin to measure entrance or exit doses, or inside internal body cavities, therefore requiring simple and unobtrusive detectors such as diodes and TL detectors. Reviews of methods for in-vivo dosimetry were published by Essers and Mijnheer (1999) and van Dam and Marinello (2006).

The potential of using optical-fiber OSL systems for in-vivo dosimetry in external beam radiotherapy has been demonstrated by Aznar *et al.* (2004). In addition to tests to evaluate the influence factors, the authors performed measurements on an anthropomorphic phantom subjected to a simulated treatment and on a patient undergoing a head and neck IMRT treatment. The results from phantom irradiation showed agreement within 1% with the planned absorbed doses using either the RL or the OSL signal. In the case of the patient measurement, the difference between the absorbed dose measured by the OSL probe and the one calculated by the treatment planning system was $(0.09 \pm 0.05)$ Gy. It should be remembered that, in addition to the total absorbed dose, the optical-fiber OSL system also provides real-time data on the dose rate during the treatment. This feature has shown to improve the ability of the system to detect treatment errors in brachytherapy (see Section 5.6.2).

OSLDs may also be used in place of TLDs in postal audit programs such as the ones operated by the International Atomic Energy Agency (IAEA) and the World Heath

Organization (WHO) (Izewska and Andreo, 2000), the ESTRO-QUALity assurance network (EQUAL) (Ferreira *et al.*, 2000), and the Radiological Physics Center (RPC) of the University of Texas M. D. Anderson Cancer Center (Kirby, Hanson and Johnston, 1992). In these programs, TLD capsules containing a TL material in powder form are irradiated inside appropriate holders or mini-phantoms by the participating hospitals. These capsules are then returned to the respective laboratories for readout with the purpose of verifying the doses delivered by radiotherapy-treatment machines. The uncertainties associated with the TLD reading per capsule from the IAEA/WHO program is 0.48% (each capsule is divided in four 35 mg aliquots). As discussed in Section 5.3.2.3, the precision associated with the reading of a single OSL detector can be as low as 0.7% using an appropriate reader. Jursinic (2010) suggests that a precision of 0.5% can be achieved using commercial readers, if the detectors are appropriately characterized. The RPC has recently implemented an OSL-based system for remote auditing (Aguirre *et al.*, 2009).

### 5.6.2 Brachytherapy

Because of their high sensitivity, small size and chemical stability, $Al_2O_3$:C OSL detectors are suitable for brachytherapy dosimetry providing simpler preparation and readout procedures than TLDs. A brief review of TLD dosimetry in brachytherapy can be found in Lambert *et al.* (2007) and references therein.

Andersen *et al.* (2009a) investigated a fiber-coupled $Al_2O_3$:C system for in-vivo dose verification during $^{192}$Ir brachytherapy. The system described is capable of real-time measurement of the dose rate using the RL signal and an independent estimation of the absorbed dose at the end of the irradiation using the OSL signal. The probe consisted of a 2 mm $\times$ 0.5 mm $\times$ 0.5 mm $Al_2O_3$:C single crystal connected to a PMMA optical fiber of 1 mm outer diameter. The irradiations were carried out using 1 mm stainless-steel brachytherapy needles inserted in a solid-water block submersed in a water tank.

The authors performed Monte-Carlo calculations, taking into account the dosimetry probe and the brachytherapy source and derived the ratio between the mean dose to $Al_2O_3$:C and the mean dose to an identical volume of water $D(Al_2O_3:C)/D(H_2O)$ in the absence of perturbation. The value of $D(Al_2O_3:C)/D(H_2O)$ was $0.91 \pm 0.01$ at 10 mm and varies by only $\pm 2\%$ in the 2–20 mm range, but increased for larger distances (e.g., $1.03 \pm 0.06$ at 100 mm) due to changes in the photon energy spectrum.

The performance of the system was similar using both RL and OSL signals, with no significant deviation from linearity in the 0 to 4.3 Gy dose range and with a reproducibility of 1.3% (experimental standard deviation of the difference between reference and measured dose for various dose levels). The measurement accuracy worsened as the distance to the source increased, the difference between the measurement and the expected value calculated by the treatment planning system increasing at 40 mm to $(2.4 \pm 1.4)\%$ for the OSL signal and $(10 \pm 1)\%$ for the RL signal. The large discrepancy in the RL signal was attributed to the influence of the stem effect. The authors estimate that the combined standard deviation of the system is 8% for the RL signal and 5% for the OSL signal (Andersen *et al.*, 2009a).

The feasibility of using optical-fiber systems to detect errors in brachytherapy was demonstrated by in-vivo measurements for five patients undergoing pulsed-dose-rate (PDR) brachytherapy (Andersen *et al.*, 2009b). In this study, the optical-fiber probes were placed in interstitial or intracavitary needles (Figure 5.37) and the absorbed dose was determined

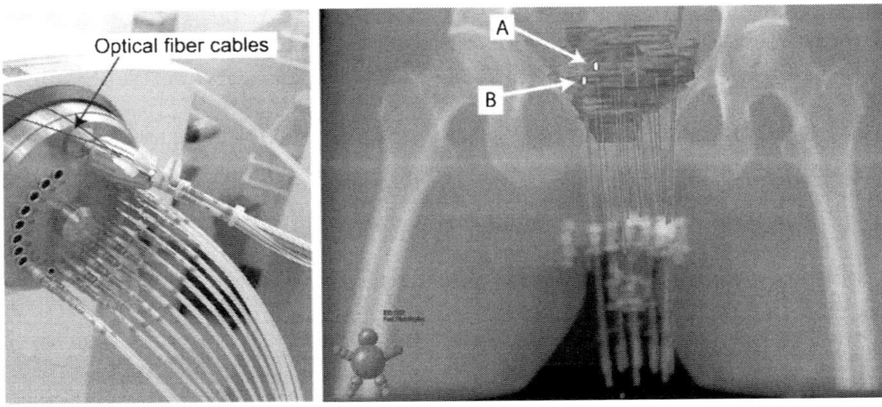


*Figure 5.37* Afterloader with $^{192}Ir$ source and guide tubes used in the brachytherapy treatment. The right-hand figure indicates the position of the OSL probes (A and B). Reprinted from Medical Physics, Andersen, C. E., et al., Time-resolved in vivo luminescence dosimetry for online error detection in pulsed dose-rate brachytherapy. Vol. 36, 5033–5043. Copyright (2009) with permission from American Association of Physicists in Medicine.

using both the RL and the OSL signal from the $Al_2O_3$:C probes. In addition, real-time information on the dose rate was provided by the RL signal. The authors showed that the dose rate indicated by the OSL probes correctly followed the dose rates given by the treatment-planning system, and that the ability to detect errors was greatly increased by having real-time information instead of only the total absorbed dose at the probe's position. The system was demonstrated to offer the precision and sensitivity required for this application. The main advantage of this system is the ability to fit the probe in standard guide tubes and applicators.

### 5.6.3 Measurement of Dose Profiles in X-ray Computed Tomography (CT)

Evaluation of the CTDI and other CTDI-related indexes such as the $CTDI_w$ and the $CTDI_{vol}$ (AAPM, 2008) are based on the $CTDI_{100}$, measured using a 100 mm pencil ionization chamber and a single axial scan rotation of the X-ray tube, and defined as:

$$\mathrm{CTDI}_{100} = \frac{1}{NT} \int_{-50\,\mathrm{mm}}^{+50\,\mathrm{mm}} D(z)\mathrm{d}z. \tag{5.15}$$

The $CTDI_{100}$ represents the accumulated multiple scan dose at the center of a 100 mm scan (AAPM, 2008). However, in addition to underestimating the CTDI because of the limited integration range, the pencil ionization chamber does not capture details of the dose distribution *D(z)* which can be useful when interpreting the results from the measurements (e.g., Dixon and Ballard, 2007).

Landauer Inc. has introduced a service to measure dose profiles in X-ray CT using a 150 mm OSL strip of $Al_2O_3$:C made of the same detector material used in the Luxel

***Table 5.6*** *OSL relative response $r_Q$ obtained by comparing the OSL indicated value (in absorbed dose to air) to the dose to air measured using a point ionization chamber.*

| Phantom (position) | 100 kVp | 120 kVp | 140 kVp |
|---|---|---|---|
| Body phantom (center) | 2.73 ± 0.04 | 2.67 ± 0.11 | 2.48 ± 0.11 |
| Body phantom (periphery) | 2.63 ± 0.15 | 2.55 ± 0.03 | 2.35 ± 0.03 |
| Head phantom (center) | 2.67 ± 0.02 | 2.56 ± 0.03 | 2.512 ± 0.006 |
| Head phantom (periphery) | 2.68 ± 0.06 | 2.52 ± 0.07 | 2.33 ± 0.05 |

As discussed in Section 5.3.1.5, the field-specific correction factors $k_Q = r_Q$ correct for the energy response of $Al_2O_3$:C when calculating the dose to air. Data from Yukihara *et al.* (2009).

dosimeters (Bauhs *et al.*, 2008). By scanning the OSL strip with a laser beam, the dose profile $D(z)$ can be determined and the $CTDI_{100}$ can be calculated. An equivalent system based on LEDs and the POSL technique was described by Yukihara *et al.* (2009), who also compared the OSL dose profiles with point ionization chamber measurements, investigated the correction factors for energy dependence to be used in different experimental conditions, and investigated the accuracy of the system in determining the $CTDI_{100}$. Further investigations on the accuracy of the system and on the efficiency of the $CTDI_{100}$ were carried out by Ruan *et al.* (2010).

Yukihara *et al.* (2009) showed that the correction factors due to the energy dependence of the $Al_2O_3$:C OSL strips depend on the X-ray tube potential ($kV_p$), type of phantom and position of the detector on the phantom (center or periphery). Table 5.6 shows the relative response for the head and body CT phantoms (AAPM, 1990), for three tube potentials, and the center and periphery positions. The average relative response is ~2.55, but there is a difference of 17% between the maximum and minimum values. This indicates that field-specific correction factors should be applied if accurate results are to be achieved.

Using field-specific correction factors, Yukihara *et al.* (2009) demonstrated the agreement between the OSL dose profiles and ionization chamber point measurements. One example of this comparison can be seen in Figure 5.38. The $CTDI_{100}$ values calculated using the OSL strips were found to agree within ±5% with the $CTDI_{100}$ values obtained using a 100 mm pencil ionization chamber (Yukihara *et al.*, 2009). These preliminary results on the agreement between the $CTDI_{100}$ values obtained using OSL and a pencil ionization chamber were further confirmed by more complete investigations using different X-ray-tube voltages and different beam widths by Ruan *et al.* (2010).

Based on these results, Ruan *et al.* (2010) evaluated the CTDI efficiency, defined as the ratio between the $CTDI_{100}$ and the $CTDI_{450}$, using the OSL dose profiles for a variety of beam widths, types of phantom, detector position in the phantom and X-ray tube potential. The CTDI efficiency indicates to what extend the $CTDI_{100}$ underestimates the CTDI. Their results indicate that the $CTDI_{100}$ is between 56 and 86% of the CTDI value, depending on the type of CT phantom (head or body) and position of the detector (center or periphery). These results are comparable to other results from the literature using ionization chamber measurements by Dixon and Ballard (2007), or using Monte-Carlo calculations by Boone (2007).

Measurements in X-ray CT using an optical-fiber systems with KBr:Eu detector are also reported by Klein (2008).

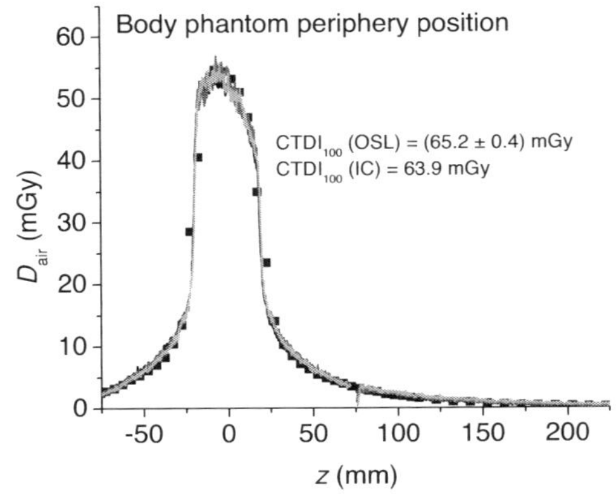


***Figure 5.38*** *Three dose profiles measured using OSL strips (dose-to-air versus position in the phantom parallel to the axis) in comparison with ionization chamber point measurements. The graph also shows the values of the $CTDI_{100}$ calculated based on the reading of a 100 mm pencil ionization chamber and on the OSL dose profile. Reprinted from Physics in Medicine and Biology, Yukihara, E. G., et al., An Optically Stimulated Luminescence system to measure dose profiles in X-ray Computed Tomography.*

### 5.6.4 Proton Therapy

OSLDs can in principle be used in place of TLDs for quality assurance and dose verification in proton therapy, taking into account variations in the relative luminescence efficiency, as discussed in Section 5.5.2.3. In addition to the characterization described in the latter section, experiments have been carried out to demonstrate the performance of $Al_2O_3$:C OSLDs in proton therapy, particularly for measurements of the spread out Bragg peak (SOBP) and output factors. The $Al_2O_3$:C dose response to protons is linear for doses up to a few grays, but supralinearity has been observed for higher doses (Reft, 2009; Sawakuchi *et al.*, 2008b).

Figure 5.39 shows an example of OSL measurements at various water equivalent depths in solid water for experimental conditions. In this case the detectors are irradiated with a broad proton energy spectrum and the resultant luminescence efficiency is a complicate combination of material properties (reduced luminescence efficiency due to particle LET) and partial volume irradiation effects (particularly for the low-energy protons present in the spectrum).

Measurements in proton dosimetry using optical-fiber systems are also reported by Edmund *et al.* (2007) and Klein (2008). The results from Edmund *et al.* (2007) are based on $Al_2O_3$:C probes and show that the degree of supralinearity in the proton response depends on the choice of signal (initial OSL intensity or total OSL area), with no supralinearity observed in case of the total OSL area. The dependence of the OSL response on the proton energy was also observed to depend on the choice of signal. Klein (2008) used a system based on KBr:Eu as the probe. Due to the fast response of this material, Klein was able to capture the rapid variations in dose rate during a single proton spill lasting ~200 ms in the RL signal. This is illustrated in Figure 5.40.

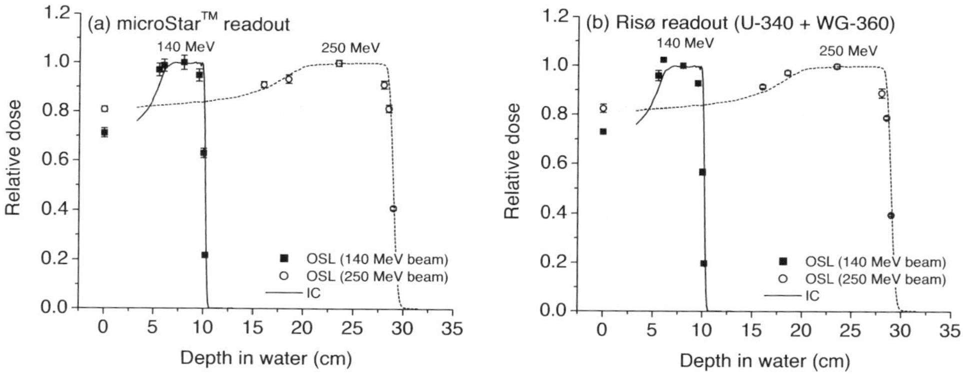


***Figure 5.39*** *Depth-dose profile (dose versus water equivalent depth), normalized to reference depth, for $Al_2O_3$:C Luxel detectors irradiated at different depths in solid water using a SOBP proton beams (SOBP created by protons of nominal energy of 140 and 250 MeV). Readout was carried out with green stimulation using (a) a microStar reader or (b) a Risø TL/OSL reader (Hoya U-340 and WG-360 filters in front of the PMT) using the total OSL area for the analysis. The data represent the mean and experimental standard deviation values for three separate experiments. The full lines represent measurements with a parallel-plate ionization chamber obtained for reference. Data provided by Dr. Gabriel O. Sawakuchi, Carleton University.*

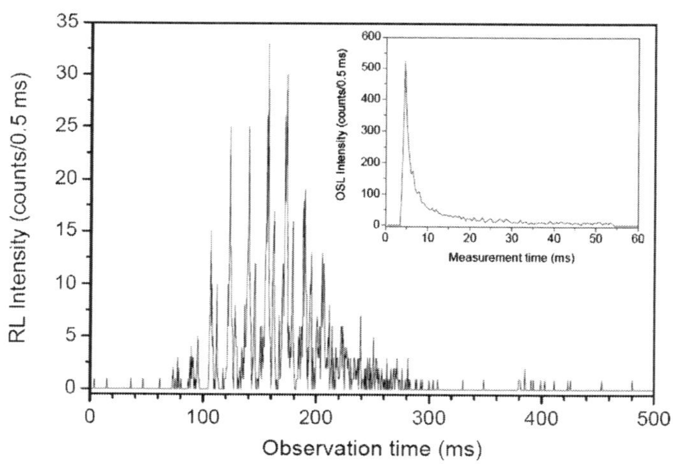


***Figure 5.40*** *RL signal from KBr:Eu probe used as an optical-fiber probe during a single proton spill. The observed RL signal was used to trigger the OSL readout shown in the inset. Reprinted from Klein, D. M., Development and characterization of remote radiation dosimetry systems using optically stimulated luminescence of $Al_2O_3$:C and KBr:Eu. PhD Dissertation. Copyright (2008) with permission from Oklahoma State University.*

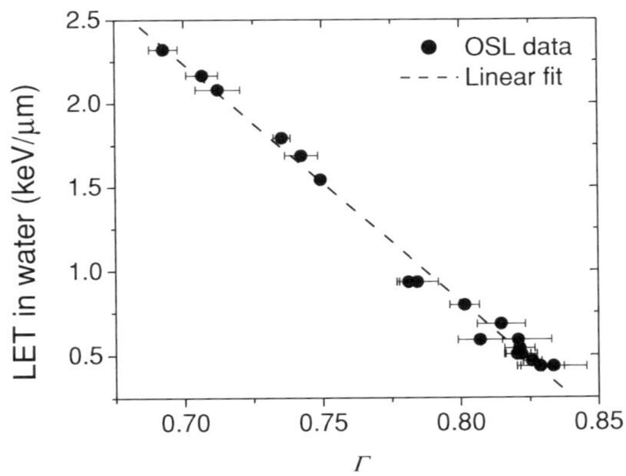


***Figure 5.41*** *Variation of the shape of the OSL curve, expressed by the parameter $\Gamma$ (see text) with average LET in a proton field. Data provided by Dr. Gabriel O. Sawakuchi, Carleton University.*

A new finding in OSL dosimetry of proton beams is the possibility of using OSL detectors for determination of the average LET of the proton field. Sawakuchi *et al.* (2010) correlated variations in the shape of the OSL decay curve in various proton conditions with the average LET of the radiation field. The shape of the OSL decay curve was expressed using a shape factor $\Gamma$, defined as the ratio between the area of the OSL decay curve after proton irradiation and reference beta irradiation ($^{90}Sr/^{90}Y$ beta source) when both OSL curves have been normalized to the initial OSL intensity, that is: $\Gamma = (S_a/S_i)_H/(S_a/S_i)_R$ where $S_a$ is the total OSL area and $S_i$ is the initial OSL intensity. Figure 5.41 shows that this shape factor exhibits a linear correlation with the average LET of the radiation field. Therefore, if the shape factor is calculated for an unknown condition, the average LET of the radiation field can be obtained using calibration data similar to that presented in Figure 5.41.

Dose profiles can also be measured in proton therapy using OSL strip detectors identical to those used for measurements in X-ray CT (Section 5.6.4). Figure 5.42 shows the dose profile measuring the same system described before (Yukihara *et al.*, 2009), but adapted for measurement of dose profiles in proton therapy by using a 1 mm aperture in contact with the OSL strip detector. The OSL strip detector was then scanned below the aperture to obtain the dose profile shown in Figure 5.42. These data were obtained for demonstration of proof of concept. A commercial reader for this type of application remains to be developed.

### 5.6.5 Fluoroscopy (Patient and Staff Dosimetry)

OSL detectors can be used for several purposes in interventional fluoroscopy. Optical-fiber OSL systems can be used as a real-time monitoring device with the purpose of alerting the interventionist of high absorbed-dose levels to the patient's skin that could lead to unwanted tissue reactions. OSL detectors can also be used as passive detectors to determine entrance doses to the patient, physician or supporting staff with the purpose of

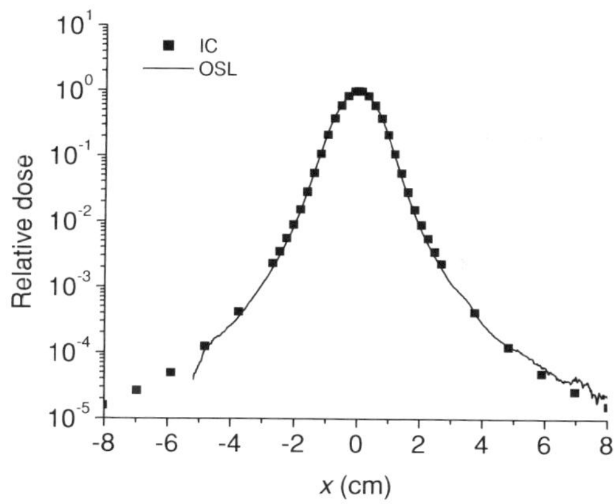


***Figure 5.42*** *Lateral dose profile measurement of a proton pencil beam using an ionization chamber (IC) and an OSL strip detector (Yukihara et al., 2010).*

determining exposure levels for procedure optimization or for correlating exposure levels with particular procedures and workloads.

An optical-fiber OSL system based on KBr:Eu was tested in conditions used in fluoroscopy by Klein (2008). This study indicated that OSL signal under periodic readout during irradiation is linear with exposure rate and the angle dependence of the OSL signal was small. In pulsed systems (~5–8 ms pulse durations at rates between 7.5 and 30 pulses per second), the fast response of the KBr:Eu material required careful synchronization between the radiation pulses and the OSL readout to avoid the introduction of artifacts in the OSL measurements (e.g., occurrence of radiation pulse during an OSL readout).

Gasparian *et al.* (2010) used 76 $Al_2O_3$:C single-crystal OSL detectors to map the entrance surface air kerma (which includes backscatter from the phantom) on an anthropomorphic phantom used to simulate the physician during a fluoroscopy procedure. The detectors were read immediately after exposure using a portable POSL reader that allowed detection of doses as low as 0.2–0.4 μGy. This sensitivity was sufficient for measurement of the entrance surface air kerma for a single simulated fluoroscopy procedure in the absence of protective apparel (e.g. lead apron). This study demonstrated the convenience and feasibility of using an OSL dosimetry system for dose mapping of staff exposures.

### 5.6.6 Mammography

On-phantom and in-vivo measurements in mammography have been carried out using optical-fiber OSL systems based on $Al_2O_3$:C probes (Aznar *et al.*, 2005). The response of the system was observed to be linear for air kerma values from 4.5 to 30 mGy and the minimum detectable dose was determined to be ~50 μGy using the RL signal and ~200 μGy using the OSL signal. The response of the $Al_2O_3$:C increased ~18% with an increase in X-ray tube potential from 23 to 35 keV (HVL in the range of 0.3–0.41 mm Al). Irradiation in a PMMA phantom used for quality control in mammography and in-vivo indicated that the characteristic size and shape of the OSL probes did not pose a problem for the identification of structures in the mammogram (Aznar *et al.*, 2005).

### 5.6.7 Out-of-field Dose Assessment in Radiotherapy

A point that has been raised in modern radiotherapy is the impact of new dose-delivery techniques on the absorbed dose outside the treatment target volume and, consequently, on the risks for secondary cancer induction (Kry *et al.*, 2005; Xu, Bednarz and Paganetti, 2008). The concern is that beam intensity modulation results in longer irradiation periods and larger radiation scattering, increasing the out-of-field doses and the risks for a secondary cancer. Neutron production from nuclear reactions is also a concern for high-energy photon and proton beams.

Three surveys using TLDs and OSLDs compared by d'Errico (2006) showed that the doses outside the treatment volume can be considerably higher in the case of IMRT than in the case of conventional external-beam radiation therapy (Meeks *et al.*, 2002; Mutic and Low, 1998; Verellen and Vanhavere, 1999). This is caused by the fact that in IMRT a larger portion of the beam is blocked by the multi-leaf collimator, therefore resulting in longer irradiation periods, more scattered radiation and consequently more exposure to secondary radiation.

Meeks *et al.* (2002) used $Al_2O_3$:C OSLDs to survey the out-of-field doses received by patients undergoing an entire tomotherapy treatment for intracranial or head and neck lesions with an average dose for the treatment course of $\sim$60 Gy. The ratio between the dose to OSLDs and prescribed doses varied between 0.6% and 3.6% depending on the distance between the OSLD and target centroid. Reft, Runkel-Muller and Myrianthopoulos (2006) also performed in-vivo and phantom studies of the out-of-field doses during prostate IMRT treatments using $Al_2O_3$:C OSLDs. One of the proposed advantages of $Al_2O_3$:C for this application is the lower neutron response when compared to TLD-700 (Klemic, Azziz and Marino, 1996). Although this is in fact true for $Al_2O_3$:C single crystals, the neutron response of $Al_2O_3$:C Luxel detectors may be different than $Al_2O_3$:C single crystals due to the presence of plastic and the possibility of neutrons generating recoil protons.

With the introduction of OSLN, which is a neutron-sensitive OSL detector (see Section 3.6), it is also possible to use OSLN and OSLDs in combination to determine the photon and neutron response in the same way as a $^{6}$LiF:Mg,Ti and $^{7}$LiF:Mg,Ti pair of detectors. The first application of OSLN in medical dosimetry was described by Sykora (2010).

### 5.6.8 Dose Mapping

Although OSL point detectors have been used for dose mapping in X-ray CT (Peakheart, 2006) and fluoroscopy (Section 5.6.5), this approach is very time-demanding and labor-intensive, requiring the positioning and readout of a large number of individual detectors. Spatial resolution is very poor and limited to the size of the detectors. As discussed in the introduction of this chapter, detectors with high spatial resolution, accuracy and precision are required to meet the challenges encountered in modern applications of radiation in medicine.

The optical nature of the OSL technology makes it suitable for one- and two-dimensional dose mapping using either linear detector strips, such as those used in the measurement of dose profiles in X-ray CT (Section 5.6.3), or plastic sheets containing dosimetric material. In fact, as already discussed in Chapter 2, imaging systems are commercially available using photostimulable phosphors and are widely used in computed radiography. Although these detectors were not developed for dosimetry per se, studies have tried to extend the range of applicability of these imaging plates for dosimetry applications. Some studies looked at the light output from imaging plates for dosimetry applications, but not necessarily for dose mapping (Ariga *et al.*, 2007; Boukhair *et al.*, 2001). Others have tried to address the problem of quantitative dosimetry (Floyd *et al.*, 1990). The main problem of this technique for dose mapping is the high effective atomic number of currently available photostimulable phosphors (see Table 2.2) and fading of the signal.

The development of one- and two-dimensional OSL detectors for dose mapping is currently limited by the availability of OSL materials. Encouraging results were obtained by Idri *et al.* (2004), who demonstrated the feasibility of using OSL detectors based on SrS:Ce,Sm for quality control in IMRT. The effective atomic number of SrS:Ce,Sm is still relatively high (see Table 2.2), which may cause a dependence of the response on the contribution from scattered low-energy photons. Although $Al_2O_3$:C has been used for measurement of dose profiles in X-ray CT, the lifetime of the main luminescence centers ($\sim$35 ms) is too slow for an efficient measurement of 2D dose profiles using laser spot scanning, since the laser spot must stay in the same position (pixel) for a duration corresponding to at least three times the luminescence lifetime (von Seggern, 1999).

Based on currently available applications, dose mapping is an area in which the OSL technology may contribute by providing accurate dose estimates with high spatial resolution and accuracy. However, this will require significant efforts in terms of material development.

### 5.6.9 Final Remarks on Clinical Applications

Application of OSL in medical dosimetry is still incipient, but has grown considerably in recent years. On the one hand, the similarity between the physical phenomena behind OSLDs and TLDs has provided a bridge between the well-known TL technique and the more-recent OSL technique. On the other hand, lack of knowledge regarding the properties of specific OSL materials and the effects of the various influence factors on the OSL signal present the main obstacles for a wider adoption of the OSL technology. Given the convenience and potential precision of the technique, it may be only a matter of time until protocols for OSL dosimetry become more established and medical physicists become comfortable with the routine use of OSLDs.

In parallel, new OSL dosimetry systems for passive and real-time dosimetry may be introduced, depending on technological advances and market demand. In some cases, development of commercial dosimetry systems for specialized applications may be hampered by a lack of broad applicability. Nevertheless, the fast developments in radiation diagnostic and treatment modalities will continue to place a continuous and ever-growing pressure for the development of dose and dose-rate monitoring and dosimetry instruments. It is difficult to predict at this point whether or not OSL will become widely adopted in medical dosimetry, but we believe there will be a range of applications in which OSL dosimetry will offer convenient solutions to practical problems in medical dosimetry.

# 6

# Other Applications and Concepts

## 6.1 Introduction

In addition to the applications discussed in previous chapters, OSL has also generated interest in a number of areas ranging from retrospective and accident dosimetry, including triage in the case of a radiological or nuclear terrorist attack, to UV dosimetry and security (e.g., detection of illicit radioactive materials). Research in some of these areas is recent and, as a result, the viability of the OSL technique remains to be fully demonstrated in many cases.

This chapter discusses these OSL applications, identifying the main problems to be addressed, the rationale for using OSL and the main results from the literature. Retrospective dosimetry overlaps to a certain extend with luminescence dating, which is not the subject of this book. However, it is discussed here because of the relevance of the methodology for other (i.e., "non-dating") applications.

## 6.2 Retrospective and Accident Dosimetry

The widespread use of radiation and radioactive materials has unfortunately led to a variety of incidents in which individuals have received unjustified doses of ionizing radiation. In the past, these incidents have included accidental exposure of workers, criticality accidents, dose misadministration with resultant over-exposure of radiotherapy patients, and the release of radioactive material in the environment due to breach of sealed sources, nuclear reactor accidents, nuclear weapon tests or mismanagement of nuclear waste. (Excellent, detailed descriptions of some of these events can be found in reports from the International Atomic Energy Agency[1]). The number of individuals directly or indirectly affected, the

[1] http://www-pub.iaea.org/mtcd/publications/accres.asp.

range of absorbed doses received by individuals, and the type of radiation and dominant pathway of exposure (e.g., external versus internal exposure) depends on the specific type of incident. However, the common denominator in these situations is the absence of appropriate dosimeters. Consequently, one must rely on methods of retrospective dosimetry for an assessment of the doses received by individuals or population groups.

In addition to such cases, the potential of a terrorist attack to nuclear installations or using nuclear or radiological materials is also of significant concern in many large population centers around the globe. For unintentional events the possibility of acute exposure of a large number of individuals is considered to be improbable, at least from historical records (González, 2007). In contrast, possible terrorist attacks have the likelihood of taking place in an urban environment, with the consequential potential for mass casualties. Potential terrorism scenarios have been amply discussed in the literature and include the use of radiation exposure devices (e.g., a covert irradiator), the detonation of radioactive dispersion devices (conventional explosives mixed with radioactive material, popularly known as "dirty bombs"), the dispersion of radioactive material, site-specific radioactive contamination, radioactive contamination of food and water supplies, attack to a nuclear facility and the detonation of an improvised or sophisticated nuclear device (González, 2007; ICRP, 2006; Mettler and Voelz, 2002; NCRP, 2001; Waselenko *et al.*, 2004). In some of these cases, triage for radiation exposure will be essential to prevent the number of individuals seeking medical attention from overwhelming the medical response (Moulder, 2004).

Retrospective dosimetry, defined in the broader sense of assessment of past exposure to ionizing radiation, plays an important role during all phases of an emergency exposure situation. In the early and intermediate phases, in which acute exposures and deterministic effects are the highest concerns, dosimetric data may assist in triage, allowing the resources to be focused on those in need of immediate medical attention, and in supporting medical treatment decisions by alerting the physicians of the likely progression of the clinical condition. The data may also be used to inform the public and allow authorities to plan further actions. In the intermediate and late phases, in which stochastic effects are of most concern, the data can assist in the process of counseling individuals for potential long-term effects, and reassuring the population who were not significantly exposed. Furthermore, since large populations of exposed individuals are rare, the reconstruction of individual doses in such a mass-casuality event will be essential for longitudinal, epidemiological studies of the long-term effects of radiation exposure (Simon *et al.*, 2006).

There are several methods that can provide retrospective assessment of exposure to ionizing radiation (Alexander *et al.*, 2007; Simon *et al.*, 2007). Early assessment can be made based on clinical signs and symptoms and their time evolution (AFRRI, 2003) and on lymphocyte depletion kinetics (Goans *et al.*, 1997). Cytogenetic analysis (IAEA, 2001; ICRU, 2002) and physical methods such as electron paramagnetic resonance (EPR) of biopsied dental enamel (IAEA, 2002; ICRU, 2002) have been used to obtain more refined data in a variety of situations.

However, given the technical limitations and the potential complexity of the situations, no single assay can address all potential scenarios in an emergency exposure situation, particularly in cases of mass casualties (Alexander *et al.*, 2007). Dose assessment based solely on clinical signs and symptoms are problematic because of confounding factors such as high variability between individuals, stress and psychological reactions (AFRRI, 2003). Lymphocyte depletion analysis requires periodic measurements and is limited to acute exposures (Goans *et al.*, 1997). Cytogenetic analyses are limited by the time required to process and analyze the sample and by the capacity of the laboratories to process samples (Alexander *et al.*, 2007; Joint Interagency Working Group, 2005). EPR currently requires teeth biopsy for in-vitro measurements, although research is underway to develop instrumentation and methodology for in-vivo tooth enamel measurements (Swartz *et al.*, 2007) and also in-vitro measurements of fingernails (Trompier *et al.*, 2007). Moreover, the demand for dose estimates in a mass casualty event can overwhelm the few laboratories specialized in these biological and physical dosimetry methods (González, 2007).

In this context, luminescence techniques (TL and OSL) may complement the range of techniques that can be used in retrospective and accident dosimetry. TL was used for reconstruction of doses resulting from the Hiroshima and Nagasaki atomic bomb explosions (Hashizume *et al.*, 1967; Ichikawa, Higashimura and Shidei, 1966), from the radioactive fallout of nuclear tests detonated at the Nevada Test Site (Haskell, Kaipa and Wrenn, 1988; Haskell et al., 1994), from the fallout of the Chernobyl accident (Hütt *et al.*, 1993), from the release of radioactive material on the Techa and Iset river systems by the Mayak plutonium production facility (Bougrov *et al.*, 1997) and from the Radiological Accident in Tammiku (IAEA, 1998). The OSL technique has been applied to retrospective dosimetry using environmental materials from the Chernobyl accident area (Bailiff, 1999; Bailiff *et al.*, 2004, 2005; Banerjee, Bøtter-Jensen and Murray, 2000; Ramzaev *et al.*, 2008) and settlements in the Semipalatinsk Nuclear Test site (Göksu *et al.*, 2006). Research is also underway to develop OSL for individual assessment of radiation exposure using dental enamel (DeWitt *et al.*, 2010; Godfrey-Smith and Pass, 1997; Godfrey-Smith, 2008; Yukihara *et al.*, 2007), dental ceramics (Bailiff *et al.*, 2002; Veronese *et al.*, 2010) and electronic components from personal electronic devices (Bassinet, Trompier and Clairand, 2010; Göksu, 2003; Inrig, Godfrey-Smith and Khanna, 2008; Inrig, Godfrey-Smith and Larsson, 2010; Larsson *et al.*, 2005; Mathur *et al.*, 2007; Woda *et al.*, 2009; Woda, Greilich and Beerten, 2010). Although this research is in its early stages, the potential of addressing some of the challenges involved in a mass casualty event have drawn attention to the OSL technique (Alexander *et al.*, 2007; Joint Interagency Working Group, 2005; Simon *et al.*, 2007). A recent review on the potential application of luminescence techniques in retrospective dosimetry and emergency response can be found in Woda *et al.* (2009).

In this section, we review the basic considerations, analytical protocols and material properties related to the application of OSL in retrospective and accident dosimetry. Herein, the terms "accident dosimetry" and "retrospective dosimetry" are used to refer to the evaluation of radiation exposures received in the past by individuals or groups from the population in cases where dosimeters were found to be inadequate or, more usually, completely absent. "Accident dosimetry" is used in the more specific problem of assessment of radiation exposure immediately or shortly after a radiological or nuclear event, the objectives being to assist in the triage of the population, medical management of the victims, and counseling or reassurance of individuals that do not require immediate medical attention. Although the term "accident" implies an unintentional event, it is used here in a broader sense which includes malevolent incidents such as terrorist attacks. "Retrospective dosimetry," on the other hand, refers here to the reconstruction of absorbed doses to populations long after the event is over (atomic bomb use and tests, radiation accidents not realized at the time, etc.). In addition to the papers cited in this chapter, we also recommend Bøtter-Jensen, McKeever and Wintle (2003) and Aitken (1985, 1998).

### 6.2.1 Basic Considerations

The goal when applying OSL to retrospective or accident dosimetry is to estimate the dose to individuals or populations using natural or artificial OSL materials present in the environment or carried by a person, such as minerals in bricks, tiles, porcelain, electronic components, dental enamel and dental ceramics. These materials can be sampled for analysis in a laboratory, or possibly probed in situ using portable OSL readers. The latter is not yet a reality, but portable OSL readers for field analysis are already under development for luminescence dating applications, and prototype portable optical fiber systems already exist for medical and environmental dosimetry (see Section 2.5).

The primary quantity that can be obtained from luminescence measurements is the total absorbed dose $D_{\mathrm{m}}$ to the medium that constitutes the luminescent material, which has two components: the accident dose $D_x$ (which is the object of the study) and the background dose $D_{\mathrm{BG}}$, due to natural sources (ICRU, 2002). Therefore, except when the background dose is significantly lower than the accident dose, $D_{\mathrm{BG}}$ needs to be estimated and appropriately subtracted from $D_{\mathrm{m}}$ to obtain $D_x$, thus:

$$D_x = D_{\mathrm{m}} - D_{\mathrm{BG}}. \tag{6.1}$$

For practical purposes, the "accident" dose $D_x$ absorbed by the luminescence material needs to be translated into a quantity of interest for the particular problem, for example the air kerma at a reference location, the absorbed dose to a specific organ or the effective dose. In the case of environmental materials, this has been done by using conversion factors derived from computational models (Bailiff *et al.*, 2004, 2005; Göksu *et al.*, 2006). In the case of luminescent materials that are carried by a person, computational models or calibration procedures similar to those used in personal dosimetry can be employed.

The material property requirements for using the luminescence techniques in retrospective and accident dosimetry are: (i) appropriate sensitivity to the dose levels of interest, (ii) negligible or predictable fading in the period between exposure and measurement, (iii) the absence of signal saturation due to background accumulated over geological times and (iv) the possibility of estimating the accumulated dose due to natural background (Haskell, 1993a). Minerals that are known to satisfy some of these criteria, most notably quartz, feldspar and $Al_2O_3$, can be found in bricks, glazed and unglazed tiles, roof tiles, and porcelain from sanitary and electrical fixtures (Bailiff, 1997).

There are several advantages to using luminescence materials in retrospective and accident dosimetry (Bailiff, 1999; Haskell, 1993a, b):

- Environmental materials are ubiquitous and can provide location-specific dose estimates which could be used to validate or refine models of transport and exposure, or serve as a benchmark in dose reconstruction studies.
- Luminescence materials work as integrating dosimeters, providing an assessment of the total dose since the onset of exposure; this information is particularly relevant in the cases of insufficient or absent initial monitoring data, changes in source configuration due to decontamination efforts or meteorological conditions, and uncertainties in the concentration or dispersion of short-lived radioisotopes (Göksu and Bailiff, 2006), all of which may not be registered in surveys performed at a later time after the accident.

- The stability of the luminescence signal of some materials allows the assessment of doses even decades after the accident has occurred, which would be particularly useful in those cases of accidents that are not discovered immediately.
- Materials with unstable signals could also be used if the time period between exposure and readout is short compared to the decay timescale. Alternatively, fading can be characterized and suitable corrections made. Signals that are unstable over long periods of time may have an advantage over stable signals if the materials are analyzed shortly after exposure, since these unstable signals reflect only the most recent events.
- Information on the time-averaged energy and angular distribution of the incident radiation, that is, the source configuration, can be obtained by analyzing the depth-dose profile in bricks (Meckbach *et al.*, 1996).

In addition to the above, OSL offers the following possibilities:

- The readout apparatus is relatively simple, making easier to develop field instruments.
- There is the possibility of developing optical fiber probes for *in situ* analysis of the luminescent materials.
- The readout does not involve heating, therefore making it applicable to materials that would decompose or result in spurious signals when heated.

Some of these possibilities have not yet been explored.

In its present stage of development, OSL is limited to the assessment of external gamma exposures and the results may be difficult to interpret in cases of partial body exposure. Another limitation is that the samples should not be exposed to light from the time of exposure until the measurement is performed.

### 6.2.2 Methodological Aspects

Before discussing the characteristics of OSL materials that may be useful for accident and retrospective dosimetry, it is helpful to introduce available analytical procedures to estimate the dose to the luminescent material. These procedures have been developed largely for luminescence dating of sediments and archeological artifacts, mainly using quartz and feldspars (Aitken, 1998; Wintle, 1997), and were accompanied by the development of automated readers to implement them (Bøtter-Jensen, 1997; Bøtter-Jensen *et al.*, 2000; Bøtter-Jensen, Thomsen and Jain, 2010). These analytical procedures also provide a conceptual framework for the development of procedures for other materials.

In luminescence dating the absorbed dose to the luminescence material $D_m$ is referred to as "equivalent dose, $D_e$," meaning the laboratory dose from a standard radiation source (e.g., $^{60}$Co or $^{90}$Sr/$^{90}$Y source) that would produce a luminescence signal in the material equivalent to that produced by the cumulative dose from artificial and natural sources. (The term "equivalent dose" should not be confused with the concept of dose equivalent used in radiation protection.)

Several analytical procedures have been proposed to estimate $D_e$ using either multiple or single aliquots. In multiple aliquot procedures, the sample material is divided into several aliquots, one being used to measure the "as-received" OSL signal $S$, whereas the others are used to construct the calibration curve (also known as "growth curve"). In this case, sensitivity changes in the material do not pose a problem because the aliquots are used only once and not re-used. However, sufficient material is required to prepare a number of aliquots and a normalization process needs to be employed to correct for variations in mass or sensitivity among different aliquots (Aitken, 1998). In single-aliquot procedures all the information (i.e., the "as-received" signal and the data for the calibration curve) are obtained using one aliquot, which is repeatedly irradiated and the OSL read after each irradiation. Single-aliquot procedures require significantly less material than multiple-aliquot techniques and therefore are preferred in retrospective and accident dosimetry. However, sensitivity changes induced during repeated irradiations and readouts in the procedure need to be monitored and corresponding corrections made.

Both multiple- and single-aliquot procedures can be classified as either "additive-dose" or "regenerative-dose" procedures, depending on how the calibration curve is determined. In additive-dose procedures, the calibration curve is determined using doses added to the "as-received" (accident plus background) dose, with the equivalent dose $D_e$ being determined by extrapolation of the calibration curve to zero OSL signal (Figure 6.1a). This procedure has the disadvantage of relying on an extrapolation, which is highly dependent on the model used to describe the dose response. Regenerative-dose procedures, on the other hand, rely on interpolation of the dose response curve. In these procedures, the calibration curve is obtained by a sequence of repeated cycles that include bleaching the OSL signal (i.e., erasing it by illumination), irradiation of the sample with a known dose, and readout of the resulting "regenerated" signal. With an appropriate choice of the regenerative doses, $D_e$ can be found by interpolation (Figure 6.1b). However, one has to determine if the sample sensitivity to the regenerative dose is similar to the sensitivity to the original accident plus background dose, and correct for those changes if they are shown to occur. In the case of quartz the sensitivity can change as much as 300% throughout the procedure of estimating $D_e$ (Murray and Wintle, 2000).

The most robust and widely adopted procedure for the estimation of the equivalent dose $D_e$ in luminescence dating and retrospective dosimetry is the single-aliquot regenerative (SAR) dose procedure outlined in Table 6.1 (Murray and Wintle, 2000, 2002). The main idea of the SAR procedure is to use a test dose to monitor the sensitivity of the sample and then to correct for sensitivity changes. The calibration curve is determined by the "sensitivity-corrected OSL" given by $S_i/T_i$, where $S_i$ ($i \neq 0$) is the OSL signal produced by a known $i$th dose and $T_i$ is the OSL signal produced by the $i$th test dose. The equivalent dose $D_e$ can then be obtained by interpolating the value $S_0/T_0$, which is the ratio between the "as-received" OSL signal ($S_0$) and the OSL signal produced by the first test dose ($T_0$). An example of the calibration curve and the interpolation method is shown in Figure 6.2.

Additional tests can be included to check whether the sensitivity correction is working properly and whether there is evidence of a phenomenon called "recuperation," in which an OSL signal is observed for an unirradiated sample due to thermal transfer of charges to optically sensitive traps during heating of the sample used in the procedure. Basic requirements, tests of the procedure and procedure modifications were reviewed by Wintle and Murray (2006).

The SAR procedure can be easily carried out using automated OSL readers, such as the Risø TL/OSL readers described in Chapter 2. The reader can be used for automated readout of the "as-received" signal (called the "natural" signal in Figure 6.1 and Figure 6.2) and determination of the calibration curve. The uncertainties can be reduced by the measurement of more aliquots if necessary, although with a reduction of throughput. Further technological

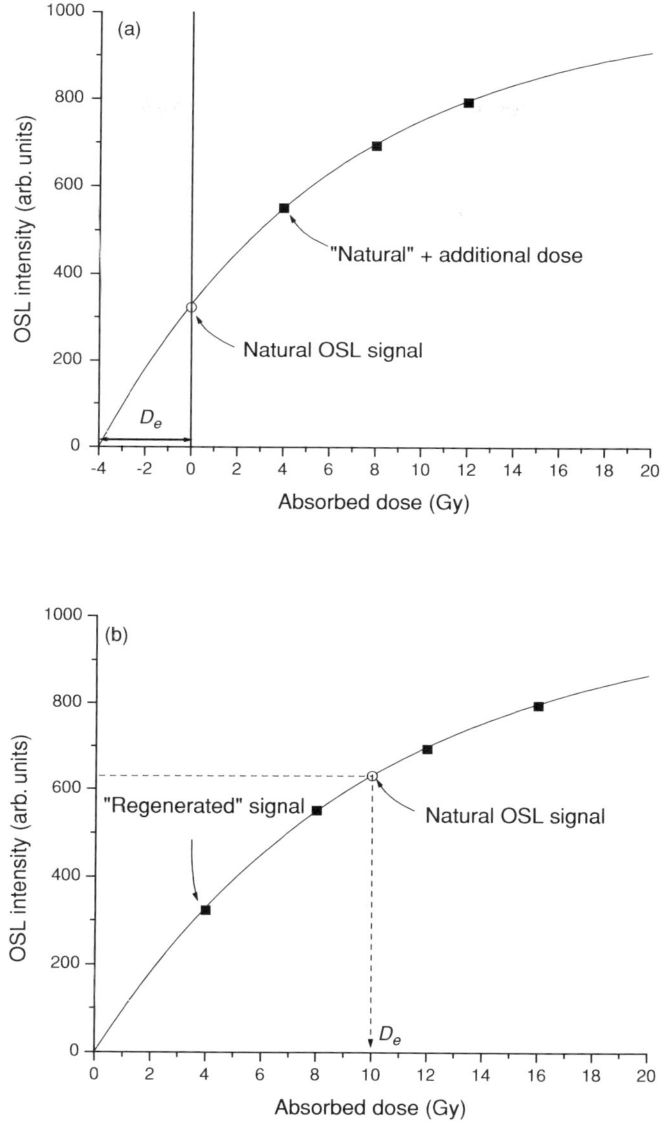


***Figure 6.1*** *Illustration of calibration curves and determination of the equivalent dose $D_e$ using the (a) additive dose procedures and (b) regenerative dose procedures. Here "natural" dose refers to the accident dose plus background dose.*

advances have also made possible the application of the SAR procedure to single grains (Bøtter-Jensen *et al.*, 2002, 2003), for example to obtain a histogram of equivalent dose based on a large number of measurements and improve the accuracy of the dose estimations by identifying well-bleached grains (see Thomsen, Murray and Bøtter-Jensen, 2005 and references therein).

**Table 6.1** *Single-aliquot regenerative-dose (SAR) procedure for quartz (Murray and Wintle, 2000)*

| Step | Treatment | Observed |
|---|---|---|
| 1 | Give dose, $D_i$[a] | — |
| 2 | Pre-heat (160-300 °C for 10s)[b] | — |
| 3 | Stimulate for 100s at 125 °C[c] | $S_i$[d] |
| 4 | Give test dose, $D_t$[e] | — |
| 5 | Heat to 160 °C[b] | — |
| 6 | Stimulate for 100s at 125 °C | $T_i$ |
| 7 | Return to step 1 | |

$L_i$ is the signal observed after dose $D_i$, and $T_i$ is the OSL signal observed after a test dose $D_t$.
[a] For the readout of the "accident" dose, $i = 0$ and $D_0 = 0$.
[b] The pre-heat is used to eliminate OSL components that are thermally unstable in the time scale of interest.
[c] The elevated stimulation temperature is used to prevent charge trapping at the traps associated with the 110 °C TL peak in quartz.
[d] For low doses, the OSL signal should be calculated for short integration times for better signal-to-noise ratio (Banerjee, Bøtter-Jensen and Murray, 2000).
[e] The test dose is fixed and approximately 10–20% of the equivalent dose $D_e$.

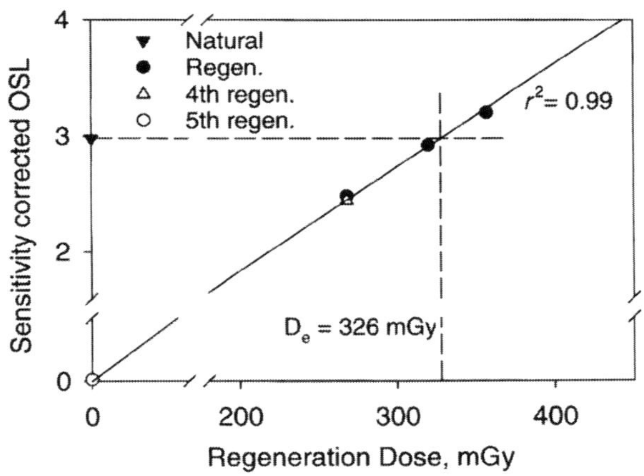


**Figure 6.2** *Example of the single-aliquot regenerative-dose procedure applied to quartz extracts from heated bricks from buildings in the area of the Chernobyl accident. In addition to the regeneration signals obtained for the calibration curve, the graph also includes a regeneration dose (4th regen.) identical to the first regeneration dose to test for the sensitivity correction procedure, and a zero dose (5th regen.) to test for recuperation of the OSL signal. "Natural" dose refers to the accident dose plus background dose. Reprinted from Applied Radiation and Isotopes, Banerjee, D., Bøtter-Jensen, L. and Murray, A.S., Retrospective dosimetry: estimation of the dose to quartz using the single-aliquot regenerative-dose protocol. Vol. 52, 831–844. Copyright (2000) with permission from Elsevier.*

Although the SAR procedure is only widely accepted for quartz, it provides a conceptual framework for the development of procedures appropriate to other materials. One of the basic requirements for the procedure to work satisfactorily is the linear relationship between the regenerated signal $S_i$ and the test dose signal $T_i$ with intercept close to the origin. This has been demonstrated for heated and unheated quartz from geological and archeological sites (Wintle and Murray, 2006). Wintle and Murray (2006) discuss other underlying assumptions of the SAR procedure.

### 6.2.3 Building Materials

The use of luminescence (both TL and OSL) from building materials for dose reconstruction, particularly using heated quartz extracts from fired bricks and tiles, is relatively well developed, as reviewed in the ICRU Report 68: "Retrospective Assessment of Exposures to Ionizing Radiation" (ICRU, 2002). Accident doses as low as $\sim$10 mGy on a natural background of 50 mGy have been detected using heated quartz (Banerjee, Bøtter-Jensen and Murray, 2000). Other building materials that have been investigated for application in accident and retrospective dosimetry include unheated quartz found in calcium silicate bricks (Bailiff and Mikhailik, 2004), mortar (Jain *et al.*, 2004) and concrete (Thomsen *et al.*, 2002, 2003).

Radiation-induced luminescence from minerals in building materials enables the indirect determination of external gamma dose to organs or tissue (ICRU, 2002). In general, the dose or kerma in air at a reference location is determined as an intermediate step. The information is then combined with models of residence time from individuals or population groups to obtain broad estimates of the dose quantity of interest.

This framework is illustrated in Figure 6.3 for the case of doses from fallout. Mineral samples extracted from a particular building location at a height $h$ and depth $d$ on a wall are evaluated using luminescence techniques to determine the total absorbed dose $D_{\mathrm{m}}$ to the medium. Estimation of the background dose to the medium, $D_{\mathrm{BG}}$, allows the absorbed dose to the luminescence material due to the accident $D_x$ to be determined. The dose $D_x$ can then be converted to dose in air at a reference location using conversion factors calculated using computational models and assuming a source term for the radiation exposure, usually a uniform distribution of isotopes on the soil surface due to the fallout, as illustrated in Figure 6.3 (Bailiff *et al.*, 2004, 2005). The dose at the reference location (RL) can then be determined from:

$$ {}_{\mathrm{RL}}D_x = D_x C_{\mathrm{RL}} F_{\mathrm{h}}. \tag{6.2} $$

In this expression, ${}_{\mathrm{RL}}D_x$ is the accident dose at a reference location, $C_{\mathrm{RL}}$ is a factor that converts dose in the luminescent material at a particular location to the dose at a reference location, and $F_{\mathrm{h}}$ is a factor to account for possible heterogenous distribution of fallout associated, for example, with the presence of buildings (Bailiff *et al.*, 2004; Bailiff and Slim, 2008). Bailiff and Slim (2008) describe a reference database of conversion factors from the accident dose to kerma in air for use in dose reconstruction.

There are various approaches to estimate the natural background dose $D_{\mathrm{BG}}$. For samples of known age that were reset (e.g., fired bricks), $D_{BG}$ can be estimated by:

$$ D_{\mathrm{BG}} = \mathrm{Age} \times \left(\dot{D}'_{\alpha} + \dot{D}'_{\beta} + \dot{D}_{\gamma} + \dot{D}_{c}\right), \tag{6.3} $$

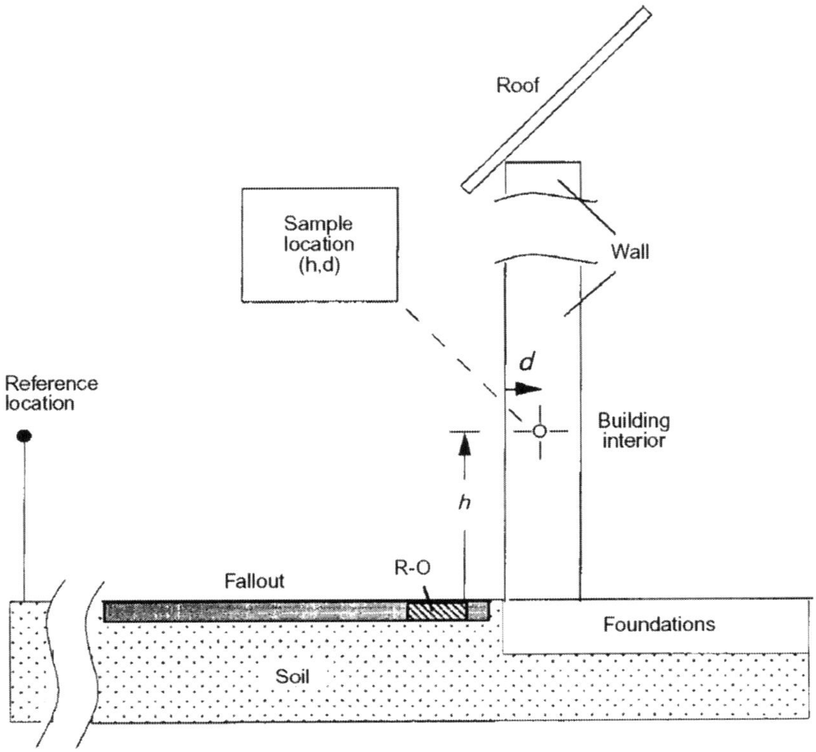


*Figure 6.3* Cross-section of a building illustrating the use of luminescence techniques using building materials for retrospective dosimetry. The dose to the medium of a sample at height h and depth d in the wall, determined by luminescence methods, is used to determine the dose quantity of interest at a reference location. The area indicated R-O has a higher activity due to run-off from the roof. Reprinted from Radiation Protection Dosimetry, Bailiff, I. K., The development of retrospective luminescence dosimetry for dose reconstruction in areas downwind of chernobyl. Vol. 84, 411–419.

where $\dot{D}'_\alpha$ and $\dot{D}'_\beta$ are the effective alpha and beta dose rates from natural radioisotopes, which take into account attenuation coefficients appropriate for the grain sizes in the sample and reduced luminescence yield in the case of high-LET particles; $\dot{D}_\gamma$ is the gamma dose rate from natural radioisotopes and $\dot{D}_c$ is the dose rate from cosmic radiation. The dose rates from natural radioisotopes, mostly potassium, rubidium and isotopes from the uranium and thorium decay chains, can be estimated from a separate measurement of the radioisotope content (Aitken, 1998).

$D_{BG}$ can also be estimated using samples from a heavily shielded location (e.g., basement of the building or interior wall) in the same building, although the radioisotopic composition and the contribution from cosmic radiation and natural gamma rays may not be the same (Ramzaev *et al.*, 2008). Another method is to use a high-sensitivity detector with a photon energy response similar to the sample material (e.g. $Al_2O_3$:C in the case of quartz) in order to measure the background dose rate in situ. This is not always practical, however, since

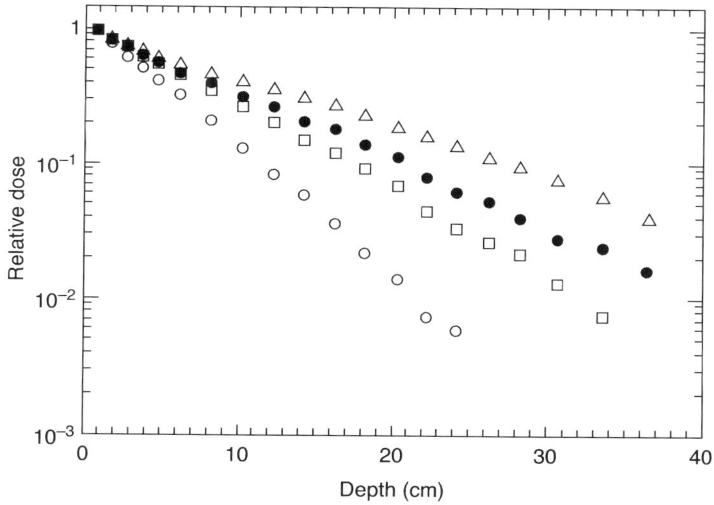


***Figure 6.4*** *Calculated depth-dose profiles in a brick wall for various energies: 140 keV (open circles), 364 keV (open squares), 662 keV (full circle) and 1600 keV (open triangles). The calculation was carried out for a brick at a height of 1 m, with a isotropic distribution of sources on the ground. Reprinted from Radiation Proection Dosimetry, Meckbach, R., et al., Calculation and measurement of depth dose distributions in bricks. Vol. 66, 183–186. Copyright (1996) with permission from Oxford University Press.*

one needs access to the site, long enough time for the dosimeters to be in place in order to accumulate sufficient dose, and the absence of contamination from the fallout.

The depth-dose profile along the depth of a brick can also provide an estimate of $D_{BG}$. Monte-Carlo calculations and measurements have shown that the doses $D_e$ decrease exponentially with depth in the brick for a variety of energies and source configurations (Meckbach *et al.*, 1996), as shown in Figure 6.4. Therefore, the presence of a constant level in the depth-dose profile may indicate the background dose. This is illustrated for the depth-dose profile for a brick from the Semipalatinsk region shown in Figure 6.5. The data in Figure 6.5 were fitted with a function of the form:

$$D_m = D_{BG} + D_0 e^{-\alpha t}$$

where $D_0$ is the dose at the surface, $t$ is the depth in the brick and $D_{BG}$ is the constant dose level. A different approach was employed to obtain $D_{BG}$ for bricks from the Semipalatinsk region, where the depth-dose profile calculated using Monte-Carlo radiation transport code for a chosen source configuration was matched interactively with the experimental depth-dose profile, assuming $D_{BG}$ to be constant throughout the brick (Göksu *et al.*, 2006).

Although the uncertainties in the dose estimations are important for dose reconstruction, the uncertainties related to the evaluation of $D_x$ represent only a small contribution to the final uncertainty in the dose quantity of interest. The other major contributions to the uncertainty come from converting $D_x$ to ${}_{RL}D_x$, and from converting ${}_{RL}D_x$ to the dose quantity of interest using population or individual behavioral models.

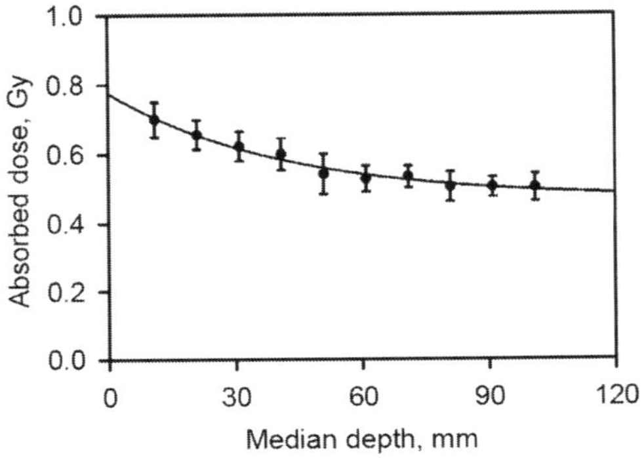


***Figure 6.5*** *Total absorbed dose $D_m$ as a function of median depth for a brick from the Semipalatinsk region used as part of an interlaboratory comparison study. Reprinted from Blair, M. W., Development of Optically Stimulated Luminescence Techniques for Application to Terrestrial and Martian Studies. Ph. D. Thesis, Oklahoma State University.*

The uncertainty in the estimation of the natural background dose $D_{\mathrm{BG}}$ and its relative contribution to the total dose also imposes limitations on the minimum accident doses that can be determined. For young samples and large accident doses, the uncertainty in $D_{\mathrm{BG}}$ may be small; for old samples and small accident doses, the uncertainty in $D_{\mathrm{BG}}$ may result in relatively large uncertainty in $D_x$ (Haskell, 1993a).

The sampling methodology has a strong bearing on the dose reconstruction process. The absorbed dose to the luminescent medium $D_{\mathrm{m}}$ is affected by the presence or absence of secondary charged particle equilibrium at the sampling location, radiation backscattered by underlying materials, radiation scattered or attenuated by surrounding structures and source configuration (fallout distribution). For this reason, recommended sampling procedures include finding locations with simple irradiation geometry. For a wall, this translates into a brick at ~1 m height from the ground at a flat section extending ~1 m either side, facing horizontal undisturbed ground, preferably at a depth larger than ~2 mm (ICRU, 2002). The methodology adopted in dose reconstruction efforts in populated areas contaminated by the Chernobyl accident involved obtaining brick samples using a diamond corer or removing the whole brick. These were then sliced using a diamond blade saw for extraction of the minerals at a specific depth in the brick (Bailiff *et al.*, 2005).

Quartz previously heated during manufacture of the bricks is particularly suitable for retrospective and accident dosimetry. Firstly, heating during manufacture works as a "zeroing" event, in which trapped charges created by the natural background radiation during geological times are released from their traps and thus the stored information regarding previous dose accumulations is reset to zero. In this case, for bricks that were shielded from the accident dose, all grains or aliquots will reflect only the accumulated natural background dose since the manufacturing process. Secondly, heating is known to cause sensitization of quartz, with the result that heated quartz from fired bricks can be more than two orders of magnitude more sensitive than unheated quartz (Bøtter-Jensen *et al.*, 1995).

The quartz is generally extracted from the brick samples using methodologies developed in luminescence dating (Aitken, 1985, 1998), which involves crushing and sieving the ceramic, extracting the quartz grains using heavy liquid separation, and etching with

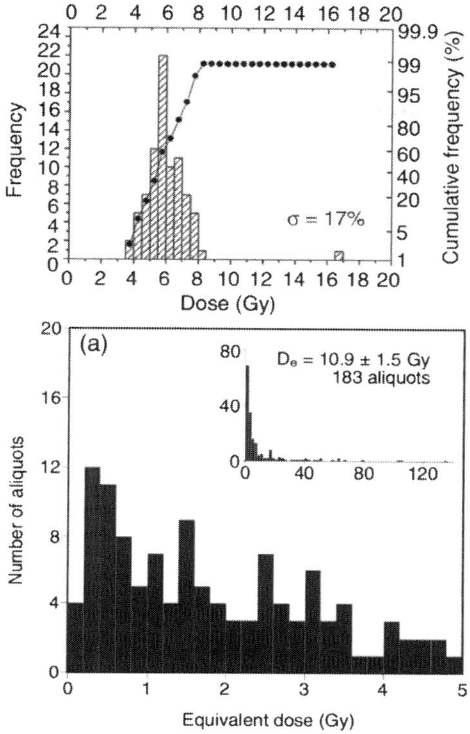


***Figure 6.6*** *Histogram of equivalent doses measured using small aliquots of (a) well-bleached quartz from a mortar extracted from a radioactive waste storage building and (b) poorly bleached quartz from unirradiated concrete. Figure (a) was reprinted from Radiation Protection Dosimetry, Jain, M., et al., Retrospective dosimetry: dose evaluation using unheated and heated quartz from a radioactive waste storage building. Vol. 101, 525–530. Copyright (2002) with permission from Oxford University Press, and Figure (b) was reprinted from Radiation Protection Dosimetry, Thomsen, K. J., et al., Retrospective dosimetry using unheated quartz: a feasibility study. Vol. 101, 345–348, Copyright (2002) with permission from Oxford University Press.*

hydrofluoric acid to remove the outer layer of the quartz grains. The latter reduces the contribution to the OSL signal from natural alpha-particle radiation (Bailiff *et al.*, 2005).

Unheated quartz extracted from mortar and concrete poses a challenge for accident dosimetry, not only because of the lower sensitivity, but also because of the difficulty of determining the background dose accumulated over geological times. In some situations, exposure to light during mixing of the concrete or mortar in the construction may work as a zeroing event. For well-bleached samples, the distributions of equivalent doses $D_e$ (i.e. the "equivalent dose distributions") obtained from multiple aliquots or, even better, multiple single grains are relatively well defined and easier to interpret (Figure 6.6a). However, the bleaching process can be highly heterogeneous, resulting in a range of doses that can be very broad (Figure 6.6b), making it difficult to estimate the accident dose without further statistical analysis.

The proposed methodology for unheated quartz is to obtain the equivalent dose for a large number of small aliquots ($\sim$100 grains or less) or single grains, and use statistical analysis to identify the well-bleached samples (see Bøtter-Jensen, McKeever and Wintle, 2003a). These samples would thus indicate the accumulated dose since the zeroing process, and could then be used to estimate the accident dose (assuming the natural background dose can be estimated). The reason for using small aliquots is the fact that most quartz grains are insensitive, with only a small percentage of the grains contributing to more than 70% of the OSL signal in an aliquot (Duller, Bøtter-Jensen and Murray, 2000). For aliquots with $\sim$100 grains or less, most of the OSL would then be due to a small number of grains only, avoiding the problem of averaging the OSL signal from several grains that were bleached to different degrees. This methodology has been tested using small aliquots and single grains in laboratory conditions using a concrete slab irradiated with gamma rays from a $^{137}$Cs source. The resultant depth-dose profile obtained from the unheated quartz samples was shown to agree with the doses predicted by Monte-Carlo calculations and with results obtained using heated quartz (Thomsen *et al.*, 2002).

In some cases, the use of TL or OSL signals that are unstable over geological times can be advantageous, since most of the signal will be related to recent events. With TL it is often easier to separate the individual TL signals that are unstable from those that are stable by selecting individual TL peaks (Bailiff and Mikhailik, 2004; Göksu and Bailiff, 2006). In this sense the OSL technique has a disadvantage in that it may be probing the trapped charge population from multiple traps, some of which may be thermally unstable, while others may be thermally stable. The thermally stable trapping centers would result in larger signals compared to the unstable centers due to the dose accumulated over geological times (assuming equal sensitivity for the different traps).

### 6.2.4 Household Materials

Several household materials have been investigated for retrospective and accident dosimetry, including porcelain, salt, and other commonly found chemicals and products. Unfortunately, these efforts have been limited and much work is still required for the OSL technique using household materials to achieve the level of reliability that is attributed to OSL when applied to building materials, particularly using quartz inclusions in bricks.

Information on the OSL properties of porcelain from household items for applications in accident dosimetry have been reported by Bøtter-Jensen *et al.* (1996). The authors suggest that $Al_2O_3$ is a major component responsible for the OSL signal in porcelain, given the similarity of the photoluminescence emission bands detected in these samples with $Al_2O_3$:C. However, detailed emission spectra are still required to confirm these suggestions. An OSL stimulation spectrum representative of the samples investigated is shown in Figure 6.7a. The range of doses that can be measured with porcelain is wide; the dose response was observed to be linear from 10 mGy up to $\sim$20 Gy, becoming sub-linear for higher doses. The minimum detectable doses are estimated to be below 10 mGy. The only caveat is the possibility of signal bleaching due to the transparency of porcelain. As seen in Figure 6.7b, sampling at a depth larger than 2 mm is required to avoid the material that could have been bleached by exposure to ambient light.

Other household chemicals such as common salt, washing powder and water softener, among others, were investigated by Thomsen, Bøtter-Jensen and Murray (2002). One of

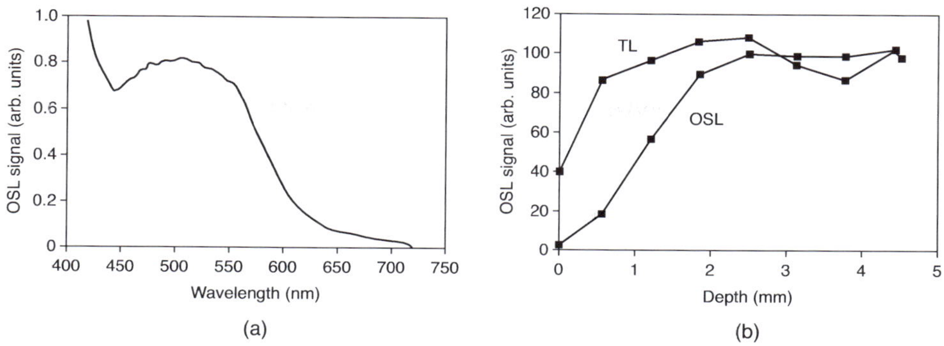


***Figure 6.7*** *(a) OSL stimulation spectra of a porcelain sample and (b) depth profile of the OSL signal from an irradiated porcelain exposed to light. The figure demonstrates that the samples at a depth of 2–3 mm should be obtained to avoid the material that was bleached due to the transparency of the porcelain. Figure (a) was reprinted from Radiation Protection Dosimetry, Bøtter-Jensen, L., et al., Luminescence properties of porcelain ceramics relevant to retrospective radiation dosimetry. Vol. 65, 369–372. Copyright (1996) with permission from Oxford University Press, and Figure (b) was reprinted from Radiation Protection Dosimetry, Bøtter-Jensen, L., et al., Luminescence properties of porcelain ceramics relevant to retrospective radiation dosimetry. Vol. 65, 369–372. Copyright (1996) with permission from Oxford University Press.*

the proposed advantages of the materials investigated is the fact that they are likely to be packaged in light-tight containers. The results obtained for the household materials are summarized in Table 6.2. In some cases, the minimum detection doses are of the order of 10 mGy or less. In the cases of common salt, Glauber salt and Omo Sensitive® washing powder, the fading after a period of two weeks was smaller than 10%. A survey of other common materials was also presented by Larsson *et al.* (2005).

The studies above demonstrate the potential of using OSL from commonly found materials in accident dosimetry. Unfortunately, investigations remain limited. If the technique is to be developed to the point of being widely accepted, systematic material characterizations and development of protocols to cope with the wide variety of behavior that is likely to be found need to be carried out.

### 6.2.5 Electronic Components

The use of OSL from electronic components contained in personal electronic devices is of great interest for accident dosimetry due to three factors: (i) they are typically positioned close to the body, (ii) they are carried by a large part of the population and (iii) they have sufficient sensitivity to address the problem of triage for acute radiation exposure (Inrig, Godfrey-Smith and Khanna, 2008). These and other items termed "fortuitous materials" have been identified as potential candidates to provide fast and accurate dose evaluation in cases of emergency (Joint Interagency Working Group, 2005).

The increasing use of smart cards or integrated-circuit cards, such as telephone, credit and ID cards, suggests their potential application in accident dosimetry. The absorbed radiation dose to exposed individuals can be estimated in principle using the radiation-induced OSL

***Table 6.2*** *OSL properties of some household materials (Thomsen, Bøtter-Jensen, and Murray, 2002)*

| Material | Major chemical component | Sensitivity (counts $s^{-1}$ $mg^{-1}$ $Gy^{-1}$) | Fading in 24 h | Fading in 2 weeks | Natural dose (mGy) | Measured/ given dose | MDD (mGy) |
|---|---|---|---|---|---|---|---|
| Bluecare® dish-washing powder | — | 1600 | 15% | 40% | $6 \pm 6$ | $1.01 \pm 0.01$ | 20 |
| Calcium chloride (frost protection) | $CaCl_2$ | 4600 | — | — | $0.1 \pm 0.2$ | $0.94 \pm 0.02$ | 1 |
| Common salt | NaCl | 2 000 000 | 4% | 5% | $1.3 \pm 0.3$ | $0.91 \pm 0.05$ | 1 |
| Dynamo Color® washing powder | — | 100 | — | — | $70 \pm 30$ | — | 100 |
| Glauber salt | $Na_2SO_4$ | 100 | 0% | 3% | $1 \pm 4$ | $0.73 \pm 0.03$ | 15 |
| Omo Sensitive® washing powder | – | 900 | 6% | 7% | $-3.0 \pm 1.5$ | $0.97 \pm 0.02$ | 5 |
| Water softener | $(NaPO_3)_x$ | 500 | 2% | 25% | $1 \pm 3$ | $1.02 \pm 0.05$ | 10 |

Minimum detectable doses (MDD) of the order of a few milligrays are observed for some materials.

signal from integrated circuits ("chips") contained in those cards. The few studies available indicate that the intensity of the OSL signal is sufficient to estimate doses from 0.1 to 10 Gy, and that the OSL signal is linear with dose, reproducible and thermally stable.

Göksu (2003) observed infrared stimulated OSL in telephone smart-card chips having a transparent cover. The OSL dose response was linear and the signal was reproducible for doses ranging from 0.25 to 5 Gy, with a natural signal of the order of 0.1 Gy. The radiation-induced IRSL (infrared stimulated OSL) signal from the chips showed significant reduction only after heating above 100 °C, attesting to its thermal stability. At 40 °C the signal remained stable for periods up to ~1.7 h, only decaying noticeably when stored at higher temperatures.

Mathur *et al.* (2007) confirmed the results above using chips from smart ID cards that were packaged within a translucent coating (see Figure 6.8) provided by the manufacturer. In this work, chips extracted directly from the ID cards had a black coating that had to be removed. Once the coating was removed, the chip did not show a signal with infrared stimulation. This is in agreement with the suggestion (Y. H. Göksu, private communication) that the OSL is coming from the globe top used to protect the chip, which is made of $SiO_2$ and glue. The IRSL signal from the top was linear over the dose range investigated, from 1 to 12 Gy from a $^{137}Cs$ gamma source. Similar results were also observed by Woda and Spötti (2009) using $SiO_2$-impregnated epoxy coverings for electronic chips from a variety of ID-card types.

In addition, Mathur *et al.* (2007) showed the potential of using blue stimulation to measure the OSL from the chip cards. Figure 6.9a shows OSL curves from one of the investigated chips irradiated with doses up to 10 Gy. The OSL response is linear and intense, with a

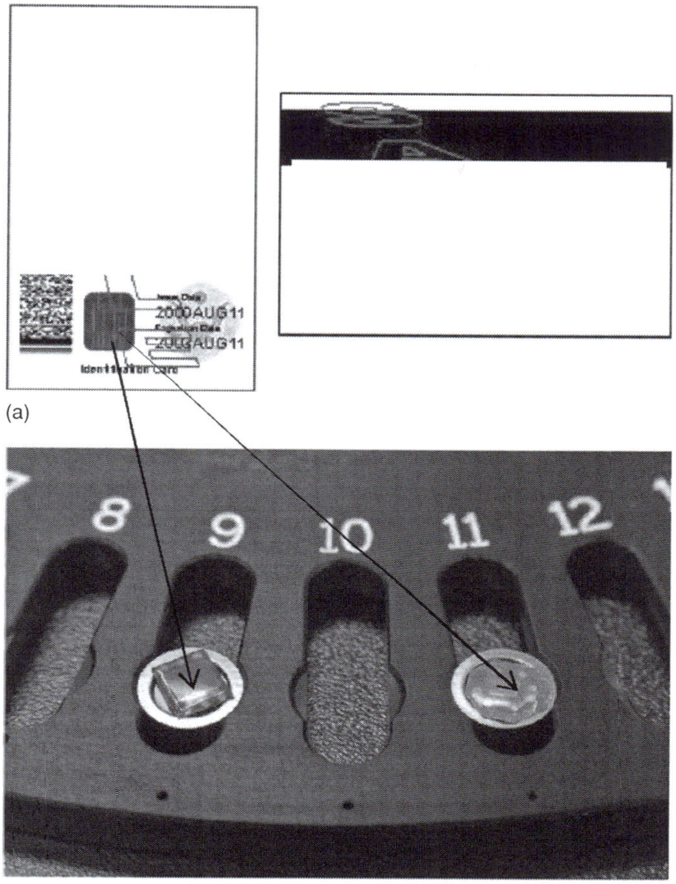

(a)

(b)

*Figure 6.8* (a) Photograph of ID card containing the (b) electronic chips investigated as a potential OSL detector for accident dosimetry. Reprinted from Radiation Measurements, Mathur, V.K, et al., Radiation sensitivity of memory chip module of an ID card. Vol. 42, 43–48. Copyright (2007) with permission from Elsevier.

minimum detectable dose estimated to be around 20 mGy. Since in these measurements no heating was used to eliminate unstable components, the signal decayed approximately 20% in the first 2 h, remaining relatively stable up to 10 h. Investigations on the thermal stability of the OSL signal for periods longer than that are needed. The signal of the "as-received" samples was also not tested; it is necessary to verify whether or not the signal for a zero dose is sufficiently small not to interfere with the evaluation of the accident dose.

The OSL of other electronic components such as resistors, capacitor, diodes, inductors and a particular type of silicon carbide substrate used in the semiconductor industry (moissanite) have also been investigated (Bassinet, Trompier and Clairand, 2010; Godfrey-Smith, 2006; Inrig, Godfrey-Smith and Khanna, 2008; Woda, Greilich and Beerten, 2010). In the case of moissanite, a radiation-induced OSL signal was detected with blue

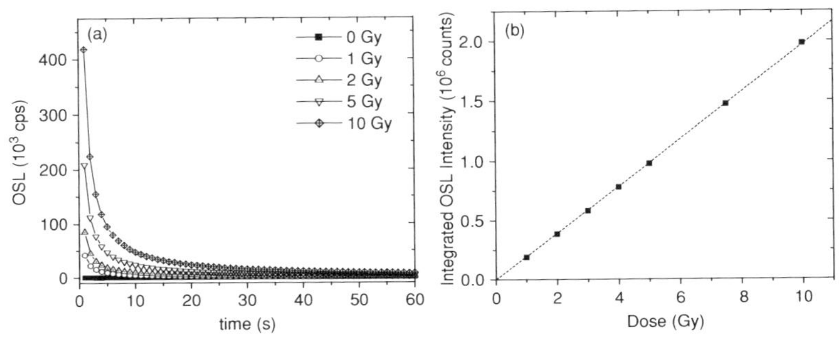


***Figure 6.9*** *(a) OSL of the memory chip modules under blue stimulation as a function of absorbed dose ($^{90}$Sr/$^{90}$Y beta source); (b) dose response curve calculated from the data presented in (b). Reprinted from Radiation Measurements, Mathur, V.K, et al., Radiation sensitivity of memory chip module of an ID card. Vol. 42, 43–48. Copyright (2007) with permission from Elsevier.*

stimulation, with a minimum detectable dose estimated around 0.7 Gy (Godfrey-Smith, 2006). The OSL dose response of surface-mount resistors using blue stimulation was linear, with minimum detection limits of the order of 10 mGy (Inrig, Godfrey-Smith and Khanna, 2008). Examples of OSL curves from a resistor are shown in Figure 6.10. The OSL signal was observed to fade by ~35% in one day, and ~50% in 10 days. However, the signal decay was relatively well-described by a function of the type:

$$I = I_c \left[ 1 - \frac{g}{100} \log_{10} \left( \frac{t}{t_c} \right) \right], \tag{6.4}$$

where $I_c$ is the signal at a time $t_c$ and $g$ is the rate of decay per decade, that is, a 10-fold increase in time since irradiation (see Figure 6.11). Fading of the OSL signal was also the major problem with electronic components encountered by Bassinet, Trompier and Clairand

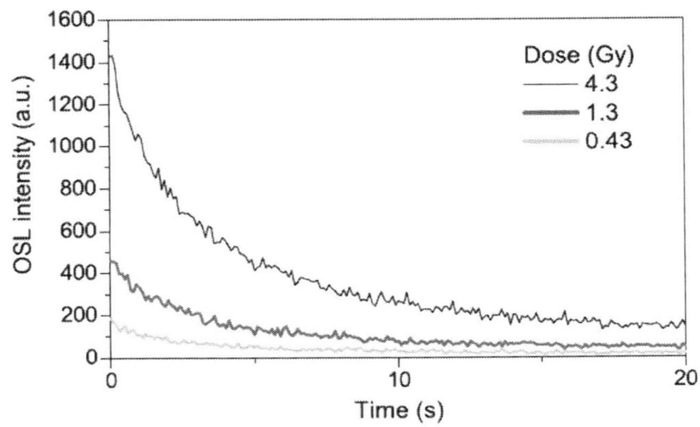


***Figure 6.10*** *OSL curves of 73 mg of a resistor, measured with blue light stimulation. Reprinted from Radiation Measurements, Inrig, E. L., Godfrey-Smith, D.I. and Khanna, S., Optically stimulated luminescence of electronic components for forensic, retrospective, and accident dosimetry. Vol. 43, 726–730. Copyright (2008) with permission from Elsevier.*

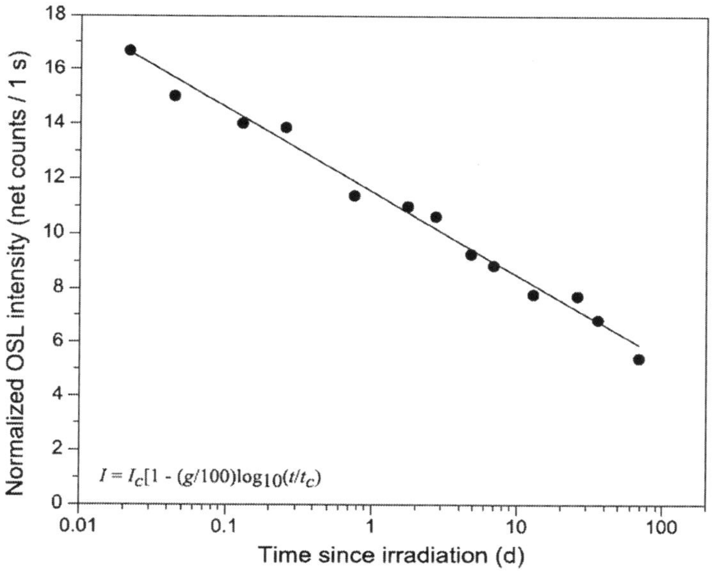


***Figure 6.11*** *Isothermal decay of OSL signal from a resistor sample irradiated with 5 Gy from a $^{60}$Co source, measured with blue light stimulation. Reprinted from Radiation Measurements, Inrig, E. L., Godfrey-Smith, D.I. and Khanna, S., Optically stimulated luminescence of electronic components for forensic, retrospective, and accident dosimetry. Vol. 43, 726–730. Copyright (2008) with permission from Elsevier.*

(2010) and by Woda and Spötti (2009). The latter described the fading as being due to a combination of thermal fading from shallow traps and anomalous fading from deeper traps.

Inrig, Godfrey-Smith and Khanna (2008) conducted trials in which cell phones were affixed on an anthropomorphic phantom and exposed to gamma sources. Resistors were then removed from the cell phones and an OSL procedure used to recover the dose absorbed. Figure 6.12 shows the OSL dose response for one of the cell phones, showing the level of the originally measured signal and the signal after correction for fading. The results of these trials showed that the recovered doses were within 20% of the dose recorded by a nearby electronic dosimeter during irradiation. Although these trials are of limited scope, they demonstrate the potential of the technique to measure doses below 1 Gy.

As in the case of household materials, the development of OSL from electronic components for accident dosimetry still requires developments in terms of material characterization and protocols.

### 6.2.6 Dental Enamel and Dental Ceramics

In the event of a radiological or nuclear event in urban areas, the number of individuals requiring evaluation for potential exposure to ionizing radiation could be very large. The radiological accident in Goiânia, Brazil, illustrates the potential scenario of a radioactive dispersion device detonation. In 1987, in the city of Goiânia two people disassembled an abandoned $^{137}$Cs teletherapy unit, rupturing the capsule containing nearly 93 g of radioactive cesium chloride salt, amounting to a total activity of 50.9 TBq (1375 Ci). As

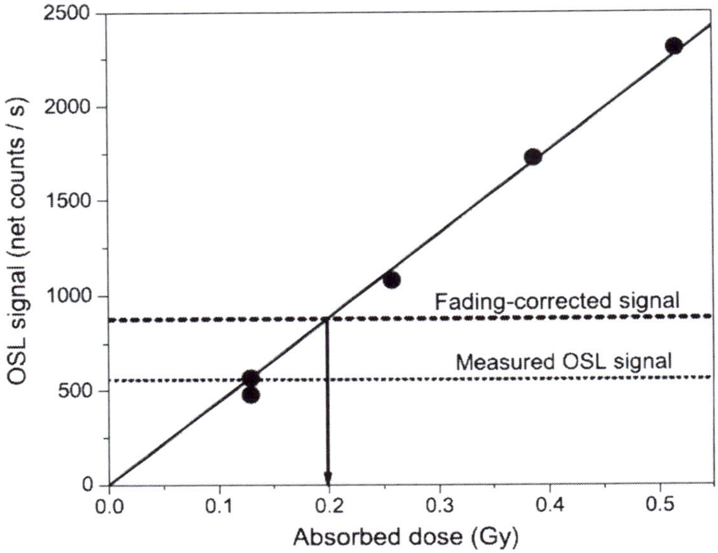


***Figure 6.12*** *Dose response curve and determination of the "accident dose" in a laboratory trial using resistors removed from a commercial cellphone. Reprinted from Radiation Measurements, Inrig, E. L., Godfrey-Smith, D.I. and Khanna, S., Optically stimulated luminescence of electronic components for forensic, retrospective, and accident dosimetry. Vol. 43, 726–730. Copyright (2008) with permission from Elsevier.*

a result, and according to cytogenetic analysis, at least 21 individuals received doses in excess of 1 Gy out of 110 people monitored due to suspected exposure to doses above 0.1 Gy. In total, 112 000 individuals were monitored for external and internal exposure (IAEA, 1988). In the case of a malicious event in a highly populated area, the numbers could be even higher, especially in the case of a nuclear attack. Waselenko *et al.* (2004) estimates that more than 100 000 people would receive doses above 1 Gy for a 1 kiloton nuclear detonation in a city with population of 2 million people. Moreover, the radiation injuries would be complicated by blast and thermal injuries (AFRRI, 2003; NCRP, 2001). Therefore, the capability to perform triage for radiation exposure and contamination in order to allow medical authorities to determine the most appropriate medical attention is an important, but currently deficient, element of emergency preparedness. The intended purpose of radition dosimeters in triage is to evaluate individual doses with high enough precision to identify persons exposed to >2 Gy, within the triage range 0.5–10 Gy while minimizing misclassification between ranges. Furthermore, the method should be able to provide a comparison dose value for other biodosimetry methods. The method, ideally, should be able to be deployed in multiple sites, including hospitals, clinics and dental surgeries, thereby easing the burden on any one location while enabling screening of large numbers of individuals.

Unfortunately in such emergency scenarios, the population directly exposed, or those simply worried about being exposed, would not be expected to be carrying dosimeters and the resources for cytogenetic and other biological dosimetry methods would have to be focused on those requiring most urgent attention. Even the electronic components described

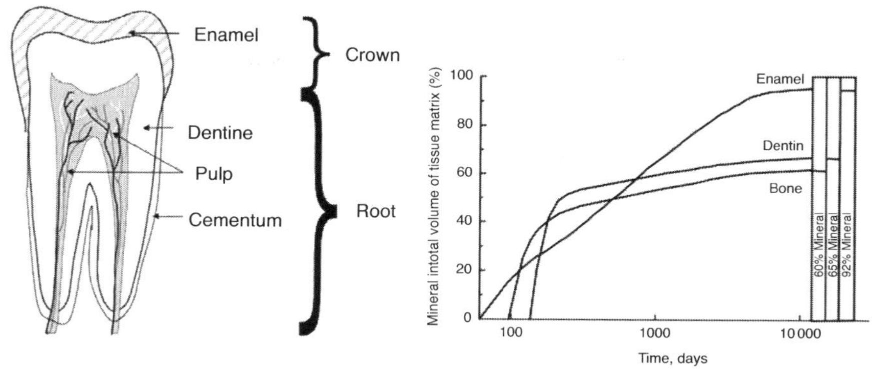


***Figure 6.13*** *(a) Anatomy and location of tissues on tooth; (b) degree of mineralization of calcified human tissue as a function of time. Reprinted from IAEA, Use of electron paramagnetic resonance dosimetry with tooth enamel for retrospective dose assessment. IAEA-TECDOC–1331. Copyright (2002) with permission from International Atomic Energy Agency, Vienna.*

in the previous section may not be available, or the location and positioning with respect to the person's body may be uncertain. In any case, assessment of an individual's absorbed dose using electronic components would be a lengthy and time-consuming process.

Therefore, there has been increasing interest in the development of techniques that could provide a fast and convenient assessment of dose to an individual in an emergency situation using samples which: (i) could be found in every case, (ii) are personal and located in a well-defined position with respect to the person's body, (e.g. biological material), (iii) have dosimetric properties that do not change over time, (iv) present a radiation-induced signal that is stable from the time of exposure until the time of measurement, or have predictable fading properties such that a fading correction could be reliably applied and (v) could be easily sampled or measured in vivo.

Dental enamel is a material that satisfies some of these requirements and, consequently, has been the subject of research using luminescence techniques. Radiation-induced radicals in dental enamel are already used in retrospective and accident dosimetry using the electron paramagnetic resonance technique with biopsied material (IAEA, 2002). Efforts to solve the technical and methodological challenges to perform measurements in vivo are also currently underway (Swartz *et al.*, 2007; Williams *et al.*, 2007).

Dental enamel is a highly mineralized tissue, containing 96% of inorganic material when fully formed (see Figure 6.13). The inorganic material is mostly hydroxyapatite, $Ca_5(PO_4)_3(OH)$, but has $K^+$, $Mg^{2+}$, $Na^+$, $Cl^-$, $HPO_4{}^{2-}$ and $CO_3{}^{2-}$ as common substitutions (Eanes, 1979; IAEA, 2002; Shi *et al.*, 2005). Atomic force and scanning electron microscopies reveal a hierarchical organization of dental enamel in which nanofibrils of hexagonal hydroxyapatite crystals, of typical diameter 30–40 nm, aggregate, forming thicker fibrils 80–130 nm in diameter; those in turn are then organized into larger fibers of ~800 nm of diameter, and so on (Cui and Ge, 2007). More information and further references on the structure and composition of the tooth from the point of view of its application in retrospective and accident dosimetry can be found in the IAEA technical document on EPR dosimetry using dental enamel (IAEA, 2002).

The possibility of using the TL of dental enamel, dentin and even bones, in accident dosimetry has been proposed by several authors (Christodoulides and Fremlin, 1971; Jasińska and Niewiadomski, 1970; Kolberg, Prydz and Dahm, 1974; Willhoit and Poland, 1968). However, TL measurements are inherently problematic in the case of biological material because of the oxidization of organic components during heating, resulting in spurious luminescence. In the case of dental enamel, heating also leads to complex chemical and structural modifications (Shi *et al.*, 2005). The TL curves of undeproteinated and deproteinated dental enamel show the presence of TL peaks even without irradiation; in deproteinated dental enamel the main effect of irradiation was to increase the intensity of a TL peak at $\sim$140 °C (Godfrey-Smith and Pass, 1997).

The measurement of OSL from dental enamel does not involve heating, so the problem of oxidization of organic material is avoided. Measurements can, in principle, be performed inside the mouth using optical fibers without removing the tooth requiring the development of equipment for in-vivo measurement. Another (although less ideal) possibility is the development of readers for analysis of extracted material, taking advantage of the simple requirements of an OSL reader.

The possibility of using the OSL of dental enamel in accident dosimetry was first investigated by Godfrey-Smith and Pass (1997) using deproteinated and undeproteinated modern human enamel samples. The authors demonstrated the presence of a radiation-induced OSL signal using green stimulation with broadband detection in the UV (Hoya U-340 fiters), and using infrared stimulation with broadband detection in the visible (Schott BG-39 filters). The lowest dose of radiation used by the authors was 120 Gy, highlighting the need to increase the technique's sensitivity by orders of magnitude for applications in retrospective and accident dosimetry. Subsequent studies confirmed the presence of a radiation-induced OSL signal from dental enamel and demonstrated the possibility of improving the sensitivity of the technique (Godfrey-Smith, 2008; Yukihara *et al.*, 2007).

The main problems identified with the use of OSL with dental enamel have been: (i) the insufficient sensitivity of dental enamel with the readers tested so far, (ii) instability of the OSL signal at room or body temperature, (iii) the sample-to-sample variability, (iv) the lack of a protocol for proper calibration of the OSL signal from the samples and (v) the possible presence of OSL signal in natural (unirradiated) samples.

Yukihara *et al.* (2007) improved the sensitivity of the technique using a reader design that uses an ellipsoidal mirror for better light collection (Markey, Bøtter-Jensen and Duller, 1997), demonstrating a minimum detectable dose of the order of a few Gy for samples freshly irradiated in laboratory (Figure 6.14). The OSL was measured using blue laser stimulation (472 nm) and detection in the UV region of the spectrum (Hoya U-340 filters). Similar detection limits were obtained by Godfrey-Smith (2008) using a Risø TL/OSL reader with blue LED stimulation and detection also in the UV region (Hoya U-340 filters). These minimum detectable doses were obtained using samples already prepared in the laboratory (disinfected, separated from dentine and crushed). More recent laboratory measurements (unpublished) have demonstrated the ability to detect doses as low as 100 mGy using whole extracted teeth (Dr. Sergii Sholom, personal communication). However, the possibility of detecting these dose levels in a realistic situation has not yet been demonstrated. Technical improvements are still necessary.

DeWitt *et al.* (2010) investigated the sample-to-sample variability in the OSL signal of dental enamel and possible techniques for normalization of the signal (e.g., sample weight

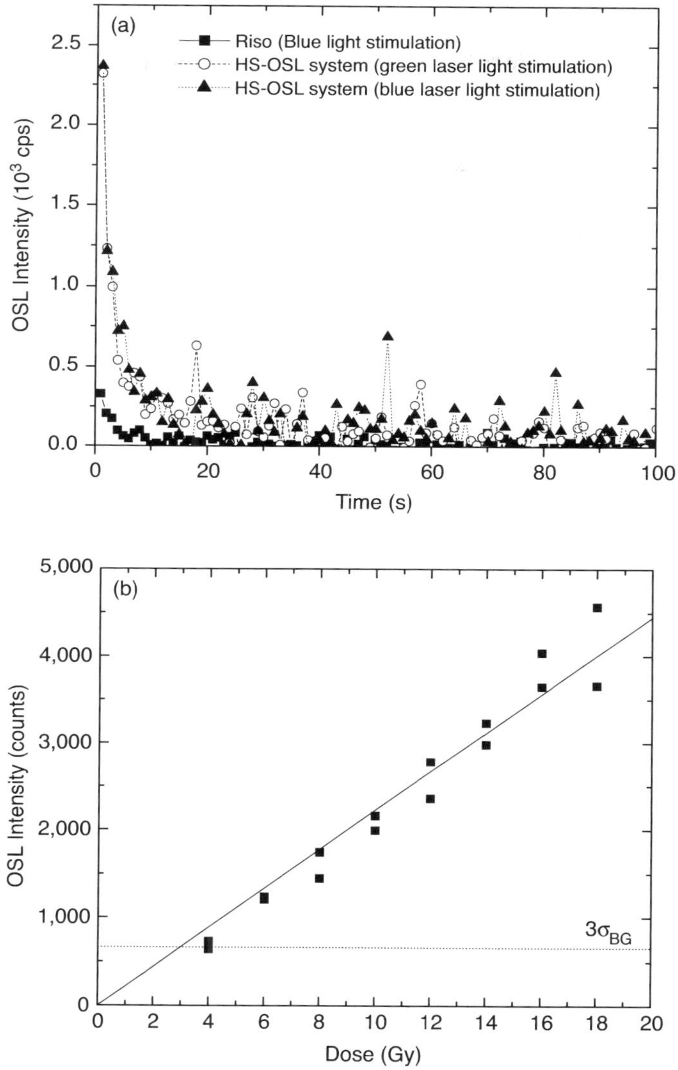


***Figure 6.14*** *(a) OSL with blue (472 nm, ~27 mW/cm²) and green (532 nm, ~23 mW/cm²) laser stimulation of a 15 mg dental enamel sample irradiated with ~40 Gy using a $^{90}Sr/^{90}Y$ beta source and measured using three different readers, where HS-OSL system refers to the system described by Yukihara et al. (2007) that uses an ellipsoidal mirror for better light collection; (b) dose response constructed with a 15 mg aliquot of dental enamel irradiated with various doses from a $^{90}Sr/^{90}Y$ beta source, measured using the HS-OSL system; the background level and respective standard deviation σ was estimated using an unirradiated aliquot. Figure (a) was reprinted from Radiation Measurements, Yukihara, E. G., et al., Optically stimulated luminescence (OSL) of dental enamel for retrospective assessment of radiation exposure. Vol. 42, 1256–1260. Copyright (2007) with permission from Elsevier, and Figure (b) was reprinted from Radiation Measurements, Yukihara, E. G., et al., Optically stimulated luminescence (OSL) of dental enamel for retrospective assessment of radiation exposure. Vol. 42, 1256–1260. Copyright (2007) with permission from Elsevier.*

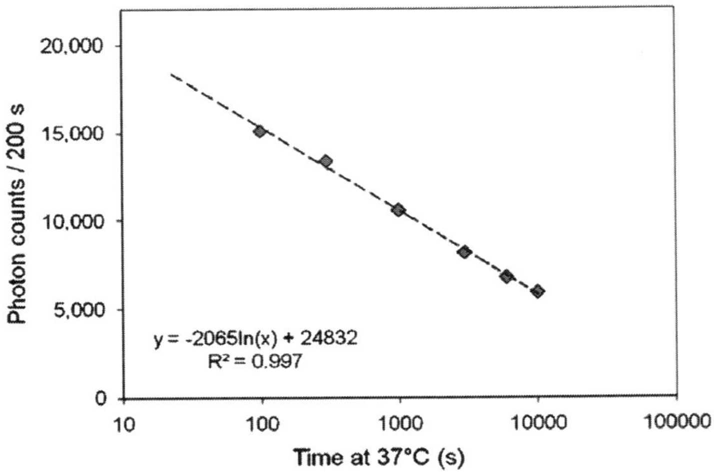


***Figure 6.15*** *OSL signal from irradiated dental enamel held at 37 °C, measured using blue LED stimulation and broadband UV detection filters (Hoya U-340). Reprinted from Radiation Measurements, Godfrey-Smith, D.I., Towards in vivo OSL dosimetry of human tooth enamel. Vol. 43, 854–858. Copyright (2008) with permission from Elsevier.*

or fragment area). Attempts to use a universal calibration curve generated from average OSL signals of different samples resulted in values that differ by a factor of 2 in some cases. One of the approaches proposed in that work, not yet tested, is the possibility of using UV illumination to generate an OSL signal that could be used to normalize the variations in sample sensitivity.

Both Godfrey-Smith (2008) and DeWitt *et al.* (2010) noted fading of the OSL signal from dental enamel over a time scale from hundreds of seconds to almost 40 days following laboratory irradiation (Figure 6.15). The sample-to-sample variability is high, but in some samples the OSL signal was observed to remain relatively stable after the first day following irradiation (DeWitt *et al.*, 2010). If the OSL signal is associated with the TL peak around 140 °C (Godfrey-Smith and Pass, 1997), it is possible that the OSL decay is due to thermally stimulated recombination of the radiation-induced trapped charges. The question then is whether or not there is also a component associated with stable trapping centers, which would result in a more stable OSL signal. If fading is associated with tunneling effects caused by the proximity between trapping and recombination centers in the hydroxyapatite nanocrystals, the only solution would be to determine whether or not fading corrections can be applied.

In addition to the above issues, in some cases an OSL signal was observed in samples freshly prepared, not irradiated in laboratory. This was observed by DeWitt *et al.* (2010) using a high-power pulsed green laser. This result indicates that an increase in stimulation power to increase the OSL signal may not always be appropriate in case of dental enamel.

The main conclusion from the above discussion is that the OSL of dental enamel has the potential to be applied in retrospective and accident dosimetry, possibly providing a non-invasive, convenient and precise method for dose assessment that could be used in triage of radiation casualties. However, there are still several questions that remain to be answered,

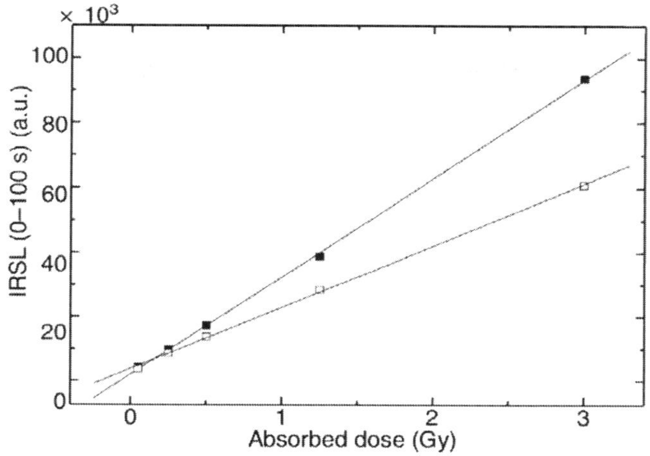


***Figure 6.16*** *IRSL dose response curves of dental ceramics at room temperature (filled squares) and at 140 °C (open squares). Reprinted from Raditation Protection Dosimetry, Bailiff, I. K., Luminescence characteristics of dental ceramics for retrospective dosimetry: a preliminary study. Vol. 101, 519–524. Copyright (2002) with permission from Oxford University Press.*

among them whether or not it is possible to increase the sensitivity of the technique even further, whether a stable OSL signal can be isolated or a correction found for fading and whether a protocol to account for sample-to-sample variability can be developed.

Dental ceramic is an alterative to dental enamel for use in retrospective and accident dosimetry in combination with the OSL technique. On the one hand, dental ceramic is also found in well-defined locations in a person's body, has dosimetric properties that do not change with time and could be easily sampled or probed for OSL measurements. On the other hand, the existence of dental ceramics in a specific person is as fortuitous as in the case of electronic dosimeters. The possibility of using the luminescence from dental ceramics for accident dosimetry has been investigated using the TL technique (Davies, 1979; Mauricio, Rosa and Cunha, 1985) and, more recently, using the OSL technique (Bailiff *et al.*, 2002). In both cases, the main advantage over dental enamel is a higher sensitivity of the luminescence signal, which currently allows the detection of doses as low as 100 mGy. However, Bailiff *et al.* also observed fading of the TL and OSL signal, with a decay of almost 50% in the first few hours, followed by a slower decay rate up to ~700 h after irradiation. Figure 6.16 illustrates the dose response of dental ceramics.

A more recent study examined the TL and OSL properties of a variety of ceramics used in dental prosthestics restoration, including feldspatic ceramics, glass ceramics, alumina-based ceramics and zirconia-based ceramics (Veronese *et al.*, 2010). Although these materials exhibit higher sensitivity to ionizing radiation than dental enamel, allowing the determination of dose in the mGy to tens of mGy range, considerable variability in the sensitivity was observed for different brands and different types of materials. Zirconia-based ceramics showed the lowest sensitivities. In addition, all samples showed some degree of fading of the TL and OSL signals. Experiments in which the samples were irradiated with 1 Gy using a $^{90}Sr/^{90}Y$ beta source and the TL or OSL signal were used to recover the dose from the original irradiation showed an underestimation of the recovered dose after 24 h. This result indicates the need for implementing fading corrections if dental ceramics are to be used in an actual accident. This will naturally be difficult because of the possible uncertainty in the time of radiation exposure, the possibility of protracted exposure, variability in the fading rate from sample to sample and the impossibility to perform extensive experiments to characterize each sample in the case of a mass casualty event. Therefore, although this approach may still be useful in the case of a few samples, further development is necessary to make this technique reliable and eventually increase the capability for analysis of many samples.

## 6.3 Environmental Monitoring

As in personal dosimetry, OSL offers an alternative to the TL technique in environmental surveillance. TLDs have been used since 1960s for environmental monitoring of gamma radiation (and beta radiation to a lesser extent) around nuclear power plants, having played an important role in the estimation of the population dose after the Three Mile Island accident (Gesell, 1982), and more recently in remediation sites (Klemic *et al.*, 1999). Characteristics that make TLDs and OSLDs attractive for environmental monitoring include their small size, low cost, reusability and the fact that they are integrating detectors. Because of the small size and low cost, they can be deployed in multiple locations. Since they are integrating detectors, they can be valuable for dose reconstruction after an accident, registering the entire event even before emergency response instruments can be deployed (Eisenbud and Gesell, 1997). Several international interlaboratory comparisons of TLDs in environmental dosimetry have been carried out, providing a wealth of information on the major challenges and sources of discrepancies (see de Planque and Gesell, 1986; Klemic *et al.*, 1995, 1999). TLDs and OSLDs can also be used for in situ measurements of the natural gamma dose rate, which is one of the components of the natural dose rate required to calculate the age in luminescence dating (Aitken, 1985, 1998).

Natural effective dose rates from external sources of radiation are typically of the order of magnitude of ~1 mSv/year (~100 nSv/h), which has contributions from terrestrial sources of gamma radiation ($^{40}K$ and nuclides from the $^{238}U$ and $^{232}Th$ series) and secondary radiation created by cosmic ray interaction with the atmosphere. The natural dose rates are influenced by radionuclide composition of the rock and soil where the measurements are made, soil moisture, possible snow cover and altitude (in case of cosmic rays) (Eisenbud and Gesell, 1997).

Because of the low dose rates in environmental monitoring, one of the challenges is the need for high-sensitivity detectors to allow for precise and accurate dosimetry, particularly when short monitoring periods are used (Gesell, 1982). The OSL from $Al_2O_3$:C provides such sensitivity, allowing doses lower than 0.5 μGy to be measured (Bøtter-Jensen *et al.*, 1997). However, as the monitoring period decreases, the dose during the monitoring period becomes comparable or lower than the dose during transit of the dosimeters from the site to the readout facility. The situation is exacerbated if the dosimeters are transported by air, due to the increased contribution from cosmic rays and the possibility that the dosimeters may be transported close to a radioactive package (Gesell, 1982). In these situations, the ability to erase the signal of the OSL detectors by exposing them to light (with the precaution of blocking short wavelengths using a yellow filter; see Chapter 3) prior to deployment and to read them on site using simple portable OSL readers is particularly attractive because it avoids complications due to the transit dose. A portable POSL reader capable of measuring doses as low as 0.2–0.4 μGy using $Al_2O_3$:C single crystals (0.9 mm thickness by 5 mm diameter) has been described for on-site measurements in diagnostic radiology (Gasparian *et al.*, 2010), which demonstrates the capability of OSL for environmental dosimetry.

It has been pointed out that the similarity between the effective atomic number of $Al_2O_3$:C ($Z_{eff} = 11.3$) and quartz ($Z_{eff} = 11.8$) is convenient for the determination of the photon dose rate to quartz in luminescence dosimetry applications in which quartz is primarily used for the determination of the geological or archeological dose. Both materials are expected to have similar responses as a function of the photon energy (Bøtter-Jensen *et al.*, 1997, 1999; ICRU, 2002).

However, for radiogical protection applications in which the quantity of interest is the absorbed dose to tissue, the over-response of $Al_2O_3$:C to low-energy photons may have to be compensated by the use of filters or by estimating the effective energy of the radiation field, as discussed in Chapter 3. In the case of nuclear facilities a significant contribution to the dose may be from photons with energy below 100 keV (Gesell, 1982). However, the effective atomic number of $Al_2O_3$:C is not as high as other TLDs used in environmental monitoring such as $CaF_2$ ($Z_{eff} = 16.9$) and $CaSO_4$ ($Z_{eff} = 15.7$). The contribution from beta rays may also have to be evaluated (Burbidge and Duller, 2003). This contribution will vary considerably depending on the type of packaging and thickness of the OSL detectors (Gesell, 1982).

Environmental monitoring may also be carried out using optical-fiber systems similar to those proposed for medical dosimetry (McKeever *et al.*, 2004; Ranchoux *et al.*, 2002). The difference between optical-fiber systems for environmental and medical dosimetry is that in environmental dosimetry there is no limitation on the size of the OSL detector. Larger crystals can be used for improved sensitivity, as demonstrated by Klein *et al.* (2005). These optical-fiber systems are suitable for remote measurements in difficult-to-access or hazardous locations for applications such as monitoring ground water around nuclear storage facilities and areas of elevated radiation levels in nuclear reactor facilities. One advantage is the fact that the signal is optical and, therefore, not affected by electromagnetic interference (Huston *et al.*, 2001). The crystal also works as an integrating detector, recording the information on the radiation field without power consumption until the OSL is probed using stimulation light. One of the problems in environmental dosimetry using optical fibers may be the attenuation of the OSL signal over the fiber length, which is typically not an issue for the short optical fibers used in medical dosimetry.

Figure 6.17 illustrates the concept of an optical fiber system for environmental dosimetry based on an $Al_2O_3$:C detector and the POSL technique. An example of POSL dose response for such system is shown in Figure 6.18. The minimum detectable dose (MDD), corresponding to three times the standard deviation of the background signal, was estimated to be ~5 μGy with a reproducibility of ~3% (Klein *et al.*, 2006). Based on this MDD, the authors estimated that a concentration of 1.85 Bq $cm^{-3}$ of $^{137}Cs$ could be detected in ~50 h. The authors also suggested that multiple optical-fiber probes could be multiplexed for readout of the dose rate at multiple locations using a single optical fiber OSL reader, as also suggested by other authors for medical applications (Huston *et al.*, 2001, 2002; Magne *et al.*, 2008). An optical fiber system based on $Al_2O_3$:C having a minimum detectable dose of ~50 μGy with a 20 m long cable has also been demonstrated by Ranchoux *et al.* (2002). The chemical stability of $Al_2O_3$:C is an attractive factor for environmental applications. Optical-fiber systems based on SrS:Ce,Sm phosphors have also been proposed for

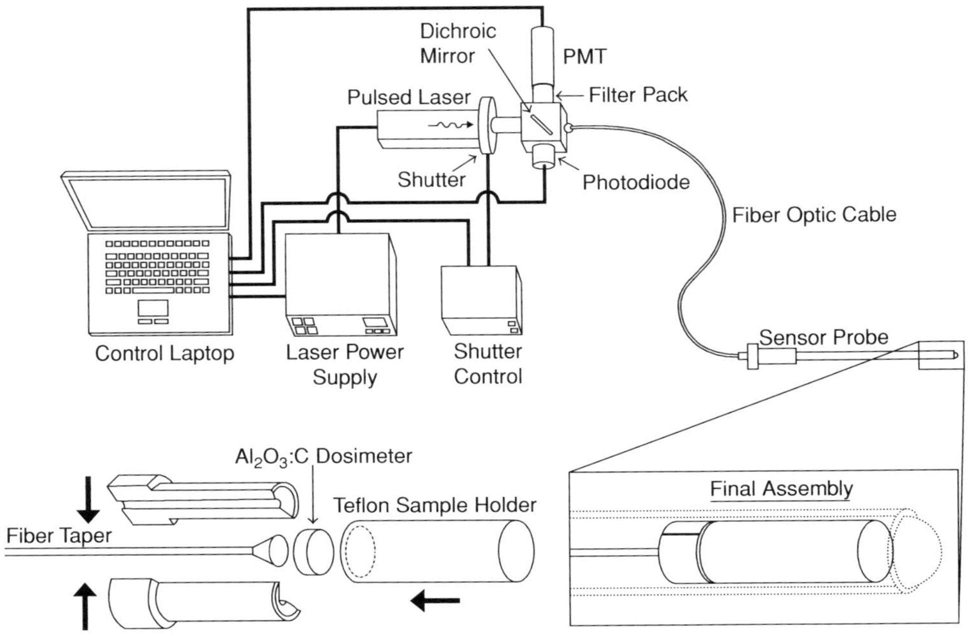


***Figure 6.17*** *Diagram of a prototype optical-fiber dosimetry system proposed for environmental measurements and detection of radioactive sources. The diagram shows the reader consisting of the laser for stimulation, the PMT for detection of the OSL signal and associated electronics. The reader is connected to the $Al_2O_3$:C OSL detector using an optical fiber. The end of the optical fiber cable is tapered for better collection of the OSL signal from the detector. Reprinted from IEEE Sensors 5, Klein, D. M., et al., An optical fiber radiation sensor for remote detection of radiological materials. 581–588. Copyright (2005) with permission from IEEE.*

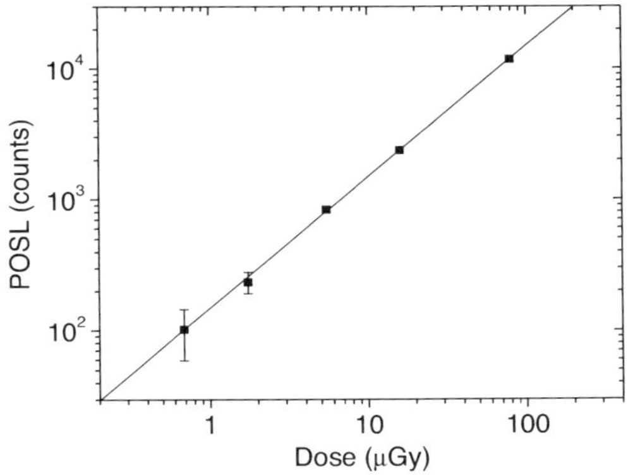


***Figure 6.18*** *Example of POSL dose response from an environmental optical fibers sytem based on $Al_2O_3$:C. Reprinted from IEEE Sensors 5, Klein, D. M., et al., An optical fiber radiation sensor for remote detection of radiological materials. 581–588. Copyright (2005) with permission from IEEE.*

applications in medicine, environmental monitoring in nuclear reactors or accelerator facilities and real-time beam monitoring (Benoit *et al.*, 2008a, 2008b, 2010).

## 6.4 UV Dosimetry

OSL and TL materials can in principle be used as passive UV dosimeters, providing a technique to assess the integrated exposure, particularly to the most harmful UVB range (280–315 nm) for different types of studies. For example, Colyott, Akselrod and McKeever (1997, 1999) describe an integrating UVB dosimeter based on an $Al_2O_3$:C OSL detector intended for biological studies on the deformation in amphibians, allowing the UVB exposure of the amphibian eggs to be quantified. Tinkler *et al.* (2000) cites three reasons motivating the research on new materials for this type of application: (i) the increasing recreational exposure of humans to sunlight, (ii) the increased solar irradiance in the UVB range at ground level due to reduction in the atmospheric ozone layer and (iii) the increased use of artificial UV sources.

The principle for using OSL materials as UV dosimeters is the light-induced signal caused by two processes: direct photoionization of defects and photo-transfer of trapped charges (Emfietzoglou and Moscovitch, 1996). Since most OSL materials are wide band-gap insulators, only photons in the vaccum UV range ($10\ \text{nm} < \lambda < 200\ \text{nm}$) have sufficient energy to cause direct excitation of electrons from the valence band to the conduction band. However, UV light with $\lambda > 200$ nm can still cause photoionization of specific defects within the band gap, provided that the photon energy is larger than the energy difference from the defect energy level to the bottom of the conduction band. The free charges created by defect photoionization can then be trapped at the trapping centers responsible for TL and OSL signals. As an example, 205 nm light can cause photoionization of $F$-centers in $Al_2O_3$:C very efficiently, filling a range of trapping centers, including those associated with the main TL and OSL signals (Lee and Crawford, 1979; Summers, 1984).

Phototransfer of charges from deep traps to the traps responsible for the TL and OSL signal have been extensively investigated in $Al_2O_3$:C (Colyott, Akselrod and McKeever, 1996; Walker *et al.*, 1996) and investigators have demonstrated that this phenomenon can be useful for UV dosimetry (Colyott, Akselrod and McKeever, 1997, 1999; Oster, Weiss and Kristianpoller, 1994). In this approach, the detector is first irradiated with a high dose of ionizing radiation to fill the trapping centers in the crystal. Annealing to an appropriate temperature is then used to empty the "main dosimetric trap" responsible for the OSL (or TL) signal of interest, zeroing it. Exposure to UV light stimulates charges from the deep traps to the "main dosimetric trap," generating an OSL signal that is proportional to the previous dose of radiation and to the accumulated UV energy incident in the detector since zeroing of the signal, i.e. annealing (Colyott, Akselrod and McKeever, 1999). Figure 6.19 shows an example of a dose response curve for $Al_2O_3$:C UV dosimeter as a function of the energy fluence from natural sunlight exposure.

One fundamental difference between exposure to UV light and to ionizing radiation is that the UV light may simultaneously cause filling and empting of the trapping center associated with the OSL signal. During exposure to ionizing radiation the detectors are usually not being stimulated by light and the thermal energy available at room temperature is typically

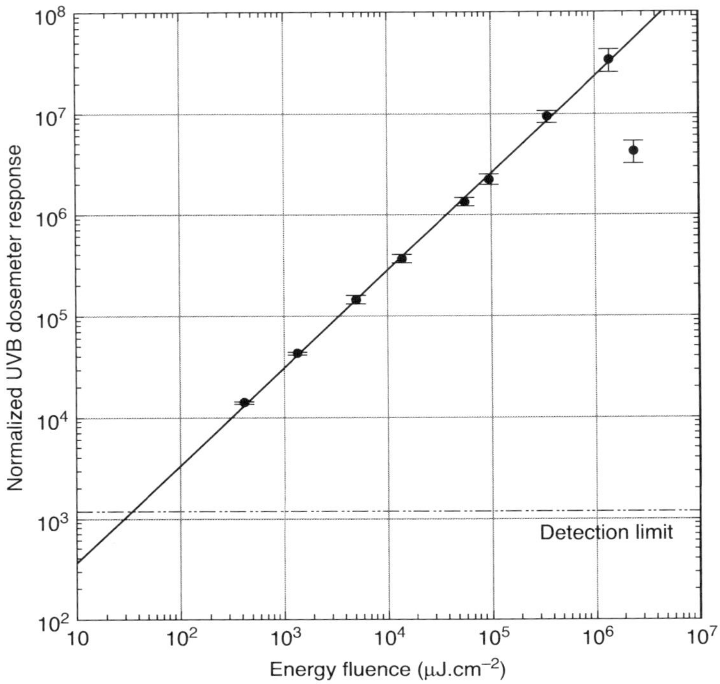


***Figure 6.19*** *Phototransferred OSL signal as a function of natural sunlight exposure from an* $Al_2O_3$*:C detector pre-irradiated with 100 Gy and annealed to 625 K for 1 h and to 675 K for 12 min to empty the main dosimetric trap. See experimental details in Colyott, Akselrod and McKeever (1999). The data demonstrate the correlation between photo-transferred OSL and UV exposure that is the basis for the use of OSL detectors in UV dosimetry. Reprinted from Raditation Protection Dosimetry, Colyott, L. E., et al., An integrating UVB dosemeter system. Vol. 85, 309–312. Copyright (1999) with permission from Oxford University Press.*

not sufficient to cause depletion of the trapping centers associated with the dosimetric TL or OSL signal. On the other hand, UV light may photoionize defects or transfer charges from deep traps to the main dosimetric trap and, at the same time, stimulate charges out of the main dosimetric trap.

Although the OSL signal may initially increase linearly with time as a function of UV exposure, the saturation levels will depend on the particular mechanisms taking place. For example, Chen, Hornyak and Mathur (1990) carried out analytical and numerical investigations on the competing mechanisms of excitation and bleaching of a TL signal, which is analogous to the OSL signal if the same trapping centers are involved. These authors showed that the saturation or equilibrium level of the TL signal is, in this case, independent of the initial occupancy of the trapping centers or exposure intensity. If the UV-induced signal originates from phototransfer of charges from deep traps, the saturation level will be a function of the initial occupancy of trapping centers and will eventually be reduced to zero as the deep traps are emptied. However, a wider variety of behavior can be expected, as discussed in the case of TL by McKeever (1991).

## 6.5 Integrated Sensors

One concept that has been explored for potential application in personal and environmental dosimetry, including the space environment, has been the integration of the OSL detector with the stimulation source and light transducer in one compact electronic package. This requires replacing the highly sensitive, but large and costly photomultiplier tubes (PMT) used in typical OSL readers with compact semiconductor photodetectors.

Advantages to this approach include:

(i) Wide dynamic range of the OSL detector, 6–7 orders of magnitude.
(ii) Possibility of resetting the OSL detector using light, avoiding saturation of the OSL material and making it possible to detect small doses after the detector has been exposed to large doses in the past.
(iii) Real-time/periodic readout.
(iv) No need for external reader since the device is self-contained.
(v) Detector works as a passive device when not powered.
(vi) Readout is adjustable by precise control of stimulation intensity, duration and frequency (tradeoff between initial OSL intensity/total OSL area).
(vii) Works with materials that show fading, depending on the readout frequency.

The basic requirements are to match the OSL emission with the peak absorption of the semiconductor photodetectors, and to maintain good separation between the stimulation light and luminescence wavelengths in order to achieve good discrimination between the stimulation light and the OSL signal.

TL investigations by Jacob *et al.* (1993) using a photodiode mounted in a Harshaw TL reader replacing the PMT indicated the possibility of detecting doses as low as a few milligrays, depending on the sample dopants. Stoebe *et al.* (1996) recorded a clear TL curve from CaF:Mn irradiated with a dose of 0.1 Gy using a commercially available GaAsP photodiode as the light detector, and estimated the minimum detectable dose of the system to be around 0.6 mGy. One disadvantage of integrated TL sensors is the problem of heating the detector material without damaging the electronic components, although one solution to this problem may be to apply short heating pulses to the TL material (Mah *et al.*, 2010).

Dusseau *et al.* (2000) described a prototype consisting of a layer of calcium sulfide doped with cerium and samarium (CaS:Ce,Sm) as the OSL material, mounted in between an infrared LED (940–950 nm peak emission) used for stimulation and a GaAsP or GaP photodiode for light detection. This prototype, intended for space dosimetry, was further developed and tested using other doped sulfides as the OSL material (BaS:Ce, or SrS:Ce) (Plattard *et al.*, 2002; Vaillé *et al.*, 2003). A version of this integrated OSL sensor is shown in Figure 6.20, along with a typical OSL curve and dose response for $^{60}$Co. The whole sensor fits in a volume of 4 mm $\times$ 4 mm $\times$ 7 mm and weights $\sim$3 g (Plattard *et al.*, 2002). For the OSL readouts the LEDs are operated at 50 mA for 20 s with a bias of 2 V voltage, resulting in complete resetting of the material after three seconds (Figure 6.20b). In between the OSL readouts, the detector works as a passive detector, the information on the absorbed dose being accumulated as trapped charges in the crystal. Characterization of the sensor indicates a repeatability of $\pm$4% and reproducibility (measured over a period of 15 days) of $\pm$15%, although the authors do not exclude the possibility of changes in the irradiation conditions (Plattard *et al.*, 2002). The detector is linear within the experimental uncertainties.

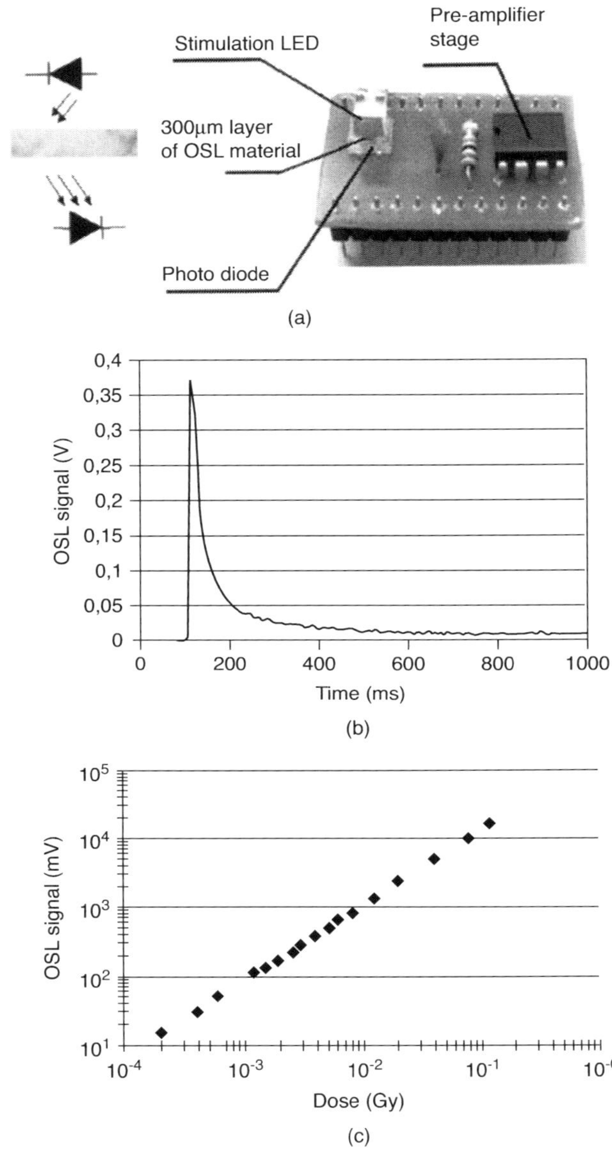


***Figure 6.20*** *(a) Integrated OSL sensor developed at the Université Montpellier II; (b) OSL curve of CaS:Ce,Sm in an integrated OSL sensor irradiated with 4 mGy; (c) OSL dose response of the integrated sensor as a function of $^{60}$Co gamma dose. Figure (a) reprinted from IEEE Transactions on Nuclear Science 50, Vaillé, J.-R., et al., Hardening of a radiation sensor based on optically stimulated luminescence. 2358. Copyright (2003) with permission from IEEE; Figures (b) reprinted from IEEE Transactions on Nuclear Science 47, Dusseau, et al., An integrated sensor using optically stimulated luminescence for in flight dosimetry. 2412–2416. Copyright (2000) with permission from IEEE; and (c) reprinted from IEEE Transactions on Nuclear Science 47, Dusseau, et al., An integrated sensor using optically stimulated luminescence for in flight dosimetry. 2412–2416. Copyright (2000) with permission from IEEE.*

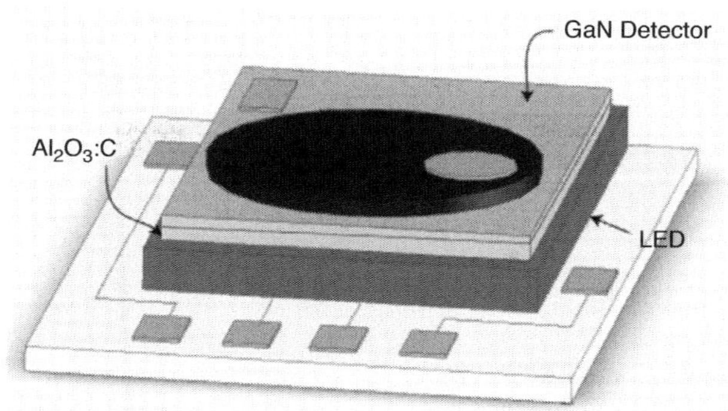


*Figure 6.21* Concept of an integrated OSL sensor consisting of $Al_2O_3$:C as the OSL detector, an LED and a GaN photodetector. Reprinted from Radiation Protection Dosimetry, Chen, T., Poochinda, K. and Stoebe, T.G., Innovational radiation sensor by integrating $Al_2O_3$:C optically stimulated luminescent dosemeter and GaN detectors. Vol. 119, 380–385. Copyright (2006) with permission from Oxford University Press.

Some of the problems that have been mentioned with respect to these integrated sensors are the radiation damage to the LEDs and photodiodes and the instability of the trapped charges, which in turn cause temperature and dose rate dependencies of the OSL signal. Vaillé *et al.* (2003) introduced a feedback loop to correct for degradation in the sensor performance due to radiation damage in the LEDs observed after irradiation with 12 MeV electrons at fluences of $\sim 10^{11}$–$10^{12}$ electrons/cm$^2$. The sensor electronics were further improved to avoid accidental triggering of the readout process (Vaillé *et al.*, 2005). The influence of signal fading was reduced by the use of an algorithm that takes into account the lifetime of the trapped charges as given by the Arrhenius law $[\tau^{-1} = s\exp(-E/kT)]$ (Garcia *et al.*, 2007). Finally, irradiation with 9 MeV electrons up to a dose of 500 Gy did not induce significant degradation of the photodiodes (Plattard *et al.*, 2002).

Chen, Poochinda and Stoebe (2006) proposed to use $Al_2O_3$:C single crystals as the substrate for growing GaN-based p-i-n photodiodes by metal-organic chemical vapor deposition (MOCVD) (Poochinda *et al.*, 2004), placing an LED on the other side of the $Al_2O_3$:C crystal for stimulation and thereby creating an integrated OSL sensor. The concept proposed is illustrated in Figure 6.21. The simulated responsivity (ratio of cathode current to optical power) of the GaN p-i-n photodiode shows a cutoff above ~400 nm, which would be ideal to eliminate detection of the stimulation light at longer wavelengths, while still detecting part of the $F$-center emission band in $Al_2O_3$:C and most of the UV emission band. Unfortunately, a working device is still not available.

At this point the research on integrated OSL dosimeters is very limited, making it difficult to compare potential advantages or disadvantages with respect to other types of active detectors, such as electronic dosimeters. For the latter, despite several models being available in the market (Bolognese-Milsztajn *et al.*, 2004), there are still concerns regarding energy and dose-rate dependencies (Ginjaume *et al.*, 2007; McCaffrey, 2008). Therefore, more research is necessary to answer the question of whether integrated OSL dosimeters may or may not offer a clear advantage over other types of active radiation sensors. Nevertheless, the challenging needs in neutron, space and hadron radiotherapy dosimetry, as well as homeland security, should provide more than sufficient scientific and, perhaps, commercial motivation.

## 6.6 Passive/Active Devices

The concept of having a passive and active type of detector in the same device can also take advantage of the OSL signal from commercial scintillators. An ideal scintillator is free of trapping centers, allowing the electron–hole pairs created by exposure to ionizing radiation to immediately recombine, generating a light pulse which is proportional to the energy of the original photon. However, defects that act as trapping centers cannot be completely eliminated, as evidenced by the presence of TL peaks with known characteristics in commercial scintillators (e.g., Dorenbos *et al.* 1994a, b).

OSL can take advantage of these trapping centers by stimulating the localized charge using light. In the absence of optical stimulation, the scintillator material works as an active device, most of the energy deposited by the radiation generating scintillation pulses that can be used for detection of radiation sources or for gamma spectroscopy. However, the crystal also works as a passive detector, since a fraction of the electrons and holes are captured by trapping centers and can later be released by optical stimulation.

Hazelton *et al.* (2010) investigated the OSL from well-known fast scintillators, the cerium-doped oxyorthosilicates $Y_2SiO_5$ (YSO), $Gd_2SiO_5$ (GSO) and $Lu_2SiO_5$ (LSO). The authors found the optimum combination of detection optical filters and stimulation light to obtain an OSL signal from these scintillators. Figure 6.22 shows an example of

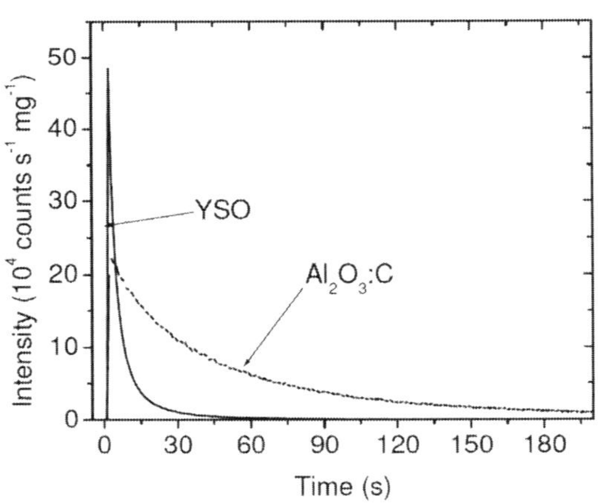


***Figure 6.22*** *Example of an OSL curve for a YSO scintillator (approximately 10 mm × 10 mm × 2 mm) in comparison to a $Al_2O_3$:C single crystal (5 mm diameter by 0.9 mm thickness). The samples were irradiated with 10 mGy beta dose prior to the OSL readout. The OSL readout was carried out using green LED stimulation (~100 mW/cm$^2$, Schott OG-515 filters in front of the LED) with detection using a PMT and a combination of Schott B-370 and Schott BG-39 filters in front of the PMT. See Hazelton et al. (2010) for more details. Reprinted from Raditation Measurements, Hazelton, J.R., et al., Feasibility of using oxyorthosilicates as optically stimulated luminescence detectors. Vol. 45, 681–683. Copyright (2010) with permission from Elsevier.*

an OSL curve from YSO compared to that from $Al_2O_3$:C, both crystals irradiated with approximately the same dose of beta rays. GSO did not show a reasonable OSL signal and LSO showed a large signal due to self-irradiation from the presence of the beta emitting $^{176}$Lu (2.6% natural abundance).

The results shown in Figure 6.22 open the possibility of developing a type of device that works as a passive integrating detector, recording the information on the ionizing radiation field without power consumption. This information can be read using the OSL technique at regular time intervals, as needed. If elevated radiation levels are detected, the device can then be operated as an active detector in scintillation mode for better discrimination against natural background or for gamma spectroscopy. It remains to be demonstrated whether or not this concept could be useful for applications in security. In this sense the OSL device is a "sentinel" for an increase in radiation exposure, reverting to active, real-time mode once the alarm has been raised.

## 6.7 Other Potential Security Applications

A continuing requirement for security agencies throughout the world is the need for sensitive detection methods for special nuclear materials (SNMs) such as $^{239}$Pu or $^{235}$U, at border controls and other locations. An often proposed, but not yet developed concept has been to use OSL (and/or TL) as a crude discriminator to determine the energy of the radiation source and thereby identify the source radionuclide. The idea, first proposed by Spurný (1980), is to find a set of materials with different effective atomic numbers and, as a consequence, different mass energy absorption coefficients and relative energy responses. Each material will thus respond differently when exposed to the same energies and doses such that the ratio of the OSL signals from the materials is then a "fingerprint" of the radionuclide to which they were exposed.

This is illustrated in Figure 6.23 wherein the three main gamma emissions from $^{239}$Pu and $^{235}$U are illustrated on the energy axis and are superimposed on the relative energy responses, with respect to water, of three common luminescence materials, in this case TLD materials, $CaF_2$, LiF and $Li_2BO_4$. Thus, if each of the three TLDs noted in the figure were exposed to these radionuclides, one can expect that their relative TL responses would vary depending upon the radioisotope to which they were exposed. The TL responses relative to each other will, of course, also be dictated by the relative TL efficiencies of the individual materials, not just the ionization efficiency as represented in the figure. However, one can speculate that calibration of a set of, say, three of four TLD materials against the sources of interest over this energy range would produce a set of relative responses which could then be used as a "look-up table" in order to identify the isotope in an unknown exposure. In this way, without full spectroscopic measurement, one may be able to use a high-sensitivity, integrating dosimetry system such as this to identify the source from the subsequent ratios of the signals. For OSL one would have to find a corresponding set of suitable OSL materials with significantly different effective atomic numbers. If this is achievable, the combination of this concept with that of the "sentinel," passive/active sensors described in the previous section may result in a useful radiological detection device for security purposes.

One can also speculate upon the applicability of OSL in forensic work relating to the illicit movement and storage of radioactive substances. The potential to use OSL in forensic

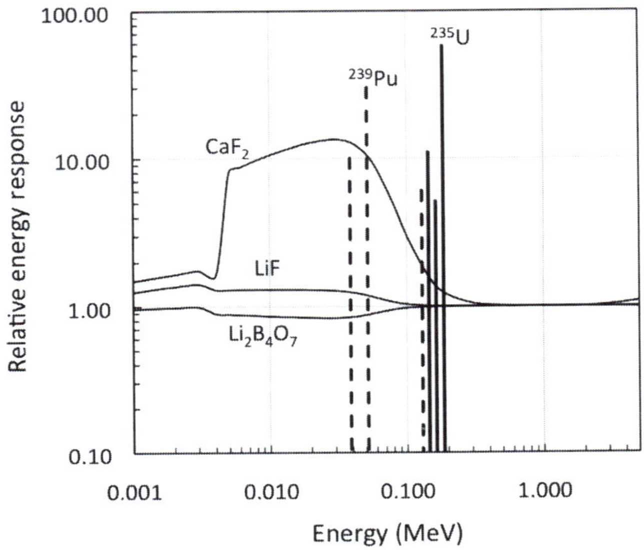


*Figure 6.23 Energy responses (relative to $H_2O$) of $CaF_2$, LiF and $Li_2BO_4$. The vertical lines are the three main gamma energies from $^{239}Pu$ (dashed) and $^{235}U$ (full). The heights of the lines represent the relative intensities of the emissions.*

analysis is related to the use of OSL in retrospective dosimetry (Section 6.2.3). For example, one might ask the question whether or not an illegal radioactive source has ever been present in a room after having been removed from that room. As already noted, a typical room contains many common materials that can be used as OSL dosimeters (Bøtter-Jensen *et al.*, 2003) and exposure of any of these items to radiation may render the material useful as an OSL dosimeter. Thus, even though the radioactive source may have already been removed from the room, its presence may have been recorded by the latent OSL signal in the surrounding materials. Furthermore, any additional intelligence which may be available concerning the type of source and its shielding may enable estimates to be made of the length of time the source was in the room. By examining materials from different locations, it may also prove possible to determine the location of the source within the room. The efficacy of the use of OSL to address these issues depends on the strength and type of the source in question, the amount of shielding and the period over which the source was present. No such applications have yet been described in the literature.

Similarly, the same procedures may be used, in principle, to determine if a source has been present in a cave. This is a much more difficult problem, however, since the cave walls will have accumulated a natural background radiation exposure level which may be much higher than that from the illicit source. In contrast, the natural background dose accumulated by modern building or household items will be much less than that accumulated by natural rock minerals.

# References


AAPM (1990) *AAPM Report No. 31: Standardized Methods for Measuring Diagnostic X-ray Exposures*, American Institute of Physics, New York.

AAPM (1991) Recommendations on performance characteristics of diagnostic exposure meters, Report of AAPM Diagnostic X-Ray Imaging Group No. 6. *Med. Phys.*, **19**, 231–241.

AAPM (1993) *AAPM Report No. 39: Specification and Acceptance Testing of Computer Tomography Scanners*, American Institute of Physics, New York.

AAPM (1995) *AAPM Report No. 54: Stereotactic Radiosurgery*, American Association of Physicists in Medicine, College Park.

AAPM (2002) *AAPM Report No. 74: Quality Control in Diagnostic Radiology*, American Association of Physicists in Medicine, College Park.

AAPM (2005) *AAPM Report No. 87: Diode in Vivo Dosimetry for Patients Receiving External Beam Radiation Therapy*, American Association of Physicists in Medicine, College Park.

AAPM (2008) *AAPM Report No. 96: The Measurement, Reporting, and Management of Radiation Dose in CT*, American Association of Physicists in Medicine, College Park.

AFRRI (2003) *Medical Management of Radiological Casualties Handbook*, Armed Forces Radiobiology Research Institute, Bethesda, Maryland.

Agersnap Larsen, N., Bøtter-Jensen, L. and McKeever, S.W.S. (1999) Thermally stimulated conductivity and thermoluminescence from $Al_2O_3$:C. *Radiat. Prot. Dosim.*, **84**, 87–90.

Agersnap Larsen, N., Bulur, E., Bøtter-Jensen, L. and McKeever, S.W.S. (2000) Use of the LM-OSL technique for the detection of partial bleaching in quartz. *Radiat. Meas.*, **32**, 419–425.

Aguirre, J., Alvarez, P., Followill, D. *et al.* (2009) SU-FF-T-306: Optically stimulated light dosimetry: commissioning of an optically stimulated luminescence (OSL) system for remote dosimetry audits, the Radiological Physics Center Experience. *Med. Phys.*, **36**, 2591–2592.

Aguirre, J.F., Tailor, R., Ibbott, G.S. *et al.* (2002) Thermoluminescence Dosimetry as a Tool for the Remote Verification of Output for Radiotherapy Beams: 25 years of Experience. *International Symposium on Standards and Codes of Practice in Medical Radiation Dosimetry, IAEA*, pp. 191–199.

Aitken, M.J. (1985) *Thermoluminescence Dating*, Academic Press, Orlando.

Aitken, M.J. (1998) *An Introduction to Optical Dating*, Oxford University Press, Oxford.

Akselrod, M.S., Agersnap Larsen, N. and McKeever, S.W.S. (2000) A procedure for the distinction between static and dynamic radiation exposures of personal radiation badges using pulsed optically stimulated luminescence. *Radiat. Meas.*, **32**, 215–225.

Akselrod, M.S., Agersnap Larsen, N., Whitley, V.H. and McKeever, S.W.S. (1998b) Thermal quenching of F-center luminescence in $Al_2O_3$:C. *J. Appl. Phys.*, **84**, 3364–3373.

Akselrod, A.E. and Akselrod, M.S. (2002) Correlation between OSL and the distribution of TL traps in $Al_2O_3$:C. *Radiat. Prot. Dosim.*, **100**, 217–220.

Akselrod, M.S. and Akselrod, A.E. (2006) New $Al_2O_3$:C,Mg crystals for radiophotoluminescent dosimetry and optical imaging. *Radiat. Prot. Dosim.*, **119**, 218–221.

Akselrod, M.S., Akselrod, A.E., Orlov, S.S. *et al.* (2003) Fluorescent Aluminum Oxide Crystals for Volumetric Optical Data Storage and Imaging Applications. *J. Fluoresc.*, **13**, 503–511.

Akselrod, M.S., Bøtter-Jensen, L. and McKeever, S.W.S. (2007) Optically stimulated luminescence and its use in medical dosimetry. *Radiat. Meas.*, **41**, S78-S99.

Akselrod, M.S. and Gorelova, E.A. (1993) Deep traps in highly sensitive $\alpha$-$Al_2O_3$:C TLD crystals. *Nucl. Tracks Radiat. Meas.*, **21**, 143–146.

Akselrod, M.S. and Kortov, V.S. (1990) Thermoluminescent and exoemission properties of new high-sensitivity TLD $\alpha$-$Al_2O_3$:C crystals. *Radiat. Prot. Dosim.*, **33**, 123–126.

Akselrod, M.S., Kortov, V.S. and Gorelova, E.A. (1993) Preparation and properties of $\alpha$-$Al_2O_3$:C. *Radiat. Prot. Dosim.*, **47**, 159–164.

Akselrod, M.S., Kortov, V.S., Kravetsky, D.J. and Gotlib, V.I. (1990) Highly sensitive thermoluminescent anion-defect $\alpha$-$Al_2O_3$:C single crystal detectors. *Radiat. Prot. Dosim.*, **33**, 119–122.

Akselrod, M.S., Lucas, A.C., Polf, J.C. and Mckeever, S.W.S. (1998a) Optically stimulated luminescence of $Al_2O_3$. *Radiat. Meas.*, **29**, 391–399.

Akselrod, M.S. and McKeever, S.W.S. (1999) A radiation dosimetry method using pulsed optically stimulated luminescence. *Radiat. Prot. Dosim.*, **81**, 167–176.

Alagu Raja, E., Menon, S., Dhabekar, B. *et al.* (2009) Investigation of defect centres responsible for TL/OSL in $MgAl_2O_4$:$Tb^{3+}$. *J. Lumin.*, **129**, 829–835.

Albrecht, H.O. and Mandeville, C.E. (1956) Storage of energy in beryllium oxide. *Phys. Rev.*, **101**, 1250–1252.

Alexander, G.A., Swartz, H.M., Amundson, S.A. *et al.* (2007) BiodosEPR-2006 Meeting: Acute dosimetry consensus committee recommendations on biodosimetry applications in events involving uses of radiation by terrorists and radiation accidents. *Radiat. Meas.*, **42**, 972–996.

Almond, P.R. (2009) A historical perspective: a brief history of dosimetry, calibration protocols, and the need for accuracy, in *Clinical Dosimetry Measurements in Radiotherapy* (eds D.W.O. Rogers and J.E. Cygler), Medical Physics Publishing, Madison, Wisconsin.

Almond, P.R., Biggs, P.J., Coursey, B.M. *et al.* (1999) AAPM's TG-51 protocol for clinical reference dosimetry of high-energy photon and electron beams. *Med. Phys.*, **26**, 1847–1870.

Andersen, C.E., Aznar, M.C., Bøtter-Jensen, L. *et al.* (2003) Development of optical fibre luminescence techniques for real time *in vivo* dosimetry in radiotherapy. *International Symposium on Standards and Codes of Practice in Medical Radiation Dosimetry, Vienna, 2003, IAEA*, pp. 353–360.

Andersen, C.E., Bøtter-Jensen, L. and Murray, A.S. (2003) A mini X-ray generator as an alternative to a $^{90}Sr/^{90}Y$ beta source in luminescence dating. *Radiat. Meas.*, **37**, 557–561.

Andersen, C.E., Edmund, J.M., Damkjær, S.M.S. and Greilich, S. (2008) Temperature coefficients for in vivo RL and OSL dosimetry using $Al_2O_3$:C. *Radiat. Meas.*, **43**, 948–953.

Andersen, C.E., Edmund, J.M., Medin, J. *et al.* (2007) Medical proton dosimetry using radioluminescence from aluminium oxide crystals attached to optical fiber cables. *Nucl. Instrum. Methods Phys. Res. A*, **580**, 466–468.

Andersen, C.E., Marckmann, C.J., Aznar, M.C. *et al.* (2006) An algorithm for real-time dosimetry in intensity-modulated radiation therapy using the radioluminescence signal from $Al_2O_3$:C. *Radiat. Prot. Dosim.*, **120**, 7–13.

Andersen, C.E., Nielsen, S.K., Greilich, S. *et al.* (2009a) Characterization of a fiber-coupled $Al_2O_3$:C luminescence dosimetry system for online in vivo dose verification during $^{192}Ir$ brachytherapy. *Med. Phys.*, **36**, 708–718.

Andersen, C.E., Nielsen, S.K., Lindegaard, J.C. and Tanderup, K. (2009b) Time-resolved in vivo luminescence dosimetry for online error detection in pulsed dose-rate brachytherapy. *Med. Phys.*, **36**, 5033–5043.

ANSI/HPS (1999) American National Standard – Personnel Neutron Dosimeters (Neutron Energies Less Than 20 MeV). ANSI/HPS N13.52-1999, American National Standards Institute/Health Physics Society.

ANSI/HPS (2001) Personnel Dosimetry Performance – Criteria for Testing. American National Standard ANSI/HPS N13.11-2001, American National Standards Institute/Health Physics Society.

Antonov-Romanovsky, V.V., Keirum-Markus, I.F., Poroshina, M.S. and Trapeznikova, Z.A. (1955) Dosimetry of Ionizing Radiation with the Aid of Infrared Sensitive Phosphors. *Conference of the Academy of Sciences of the U.S.S.R. on the Peaceful Uses of Atomic Energy, Moscow*, USAEC Report AEC-tr-2435 (Pt. 1), 239–250.

Apáthy, I., Deme, S., Fehér, I. *et al.* (2002) Dose measurements in space by the Hungarian Pille TLD system. *Radiat. Meas.*, **35**, 381–391.

Appleby, G.A., Edgar, A., Williams, G.V.M. and Bos, A.J.J. (2006) Photostimulated luminescence from $BaCl_2:Eu^{2+}$ nanocrystals in lithium borate glasses following neutron irradiation. *Appl. Phys. Lett.*, **89**, 101902.

Appleby, G.A., Zimmermann, J., Hesse, S. *et al.* (2009a) Sensitization and radiation hardening of the photostimulable X-ray storage phosphor $CsBr:Eu^{2+}$. *J. Mater. Sci. – Mater. Electron.*, **20**, 54–58.

Appleby, G.A., Zimmermann, J., Hesse, S. *et al.* (2009b) Sensitization of the photostimulable x-ray storage-phosphor CsBr: Eu2+ following room-temperature hydration. *J. Appl. Phys.*, **105**, 073511–073511-5.

Archambault, L., Beddar, A.S., Gingras, L. *et al.* (2006) Measurement accuracy and Cerenkov removal for high performance, high spatial resolution scintillation dosimetry. *Med. Phys.*, **33**, 128–135.

Ariga, E., Ito, S., Deji, S. *et al.* (2007) Development of dosimetry using detectors of diagnostic digital radiography systems. *Med. Phys.*, **34**, 166–174.

Arnfield, M.R., Gaballa, H.E., Zwicker, R.D. *et al.* (1996) Radiation-induced light in optical fibers and plastic scintillators: Application to brachytherapy dosimetry. *IEEE Trans. Nucl. Sci*, **43**, 2077–2084.

Attix, F.H. (2004) *Introduction to Radiological Physics and Radiation Dosimetry*, Wiley-VCH, Weinheim.

Aznar, M.C., Andersen, C.E., Bøtter-Jensen, L. *et al.* (2004) Real-time optical-fibre luminescence dosimetry for radiotherapy: physical characteristics and applications in photon beams. *Phys. Med. Biol.*, **49**, 1655–1669.

Aznar, M.C., Hemdal, B., Medin, J. *et al.* (2005) *In vivo* absorbed dose measurements in mammography using a new real-time luminescence technique. *Br. J. Radiol.*, **78**, 328–334.

Badavi, F.F., Nealy, J.E., deAngelis, G. et al. (2005) Radiation environment and shield modeling validation for CEV design. Space 2005, Long Beach, CA, 30 Aug–1 Sep 2005, American Institute of Aeronautics and Astronautics, AIAA 2005–6651, http://pdf.aiaa.org/preview/CDReadyMSPACE05_1181/PV2005_6651.pdf (acessed on 5 May 2010).

Badhwar, G.D. (1997) The radiation environment in low-Earth orbit. *Radiat. Res.*, **148**, S3–S10.

Badhwar, G.D. and Cucinotta, F.A. (2000) Comparison of a depth dependence of dose and linear energy transfer spectra in aluminum and polyethylene. *Radiat. Res.*, **153**, 1–8.

Badhwar, G.D. and O'Neill, P.M. (1996) Galactic cosmic ray model and its application. *Adv. Space Res.*, **17**, 7–17.

Bailey, R.M. (2001) Towards a general kinetic model for optically and thermally stimulated luminescence of quartz. *Radiat. Meas.*, **33**, 17–45.

Bailiff, I.K. (1997) Retrospective dosimetry with ceramics. *Radiat. Meas.*, **27**, 923–941.

Bailiff, I.K. (1999) The development of retrospective luminescence dosimetry for dose reconstruction in areas downwind of chernobyl. *Radiat. Prot. Dosim.*, **84**, 411–419.

Bailiff, I.K., Correcher, V., Delgado, A. *et al.* (2002) Luminescence characteristics of dental ceramics for retrospective dosimetry: a preliminary study. *Radiat. Prot. Dosim.*, **101**, 519–524.

Bailiff, I.K. and Mikhailik, V.B. (2003) Spatially-resolved measurement of optically stimulated luminescence and time-resolved luminescence. *Radiat. Meas.*, **37**, 151–159.

Bailiff, I.K. and Mikhailik, V.B. (2004) The use of calcium silicate bricks for retrospective dosimetry. *Radiat. Meas.*, **38**, 91–99.

Bailiff, I.K. and Slim, H.A. (2008) Development of reference database for gamma dose assessment in retrospective luminescence dosimetry. *Radiat. Meas.*, **43**, 859–863.

Bailiff, I.K., Stepanenko, V.F., Göksu, H.Y. *et al.* (2004) Comparison of retrospective luminescence dosimetry with computational modeling in two highly contaminated settlements downwind of the Chernobyl NPP. *Health Phys.*, **86**, 25–41.

Bailiff, I.K., Stepanenko, V.F., Göksu, H.Y. *et al.* (2005) Retrospective luminescence dosimetry: development of approaches to application in populated areas downwind of the Chernobyl NPP. *Health Phys.*, **89**, 233–246.

Bandyopadhyay, P.K., Russell, G.W. and Chakrabarti, K. (1999) Optically stimulated luminescence in KCl:Cu x-irradiated at room temperature. *Radiat. Meas.*, **30**, 51–57.

Banerjee, D., Bøtter-Jensen, L. and Murray, A.S. (2000) Retrospective dosimetry: estimation of the dose to quartz using the single-aliquot regenerative-dose protocol. *Appl. Radiat. Isotopes*, **52**, 831–844.

Baril, M.R. (2004) CCD imaging of the infra-red stimulated luminescence of feldspars. *Radiat. Meas.*, **38**, 81–86.

Barth, J., Dyer, S.C. and Sassinopulos, E.G. (2003) Space, atmospheric, and terrestrial radiation environments. *IEEE Trans. Nucl. Sci*, **50**, 466–482.

Bassinet, C., Trompier, F. and Clairand, I. (2010) Radiation accident dosimetry on electronic components by OSL. *Health Phys.*, **98**, 440–445.

Bauhs, J.A., Sturchio, G., Vrieze, T.J. and McCollough, C.H. (2007) Accuracy of CT Optically Stimulated Luminescent (OSL) dosimeters for the measurement of CT Dose Index (CTDI). Radiological Society of North America, Chicago, November 25–30.

Bauhs, J.A., Vrieze, T.J., Primak, A.N. *et al.* (2008) CT dosimetry: Comparison of Measurement Techniques and Devices. *RadioGraphics*, **28**, 245–253.

Becquerel, E. (1843) Des effets produits sur les corps par les rayons solaires. *Ann. de Chim. et de Phys.*, **9**, 257–322.

Becquerel, H. (1883) Maxima et minima d'extinction de la phosphorescence sous l'influence des radiations infra-rouges. *Comptes Rendus*, **96**, 1853–1856.

Beddar, A.S., Mackie, T.R. and Attix, F.H. (1992a) Water-equivalent plastic scintillation detectors for high-energy beam dosimetry: II. Properties and measurements. *Phys. Med. Biol.*, **37**, 1901–1913.

Beddar, A.S., Mackie, T.R. and Attix, F.H. (1992b) Cerenkov light generated in optical fibres and other light pipes irradiated by electron beams. *Phys. Med. Biol.*, **37**, 925–935.

Beddar, A.S., Mackie, T.R. and Attix, F.H. (1992c) Water-equivalent plastic scintillation detectors for high-energy beam dosimetry: I. Physical characteristics and theoretical considerations. *Phys. Med. Biol.*, **37**, 1883–1990.

Beddar, A.S. (2006) Water equivalent plastic scintillation detectors in radiation therapy. *Radiat. Prot. Dosim.*, **120**, 1–6.

Beierholm, A.R., Andersen, C.E., Lindvold, L.R. *et al.* (2008) A comparison of BCF-12 organic scintillators and $Al_2O_3$:C crystals for real-time medical dosimetry. *Radiat. Meas.*, **43**, 898–903.

Bell, J., Friedman, L., Hubbard, G.S. *et al.* (2008) Beyond the Moon: A new roadmap for human space exploration in the 21st century. The Planetary Society. http://planetary.org/special/roadmap/beyond_the_moon.pdf (accessed on 1 May 2010).

Bender, J.B. and Merriman, M. (2005) *Regimes of Description: In the Archive of the Eighteenth Century*, Stanford University Press, Stanford.

Benoit, D., Dusseau, L., Glaser, M. *et al.* (2010) Performance studies of an optical fiber OSL/RL dosimetry system in pulsed high-intensity radiation beams. *Radiat. Meas.*, **45**, 688–690.

Benoit, D., Garcia, P., Matias-Vaillé, S. *et al.* (2008b) Real-time fibered optically stimulated luminescence dosimeter based on SrS:Ce,Sm phosphor. *IEEE Trans. Nucl. Sci.*, **55**, 2154–2160.

Benoit, D., Ravotti, F., Garcia, P. *et al.* (2008a) Characterization of an optically stimulated luminescence (OSL) material for thermal neutron detection: SrS : Ce,Sm,B. *Phys. Status Solidi A*, **205**, 1196–1202.

Benton, E.R. and Benton, E.V. (2001) Space radiation dosimetry in low-Earth orbit and beyond. *Nucl. Instrum. Methods Phys. Res. B*, **184**, 255–294.

Benton, E.R., Benton, E.V. and Frank, A.L. (2002a) Passive dosimetry aboard the Mir Orbital Station: internal measurements. *Radiat. Meas.*, **35**, 439–455.

Benton, E.R., Benton, E.V. and Frank, A.L. (2002b) Passive dosimetry aboard the Mir Orbital Station: external measurements. *Radiat. Meas.*, **35**, 457–471.

Benton, E.R., Frank, A.L. and Benton, E.V. (2000) TLD efficiency of $^7$LiF for doses deposited by high-LET particles. *Radiat. Meas.*, **32**, 211–214.

Benton, E.R., Frank, A.L., Gaza, R. *et al.* (2005) US results from the BRADOS/Space intercomparison experiment. Unpublished final report, Eril Research.

Berger, T. and Hajek, M. (2008a) On the linearity of the high-tempeture emission from $^7$LiF:Mg,Ti (TLD-700). *Radiat. Meas.*, **43**, 1467–1473.

Berger, T. and Hajek, M. (2008b) TL efficiency – overview and experimental results over the years. *Radiat. Meas.*, **43**, 146–156.

Berger, T., Hajek, M., Fugger, M. and Vana, N. (2006b) Efficiency corrected dose verification with thermoluminescence dosemeters in heavy ion beams. *Radiat. Prot. Dosim.*, **120**, 361–364.

Berger, T., Hajek, M., Summerer, L. *et al.* (2006a) The efficiency of various thermoluminescence dosemeter types to heavy ions. *Radiat. Prot. Dosim.*, **120**, 365–368.

Berger, T., Reitz, G., Hajek, M. and Vana, N. (2006c) Comparison of various techniques for the exact determination of absorbed dose in heavy ion fields using passive detectors. *Adv. Space Res.*, **37**, 1716–1721.

Bernhardt, R. and Herforth, L. (1974) Radiation Dosimetry by Optically Stimulated Phosphorescence of $CaF_2$:Mn. *Proceedings of the Fourth International Conference on Luminescence Dosimetry, Krakow, Poland, 1974*, (ed. T. Niewiadomski), pp. 1091–1104.

Bilski, P. (2006a) Response of various LiF thermoluminescent detectors to high energy ions – Results of the ICCHIBAN experiment. *Nucl. Instrum. Methods Phys. Res. B*, **251**, 121–126.

Bilski, P. (2006b) Dosimetry of densely ionising radiation with three LiF phosphors for space applications. *Radiat. Prot. Dosim.*, **120**, 397–400.

Blair, M.W. (2005) Development of Optically Stimulated Luminescence Techniques for Application to Terrestrial and Martian Studies. Ph. D. Dissertation, Oklahoma State University.

Blair, M.W., Jacobsohn, L.G., Bennett, B.L. *et al.* (2009) Luminescence and structural properties of oxyorthosilicate and $Al_2O_3$ nanophosphors. *Phys. Status Solidi A*, **206**, 904–909.

Blair, M.W., Jacobsohn, L.G., Tornga, S.C. *et al.* (2010) Nanophosphor aluminum oxide: luminescence response of a potential dosimetric material. *J. Lumin.*, **130**, 825–831.

Blasse, G. and Grabmaier, B.C. (1994) *Luminescent Materials*, Springer, Heidelberg.

Boas, M.L. (2005) *Mathematical Methods in the Physical Sciences*, Wiley.

Bolognese-Milsztajn, T., Ginjaume, M., Luszik-Bhadra, M. *et al.* (2004) Active personal dosiemters for individual monitoring and other new developments. *Radiat. Prot. Dosim.*, **112**, 141–168.

Boone, J.M. (2007) The trouble with $CTDI_{100}$. *Med. Phys.*, **34**, 1364–1371.

Borio, R., Chiocchini, S., Demaio, U. *et al.* (1990) Effects of cooling rate during annealing procedure in BeO. *Radiat. Prot. Dosim.*, **33**, 197–200.

Bortfeld, T. (2006) IMRT: a review and preview. *Phys. Med. Biol.*, **2006**, R363–R379.

Bortolot, V.J. (2000) A new modular high capacity OSL reader system. *Radiat. Meas.*, **32**, 751–757.

Bos, A.J.J. (2001) High sensitivity thermoluminescence dosimetry. *Nucl. Instrum. Methods Phys. Res. B*, **184**, 3–28.

Bos, A.J.J., Prokić, M. and Brouwer, J.C. (2006) Optically and thermally stimulated luminescence characteristics of $MgO:Tb^{3+}$. *Radiat. Prot. Dosim.*, **119**, 130–133.

Bos, A.J.J. and Wallinga, J. (2009) Optically stimulated luminescence signals under various stimulation modes assuming first-order kinetics. *Phys. Rev. B*, **79**, 195118.

Bos, A.J.J., Winkelman, A.J.M., Le Masson, N.J.M. *et al.* (2002) A TL/OSL emission spectrometer extension of the Risø reader. *Radiat. Prot. Dosim.*, **101**, 111–114.

Bøtter-Jensen, L. (1997) Luminescence techniques: instrumentation and methods. *Radiat. Meas.*, **27**, 749–768.

Bøtter-Jensen, L. (2000) Development of Optically Stimulated Luminescence Techniques using Natural Minerals and Ceramics, and their Application to Retrospective Dosimetry. Risø-R-1211(EN), DSc. Thesis, Risø National Laboratory.

Bøtter-Jensen, L., Agersnap Larsen, N., Markey, B.G. and McKeever, S.W.S. (1997) $Al_2O_3$:C as a sensitive OSL dosemeter for rapid assessment of environmental photon dose rates. *Radiat. Meas.*, **27**, 295–298.

Bøtter-Jensen, L., Agersnap Larsen, N., Mejdahl, V. *et al.* (1995) Luminescence sensitivity changes in quartz as a results of annealing. *Radiat. Meas.*, **24**, 535–541.

Bøtter-Jensen, L., Andersen, C.E., Duller, G.A.T. and Murray, A.S. (2003) Developments in radiation, stimulation and observation facilities in luminescence measurements. *Radiat. Meas.*, **37**, 535–541.

Bøtter-Jensen, L., Banerjee, D., Jungner, H. and Murray, A.S. (1999) Retrospective assessment of environmental dose rates using optically stimulated luminescence from $Al_2O_3$:C and quartz. *Radiat. Prot. Dosim.*, **84**, 537–542.

Bøtter-Jensen, L., Bulur, E., Duller, G.A.T. and Murray, A.S. (2000) Advances in luminescence instrument systems. *Radiat. Meas.*, **32**, 523–528.

Bøtter-Jensen, L., Bulur, E., Murray, A.S. and Poolton, N.R.J. (2002) Enhancements in luminescence measurement techniques. *Radiat. Prot. Dosim.*, **101**, 119–124.

Bøtter-Jensen, L., Markey, B.G., Poolton, N.R.J. and Jungner, H. (1996) Luminescence properties of porcelain ceramics relevant to retrospective radiation dosimetry. *Radiat. Prot. Dosim.*, **65**, 369–372.

Bøtter-Jensen, L., McKeever, S.W.S. and Wintle, A.G. (2003) *Optically Stimulated Luminescence Dosimetry*, Elsevier, Amsterdam.

Bøtter-Jensen, L. and Murray, A.S. (1999) Developments in optically stimulated luminescence techniques for dating and retrospective dosimetry. *Radiat. Prot. Dosim.*, **84**, 307–315.

Bøtter-Jensen, L. and Murray, A.S. (2001) Optically stimulated luminescence techniques in retrospective dosimetry. *Radiat. Phys. Chem.*, **61**, 181–190.

Bøtter-Jensen, L., Poolton, N.R.J., Willumsen, F. and Christiansen, H. (1994) A compact design for monochromatic OSL measurements in the wavelength range 380-1020nm. *Radiat. Meas.*, **23**, 519–522.

Bøtter-Jensen, L., Thomsen, K.J. and Jain, M. (2010) Review of optically stimulated luminescence (OSL) instrumental developments for retrospective dosimetry. *Radiat. Meas.*, **45**, 253–257.

Bougrov, N.G., Göksu, H.Y., Haskell, E.H. *et al.* (1997) Issues in the reconstruction of environmental doses on the basis of thermoluminescence measurements in the Techa riverside. *Health Phys.*, **75**, 574–583.

Boukhair, A., Heilmann, C., Nourreddine, A. *et al.* (2001) Fast neutron and $\gamma$-ray dosimetry with imaging plates. *Radiat. Meas.*, **34**, 513–516.

Boyle, R. (1664) *Experiments and Considerations Touching Colors*, Transactions of the Royal Society, London, pp. 413–421.

Boyle, R. (1680) *The Aerial Noctiluca: Or, Some New Phenomena, and a Process of a Factitious Self-Shining Substance*, N. Ranew, London.

Bransden, B.H. and Joachain, C.J. (1983) *Physics of Atoms and Molecules*, Prentice Hall, Harlow.

Bräunlich, P. (ed.) (1979) *Thermally Stimulated Relaxation in Solids*, Springer-Verlag, Berlin.

Bräunlich, P., Schaffer, D. and Scharmann, A. (1967) A simple model for thermoluminescence and thermally stimulated conductivity of inorganic photoconducting phosphors and experiments pertaining to infra-red stimulated luminescence. Proceedings of the First International Conference on Luminescence Dosimetry, Stanford, June, 1965, USAEC, pp. 57–73.

Bray, H.E., Bailey, R.M. and Stokes, S. (2002) Quantification of cross-irradiation and cross-illumination using a Risø TL/OSL DA-15 reader. *Radiat. Meas.*, **35**, 275–280.

Bube, R.H. (1951) A comparison study of photoconductivity and luminescence. *Phys. Rev.*, **83**, 393–396.

Bube, R.H. (1960) *Photoconductivity of Solids*, John Wiley & Sons, Inc., New York.

Bulur, E. (1996) An alternative technique for optically stimulated luminescence (OSL) experiment. *Radiat. Meas.*, **26**, 701–709.

Bulur, E. and Göksu, H.Y. (1997) IR stimulated luminescence from ZnS and SrS based storage phosphors: A re-examination with linear modulation technique. *Phys. Status Solidi A*, **161**, R9–R10.

Bulur, E. and Goksu, H.Y. (1999) Infrared (IR) stimulated luminescence from feldspars with linearly increasing excitation light intensity. *Radiat. Meas.*, **30**, 505–512.

Bulur, E. (2000) A simple transformation for converting CW-OSL curves to LM-OSL curves. *Radiat. Meas.*, **32**, 141–145.

Bulur, E., Bøtter-Jensen, L. and Murray, A.S. (2000) Optically stimulated luminescence from quartz measured using linear modulation technique. *Radiat. Meas.*, **32**, 407–411.

Bulur, E., Bøtter-Jensen, L. and Murray, A.S. (2001) LM-OSL signals from some insulators: an analysis of the dependency of the detrapping probability on stimulation light intensity. *Radiat. Meas.*, **33**, 715–719.

Bulur, E., Duller, G.A.T., Solongo, S. *et al.* (2002) LM-OSL from single grains of quartz: a preliminary study. *Radiat. Meas.*, **35**, 79–85.

Bulur, E. and Göksu, H.Y. (1998) OSL from BeO ceramics: new observations from an old material. *Radiat. Meas.*, **29**, 639–650.

Bulur, E. and Yeltik, A. (2010a) Time-resolved OSL from BeO ceramics: a preliminary study. Submitted.

Bulur, E. and Yeltik, A. (2010b) Optically stimulated luminescence from BeO ceramics: An LM-OSL study. *Radiat. Meas.*, **45**, 29–34.

Burbidge, C.I. and Duller, G.A.T. (2003) Combined gamma and beta dosimetry, using $Al_2O_3$:C, for in situ measurements on a sequence of archaeological deposits. *Radiat. Meas.*, **37**, 285–291.

Burgkhardt, B. and Piesch, E. (1988) Field calibration technique for albedo neutron dosemeters. *Radiat. Prot. Dosim.*, **23**, 121–126.

Bushberg, J.T., Seibert, J.A., Leidholdt, J., E.M. and Boone, J.M. (1994) *The Essential Physics of Medical Imaging*, Williams & Wilkins, Baltimore.

Butts, J.J. and Katz, R. (1967) Theory of RBE for heavy ion bombardment of dry enzymes and viruses. *Radiat. Res.*, **30**, 855–871.

Chadwick, M.B., Oblozinsky, P., Herman, M. *et al.* (2006) ENDF/B-VII.0: Next generation evaluated nuclear data library for nuclear science and technology. *Nuclear Data Sheets*, **107**, 2931–3059.

Chatterjee, A. and Schaefer, H.J. (1976) Microdosimetric structure of heavy ion tracks in tissue. *Radiat. Environ. Biophys.*, **13**, 215–227.

Chen, R., Hornyak, W.F. and Mathur, V.K. (1990) Competition between excitation and bleaching of thermoluminescence. *J. Phys. D: Appl. Phys.*, **23**, 724–728.

Chen, R. and Leung, P.L. (2001a) Nonlinear dose dependence and dose-rate dependence of optically stimulated luminescence and thermoluminescence. *Radiat. Meas.*, **33**, 475–481.

Chen, R. and Leung, P.L. (2001b) Dose dependence and dose-rate dependence of the optically stimulated luminescence signal. *J. Appl. Phys.*, **89**, 475–481.

Chen, R. and Leung, P.L. (2003) The decay of OSL signals as stretched-exponential functions. *Radiat. Meas.*, **37**, 519–526.

Chen, R. and McKeever, S.W.S. (1994) Characterization of nonlinearities in the dose dependence of thermoluminescence. *Radiat. Meas.*, **23**, 667–673.

Chen, R. and McKeever, S.W.S. (1997) *Theory of Thermoluminescence and Related Phenomena*, World Scientific Publishing Co., Singapore.

Chen, R., McKeever, S.W.S. and Durrani, S.A. (1981) Solution of the kinetic equations governing trap filling. Consequences concerning dose dependence and dose-rate effects. *Phys. Rev. B*, **24**, 4931–4944.

Chen, R., Pagonis, V. and Lawless, J.L. (2006) The nonmonotonic dose dependence of optically stimulated luminescence in $Al_2O_3$:C: Analytical and numerical simulation results. *J. Appl. Phys.*, **99**, 033511.

Chen, T., Poochinda, K. and Stoebe, T.G. (2006) Innovational radiation sensor by integrating $Al_2O_3$:C optically stimulated luminescent dosemeter and GaN detectors. *Radiat. Prot. Dosim.*, **119**, 380–385.

Chen, S.W., Wang, X.T., Chen, L.X. *et al.* (2009) Monte Carlo evaluations of the absorbed dose and quality dependence of $Al_2O_3$ in radiotherapy photon beams. *Med. Phys.*, **36**, 4421–4424.

Chithambo, M.L. and Galloway, R.B. (2000a) A pulsed light-emitting-diode system for stimulation of luminescence. *Meas. Sci. Technol.*, **11**, 418–424.

Chithambo, M.L. and Galloway, R.B. (2000b) On luminescence lifetimes in quartz. *Radiat. Meas.*, **32**, 621–626.

Chithambo, M.L. and Ogundare, F.O. (2009) Luminescence lifetime components in quartz: Influence of irradiation and annealing. *Radiat. Meas.*, **44**, 453–457.

Christodoulides, C. and Fremlin, J.H. (1971) Thermoluminescence of biological material. *Nature*, **232**, 257–258.

Chruścińska, A. (2006) Modelling the optical bleaching of a complex TL signal. *J. Phys. D.: Appl. Phys.*, **39**, 2321–2329.

Chruścińska, A. (2009) Modelling the thermal bleaching of OSL signal in the case of a competition between recombination centres. *Radiat. Meas.*, **44**, 329–337.

Clark, R.J. and Bailiff, I.K. (1998) Fast time-resolved luminescence emission spectroscopy in some feldspars. *Radiat. Meas.*, **29**, 553–560.

Clarke, M.L. and Rendell, H.M. (1997) Infra-red stimulated luminescence spectra of alkali feldspars. *Radiat. Meas.*, **27**, 221–236.

Clift, M.A., Johnston, P.N. and Webb, D.V. (2002) A temporal methods of avoiding the Cerenkov radiation generated in organic scintillator dosimeters by pulsed mega-voltage electron and photon beams. *Phys. Med. Biol.*, **47**, 1421–1433.

Clift, M.A., Sutton, R.A. and Webb, D.V. (2000) Dealing with Cerenkov radiation generated in organic scintillator dosimeters by bremsstrahlung beams. *Phys. Med. Biol.*, **45**, 1165–1182.

Colyott, L.E., Akselrod, M.S. and McKeever, S.W.S. (1996) Phototransferred thermoluminescence in $\alpha$-$Al_2O_3$:C. *Radiat. Prot. Dosim.*, **65**, 263–266.

Colyott, L.E., Akselrod, M.S. and McKeever, S.W.S. (1997) An integrating ultraviolet-B dosemeter using phototransferred thermoluminescence from $\alpha$-$Al_2O_3$:C. *Radiat. Prot. Dosim.*, **72**, 87–94.

Colyott, L.E., McKeever, S.W.S. and Akselrod, M.S. (1999) An integrating UVB dosemeter system. *Radiat. Prot. Dosim.*, **85**, 309–312.

Cucinotta, F.A. and Durante, M. (2006) Cancer risk from exposure to galactic cosmic rays: implications for space exploration by human beings. *Lancet Oncol.*, **7**, 431–435.

Cucinotta, F.A., Kim, M.-H.Y. and Ren, L. (2005) Managing Lunar and Mars Mission Radiation Risks. Part I: Cancer Risks, Uncertainties, and Shielding Effectiveness. NASA/TP-2005-213164, NASA Center for AeroSpace Information, Hanover, MD, http://marsjournal.org/contents/2006/0004/files/Cucinotta2005.pdf (accessed on 5 May 2010).

Cui, F.Z. and Ge, J. (2007) New observations of the hierarchical structure of human enamel, from nanoscale to microscale. *J. Tissue Eng. Regen. Med.*, **1**, 185–191.

Currie, L.A. (1968) Limits for qualitative detection and quantitative determination – application to radiochemistry. *Anal. Chem.*, **40**, 586–593.

Curtis, S.D., Atwell, W., Beever, R. and Hardy, A. (1986) Radiation environments and absorbed dose estimations on manned space missions. *Adv. Space Res.*, **6**, 269–274.

Cygler, J.E. and Yukihara, E.G. (2009) Optically stimulated luminescence (OSL) dosimetry in radiotherapy, in *Clinical Dosimetry Measurements in Radiotherapy (AAPM 2009 Summer School)* (eds D.W.O. Rogers and J.E. Cygler), Medical Physics Publishing.

Damkjær, S.M.S., Andersen, C.E. and Aznar, M.C. (2008) Improved real-time dosimetry using the radioluminescence signal from $Al_2O_3$:C. *Radiat. Meas.*, **43**, 893–897.

Danzer, J., Dudney, C., Seibert, R. *et al.* (2007) TH-C-M100E-02: Optically Stimulated Luminescence of Aluminum Oxide Detectors for Radiation Therapy Quality Assurance. *Med. Phys.*, **34**, 2628–2629.

Das, R.K., Keleti, D., Zhu, Y.M. *et al.* (1997) Validation of Monte Carlo dose calculations near I-125 sources in the presence of bounded heterogeneities. *Int. J. Radiat. Oncol. Biol. Phys.*, **38**, 843–853.

Davies, J.E. (1979) On the use of dental ceramics as a possible second-line approach to accident irradiation dosimetry. *Radioprotection*, **14**, 89–97.

Deboer, S.F., Beddar, A.S. and Rawlinson, J.A. (1993) Optical filtering and spectral measurements of radiation-induced light in plastic scintillation dosimetry. *Phys. Med. Biol.*, **38**, 945–958.

Demuro, M., Roberts, R.G., Froese, D.G. *et al.* (2008) Optically stimulated luminescence dating of single and multiple grains of quartz from perennially frozen loess in western Yukon Territory, Canada: Comparison with radiocarbon chronologies for the late Pleistocene Dawson tephra. *Quat. Geochronol.*, **3**, 346–364.

Denby, P.M., Bøtter-Jensen, L., Murray, A.S. *et al.* (2006) Application of pulsed OSL to the separation of the luminescence components from a mixed quartz/feldspar sample. *Radiat. Meas.*, **41**, 774–779.

d'Errico, F. (2006) Dosimetric issues in radiation protection of radiotherapy patients. *Radiat. Prot. Dosim.*, **118**, 205–212.

DeWitt, R., Klein, D.M., Yukihara, E.G. *et al.* (2010) Optically stimulated luminescence (OSL) of tooth enamel and its potential use in post-exposure triage. *Health Phys.*, **98**, 432–439.

Di Bartolo, B. (1968) *Optical Interactions in Solids*, John Wiley & Sons, Inc., New York.

Ding, X. (2002) Energy spectra, angular spread, fluence profiels and dose distributions of 6 and 18 MV photon beams: results of Monte Carlo simulations for a Varian 2100EX accelerator. *Phys. Med. Biol.*, **47**, 1025–1046.

Dixon, R.L. (2006) Restructuring CT dosimetry – A realistic strategy for the future requiem for the pencil chamber. *Med. Phys.*, **33**, 3973–3976.

Dixon, R.L. and Ballard, A.C. (2007) Experimental validation of a versatile system of CT dosimetry using a conventional ion chamber: Beyond $CTDI_{100}$. *Med. Phys.*, **34**, 3399–3413.

Doke, T., Hayashi, T., Nagaoka, S. *et al.* (1995) Estimation of dose equivalent in STS-47 by a combination of TLDs and CR-39. *Radiat. Meas.*, **24**, 75–82.

Dorenbos, P., Van Eijk, C.W.E., Bos, A.J.J. and Melcher, C.L. (1994a) Afterglow and thermoluminescence properties of $Lu_2SiO_5$:Ce scintillation crystals. *J. Phys.: Condens. Matter*, **6**, 4167–4180.

Dorenbos, P., Van Eijk, C.W.E., Bos, A.J.J. and Melcher, C.L. (1994b) Scintillation and thermoluminescence properties of $Lu_2SiO_5$:Ce fast scintillation crystals. *J. Lumin.*, **60 & 61**, 979–982.

Dotzler, C., Williams, G.V.M., Edgar, A. and Appleby, G.A. (2007) Dosimetric properties of $RbCdF_3$:$Mn^{2+}$. *Radiat. Meas.*, **42**, 586–589.

Douguchi, Y., Nanto, H., Sato, T. *et al.* (1999) Optically stimulated luminescence in Eu-doped KBr phosphor ceramics. *Radiat. Prot. Dosim.*, **84**, 143–148.

Dudkin, V.E., Potapov, Y.V., Akapova, A.B. *et al.* (1996) Measurements of fast and intermediate neutron energy spectra on Mir space station in the second half of 1991. *Radiat. Meas.*, **26**, 535–539.

Duller, G.A.T. (1997) Behavioural studies of stimulated luminescence from feldspars. *Radiat. Meas.*, **27**, 663–694.

Duller, G.A.T., Bøtter-Jensen, L. and Markey, B.G. (1997) A luminescence imaging system based on a CCD camera. *Radiat. Meas.*, **27**, 91–99.

Duller, G.A.T., Bøtter-Jensen, L. and Murray, A.S. (2000) Optical dating of single sand-sized grains of quartz: sources of variability. *Radiat. Meas.*, **32**, 453–457.

Dusseau, L., Plattard, D., Vaillé, J.-R. *et al.* (2000) An integrated sensor using optically stimulated luminescence for in flight dosimetry. *IEEE Trans. Nucl. Sci*, **47**, 2412–2416.

Dusseau, L., Polge, G., Albert, L. *et al.* (1998) Irradiated integrated circuits dose-attenuation mapping using optically stimulated phosphor for packaging dosimetry. *IEEE Trans. Nucl. Sci*, **45**, 2695–2699.

Dusseau, L., Ranchoux, G., Polge, G. *et al.* (1999) High energy electron dose-mapping using optically stimulated luminescent films. *IEEE Trans. Nucl. Sci*, **46**, 1757–1761.

Eanes, E.D. (1979) Enamel Apatite: Chemistry, Structure and Properties. *J. Dent. Res.*, **58**(B), 829–834.

Edmund, J.M. and Andersen, C.E. (2007) Temperature dependence of the $Al_2O_3$:C response in medical luminescence dosimetry. *Radiat. Meas.*, **42**, 177–189.

Edmund, J.M., Andersen, C.E., Greilich, S. *et al.* (2007) Optically stimulated luminescence from $Al_2O_3$:C irradiated with 10-60 MeV protons. *Nucl. Instrum. Methods Phys. Res. A*, **580**, 210–213.

van Eijk, C.W.E. (2004) Neutron detection and neutron dosimetry. *Radiat. Prot. Dosim.*, **110**, 5–13.

Eisenbud, M. and Gesell, T. (1997) *Environmental Radioactivity*, Academic Press, London.

Emfietzoglou, D. and Moscovitch, M. (1996) Phenomenological study of light-induced effects in $\alpha$-$Al_2O_3$:C. *Radiat. Prot. Dosim.*, **65**, 259–262.

Erfurt, G., Krbetschek, M.R., Trautmann, T. and Stolz, W. (2000) Radioluminescence (RL) behavior of $Al_2O_3$:C-potential for dosimetric applications. *Radiat. Meas.* **32**, 735–739.

Essers, M. and Mijnheer, B.J. (1999) In vivo dosimetry during external photon beam radiotherapy. *Int. J. Radiat. Oncol. Biol. Phys.* **43**, 245–259.

Evans, B.D. (1995) A review of the optical properties of anion lattice vacancies, and electrical conduction in $\alpha$-$Al_2O_3$: their relation to radiation-induced electrical degradation. *J. Nucl. Mater.*, **219**, 202–223.

Evans, B.D., Pogatshnik, G.J. and Chen, Y. (1994) Optical properties of lattice defects in $\alpha$-$Al_2O_3$. *Nucl. Instrum. Methods Phys. Res. B*, **91**, 258–262.

Fantuzzi, E., Alves, J.G., Ambrosi, P. *et al.* (2004) Implementation of standards for individual monitoring in Europe. *Radiat. Prot. Dosim.*, **112**, 3–44.

Fattahi, M., Singarayer, J.S., Bailey, R.M. and Stokes, S. (2006) The linearly modulated IRSL red emission from feldspars. *Geochronometria*, **25**, 19–28.

Ferreira, I.H., Dutreix, A., Bridier, A. *et al.* (2000) The ESTRO-QUALity assurance network (EQUAL). *Radiother. Oncol.*, **55**, 273–284.

Fischer, R.E., Tadic-Galeb, B. and Yoder, P.R. (2008) *Optical System Design*, McGraw Hill, New York.

Fleischer, R.L., Price, P.B. and Walker, R.M. (1975) *Nuclear Tracks in Solids*, University of California Press, Berkeley, USA.

Floyd, C.E., Chotas, H.G., Dobbins III, J.T. and Ravin, C.E. (1990) Quantitative radiographic imaging using a photostimulable phosphor system. *Med. Phys.*, **17**, 454–459.

Flynn, R.T., Barbee, D.L., Mackie, T.R. and Jeraj, R. (2007) Comparison of intensity modulated x-ray therapy and intensity modulated proton therapy for selective subvolume boosting: a phantom study. *Phys. Med. Biol.*, **52**, 6073–6091.

Flynn, R.T., Bowen, S.R., Bentzen, S.M. *et al.* (2008) Intensity-modulated x-ray (IMXT) versus proton (IMPT) therapy for theragnostic hypoxia-based dose painting. *Phys. Med. Biol.*, **53**, 4153–4167.

Fontbonne, J.M., Iltis, G., Ban, G. *et al.* (2002) Scintillating fiber dosimeter for radiation therapy accelerator. *IEEE Trans. Nucl. Sci*, **49**, 2223–2227.

Ford, R.M., Hanify, R.D. and Perks, C.A. (2004) *Depletion of the Signal from Optically Stimulated Luminescence Dosemeters*. 11th International Congress of the International Radiation Protection Association, Madrid.

Fowler, J.F. (1963) Solid state dosimetry. *Phys. Med. Biol.*, **8**, 1–32.

Frelin, A.M., Fontbonne, J.M., Ban, G. *et al.* (2005) Spectral discrimination of Cerenkov radiation in scintillating dosimeters. *Med. Phys.*, **32**, 3000–3006.

Fröhlich-Schlapp, M., Ioffe, A., Conrad, H. *et al.* (2005) Novel materials and concepts for neutron image plates. *Nucl. Instrum. Methods Phys. Res. A*, **551**, 46–51.

Galloway, R.B. (2002) Luminescence lifetimes in quartz: dependence on annealing temperature prior to beta irradiation. *Radiat. Meas.*, **35**, 67–77.

Garcier, Y., Cordier, G., Pauron, C. and Fazileabasse, J. (2007) Intercomparison of passive dosimetry technology at EDF facilities in France. *Radiat. Prot. Dosim.*, **124**, 107–114.

Garcia, P., Vaillé, J.R., Benoit, D. *et al.* (2006) Simultaneous evaluation of TID and displacement damage dose using a single OSL sensor. *IEEE Trans. Nucl. Sci*, **53**, 3713–3717.

Garcia, P., Vaillé, J.-R., Benoit, D. *et al.* (2007) Study of the thermal behavior of the OSL integrated sensor response. *IEEE Trans. Nucl. Sci*, **54**, 2272–2275.

Gasparian, P.B.R. (2009) Methodological developments for application of optically stimulated luminescence (OSL) in medical dosimetry. M. S. Thesis, Oklahoma State University.

Gasparian, P.B.R., Ruan, C., Ahmad, S. *et al.* 2010. Demonstrating the use of optically stimulated luminescence dosimeters (OSLDs) for measurement of staff radiation exposure in interventional fluoroscopy and helmet output factors in radiosurgery. *Radiat. Meas.* **45**, 677–680.

Gaza, R., Bulur, E., McKeever, S.W.S. and Soares, C.G. (2006a) Experimental determination of the dose deposition profile of a $^{90}$Sr beta source. *Radiat. Prot. Dosim.*, **120**, 33–37.

Gaza, R., Cooper, T., Hussein, H. *et al.* (2008) *Orion Spacecraft: Crew Radiation Protection Strategies*. 37$^{th}$ COSPAR Scientific Assembly, Montreal, Canada, July 2008.

Gaza, R. and McKeever, S.W.S. (2006) A real-time, high-resolution optical fibre dosemeter based on optically stimulated luminescence (OSL) of KBr:Eu, for potential use during the radiotherapy of cancer. *Radiat. Prot. Dosim.*, **120**, 14–19.

Gaza, R., McKeever, S.W.S. and Akselrod, M.S. (2005) Near-real-time radiotherapy dosimetry using optically stimulated luminescence of $Al_2O_3$:C: Mathematical models and preliminary results. *Med. Phys.*, **32**, 1094–1102.

Gaza, R., McKeever, S.W.S., Akselrod, M.S. *et al.* (2004) A fiber-dosimetry method based on OSL from $Al_2O_3$:C for radiotherapy applications. *Radiat. Meas.*, **38**, 809–812.

Gaza, R., Yukihara, E.G. and McKeever, S.W.S. (2004) The response of thermally and optically stimulated luminescence from $Al_2O_3$:C to high-energy heavy charged particles. *Radiat. Meas.*, **38**, 417–420.

Gaza, R., Yukihara, E.G., McKeever, S.W.S. *et al.* (2006b) Ionization density dependence of the optically stimulated luminescence dose response of $Al_2O_3$:C to low-energy charged particles. *Radiat. Prot. Dosim.*, **119**, 375–379.

Gaza, R., Yukihara, E.G. and McKeever, S.W.S. (2006) The use of optically stimulated luminescence from $Al_2O_3$:C in the dosimetry of high-energy heavy charged particle fields. *Radiat. Prot. Dosim.* **120**, 354–357.

Gaza, R., Yukihara, E.G. and McKeever, S.W.S. (2008) OSU $Al_2O_3$:C OSL detectors, in *Results from the ICCHIBAN-3 and ICCHIBAN-4 experiments to intercompare the response of space radiation dosimeters, HIMAC-128* (eds Y. Uchihori and E.R. Benton), National Institute of Radiological Sciences, Chiba, Japan.

Geiß, O.B., Krämer, M. and Kraft, G. (1998) Efficiency of thermoluminescent detectors to heavy charged particles. *Nucl. Instrum. Methods Phys. Res. B*, **142**, 592–598.

Gesell, T.F. (1982) Environmental monitoring with thermoluminescence dosimetry. *IEEE Trans. Nucl. Sci*, **NS-29**, 1225–1232.

Ginjaume, M., Bolognese-Milsztajn, T., Luszik-Bhadra, M. *et al.* (2007) Overview of active personal dosemeters for individual monitoring in the European Union. *Radiat. Prot. Dosim.*, **125**, 261–266.

Goans, R.E., Holloway, E.C., Berger, M.E. and Ricks, R.C. (1997) Early dose assessment following severe radiation accidents. *Health Phys.*, **72**, 513–518.

Godfrey-Smith, D.I. (2006) Applicability of moissanite, a monocrystalline form of silicon carbide, to retrospective and forensic dosimetry. *Radiat. Meas.*, **41**, 976–981.

Godfrey-Smith, D.I. (2008) Towards in vivo OSL dosimetry of human tooth enamel. *Radiat. Meas.*, **43**, 854–858.

Godfrey-Smith, D.I. and Pass, B. (1997) A new method of retrospective radiation dosimetry: optically stimulated luminescence in dental enamel. *Health Phys.*, **72**, 744–749.

Göksu, H.Y. (2003) Telephone chip-cards as individual dosimeters. *Radiat. Meas.* **37**, 617–620.

Göksu, H.Y. and Bailiff, I.K. (2006) Luminescence dosimetry using building materials and personal objects. *Radiat. Prot. Dosim.*, **119**, 413–420.

Göksu, H.Y., Stepanenko, V.F., Bailiff, I.K. and Jungner, H. (2006) Intercomparison of Luminescence Measurements of Bricks from Dolon' Village: Experimental Methodology and Results of European Study Group. *J. Radiat. Res.*, **47**, A29–A37.

González, A.J. (2007) An international perspective on radiological threats and the need for retrospective biological dosimetry of acute radiation overexposures. *Radiat. Meas.*, **42**, 1053–1062.

Goossens, O., Vanhavere, F., Leys, N. *et al.* (2006) Radiation dosimetry for microbial experiments in the International Space Station using different track-etch and luminescent detectors. *Radiat. Prot. Dosim.*, **120**, 433–437.

Gorbunov, S.V., Kruzhalov, A.V. and Springis, M.J. (1987) Optical properties of the F-center in beryllium oxide. *Phys. Status Solidi B*, **141**, 293–301.

Greilich, S., Glasmacher, U.A. and Wagner, G.A. (2002) Spatially resolved detection of luminescence: a unique tool for archaeochronometry. *Naturwissenschaften*, **89**, 371–375.

Greilich, S. and Wagner, G.A. (2006) Development of a spatially resolved dating technique using HR-OSL. *Radiat. Meas.*, **41**, 738–743.

Grimmeiss, H.G. and Ledebo, L.-Å. (1975a) Photo-ionization of deep impurity levels in semiconductors with non-parabolic bands. *J. Phys. C: Solid State Phys.*, **8**, 2615–2626.

Grimmeiss, H.G. and Ledebo, L.-Å. (1975b) Spectral distribution of photoionization cross sections by photoconductivity measurements. *J. Appl. Phys.*, **46**, 2155–2162.

Grusell, E., Montelius, A., Brahme, A. *et al.* (1994) A general solution to charged particle beam flattening using an optimized dual-scattering-foil technique, with application to proton therapy beams. *Phys. Med. Biol.*, **39**, 2201–2216.

Guimarães, C.C. and Okuno, E. (2003) Blind performance testing of personal and environmental dosimeters based on TLD-100 and natural $CaF_2$:NaCl. *Radiat. Meas.*, **37**, 127–132.

Hajek, M., Berger, T., Fugger, M. *et al.* (2006) BRADOS – Dose distribution in the Russian segment of the International Space Station. *Adv. Space Res.*, **37**, 1664–1667.

Hamamatsu Photonics, K.K. (2007) *Photomultiplier Tubes: Basics and Applications*, Hamamatsu Photonics K. K.

Han, Y., Shin, E.H., Lim, C. *et al.* (2008) Dosimetry in an IMRT phantom designed for a remote monitoring program. *Med. Phys.*, **35**, 2519–2527.

Handee, W.R., Ibbott, G.S. and Handee, E.G. (2005) *Radiation Therapy Physics*, John Wiley & Sons, Hoboken.

Harvey, E.N. (1957) *A History of Luminescence from Earliest Times Until 1900*, The American Philosophical Society, Philadelphia.

Hashizume, T., Maruyama, T., Shiragai, A. *et al.* (1967) Estimation of the air dose from the atomic bombs in Hiroshima and Nagasaki. *Health Phys.*, **13**, 149–161.

Haskell, E.H., Kaipa, P.L. and Wrenn, M.E. (1988) Pre-dose TL characteristics of quartz inclusions removed from bricks exposed to fallout radiation from atmospheric testing at the Nevada Test Site. *Nucl. Tracks Radiat. Meas.*, **14**, 114–120.

Haskell, E.H. (1993a) Retrospective accident dosimetry using environmental materials. *Radiat. Prot. Dosim.*, **47**, 297–303.

Haskell, E.H. (1993b) Accident dosimetry using environmental materials: the role of thermoluminescence. *Nucl. Tracks Radiat. Meas.*, **21**, 87–93.

Haskell, E.H., Bailiff, I.K., Kenner, G.H. *et al.* (1994) Thermoluminescence measurements of gamma-ray doses attributed to fallout from the Nevada Test Site using building bricks as natural dosimeters. *Health Phys.*, **66**, 380–391.

Hazelton, J.R., Yukihara, E.G., Jacobsohn, L.G. *et al.* (2010) Feasibility of using oxyorthosilicates as optically stimulated luminescence detectors. *Radiat. Meas.*, **45**, 681–683.

Hecht, E. (2001) *Optics*, Addison Wesley.

Heirtzler, J.R. (2002) The future of the South Atlantic anomaly and implications for radiation damage in space. *J. Atmospheric and Solar-Terrestrial Phys.*, **64**, 1701–1708.

Henderson, B. and Imbusch, G.F. (1989) *Optical Spectroscopy of Inorganic Solids*, Claredon Press, Oxford.

Henniger, J., Horlbeck, B., Hübner, K. and Prokert, K. (1982) The evaluation of $CaF_2$:Mn-polyethylene detectors with the aid of the optically stimulated luminescence (OSL). *Nucl. Instrum. Methods*, **204**, 209–212.

Heynderickx, D. (2002) Radiation belt modelling in the framework of space weather effects and forecasting. *J. Atmospheric and Solar-Terrestrial Phys.*, **64**, 1687–1700.

Hoffmann, W. and Prediger, B. (1984) Heavy particle dosimetry with high temperature peaks of $CaF_2$:Tm and $^7$LiF phosphors. *Radiat. Prot. Dosim.*, **6**, 149–152.

Hogstrom, K.R. and Almond, P.R. (2006) Review of electron beam therapy physics. *Phys. Med. Biol.*, **51**, R455–R489.

Horowitz, Y.S. (ed.), (1983) *Thermoluminescence and Thermoluminescent Dosimetry*, CRC Press, Boca Raton.

Horowitz, Y.S. (2001) Theory of thermoluminescence gamma dose response: the unified interaction model. *Nucl. Instrum. Methods Phys. Res. B*, **184**, 68–84.

Horowitz, Y.S. (2003) The concept of LET in thermoluminescence dosimetry – is it meaningful or helpful? *Radiat. Prot. Dosim.*, **106**, 195–196.

Horowitz, Y.S. and Olko, P. (2004) The effects of ionization density on the thermoluminescence response (efficiency) of LiF:Mg,Ti and LiF:Mg,Cu,P. *Radiat. Prot. Dosim.*, **109**, 331–348.

Horowitz, Y.S. and Kalef-Ezra, J. (1981) Relative thermoluminescent response of LiF-TLD to $^{252}$Cf fission fragments. *Nucl. Instrum. Methods Phys. Res.*, **187**, 519–525.

Horowitz, Y.S., Oster, L. and Datz, H. (2007) The thermoluminescence dose response and other characteristics of the high-temperature TL in LiF:Mg,Ti (TLD-100). *Radiat. Prot. Dosim.*, **124**, 191–205.

Hu, S., Kim, M.-H.Y., McClellan, G.E. and Cucinotta, F.A. (2009) Modeling the acute health effects of astronauts from exposure to large solar particle events. *Health Phys.*, **96**, 465–476.

Hubbell, J.H. and Seltzer, S.M. (2004) Tables of X-ray mass attenuation coefficients and mass energy-absorption coefficients (version 1.4). [Online] Available: http://physics.nist.gov/xaamdi [2010, January 20]. National Institute of Standards and Technology, Gaithesburg, MD.

Huntley, D.J., Godfrey-Smith, D.I. and Haskell, E.H. (1991) Light-induced emission spectra from some quartz and feldspars. *Nucl. Tracks Radiat. Meas.*, **18**, 127–131.

Huntley, D.J., Godfrey-Smith, D.I. and Thewalt, M.L.W. (1985) Optical dating of sediments. *Nature*, **313**, 105–107.

Huntley, D.J. and Lamothe, M. (2001) Ubiquity of anomalous fading in K-feldspars and the measurement and correction for it in optical dating. *Can. J. Earth Sci.*, **38**, 1093–1106.

Huston, A.L., Justus, B.L., Falkenstein, P.L. *et al.* (2001) Remote optical fiber dosimetry. *Nucl. Instrum. Methods Phys. Res. B*, **184**, 55–67.

Huston, A.L., Justus, B.L., Falkenstein, P.L. *et al.* (2002) Optically stimulated luminescence glass optical fibre dosemeters. *Radiat. Prot. Dosim.*, **101**, 23–26.

Hütt, G., Brodski, L., Bailiff, I.K. *et al.* (1993) Accident dosimetry using environmental materials collected from regions downwind of Chernobyl: a preliminary evaluation. *Radiat. Prot. Dosim.*, **47**, 307–311.

Hütt, G., Jaek, I. and Tchonka, J. (1988) Optical dating: K-feldspars optical response stimulation spectra. *Quat. Sci. Rev.*, **7**, 381–385.

IAEA (1988) *The Radiological Accident in Goiania*, International Atomic Energy Agency, Vienna.

IAEA (1998) The radiological accident in Tammiku. STI/PUB/1053, International Atomic Energy Agency, Vienna.

IAEA (2000) Absorbed Dose Determination in External Beam Radiotherapy. Technical Reports Series No. 398, International Atomic Energy Agency, Vienna.

IAEA (2001) Cytogenetic Analysis for Radiation Dose Assessment: A Manual. Technical Reports Series No. 405, International Atomic Energy Agency, Vienna.

IAEA (2002) Use of electron paramagnetic resonance dosimetry with tooth enamel for retrospective dose assessment. IAEA-TECDOC-1331, International Atomic Energy Agency, Vienna.

IAEA (2007) Dosimetry in Diagnostic Radiology: An International Code of Practice. Technical Reports Series No. 457, International Atomic Energy Agency, Vienna.

Ibbott, G.S., Molineu, A. and Followill, D.S. (2006) Independent evaluations of IMRT through the use of an anthropomorphic phantom. *Technol. Cancer Res. Treat.*, **5**, 481–487.

Ichikawa, Y., Higashimura, T. and Shidei, T. (1966) Thermoluminescence dosimetry of gamma rays from atomic bombs in Hiroshima and Nagasaki. *Health Phys.*, **12**, 395–405.

ICRP (1991) ICRP Publication 60: 1990 Recommendations of the International Comission on Radiological Protection. *Ann. ICRP*, 21.

ICRP (1996) ICRP Publication 74: Conversion coefficients for use in radiological protection against external radiation. *Ann. ICRP*, 26.

ICRP (1997) ICRP Publication 75: General principles for the radiation protection of workers. *Ann. ICRP*, **27**, 1–61.

ICRP (2000a) ICRP Publication 86: Prevention of accidental exposures to patients undergoing radiation therapy. *Ann. ICRP*, **30**, 1–70.

ICRP (2000b) ICRP Publication 87: Managing patient dose in computed tomography. *Ann. ICRP*, **30**, 1–45.

ICRP (2001) ICRP Publication 85: Avoidance of radiation injuries from medical interventional procedures. *Ann. ICRP*, **30**, 1–67.

ICRP (2004) ICRP Publication 93: Managing patient dose in digital radiography. *Ann. ICRP*, **34**, 1–73.

ICRP (2005) ICRP Publication 97: Prevention of high-dose-rate brachytherapy accidents. *Ann. ICRP*, **35**, 1–52.

ICRP (2006) ICRP Publication 96: protecting people against radiation exposure in the event of a radiological attack. *Ann. ICRP*, **35**, 1–110.

ICRP (2007) ICRP Publication 103: The 2007 recommendations of the international commission on radiological protection. *Ann. ICRP*, **37**, 1–332.

ICRU (1970) *ICRU Report 16: Linear Energy Transfer*, International Comission on Radiation Units and Measurements, Washington, DC.

ICRU (1976) *ICRU Report 24: Determination of Absorbed Dose in a Patient Irradiated by Beams of X or Gamma Rays in Radiotherapy Procedures*, International Commission on Radiation Units and Measurements, Bethesda, MD.

ICRU (1993) *ICRU Report 51: Quantities and Units in Radiation Protection Dosimetry*, International Comission on Radiation Units and Measurements, Bethesda, MD.

ICRU (1998a) *Fundamental Quantities and Units for Ionizing Radiation. ICRU Report 60*, International Comission on Radiation Units and Measurements, Bethesda, MD.

ICRU (1998b) *Clinical Proton Dosimetry Part I: Beam Production, Beam Delivery and Measurement of Absorbed Dose, ICRU Report 59*, International Commission on Radiation Units and Measurements, Bethesda.

ICRU (1998c) *ICRU Report 59: Clinical Proton Dosimetry Part I: Beam Production, Beam Delivery and Measurement of Absorbed Dose*, International Commission on Radiation Units and Measurements, Bethesda, MD.

ICRU (1998d) *ICRU Report 60: Fundamental Quantities and Units for Ionizing Radiation*, International Comission on Radiation Units and Measurements, Bethesda, MD.

ICRU (2001) ICRU Report 66: Determination of operational dose equivalent quantities for neutrons. *Journal of the ICRU*, **1**, 1–93.

ICRU (2002) ICRU Report 68: retrospective assessment of exposure to ionizing radiation. *Journal of the ICRU*, 2.

ICRU (2005) ICRU Report 74: Patient dosimetry for x rays used in medical imaging. *Journal of the ICRU*, **5**, 1–113.

ICRU (2006) ICRU Report 76: measurement quality assurance for ionizing radiation dosimetry. *Journal of the ICRU*, **6**, 1–50.

Idri, K., Santoro, L., Charpiot, E. *et al.* (2004) Quality control of intensity modulated radiation therapy with optically stimulated luminescent films. *IEEE Trans. Nucl. Sci*, **51**, 3638–3641.

IEC (2006) *International Standard IEC 61066: Thermoluminescence Dosimetry Systems for Personal and Environmental Monitoring*, International Electrotechnical Commission, Geneva.

IEC (2007) International Standard 62387-1: Radiation protection instrumentation – Passive integrating dosimetry systems for environmental and personal monitoring. Part 1: General characteristics and performance requirements. International Electrotechnical Commission, Geneva.

Inkson, J.C. (1981) Deep impurities in semiconductors: II. The optical cross section. *J. Phys. C: Solid State Phys.*, **14**, 1093–1101.

Inrig, E.L., Godfrey-Smith, D.I. and Khanna, S. (2008) Optically stimulated luminescence of electronic components for forensic, retrospective, and accident dosimetry. *Radiat. Meas.*, **43**, 726–730.

Inrig, E.L., Godfrey-Smith, D.I. and Larsson, C.L. (2010) Fading corrections to electronic component substrates in retrospective accident dosimetry. *Radiat. Meas.*, **45**, 608–610.

Intensity Modulated Radiation Therapy Collaborative Working Group (2001) Intensity-modulated radiotherapy: Current status and issues of interest. *Int. J. Radiat. Oncol. Biol. Phys.*, **51**, 880–914.

ISO (1996a) X and gamma reference radiation for calibrating dosemeters and doserate meters and for determining their response as a function of photon energy – Part 1: Radiation characteristics and production methods. International Standard ISO 4037-1, International Organization for Standardization, Geneva.

ISO (1996b) X and gamma reference radiation for calibrating dosemeters and doserate meters and for determining their response as a function of photon energy – Part 3: Calibration of area and personal dosemeters and the measurement of their response as a function of energy and angle of incidence. International Standard ISO 4037-3, International Organization for Standardization, Geneva.

ISO (1996c) X and gamma reference radiation for calibrating dosemeters and doserate meters and for determining their response as a function of photon energy – Part 4: Calibration of area and personal dosemeters in low energy X reference radiation fields. International Standard ISO 4037-4, International Organization for Standardization, Geneva.

ISO (1997) International Standard ISO 4037-2: X and gamma reference radiation for calibrating dosemeters and doserate meters and for determining their response as a function of photon energy – Part 2: Dosimetry for radiation protection over the energy ranges 8 keV to 1,3 MeV and 4 MeV to 9 MeV. International Organization for Standardization, Geneva.

ISO (1999) International Standard ISO 4037-3: X and gamma reference radiation for calibrating dosemeters and doserate meters and for determining their response as a function of photon energy – Part 3: Calibration of area and personal dosemeters and the measurement of their response as a function of energy and angle of incidence. International Organization for Standardization, Geneva.

ISO (2000a) Reference neutron irradiations – Part 3: Calibration of area and personal dosimeters and determination of their response as a function of neutron energy and angle of incidence. International Standard ISO 8529-3, International Organization for Standardization, Geneva.

ISO (2000b) Reference neutron irradiations – Part 2: Calibration fundamentals related to basic quantities characterizing the radiation field. International Standard ISO 8529-2, International Organization for Standardization, Geneva.

ISO (2001) International Standard ISO 8529-1: Reference neutron irradiations – Part 1: Characteristics and methods of production. International Organization for Standardization, Geneva.

ISO (2004) International Standard ISO 6980-2: Reference beta particle radiations – Part 2: calibration fundamentals related to basic quantities characterizing the radiation field. International Standardization Organization, Geneva.

ISO (2006a) International Standard ISO 6980-2: Reference beta particle radiations – Part 3: calibration of area and personal dosemeters and determination of response as a function of energy and angle of incidence. International Standardization Organization, Geneva.

ISO (2006b) International Standard ISO 6980-1: Reference beta particle radiations – Part 1: methods of production. International Standardization Organization, Geneva.

ISO/IEC (2007) ISO/IEC Guide 99: International vocabulary of metrology – Basic and general concepts and associated terms (VIM). International Organization for Standardization, Geneva.

ISO/IEC (2008) ISO/IEC Guide 98-3: Guide to the expression of uncertainty in measurement (GUM:1995). International Organization for Standardization, Geneva.

ISO/IEC GUIDE 99 (2007) International vocabulary of metrology – Basic and general concepts and associated terms (VIM). International Organization for Standardization, Geneva.

Izewska, J. and Andreo, P. (2000) The IAEA/WHO TLD postal programme for radiotherapy hospitals. *Radiother. Oncol.*, **54**, 65–72.

Izewska, J., Georg, D., Bera, P. *et al.* (2007) A methodology for TLD postal dosimetry audit of high-energy radiotherapy photon beams in non-reference conditions. *Radiother. Oncol.*, **84**, 67–74.

Izewska, J., Hultqvist, M. and Bera, P. (2008) Analysis of uncertainties in the IAEA/WHO TLD postal dose audit system. *Radiat. Meas.*, **43**, 959–963.

Jackson, J.D. (1999) *Classical Electrodynamics*, 3rd edn, John Wiley & Sons, New York.

Jacob, M., Spallek, R., Rassow, J. and Hoffmann, W. (1993) Infrared thermoluminescence glow curves of rare earth doped calcium fluoride. *Radiat. Prot. Dosim.*, **48**, 265–272.

Jacobs, Z., Wintle, A.G. and Duller, G. (2006) Evaluation of SAR procedures for $D_e$ determination using single aliquots of quartz from two archaeological sites in South Africa. *Radiat. Meas.*, **41**, 520–533.

Jahn, A., Sommer, M. and Henniger, J. (2009) 2D-OSL-dosimetry using beryllium oxide. *Radiat. Meas.*, **45**, 674–676.

Jain, M., Bøtter-Jensen, L., Murray, A.S. and Jungner, H. (2002) Retrospective dosimetry: dose evaluation using unheated and heated quartz from a radioactive waste storage building. *Radiat. Prot. Dosim.*, **101**, 525–530.

Jain, M., Bøtter-Jensen, L. and Thomsen, K.J. (2007) High local ionization density effects in x-ray excitations deduced from optical stimulation of trapped charge in $Al_2O_3$:C. *J. Phys.: Condens. Matter*, **19**, 116201.

Jain, C., Murray, A.S., Botter-Jensen, L. and Wintle, A.G. (2005) A single-aliquot regenerative-dose method based on IR (1.49 eV) bleaching of the fast OSL component in quartz. *Radiat. Meas.*, **39**, 309–318.

Jain, M., Thomsen, K.J., Bøtter-Jensen, L. and Murray, A.S. (2004) Thermal transfer and apparent-dose distributions in poorly bleached mortar samples: results from single grains and small aliquots of quartz. *Radiat. Meas.*, **38**, 101–109.

Jasińska, M. and Niewiadomski, T. (1970) Thermoluminescence of Biological Materials. *Nature*, **227**, 1159–1160.

Joint Interagency Working Group (2005) Technology assessment and roadmap for the Emergency Radiation Dose Assessment Program, UCRL-TR-215887, Department of Homeland Security.

Jordan, K.J. (1996) Evaluation of ruby as a fluorescent sensor for optical fiber-based radiation dosimetry, in *Fluorescence Detection IV* (eds. E.R. Menzel and A. Katzir), SPIE, pp. 170–178.

Jun, I. and Garrett, H.B. (2005) Comparison of high-energy trapped particle environments at the Earth and Jupiter. *Radiat. Prot. Dosim.*, **116**, 50–54.

Jursinic, P.A. (2007) Characterization of optically stimulated luminescence dosimeters, OSLDs, for clinical dosimetric measurements. *Med. Phys.*, **34**, 4594–4604.

Jursinic, P.A. (2010) Changes in optically stimulated luminescent dosimeter (OSLD) dosimetric characteristics with accumulated dose. *Med. Phys.*, **37**, 132–140.

Justus, B.L., Falkenstein, P.L., Huston, A.L. *et al.* (2004) Gated fiber-optic-coupled detector for in vivo real-time radiation dosimetry. *Appl. Opt.*, **43**, 1663–1668.

Justus, B.L., Merritt, C.D., Pawlovich, K.J. *et al.* (1999a) Optically stimulated luminescence dosimetry using doped fused quartz. *Radiat. Prot. Dosim.*, **84**, 189–192.

Justus, B.L., Pawlovich, K.J., Merritt, C.D. and Huston, A.L. (1999b) Optically and thermally stimulated luminescence characteristics of $Cu^+$-coped fused quartz. *Radiat. Prot. Dosim.*, **81**, 5–10.

Justus, B.L., Rychnovsky, S., Miller, M.A. *et al.* (1997) Optically stimulated luminescence radiation dosimetry using doped silica glass. *Radiat. Prot. Dosim.*, **74**, 151–154.

Karzmark, C.J. and Morton, R.J. (1998) *A Primer on Theory and Operation of Linear Accelerators in Radiation Therapy*, Medical Physics Publishing, Madison.

Kieffer, J. and Straaten, J. (1986) A model of ion track structure based on classical collision dynamics. *Phys. Med. Biol.*, **31**, 1201–1209.

King, J.H. (1974) Solar proton fluences for 1977–1983 space missions. *J. Spacecr. Rockets*, **11**, 401–408.

Kirby, T.H., Hanson, W.F. and Johnston, D.A. (1992) Uncertainty analysis of absorbed dose calculations from thermoluminescence dosimeters. *Med. Phys.*, **19**, 1427–1433.

Kirov, A.S., Williamson, J.F., Meigooni, A.S. and Zhu, Y. (1995) TLD, diode and monte carlo dosimetry of an Ir-192 source for high dose-rate brachytherapy. *Phys. Med. Biol.*, **40**, 2015–2036.

Kitagawa, A., Fujita, T., Muramatsu, M. *et al.* (2010) Review on heavy ion radiotherapy facilities and related ion sources. Rev. Sci. Instrum., 81, 02B909.

Kitis, G., Furetta, C. and Pagonis, V. (2009) Mixed-order kinetics model for optically stimulated luminescence. *Mod. Phys. Lett. B*, **23**, 3191–3207.

Kitis, G., Polymeris, G.S. and Kiyak, N.G. (2007) Component-resolved thermal stability and recuperation study of the LM-OSL curves of four sedimentary quartz samples. *Radiat. Meas.*, **42**, 1273–1279.

Kittel, C. (1996) *Introduction to Solid State Physics*, John Wiley & Sons, Inc., New York.

Kiyak, N.G., Polymeris, G.S. and Kitis, G. (2007) Component resolved OSL dose response and sensitization of various sedimentary quartz samples. *Radiat. Meas.*, **42**, 144–155.

Klein, D.M. (2008) Development and characterization of remote radiation dosimetry systems using optically stimulated luminescence of $Al_2O_3$:C and KBr:Eu. Ph. D. Dissertation, Oklahoma State University.

Klein, D.M. and McKeever, S.W.S. (2008) Optically stimulated luminescence from KBr:Eu as a near real-time dosimetry system. *Radiat. Meas.*, **43**, 883–887.

Klein, D.M., Yukihara, E.G., Bulur, E. *et al.* (2005) An optical fiber radiation sensor for remote detection of radiological materials. *IEEE Sensors*, **5**, 581–588.

Klein, D.M., Yukihara, E.G., Durham, J.S. *et al.* (2006) In-situ, long-term monitoring system for radioactive contaminants. *Radiat. Prot. Dosim.*, **119**, 421–424.

Klemic, G.A., Azziz, N. and Marino, S.A. (1996) The neutron response of $Al_2O_3$:C, $^7$Li:Mg,Cu,P, and $^7$LiF:Mg,Ti TLDs. *Radiat. Prot. Dosim.*, **65**, 221–226.

Klemic, G., Shobe, J., Gesell, T. and Shebell, P. (1995) Results of the tenth international intercomparison of environmental dosemeters. *Radiat. Prot. Dosim.*, **58**, 133–142.

Klemic, G., Shobe, J., Sengupta, S. *et al.* (1999) State of the art of environmental dosimetry: 11th international intercomparison and proposed performance tests. *Radiat. Prot. Dosim.*, **95**, 201–206.

Knoll, G.F. (2000) *Radiation Detection and Measurements*, John Wiley & Sons, Inc.

Kobayashi, H., Satoh, M., Kobayashi, I. and Morishima, H. (2005) Neutron imaging using an optically stimulated luminescence material: $\alpha$-$Al_2O_3$:C+$Gd_2O_3$. *IEEE Trans. Nucl. Sci*, **52**, 360–363.

Kolberg, S., Prydz, S. and Dahm, S. (1974) Thermally stimulated luminescence in dental hard tissues and bone. *Calcif. Tissue Res.*, **17**, 9–23.

Kortov, V. (2007) Materials for thermoluminescent dosimetry: current status and future trends. *Radiat. Meas.*, **42**, 576–581.

Kortov, V.S., Milman, I.I. and Nikiforov, S.V. (1999) The effect of deep traps on the main features of thermoluminescence in dosimetric $\alpha$-$Al_2O_3$ crystals. *Radiat. Prot. Dosim.*, **84**, 35–38.

Koshy, M., Paulino, A.C., Marcus, R.B. *et al.* (2004) Extra-target doses in children receiving multileaf collimator (MLC) based intensity modulated radiation therapy (IMRT). *Pediatr. Blood Cancer*, **42**, 626–630

Krbetschek, M.R., Götze, J., Dietrich, A. and Trautmann, T. (1997) Spectral information from minerals relevant for luminescence dating. *Radiat. Meas.*, **27**, 695–748.

Krbetschek, M.R. and Rieser, U. (1995) Luminescence spectra of alkalifeldspars and plagioclases. *Radiat. Meas.*, **24**, 473–477.

Kron, T. (1999) Applications of thermoluminescence dosimetry in medicine. *Radiat. Prot. Dosim.*, **85**, 333–340.

Kron, T., DeWerd, L., Mobit, P.M., J. et al. (1999) A checklist for reporting of thermoluminescence dosimetry (TLD) measurements. *Phys. Med. Biol.*, **44**, L15–L17.

Kron, T., Hamilton, C., Roff, M. and Denham, J. (2002) Dosimetric intercomparison for two australasian clinical trials using an anthropomorphic phantom. *Int. J. Radiat. Oncol. Biol. Phys.*, **52**, 566–579.

Kry, S.F., Salehpour, M., Followill, D.S. *et al.* (2005) The calculated risk of fatal secondary malignancies from intensity-modulated radiation therapy. *Int. J. Radiat. Oncol. Biol. Phys.*, **62**, 1195–1203.

Kuhns, C.K., Agersnap Larsen, N. and McKeever, S.W.S. (2000) Characteristics of LM-OSL from several different types of quartz. *Radiat. Meas.*, **32**, 413–418.

Kurjewicz, L. and Berndt, A. (2007) Measurement of Gamma Knife helmet factors using MOSFETs. *Med. Phys.*, **34**, 1007–1012.

La, S.Y., Bartram, R.H. and Cox, R.T. (1973) The $F^+$ center in reactor-irradiated aluminum oxide. *J. Phys. Chem. Solids*, **34**, 1079–1086.

Lambert, J., Nakano, T., Law, S. *et al.* (2007) In vivo dosimeters for HDR brachytherapy: A comparison of a diamond detector, MOSFET, TLD, and scintillation detector. *Med. Phys.*, **34**, 1759–1765.

Lambert, J., Yin, Y., McKenzie, D.R. *et al.* (2008) Cerenkov-free scintillation dosimetry in external beam radiotherapy with an air core light guide. *Phys. Med. Biol.*, **53**, 3071–3080.

Landauer (2006) *Optically Stimulated Luminescence Dosimetry for Computed Tomography – Measurement of CTDI and dose Profiles*, Landauer Inc., Glenwood, Il.

Landauer (2008) *MicroStar User Manual*, Landauer Inc.

Lapraz, D., Prevost, H., Idri, K. *et al.* (2006) On the PL, TSL and OSL properties of SrS : Ce,Sm phosphor. *Phys. Status Solidi A*, **203**, 3793–3800.

Larsson, C., Koslowsky, V., Gao, H. *et al.* (2005) Optically stimulated luminescence in forensics. *Appl. Radiat. Isotopes*, **63**, 689–695.

Lawless, J.L., Chen, R. and Pagonis, V. (2009) Sublinear dose dependence of thermoluminescence and optically stimulated luminescence prior to the approach to saturation level. *Radiat. Meas.*, **44**, 606–610.

Le Masson, N.J.M., Bos, A.J.J., Czapla, Z. *et al.* (2004b) Fast-neutron OSL sensitivity of thallium-doped ammonium salts. *Radiat. Prot. Dosim.*, **110**, 319–323.

Le Masson, N.J.M., Bos, A.J.J. and Van Eijk, C.W.E. (2001) Optically stimulated luminescence in hydrated magnesium sulfates. *Radiat. Meas.*, **33**, 693–697.

Le Masson, N.J.M., Bos, A.J.J., Van Eijk, C.W.E. *et al.* (2002) Optically and thermally stimulated luminescence of $KMgF_3{:}Ce^{3+}$ and $NaMgF_3{:}Ce^{3+}$. *Radiat. Prot. Dosim.*, **100**, 229–234.

Le Masson, N.J.M., Bos, A.J.J., Winkelman, A.J.M. and van Eijk, C.W.E. (2001) Optically stimulated luminescence in $KMgF_3{:}Ce^{3+}$ Comparison of dosimetric characteristics with $Al_2O_3{:}C$. *IEEE Trans. Nucl. Sci*, **48**, 1143–1147.

Le Masson, N.J.M., Czapla, Z., Bos, A.J.J. *et al.* (2004a) Luminescence and OSL study of the inorganic compounds $Tl^+$-doped $(NH_4)_2BeF_4$ and $(NH_4)_2SiF_6$. *Radiat. Meas.*, **39**, 549–552.

Lee, C.K. and Chen, R. (1995) Explanation of the superlinear behavior of thermoluminescence by considering the residual holes in the recombination centres before irradiation. *J. Phys. D: Appl. Phys.*, **28**, 408–414.

Lee, K.H. and Crawford Jr., J.H. (1977) Electron centers in single-crystal $Al_2O_3$. *Phys. Rev. B*, **15**, 4065–4070.

Lee, K.H. and Crawford Jr., J.H. (1979) Luminescence of the F center in sapphire. *Phys. Rev. B*, **19**, 3217–3221.

Leverenz, H. (1949) Luminescence of solids. *Science*, **109**, 183–195.

Leverenz, H. (1950) *An introduction to Luminescence of Solids*, Dover, New York.

Li, B. and Li, S.H. (2006a) Comparison of $D_e$ estimates using the fast component and the medium component of quartz OSL. *Radiat. Meas.*, **41**, 125–136.

Li, S.H. and Li, B. (2006b) Dose measurement using the fast component of LM-OSL signals from quartz. *Radiat. Meas.*, **41**, 534–541.

Livingstone, J., Horowitz, Y.S., Oster, L. *et al.* (2009) Experimental investigation of the 100 keV X-ray dose response of the high-temperature thermoluminescence in LiF:Mg,Ti (TLD-100): theoretical interpretation using the unified interaction model. *Radiat. Prot. Dosim.*, **138**, 320–333.

Lomax, A. (1999) Intensity modulation methods for proton radiotherapy. *Phys. Med. Biol.*, **44**, 185–205.

Lopez Ponte, M.A., Castellani, C.M., Currivan, L. *et al.* (2004) A catalogue of dosemeters and dosimetric services within Europe – an update. *Radiat. Prot. Dosim.*, **112**, 45–68.

Low, D.A., Chao, K.S.C., Mutic, S. *et al.* (1998) Quality assurance of serial tomotherapy for head and neck patient treatments. *Int. J. Radiat. Oncol. Biol. Phys.*, 42, 681–692.

Lucas, A.C. and Kaspar, B.M. (1977) The Thermoluminescence of Thulium Doped Calcium Fluoride. Proceedings of the Fifth International Conference on Luminescence Dosimetry, São Paulo, Brazil, 1977, p. 131.

Lucovsky, G. (1965) On the photoionization of deep impurity centers in semiconductors. *Solid State Commun.*, **3**, 299–302.

Magne, S., Auger, L., Bordy, J.M. *et al.* (2008) Multichannel dosemeter and $Al_2O_3$:C optically stimulated luminescence fibre sensors for use in radiation therapy: evaluation with electron beams. *Radiat. Prot. Dosim.*, **131**, 93–99.

Mah, M.L., Manfred, M.E., Kim, S.S. *et al.* (2010) Measurement of rapid temperature profiles using thermoluminescent microparticles. *IEEE Sensors*, **10**, 311–315.

Mandeville, C.E. and Albrecht, H.O. (1953) The storage of energy in silver activated potassium chloride. *Phys. Rev.*, **91**, 566–567.

Mandeville, C.E. and Albrecht, H.O. (1954) Luminescence of beryllium oxide. *Phys. Rev.*, **94**, 494–495.

Marfunin, A.S. (1979) *Spectroscopy, Luminescence and Radiation Centers in Minerals*, Springer-Verlag, Berlin.

Markey, B.G., Bøtter-Jensen, L. and Duller, G.A.T. (1997) A new flexible system for measuring thermally stimulated luminescence. *Radiat. Meas.*, **27**, 89–93.

Markey, B.G., Bøtter-Jensen, L., Poolton, N.R.J. *et al.* (1996a) A new sensitive system for measurement of thermally and optically stimulated luminescence. *Radiat. Prot. Dosim.*, **66**, 413–418.

Markey, B.G., Colyott, L.E. and McKeever, S.W.S. (1995) Time-resolved optically stimulated luminescence from $\alpha$-$Al_2O_3$:C. *Radiat. Meas.*, **24**, 457–463.

Markey, B.G., McKeever, S.W.S., Akselrod, M.S. *et al.* (1996b) The temperature dependence of optically stimulated luminescence from $\alpha$-$Al_2O_3$:C. *Radiat. Prot. Dosim.*, **65**, 185–189.

Marshall, N.W., Faulkner, K. and Warren, H. (1996) Measured scattered x-ray energy spectra for simulated irradiation geometries in diagnostic radiology. *Med. Phys.*, **23**, 1271–1276.

Masalovich, S., Ioffe, A., Schlapp, M. *et al.* (2005) Optimization of a neutron image plate detector with low $\gamma$-sensitivity. *Nucl. Instrum. Methods Phys. Res. A*, 539.

Mathur, V.K., Barkyoumb, J.H., Yukihara, E.G. and Göksu, H.Y. (2007) Radiation sensitivity of memory chip module of an ID card. *Radiat. Meas.*, **42**, 43–48.

Mauricio, C.L.P., Rosa, L.A.R. and Cunha, P.G. (1985) Accident dosimetry using the TL from dental restoration ceramics. *Radiat. Prot. Dosim.*, **11**, 185–188.

McCaffrey, J.P. (2008) Dose rate dependency of electronic personal dosemeters measuring X- and gamma-ray radiation. *Radiat. Prot. Dosim.*, **131**, 229–235.

McCollough, C.H. (2008) CT dose: how to measure, how to reduce. *Health Phys.*, **95**, 508–517.

McCollough, C.H. and Schueler, B.A. (2000) Calculation of effective dose. *Med. Phys.*, **27**, 828–837.

McKeever, S.W.S. (1985) *Thermoluminescence of Solids*, Cambridge University Press, Cambridge.

McKeever, S.W.S. (1991) Mechanisms of thermoluminescence production – some problems and a few answers. *Nucl. Tracks Radiat. Meas.*, **18**, 5–12.

McKeever, S.W.S., Agersnap Larsen, N., Bøtter-Jensen, L. and Mejdahl, V. (1997) OSL sensitivity changes during single aliquot procedures: computer simulations. *Radiat. Meas.*, **27**, 75–82.

McKeever, S.W.S. and Akselrod, M.S. (1999) Radiation dosimetry using pulsed optically stimulated luminescence of $Al_2O_3$:C. *Radiat. Prot. Dosim.*, **84**, 317–320.

McKeever, S.W.S., Akselrod, M.S., Colyott, L.E. *et al.* (1999) Characterisation of $Al_2O_3$ for use in thermally and optically stimulated luminescence dosimetry. *Radiat. Prot. Dosim.*, **84**, 163–168.

McKeever, S.W.S., Akselrod, M.S. and Markey, B.G. (1996) Pulsed optically stimulated luminescence dosimetry using $\alpha$-$Al_2O_3$:C. *Radiat. Prot. Dosim.*, **65**, 267–272.

McKeever, S.W.S., Banerjee, D., Blair, M.W. *et al.* (2003) Concepts and approaches to in situ luminescence dating of martian sediments. *Radiat. Meas.*, **37**, 527–534.

McKeever, S.W.S., Blair, M.W., Bulur, E. *et al.* (2004) Recent advances in dosimetry using the optically stimulated luminescence of $Al_2O_3$:C. *Radiat. Prot. Dosim.*, **109**, 269–276.

McKeever, S.W.S., Blair, M.W., Yukihara, E.G. and DeWitt, R. (2009) The effect of low ambient temperatures on optically stimulated luminescence (OSL) processes: relevance to OSL dating of martian sediments. *Radiat. Meas.*, **45**, 60–70.

McKeever, S.W.S., Bøtter-Jensen, L., Agersnap Larsen, N. *et al.* (1996) Optically stimluated luminescence sensitivity changes in quartz due to repeated use in single aliquot readout: experiments and computer simulations. *Radiat. Prot. Dosim.*, **65**, 49–54.

McKeever, S.W.S., Chen, C.Y. and Halliburton, L.E. (1985) Point-defects and the predose effect in natural quartz. *Nucl. Tracks Radiat. Meas.*, **10**, 489–495.

McKeever, S.W.S., Gaza, R. and Yukihara, E.G. (2004) TL and OSL efficiencies of $Al_2O_3$:C and LiF:Mg,Ti detectos exposed during ICCHIBAN-2, in *Results from the First Two*

*Intercomparison of Dosimetric Instruments for Cosmic Radiation with Heavy Ion Beams at NIRS (ICCHIBAN-1 & 2) Experiments, HIMAC-078* (eds Y. Uchihori and E.R. Benton), National Institute of Radiological Sciences, Chiba, Japan.

McKeever, S.W.S. and Morris, M.F. (1994) Computer simulations of optical bleaching of TL and OSL signals. *Radiat. Meas.*, **23**, 301–306.

McKeever, S.W.S. and Moscovitch, M. (2003) On the advantages and disadvantages of optically stimulated luminescence dosimetry and thermoluminescence dosimetry. *Radiat. Prot. Dosim.*, **104**, 263–270.

McKeever, S.W.S., Moscovitch, M. and Townsend, P.D. (1995) *Thermoluminescence Dosimetry Materials: Properties and Uses*, Nuclear Technology Publishing, Ashford.

Meckbach, R., Bailiff, I.K., Göksu, H.Y. *et al.* (1996) Calculation and measurement of depth dose distributions in bricks. *Radiat. Prot. Dosim.*, **66**, 183–186.

Meeks, S.L., Paulino, A.C., Pennington, E.C. *et al.* (2002) In vivo determination of extra-target doses received from serial tomotherapy. *Radiother. Oncol.*, **63**, 217–222.

Merckle, L.D., Powell, R.C. and Wilson, T.M. (1978) Radiationless process in KCl:$Eu^{2+}$. *J. Phys. C: Solid State Phys.*, **11**, 3103–3119.

Mettler, F.A. and Voelz, G.L. (2002) Major radiation exposure – what to expect and how to respond. *New Engl. J. Med.*, **346**, 1554–1562.

Mewaldt, R.A. (1989) The abundances of isotopes in the cosmic radiation. *AIP Conf. Proc.*, **1831**, 124–146.

Miller, S.D. (1996) Composite material dosimeters. U.S. Patent, 5,567,948, issued on 22 October 1996.

Miller, S.D. (1998) Metal oxide composite dosimeter method and material. U.S. Patent, 5,731,590, issued on 24 March 1998.

Miller, D.L., Balter, S., Cole, P.E. *et al.* (2003a) Radiation doses in interventional radiology procedures: The RAD-IR Study – Part I: Overall measures of dose. *J. Vasc. Interv. Radiol.*, **14**, 711–727.

Miller, D.L., Balter, S., Cole, P.E. *et al.* (2003b) Radiation doses in interventional radiology procedures: The RAD-IR study – Part II: Skin dose. *J. Vasc. Interv. Radiol.*, **14**, 977–990.

Miller, S.D. and Murphy, M.K. (2007) Technical performance of the Luxel $Al_2O_3$:C optically stimulated luminescence dosemeter element at radiation oncology and nuclear accident dose levels. *Radiat. Prot. Dosim.*, **123**, 435–442.

Mische, E.F. and McKeever, S.W.S. (1989) Mechanisms of supralinearity in lithium fluoride thermoluminescence dosemeters. *Radiat. Prot. Dosim.*, **29**, 159–175.

Mishra, D.R., Kulkarni, M.S., Rawat, N.S. *et al.* (2008) Non-linear light modulation OSL phenomenon. *Radiat. Meas.*, **43**, 1177–1186.

Mittani, J.C., Prokić, M. and Yukihara, E.G. (2008) Optically stimulated luminescence and thermoluminescence of terbium-activated silicates and aluminates. *Radiat. Meas.*, **43**, 323–326.

Mittani, J.C.R., Silva, A.A.R.d., Vanhavere, F. *et al.* (2007) Investigation of neutron converters for production of optically stimulated luminescence (OSL) neutron dosimeters using $Al_2O_3$:C. *Nucl. Instrum. Methods Phys. Res. B*, **260**, 663–671.

Mobit, P., Agyingi, E. and Sandison, G. (2006) Comparison of the energy-response factor of LiF and $Al_2O_3$ in radiotherapy beams. *Radiat. Prot. Dosim.*, **119**, 497–499.

Mobit, P.N., Mayles, P. and Nahum, A.E. (1996) The quality dependence of LiF TLD in megavoltage photon beams: Monte Carlo simulation and experiments. *Phys. Med. Biol.*, **41**, 387–398.

Moscovitch, M. (1993) Dose algorithms for personal thermoluminescence dosimetry. *Radiat. Prot. Dosim.*, **47**, 373–380.

Moscovitch, M. (1999) Personnel dosimetry using LiF:Mg,Cu,P. *Radiat. Prot. Dosim.*, **85**, 49–56.

Moscovitch, M., Tawil, R.A. and Svinkin, M. (1993) Light induced fading in $\alpha$-$Al_2O_3$:C. *Radiat. Prot. Dosim.*, **47**, 251–253.

Mott, N.F. and Gurney, R.W. (1940) *Electronic Processes in Ionic Crystals*, Claredon Press, Oxford.

Moulder, J.E. (2004) Post-irradiation approaches to treatment of radiation injuries in the context of radiological terrorism and radiation accidents: a review. *Int. J. Radiat. Biol.*, **80**, 3–10.

Moyers, M.F. and Nelson, G.A. (2009) Dose response of $CaF_2$:Tm to charged particles of different LET. *Med. Phys.*, **36**, 3714–3723.

Muirhead, C.R. (2008) Exposure assessment: Implications for epidemiological studies of ionizing radiation. *Radiat. Prot. Dosim.*, **132**, 134–138.

Murray, A.S. and Wintle, A.G. (2000) Luminescence dating of quartz using an improved single-aliquot regenerative-dose protocol. *Radiat. Meas.*, **32**, 57–73.

Murray, A.S. and Wintle, A.G. (2002) Retrospective dose assessment: the measurement of the dose in quartz in dating and accident dosimetry. *Radiat. Prot. Dosim.*, **101**, 301–308.

Mutic, S. and Low, D.A. (1998) Whole-body dose from tomotherapy delivery. *Int. J. Radiat. Oncol. Biol. Phys.*, **42**, 229–232.

Nanjundaswamy, R., Lepper, K. and McKeever, S.W.S. (2002) Thermal quenching of thermoluminescence in natural quartz. *Radiat. Prot. Dosim.*, **100**, 305–308.

Nanto, H., Murayama, K., Usuda, T. *et al.* (1993) Optically stimulated luminescence in KCl:Eu single crystals. *Radiat. Prot. Dosim.*, **47**, 281–284.

Nascimento, L.F. and Hornos, Y.M.M. (2010) Proposal of a Brazilian accreditation program for personal dosimetry using OSL. *Radiat. Meas.*, **45**, 51–59.

Nath, R., Anderson, L.L., Luxton, G. *et al.* (1995) Dosimetry of interstitial brachytherapy sources – recommendations of the AAPM Radiation Therapy Commtitee Task Group No. 43. *Med. Phys.*, **22**, 209–234.

Nath, R., Anderson, L.L., Meli, J.A. *et al.* (1997) Code of practice for brachytherapy physics: Report of the AAPM Radiation Therapy Committee Task Group No. 56. *Med. Phys.*, **24**, 1557–1598.

NCRP (1989) *NCRP Report 98: Guidance on Radiation Received in Space Activities*, National Council on Radiation Protection and Measurements, Bethesda, MD.

NCRP (2000) *NCRP Report 132: Radiation Protection Guidance for Activities in Low-Earth Orbit*, National Council on Radiation Protection and Measurements, Bethesda, MD.

NCRP (2001) NCRP Report 138: Management of terrorist events involving radioactive material. National Council on Radiation Protection and Measurements, Bethesda.

NCRP (2002) NCRP Report 142: Operational Radiation Safety Program for Astronauts in Low-Earth Orbit: A Basic Framework. National Council on Radiation Protection and Measurements, Bethesda, Maryland.

NCRP (2009) NCRP Report 160: Ionizing Radiation Exposure of the Population of the United States. National Council on Radiation Protection and Measurements, Bethesda, Maryland.

Nichols, E.L. and Merrit, E. (1912) *Studies in Luminescence*, Carnegie Institute of Washington, Washington D.C.

Niedermayer, M. and Bayer, A. (2010) Pulsed photon-stimulated luminescence (PPSL) as a tool in retrospective dosimetry. *Radiat. Prot. Dosim.*, **139**, 494–504.

Niroomand-Rad, A., Blackwell, C.R., Coursey, B.M. *et al.* (1998) Radiochromic film dosimetry: Recommendations of AAPM Radiation Therapy Committee Task Group 55. *Med. Phys.*, **25**, 2093–2115.

Nisbet, A., Beange, I., Vollmar, H.S. *et al.* (2004) Dosimetric verification of a commercial collapsed cone algorithm in simulated clinical situations. *Radiother. Oncol.*, **73**, 79–88.

NRC (2006) Space radiation hazards and the vision for space exploration. Report of a Workshop on the Solar System Radiation Environment and NASA's Vision for Space Exploration, National Research Council of the National Academies, Wintergreen, VA, October 2005, The National Academies Press, Washington DC.

NRC (2008) Managing space radiation risk in the new era of space exploration. Report of the Committee on the Evaluation of Radiation Shielding for Space Exploration, Aeronautics and Space Engineering Board, National Research Council of the National Academies, The National Academies Press, Washington DC.

Ohuchi, H. and Yamadera, A. (2002) Possible application of an imaging plate to space radiation dosimetry. *J. Radiat. Res.*, **43**, S71–S74.

Ohuchi, H., Yamadera, A. and Baba, M. (2003) Development of a new passive integral dosemeter for gamma ray monitoring using an imaging plate. *Radiat. Prot. Dosim.*, **107**, 239–246.

Olko, P. (2002) Microdosimetric modelling of physical and biological detectors. Habilitation Thesis, The Henryk Niewodniczański Institute of Nuclear Physics.

Olko, P., Currivan, L., Van Dijk, J.W.E. *et al.* (2006b) Thermoluminescent detectors applied in individual monitoring of radiation workers in Europe – a review based on the EURADOS questionnaire. *Radiat. Prot. Dosim.*, **120**, 298–302.

Olko, P., Marczewska, B., Czopyk, L. *et al.* (2006a) New 2-D dosimetric technique for radiotherapy based on planar thermoluminescent detectors. *Radiat. Prot. Dosim.*, **118**, 213–218.

Oster, L., Weiss, D. and Kristianpoller, N. (1994) A study of photostimulated thermoluminescence in C-doped $\alpha$-$Al_2O_3$ crystals. *J. Phys. D: Appl. Phys.*, **27**, 1732–1736.

Packman, S.C., Mauz, B., Rousseau, D.D. *et al.* (2007) Implications of broad dose distributions obtained with the single-aliquot regenerative-dose method on quartz fine-grains from loess. *Quat. Geochronol.*, **2**, 39–44.

Pagonis, V., Chen, R. and Lawless, J.L. (2006) Nonmonotonic dose dependence of OSL intensity due to competition during irradiation and readout. *Radiat. Meas.*, **41**, 903–909.

Pagonis, V., Lawless, J., Chen, R. and Andersen, C. (2009) Radioluminescence in $Al_2O_3$:C - analytical and numerical simulation results. *J. Phys. D.: Appl. Phys.*, 42, 175107.

Pai, S., Das, I.J., Dempsey, J.F. *et al.* (2007) TG-69: Radiographic film for megavoltage beam dosimetry. *Med. Phys.*, **34**, 2228–2258.

Pantelides, S.T. (1978) The electronic structure of impurities and other point defects in semiconductors. *Rev. Mod. Phys.*, **50**, 797–858.

Park, C.C., Yom, S.S., Podgorsak, M.B. *et al.* (2010) American Society for Therapeutic Radiology and Oncology (ASTRO) Emerging Technology Committee Report on Electronic Brachytherapy. *Int. J. Radiat. Oncol. Biol. Phys.*, **76**, 963–972.

Pawlicki, T., Luxton, G., Le, Q.T. *et al.* (2004) Lens dose in MLC-based IMRT treatments of the head and neck. *Int. J. Radiat. Oncol. Biol. Phys.*, **59**, 293–299.

Peakheart, D.W. (2006) Evaluation of clinically feasible dosimetry systems for CT quality assurance and dose optimization. M. S. Thesis, University of Oklahoma Health Sciences Center.

Perks, C.A., Le Roy, G. and Prugnaud, B. (2007) Introduction of the InLight Monitoring Service. *Radiat. Prot. Dosim.*, **125**, 220–223.

Perks, C.A., LeRoy, G., Yoder, C. and Passmore, C. (2004) Development of the InLight$^{TM}$ Monitoring Service for World-Wide Application. 11th International Congress of the International Radiation Protection Association, Madrid.

Perks, C.A., Faugoin, S., Passmore, C.N. and Million, M. (2008) Developments in Neutron Dosimetry. 2008 ISOE European Symposium, Turku (Finland).

Piesch, E. and Burgkhardt, B. (1985) Albedo neutron dosimetry. *Radiat. Prot. Dosim.*, **10**, 175–188.

Piesch, E. and Burgkhardt, B. (1988) Albedo dosimetry system for routine personnel monitoring. *Radiat. Prot. Dosim.*, **23**, 117–120.

de Planque, G. and Gesell, T.F. (1986) Environmental measurements with thermoluminescence dosemeters – trends and issues. *Radiat. Prot. Dosim.*, **18**, 193–200.

Plattard, D., Ranchoux, G., Dusseau, L. *et al.* (2002) Characterization of an integrated sensor using optically stimulated luminescence for in-flight dosimetry. *IEEE Trans. Nucl. Sci*, **49**, 1322–1326.

Podgorsak, E.B. (2005a) Chapter 5: Treatment machines for external beam radiotherapy, in *Radiation Oncology Physics: A Handbook for Teachers and Students* (ed. E.B. Podgorsak), International Atomic Energy Agency, Vienna.

Podgorsak, E.B. (ed.), (2005b) *Radiation Oncology Physics: A Handbook for Teachers and Students*, International Atomic Energy Agency, Vienna.

Podgorsak, E.B. and Podgorsak, M.B. (2005) Chapter 15: Special procedures and techniques in radiotherapy, in *Radiation Oncology Physics: A Handbook for Teachers and Students* (ed. E.B. Podgorsak), International Atomic Energy Agency, Vienna.

Pogatshnik, G.J., Chen, Y. and Evans, B.D. (1987) A model of lattice defects in sapphire. *IEEE Trans. Nucl. Sci*, **NS-34**, 1709–1711.

Pohl, R.W. (1939) Electron conductivity and photochemical processes in alkali-halide crystals. *Proc. Phys. Soc.*, **49**, 3–31.

Polf, J.C., McKeever, S.W.S., Akselrod, M.S. and Holmstrom, S. (2002) A real-time, fibre optic dosimetry system using $Al_2O_3$ fibres. *Radiat. Prot. Dosim.*, **100**, 301–304.

Polf, J.C., Yukihara, E.G., Akselrod, M.S. and McKeever, S.W.S. (2004) Real-time luminescence from $Al_2O_3$ fiber dosimeters. *Radiat. Meas.*, **38**, 227–240.

Polymeris, G.S., Afouxenidis, D., Tsirliganis, N.C. and Kitis, G. (2009) The TL and room temperature OSL properties of the glow peak at 110 degrees C in natural milky quartz: A case study. *Radiat. Meas.*, **44**, 23–31.

Poochinda, K., Chen, T.C., Stoebe, T.G. and Ricker, N.L. (2004) Structural, optical and electrical properties of GaN and InGaN films grown by MOCVD. *J. Cryst. Growth*, **272**, 460–465.

Poolton, N.R.J., Bøtter-Jensen, L. and Jungner, H. (1995) An optically stimulated luminescence study of porcelain related to radiation dosimetry. *Radiat. Meas.*, **24**, 543–549.

Poolton, N.R.J., Kars, R.H., Wallinga, J. and Bos, A.J.J. (2009) Direct evidence for the participation of band-tails and excited-state tunneling in the luminescence of irradiated feldspars. *J. Phys.: Condens. Matter*, **21**, 485505.

Poolton, N.R.J., Ozanyan, K.B., Wallinga, J. *et al.* (2002a) Electrons in feldspar II: a consideration of the influence of conduction band-tail states on luminescence processes. *Phys. Chem. Minerals*, **29**, 217–225.

Poolton, N.R.J., Wallinga, J., Murray, A.S. *et al.* (2002b) Electrons in feldspar I: on the wavefunction of electrons trapped at simple lattice defects. *Phys. Chem. Minerals*, **29**, 210–216.

Pradhan, A.S. and Ayyanger, K. (1977) Radiation dosimetry by photostimulated luminescence of $CaSO_4$:Dy. *Int. J. Appl. Radiat. Isot.* 28.

Pradhan, A.S. and Bhatt, R.C. (1981) Photostimulated luminescence and thermoluminescence in $CaSO_4$:Dy. *Phys. Status Solidi A*, **68**, 405–411.

Pradhan, A.S., Lee, J.I. and Kim, J.L. (2008) Recent developments of optically stimulated luminescence materials and techniques for radiation dosimetry and clinical applications. *J. Med. Phys.*, **33**, 85–99.

PTCOG (2010) Particle Therapy Co-Operative Group. http://ptcog.web.psi.ch/ (accessed on 18 May 2010).

Pustovarov, V.A., Ivanov, V.Y., Kirm, M. *et al.* (2001) Time-resolved luminescent VUV spectroscopy of F- and $F^+$-centres in single BeO crystals. *Nucl. Instrum. Methods Phys. Res. A*, **470**, 353–357.

Ramzaev, V., Bøtter-Jensen, L., Thomsen, K.J. *et al.* (2008) An assessment of cumulative external doses from Chernobyl fallout for a forested area in Russia using optically stimulated luminescence from quartz inclusions in bricks. *J. Environ. Radioact.*, **99**, 1154–1164.

Ranchoux, G., Magne, S., Bouvet, J.P. and Ferdinand, P. (2002) Fibre remote optoelectronic gamma dosimetry based on optically stimulated luminescence of $Al_2O_3$:C. *Radiat. Prot. Dosim.*, **100**, 255–260.

Ranchoux, G., Plattard, D., Polge, G. *et al.* (2001) Proton dosimetry measurements and calculations in electronic packages using optically stimulated luminescent films. *IEEE Trans. Nucl. Sci*, **48**, 1731–1734.

Randall, J.T. and Wilkins, M.H.F. (1945) Phosphorescence and Electron Traps. I. The Study of Trap Distributions. *Proc. Roy. Soc. A*, **184**, 365–389.

Ravotti, F., Benoit, D., Lefebvre, P. *et al.* (2007) Time-resolved photoluminescence and optically stimulated luminescence measurements of picosecond-excited SrS:Ce,Sm phosphor. *J. Appl. Phys.*, **102**, 123102.

Reames, D.V. (1999) Particle acceleration at teh Sun and in the heliosphere. *Space Sci. Rev.*, **90**, 413–491.

Reames, D.V. (2004) Solar energetic particle variations. *Adv. Space Res.*, **34**, 381–390.

Reft, C.S. (2009) The energy dependence and dose response of a commercial optically stimulated luminescence detector for kilovoltage photon, megavoltage photon, and electron, proton, and carbon beams. *Med. Phys.*, **36**, 1690–1699.

Reft, C.S., Runkel-Muller, R. and Myrianthopoulos, L. (2006) In vivo and phantom measurements of the secondary photon and neutron doses for prostate patients undergoing 18 MV IMRT. *Med. Phys.*, **33**, 3734–3742.

Reitz, G. (2006) Past and future application of solid-state detectors in manned spacecraft. *Radiat. Prot. Dosim.*, **120**, 387–396.

Reitz, G., Berger, T., Bilski, P. *et al.* (2009) Astronaut's organ doses inferred from measurements in a human phantom outside the International Space Station. *Radiat. Res.*, **171**, 225–235.

Rhyner, C.R. and Miller, W.G. (1970) Radiation dosimetry by optically stimulated luminescence in BeO. *Health Phys.*, **18**, 681–684.

Rieser, U., Habermann, J. and Wagner, G.A. (1999) Luminescence dating: a new high sensitivity TL/OSL emission spectrometer. *Quat. Geochronol.*, **18**, 311–315.

Rittenour, T.M., Goble, R.J. and Blum, M.D. (2005) Development of an OSL chronology for Late Pleistocene channel belts in the lower Mississippi valley, USA. *Quat. Sci. Rev.*, **24**, 2539–2554.

Rivera, T., Azorin, J., Furetta, C. *et al.* (2003) Continuous wavelength and linear modulation optically stimulated luminescence characteristics of beta-irradiated ZrO2. *Nucl. Instr. Meth. Phys. Res.* A, **514**, 146–149.

Robbins, D.E. (1997) *The Space Radiation Environment*. Symposium on Acceptability of Risk from Radiation – Application to Human Space Flight, Arlington, VA, 1996, National Council on Radiation Protection and Measurements, Symposium Proceedings No. 3, pp. 5–31.

Rodnyi, P.A. (1997) *Physical Processes in Inorganic Scintillators*, CRC Press, Boca Raton.

Rodrigues, P., Trindade, A., Peralta, L. *et al.* (2004) Application of GEANT4 radiation transport toolkit to dose calculations in anthropomorphic phantoms. *Appl. Radiat. Isotopes*, **61**, 1451–1461.

Rodríguez-Villafuerte, M., Buenfil, A.E., Gamboa-deBuen, I. *et al.* (2000) Study of the TL response of LiF:Mg,Ti to 3 and 7.5 MeV helium ions: Measurements and interpretation in terms of the track interaction model. *Nucl. Instrum. Methods Phys. Res. B*, **160**, 377–386.

Rogers, D.W.O. (2009) General characteristics of radiation dosimeters and a terminology to describe them, in *Clinical Dosimetry Measurements in Radiotherapy* (eds D.W.O. Rogers and J.E. Cygler), Medical Physics Publishing, Madison, WI, 137–145.

Rogers, D.W.O. and Cygler, J.E. (eds), (2009) *Clinical Dosimetry Measurements in Radiotherapy*, Medical Physics Publishing, Madison, Wisconsin.

Rossi, H.H. and Zaider, M. (1996) *Microdosimetry and Its Applications*, Springer-Verlag, Germany.

Rowlands, J.A. (2002) The physics of computed radiography. *Phys. Med. Biol.*, **47**, R123-R166.

Ruan, C., Yukihara, E.G., Clouse, W.J. *et al.* (2010) Determination of multi-slice computed tomography dose index (CTDI) using optically stimulated luminescence technology. *Med. Phys.*, **37**, 3560–3568.

Sanborn, E.N. and Beard, E.L. (1967) *Sulfides of Strontium, Calcium, and Magnesium in Infrared-Stimulated Luminescence Dosimetry*. Proceedings of First International Conference on Luminescence Dosimetry, Stanford, June, 1965, 183–191.

Sawakuchi, G.O. (2007) Characterization and modeling of relative luminescence efficiency of optically stimulated luminescence detectors exposed to heavy charged particles. Ph. D. Dissertation, Oklahoma State University.

Sawakuchi, G.O., Sahoo, N., Gasparian, P.B.R. *et al.* (2010) Determination of average LET of therapeutic proton beams using $Al_2O_3$:C optically stimulated luminescence (OSL) detectors. *Phys. Med. Biol.*, **55**, 4963–4976.

Sawakuchi, G.O., Yukihara, E.G., McKeever, S.W.S. and Benton, E.R. (2008a) Overlap of heavy charged particle tracks and the change in shape of optically stimulated luminescence curves of $Al_2O_3$:C dosimeters. *Radiat. Meas.*, **43**, 194–198.

Sawakuchi, G.O., Yukihara, E.G., McKeever, S.W.S. and Benton, E.R. (2008b) Optically stimulated luminescence fluence response of $Al_2O_3$:C dosimeters exposed to different types of radiation. *Radiat. Meas.*, **43**, 450–454.

Sawakuchi, G.O., Yukihara, E.G., McKeever, S.W.S. and Benton, E.R. (2008c) Relative optically stimulated luminescence and thermoluminescence efficiencies of $Al_2O_3$:C dosimeters to heavy charged particles with energies relevant to space and radiotherapy dosimetry. *J. Appl. Phys.*, **104**, 124903.

Sawyer, D.M. and Vette, J.I. (1976) The AP-8 trapped proton environment for solar maximum and solar minimum. NSSDC/WDC-A-R&S 76-06, National Space Science Data Center, National Aeronautics and Space Administration, Greenbelt, Maryland, see also http://www.spenvis.oma.be/spenvis/help/background/traprad/traprad.html#TRENV (accessed 5 May 2010).

Scarpa, G. (1970) Dosimetric use of beryllium oxide as a thermoluminescent material – a preliminary study. *Phys. Med. Biol.*, **15**, 667-&.

Schardt, D. and Heavy-Ion Therapy Collaboration (2007) Tumor therapy with high-energy carbon ion beams. *Nucl. Phys. A*, **787**, 633C-641C.

Schembri, V. and Heijmen, B.J.M. (2007) Optically stimulated luminescence (OSL) of carbon-doped aluminum oxide ($Al_2O_3$:C) for film dosimetry in radiotherapy. *Med. Phys.*, **34**, 2113–2118.

Schöner, W., Vana, N. and Fogger, M. (1999) The LET dependence of LiF:Mg,Ti dosemeters and its implication for LET measurements in mixed radiation fields. *Radiat. Prot. Dosim.*, **85**, 263–266.

Schulman, J.H., Ginther, R.J., Klick, C.C. *et al.* (1951) Dosimetry of x-rays and gamma-rays by radiophotoluminescence. *J. Appl. Phys.*, **22**, 1479–1487.

von Seggern, H. (1992) X-ray imaging with photostimulable phosphors. *Nucl. Instrum. Methods Phys. Res. A*, **322**, 467–471.

von Seggern, H. (1999) Photostimulable X-ray storage phosphors: a review of present understanding. *Braz. J. Phys*, **29**, 254–268.

Shachar, B.B. and Horowitz, Y.S. (1988) Dosimetric characteristics of the high-temperature peaks in LiF:Mg,Ti and $CaF_2$:Tm using computerized glow curve deconvolution. *Radiat. Prot. Dosim.*, **22**, 87–96.

Shah, M.R. (2009) Development and characterization of neutron-sensitive optically stimulated luminescence (OSL) detectors. M. S. thesis, Oklahoma State University.

Shi, J., Klocke, A., Zhang, M. and Bismayer, U. (2005) Thermally-induced structural modification of dental enamel apatite: decomposition and transformation of cabonate groups. *Eur. J. Mineral.*, **17**, 769–775.

Shimano, T., Iwasaki, M., Miyazawa, C. *et al.* (1989) Human tooth dosimetry for gamma-rays and dental X-rays using ESR. *Appl. Radiat. Isotopes*, **40**, 1035–1038.

Shope, T.B., Gagne, R.M. and Johnson, G.C. (1981) A method for describing the doses delivered by transmission x-ray computed tomography. *Med. Phys.*, **8**, 488–495.

Simon, S.L., Bailiff, I.K., Bouville, A. *et al.* (2007) BiodosEPR-2006 consensus committee report on biodosimetric methods to evaluate radiation doses at long times after exposure. *Radiat. Meas.*, **42**, 948–971.

Simon, S.L., Kleinerman, R.A., Ron, E. and Bouville, A. (2006) Uses of dosimetry in radiation epidemiology. *Radiat. Res.*, **166**, 125–127.

Simpson, J.A. (1983) Elemental and isotopic composition of the galactic cosmic rays. *Ann. Rev. Nucl. Part. S.*, **33**, 323–381.

Singarayer, J.S. and Bailey, R.M. (2003) Further investigations of the quartz optically stimulated luminescence components using linear modulation. *Radiat. Meas.*, **37** 451–458.

Singarayer, J.S. and Bailey, R.M. (2004) Component-resolved bleaching spectra of quartz optically stimulated luminescence: preliminary results and implications for dating. *Radiat. Meas.*, **38**, 111–118.

Smetana, F., Hajek, M., Bergmann, R. *et al.* (2008) A portable multi-purpose OSL reader for UV dosimetry at workplaces. *Radiat. Meas.*, **43**, 516–519.

Smith, B.W., Aitken, M.J., Rhodes, E.J. *et al.* (1986) Optical dating: methodological aspects. *Radiat. Prot. Dosim.*, **17**, 229–233.

Sommer, M., Freudenberg, R. and Henniger, J. (2007) New aspects of a BeO-based optically stimulated luminescence dosimeter. *Radiat. Meas.*, **42**, 617–620.

Sommer, M. and Henniger, J. (2006) Investigation of a BeO-based optically stimulated luminescence dosimeter. *Radiat. Prot. Dosim.*, **119**, 394–397.

Sommer, M., Jahn, A. and Henniger, J. (2008) Beryllium oxide as optically stimulated luminescence dosimeter. *Radiat. Meas.*, **43**, 353–356.

Sonoda, M., Takano, M., Miyahara, J. and Kato, H. (1983) Computed radiography utilizing scanning laser stimulated luminescence. *Radiology*, **148**, 833–838.

Spooner, N.A. (1994) On the optical dating signal from quartz. *Radiat. Meas.*, **23**, 593–600.

Spurny, Z., 1980. *Some new materials for TLD. Nucl. Instrum. Methods*, **175**, 71–73.

Spurny, F., Králík, M., Médioni, R. and Portal, G. (1976) A new thermoluminescent detector for 14.7 MeV neutrons. *Nucl. Instrum. Methods*, **137**, 593–594.

Stanford, N. and McCurdy, D.E. (1990) A single TLD dose algorithm to satisfy federal standards and typical field conditions. *Health Phys.*, **58**, 691–704.

Stoebe, T.G., Chen, T.C., Smith, K.R. *et al.* (1996) Solid state integrated radiation sensor development. *Radiat. Prot. Dosim.*, **66**, 427–429.

Stoneham, A.M. (1975) *Theory of Defects in Solids*, Oxford University Press, Oxford.

Straube, U., Berger, T., Reitz, G. *et al.* (2010) Operational radiation protection for astronauts and cosmonauts and correlated activities of ESA Medical Operations. *Acta Astronaut.*, **66**, 963–973.

Suchowerska, N., Lambert, J., Nakano, T. *et al.* (2007) A fibre optic dosimeter customised for brachytherapy. *Radiat. Meas.*, **42**, 929–932.

Sugano, S., Tanabe, Y. and Kamikura, H. (1970) *Multiplets of Transition-Metal Ions in Crystals*, Academic Press, New York.

Summers, G.P. (1984) Thermoluminescence in single crystal $\alpha$-$Al_2O_3$. *Radiat. Prot. Dosim.*, **8**, 69–80.

Sunta, C.M., Ayta, W.E.F., Chubaci, J.F.D. and Watanabe, S. (2005) A critical look at the kinetic models of thermoluminescence – II. Non-first order kinetics. *J. Phys. D: Appl. Phys.*, **38**, 95–102.

Sunta, C.M., Yoshimura, E.M. and Okuno, E. (1994) Supralinearity and sensitization of thermoluminescence. Part I: a theoretical treatment based on an interactive trap system. *J. Phys. D: Appl. Phys.*, **27**, 852–860.

Svelto, O. (1998) *Principles of Lasers*, Springer, New York.

Swartz, H.M., Burke, G., Coey, M. *et al.* (2007) In vivo EPR for dosimetry. *Radiat. Meas.*, **42**, 1075–1084.

Sykora, G.J. (2010) Photo- and radiochromic transformations in $Al_2O_3$:C,Mg fluorescent nuclear track detectors and high resolution imaging of radiation fields. Ph. D. Dissertation, Oklahoma State University.

Sze, S.M. and Ng, K.K. (2007) *Physics of Semiconductor Devices*, John Wiley & Sons, Inc., Hoboken, New Jersey.

Tailor, R., Tolani, N. and Ibbott, G.S. (2008) Thermoluminescence dosimetry measurements of brachytherapy sources in liquid water. *Med. Phys.*, **35**, 4063–4069.

Takahashi, K., Kohda, K. and Miyahara, J. (1984) Mechanism of photostimulated luminescennce in BaFX:$Eu^{2+}$ (X=Cl, Br) phosphors. *J. Lumin.*, **31 & 32**, 266–268.

Tello, V.M., Tailor, R.C. and Hanson, W.F. (1995) How water equivalent are water-equivalent solid materials for output calibration of photon and electron beans? *Med. Phys.*, **22**, 1177–1189.

Thoms, M., von Seggern, H. and Winnacker, A. (1991) Spatial correlation and photostimulability of defects centers in the x-ray-storage phosphor BaFBr:$Eu^{2+}$. *Phys. Rev. B*, **44**, 9240–9247.

Thomsen, K.J., Bøtter-Jensen, L., Denby, P.M. *et al.* (2006) Developments in luminescence measurement techniques. *Radiat. Meas.*, **41**, 768–773.

Thomsen, K.J., Botter-Jensen, L., Jain, M. *et al.* (2008) Recent instrumental developments for trapped electron dosimetry. *Radiat. Meas.*, **43**, 414–421.

Thomsen, K.J., Bøtter-Jensen, L. and Murray, A.S. (2002) Household and workplace chemicals as retrospective luminescence dosimeters. *Radiat. Prot. Dosim.*, **101**, 515–518.

Thomsen, K.J., Bøtter-Jensen, L., Murray, A.S. and Solongo, S. (2002) Retrospective dosimetry using unheated quartz: a feasibility study. *Radiat. Prot. Dosim.*, **101**, 345–348.

Thomsen, K.J., Jain, M., Bøtter-Jensen, L. *et al.* (2003) Variation with depth of dose distributions in single grains of quartz extracted from an irradiated concrete block. *Radiat. Meas.*, **37**, 315–321.

Thomsen, K.J., Murray, A.S. and Bøtter-Jensen, L. (2005) Sources of variability in OSL dose measurements using single grain of quartz. *Radiat. Meas.*, **39**, 47–61.

Thwaites, D.I., Mijnheer, B.J. and Mills, J.A. (2005) Chapter 12: Quality assurance of external beam radiotherapy, in *Radiation Oncology Physics: A Handbook for Teachers and Students* (ed. E.B. Podgorsak), International Atomic Energy Agency, Vienna.

Tinkler, L., Bøtter-Jensen, L., Christensen, P. and Berzina, B. (2000) Studies of aluminum nitride ceramics for application in UV dosimetry. *Radiat. Prot. Dosim.*, **92**, 299–306.

Tochilin, E., Goldstein, N. and Miller, W.G. (1969) Beryllium oxide as a thermoluminescent dosimeter. *Health Phys.*, **16**, 1–7.

Townsend, P.D. and Rowlands, J.A. (1999) Extended defect models for thermoluminescence. *Radiat. Prot. Dosim.*, **84**, 7–12.

Trompier, F., Kornak, L., Calas, C. *et al.* (2007) Protocol for emergency EPR dosimetry in fingernails. *Radiat. Meas.*, **42**, 1085–1088.

Tsujii, H., Mizoe, J., Kamada, T. *et al.* (2007) Clinical results of carbon ion radiotherapy at NIRS. *J. Radiat. Res.*, **48**, A1-A13.

Uchihori, Y. and Benton, E.R. (2004) Results from the first two interconparison of dosimetric instruments for cosmic radiation with heavy ions beams at NIRS (ICCHIBAN-1&2) experiments. Report HIMAC-078, National Institute of Radiological Sciences, Chiba, Japan.

Uchihori, Y. and Benton, E.R. (2008) Results from the ICCHIBAN-3 and ICCHIBAN-4 experiments to intercompare the response of space radiation dosimeters. Report HIMAC-128, National Institute of Radiological Sciences, Chiba, Japan.

Uchihori, Y., Benton, E.R., Yasuda, N. *et al.* (2006) *Status of the NSRL-ICCHIBAN, ICCHIBAN-7 AND -8 and Future ICCHIBAN Experiments*. 36th COSPAR Scientific Assembly, Beijing, 16–23 July 2006.

Uchihori, Y., Fujitaka, K., Yasuda, N., Benton, E.R. and the ICCHIBAN Collaboration (2002) Intercomparison of radiation instruments for cosmic rays with heavy ion beams at NIRS (ICCHIBAN project). *J. Radiat. Res.*, **43**, S81-S85.

Umisedo, N.K., Yoshimura, E.M., Gasparian, P.B.R. and Yukihara, E.G. (2010) Comparison between blue and green stimulated luminescence of $Al_2O_3$:C. *Radiat. Meas.*, **45**, 151–156.

United Nations (2000) Sources and Effects of Ionizing Radiation. Report to the General Assembly, United Nations Scientific Committee on the Effects of Atomic Radiation (UNSCEAR), United Nations, New York.

Vaillé, J.-R., Ducret, S., Idri, K. *et al.* (2003) Hardening of a radiation sensor based on optically stimulated luminescence. *IEEE Trans. Nucl. Sci*, **50**, 2358.

Vaillé, J.-R., Ravotti, F., Garcia, P. *et al.* (2005) Online dosimetry based on optically stimulated luminescence materials. *IEEE Trans. Nucl. Sci*, **52**, 2578–2582.

Van Dam, J. and Marinello, G. (2006) *Methods for in Vivo Dosimetry in External Radiotherapy*, ESTRO, Brussels.

Van Dyk, J., Barnett, R.B., Cygler, J.E. and Shragge, P.C. (1993) Comissioning and quality assurance of treatment planning computers. *Int. J. Radiat. Oncol. Biol. Phys.*, **26**, 261–273.

Vanhavere, F., Genicot, J.L., O'Sullivan, D. *et al.* (2008) DOsimetry of BIological EXperiments in SPace (DOBIES) with luminescence (OSL and TL) and track etch detectors. *Radiat. Meas.*, **43**, 694–697.

Vano, E., Gonzalez, L., Fernandez, J.M. and Haskal, Z.J. (2008) Eye lens exposure to radiation in interventional suites: Caution is warranted. *Radiology*, **248**, 945–953.

Verellen, D., Linthout, N., Berge, D.V.d. *et al.* (1997) Initial experience with intensity-modulated conformal radiation therapy for treatment of the head and neck region. *Int. J Radiat. Oncol. Biol. Phys.*, **39**, 99–114.

Verellen, D. and Vanhavere, F. (1999) Risk assessment of radiation-induced malignancies based on whole-body equivalent dose estimates for IMRT treatment in teh head and neck region. *Radiother. Oncol.*, **53**, 199–203.

Veronese, I., Galli, A., Cantone, M.C. *et al.* (2010) Study of TSL and OSL properties of dental ceramics for accidental dosimetry applications. *Radiat. Meas.*, **45**, 35–41.

Vette, J.I. (1976) The AP-8 trapped electron model environment. NSSDC/WDC-A-R&S 91-24, National Space Science Data Center, National Aeronautics and Space

Administration, Greenbelt, Maryland, see also http://www.spenvis.oma.be/spenvis/help/background/traprad/traprad.html#TRENV (accessed 5 May 2010).

Viamonte, A., da Rosa, L.A.R., Buckley, L.A. *et al.* (2008) Radiotherapy dosimetry using a commercial OSL system. *Med. Phys.*, **35**, 1261–1266.

Walker, F.D., Colyott, L.E., Agersnap Larsen, N. and McKeever, S.W.S. (1996) The wavelength dependence of light-induced fading of thermoluminescence from $\alpha$-$Al_2O_3$:C. *Radiat. Meas.*, **26**, 711–718.

Wallack, S. (1959) Dosimeter. U.S. Patent, 2,902,605.

Wang, X.L., Lu, Y.C. and Zhao, H. (2006) On the performances of the single-aliquot regenerative-dose (SAR) protocol for Chinese loess: fine quartz and polymineral grains. *Radiat. Meas.*, **41**, 1–8.

Waselenko, J.K., MacVittie, J., Blakely, W.F. *et al.* (2004) Medical management of the Acute Radiation Syndrome: Recommendations of the Strategic National Stockpile Radiation Working Group. *Ann. Intern. Med.*, **140**, 1037–1051.

Weiss, D., Horowitz, Y.S. and Oster, L. (2009) Ionization density effects following F-centre optical excitation in LiF:Mg,Ti (TLD-100): analysis via track structure theory. *J. Phys. D: Appl. Phys.*, **42**, 085113.

West, W.G., Kearfott, K.J. and Bernal, S.M. (2006) The sunlight OSL response of a commercially available $\alpha$-$Al_2O_3$:C personnel dosimetry material. *Radiat. Prot. Dosim.*, **119**, 344–349.

Whitley, V.H. and McKeever, S.W.S. (2001) Linearly modulated photoconductivity and linearly modulated optically stimulated luminescence measurements on $Al_2O_3$:C. *J. Appl. Phys.*, **90**, 6073–6083.

Whitley, V.H. and McKeever, S.W.S. (2002) Linear modulation optically stimulated luminescence and thermoluminescence techniques in $Al_2O_3$:C. *Radiat. Prot. Dosim.*, **100**, 61–66.

Wiedemann, E. (1889) Zurmechanik des leuchtens. *Ann. Phys-Berlin*, **37**, 177–248.

Wiedemann, E. and Schmidt, G.C. (1895) Überluminescenz. *Ann. Phys-Berlin*, **54**, 604–625.

Wilkens, J.J. and Oelfke, U. (2008) Direct comparison of biologically optimized spread-out Bragg peaks for protons and carbon ions. *Int. J. Radiat. Oncol. Biol. Phys.*, **70**, 262–266.

Willhoit, D.G. and Poland, A.D. (1968) Thermoluminescent characteristics of irradiated enamel and dentin. *Health Phys.*, **15**, 91–93.

Williams, B.B., Sucheta, A., Dong, R. *et al.* (2007) Experimental procedures for sensitive and reproducible in situ EPR tooth dosimetry. *Radiat. Meas.*, **42**, 1094–1098.

Williams, P.C. (2003) IMRT: delivery techniques and quality assurance. *Br. J. Radiol.*, **76**, 766–776.

Wilson, J.W., Cucinotta, F.A., Shinn, J.L. *et al.* (1999) Shielding from solar particle event exposures in deep space. *Radiat. Meas.*, **30**, 361–383.

Wintle, A.G. (1973) Anomalous fading of thermoluminescence in mineral samples. *Nature*, **245**, 143–144.

Wintle, A.G. (1975) Thermal quenching of thermoluminescence in quartz. *Geophys. J. R. Astr. Soc.*, **41**, 107–113.

Wintle, A.G. (1997) Luminescence dating: laboratory procedures and protocols. *Radiat. Meas.*, **27**, 769–817.

Wintle, A.G. (2008) Fifty years of luminescence dating. *Archaeometry*, **50**, 276–312.

Wintle, A.G. and Murray, A.S. (1997) The relationship between quartz thermoluminescence, photo-transferred thermoluminescence, and optically stimulated luminescence. *Radiat. Meas.*, **27**, 611–624.

Wintle, A.G. and Murray, A.S. (2006) A review of quartz optically stimulated luminescence characteristics and their relevance in single-aliquot regeneration dating protocols. *Radiat. Meas.*, **41**, 369–391.

Woda, C., Bassinet, C., Trompier, F. *et al.* (2009) Radiation-induced damage analysed by luminescence methods in retrospective dosimetry and emergency response. *Ann. Ist. Super. Sanita*, **45**, 297–306.

Woda, C., Greilich, S. and Beerten, K. (2010) On the OSL curve shape and preheat treatment of electronic components from portable electronic devices. *Radiat. Meas.*, **45**, 746–748.

Woda, C. and Spöttl, T. (2009) On the use of OSL of wire-bond chip card modules for retrospective and accident dosimetry. *Radiat. Meas.*, **44**, 548–553.

Xu, X.G., Bednarz, B. and Paganetti, H. (2008) A review of dosimetry studies on external-beam radiation treatment with respect to second cancer induction. *Phys. Med. Biol.*, **53**, R193-R241.

Yang, X.B., Li, H.J., Bi, Q.Y. *et al.* (2008) Influence of carbon on the thermoluminescence and optically stimulated luminescence of $\alpha$-$Al_2O_3$:C crystals. *J. Appl. Phys.* 104, 123112.

Yasuda, H. (2001) Conservative evaluation of space radiation dose equivalent using the glow curve of $^7$LiF:Mg,Ti (TLD-700). *Health Phys.*, **80**, 576–582.

Yasuda, H. (2002) Application of solid state integrating dosemeters to the determination of biologically equivalent doses in space. *Radiat. Prot. Dosim.*, **100**, 499–502.

Yasuda, H. (2006) Space radiation dosimetry by combination of integrating dosemeters. *Radiat. Prot. Dosim.*, **120**, 410–413.

Yasuda, H. (2009) Effective dose measured with a life size human phantom in a Low Earth Orbit mission. *J. Radiat. Res.*, **50**, 89–96.

Yasuda, H., Badhwar, G., Komiyama, T. and Fujitaka, K. (2000) Effective dose equivalent on the ninth Shuttle-MIR mission (STS-91). *Radiat. Res.*, **154**, 705–713.

Yasuda, H. and Fujitaka, K. (2000) Non-linearity of the high temperature peak area ratio of $^6$LiF:Mg,Ti (TLD-600). *Radiat. Meas.*, **32**, 355–360.

Yasuda, H. and Kobayashi, I. (2001) Optically stimulated luminescence from $Al_2O_3$:C irradiated with relativistic heavy ions. *Radiat. Prot. Dosim.*, **95**, 339–343.

Yasuda, H., Kobayashi, I. and Morishima, H. (2002) Decay patterns of optically stimulated luminescence from $Al_2O_3$:C for different quality radiations. *J. Nucl. Sci. Technol.*, **39**, 211–213.

Yasuda, N., Uchihori, Y., Benton, E.R. *et al.* (2006) The intercomparison of cosmic rays with heavy ion beams at NIRS (ICCHIBAN) project. *Radiat. Prot. Dosim.*, **120**, 414–420.

Yoder, R.C. and Salasky, M.R. (1997) A dosimetry system based on delayed optically stimulated luminescence. *Health Phys.*, **72**, S18-S19.

Yoshimura, E.M. (2007) Correlation of optically stimulated luminescence and thermoluminescence of $Al_2O_3$:Fe,Mg,Cr crystals. *Nucl. Instrum. Methods Phys. Res. B*, **580**, 606–609.

Yoshimura, E.M. and Yukihara, E.G. (2006a) Optically Stimulated Luminescence: searching for new dosimetric materials. *Nucl. Instrum. Methods Phys. Res. B*, **250**, 337–341.

Yoshimura, E.M. and Yukihara, E.G. (2006b) Optically stimulated luminescence of magnesium aluminate ($MgAl_2O_4$) spinel. *Radiat. Meas.*, **41**, 163–169.

Yukihara, E.G., Gasparian, P.B.R., Sawakuchi, G.O. *et al.* (2010) Medical applications of optically stimulated luminescence dosimeters (OSLDs). *Radiat. Meas.*, **45**, 658–662.

Yukihara, E.G., Gaza, R., McKeever, S.W.S. and Soares, C.G. (2004b) Optically stimulated luminescence and thermoluminescence efficiencies for high-energy heavy charged particle irradiation in $Al_2O_3$:C. *Radiat. Meas.*, **38**, 59–70.

Yukihara, E.G., Mardirossian, G., Mirzasadeghi, M. *et al.* (2008a) Evaluation of $Al_2O_3$:C Optically Stimulated Luminescence (OSL) dosimeters for passive dosimetry of high-energy photon and electron beams in radiotherapy. *Med. Phys.*, **35**, 260–269.

Yukihara, E.G., Mittani, J.C.R., Vanhavere, F. and Akselrod, M.S. (2008b) Development of new optically stimulated luminescence neutron dosimeters. *Radiat. Meas.*, **43**, 309–314.

Yukihara, E.G. and McKeever, S.W.S. (2006a) Ionization density dependence of the optically and thermally stimulated luminescence from $Al_2O_3$:C. *Radiat. Prot. Dosim.*, **119**, 206–217.

Yukihara, E.G. and McKeever, S.W.S. (2006b) Spectroscopy and optically stimulated luminescence of $Al_2O_3$:C using time-resolved measurements. *J. Appl. Phys.*, **100**, 083512.

Yukihara, E.G. and McKeever, S.W.S. (2008) Optically Stimulated Luminescence (OSL) dosimetry in Medicine. *Phys. Med. Biol.*, **53**, R351–R379.

Yukihara, E.G., Mittani, J., McKeever, S.W.S. and Simon, S.L. (2007) Optically stimulated luminescence (OSL) of dental enamel for retrospective assessment of radiation exposure. *Radiat. Meas.*, **42**, 1256–1260.

Yukihara, E.G., Ruan, C., Gasparian, P.B.R. *et al.* (2009) An Optically Stimulated Luminescence system to measure dose profiles in X-ray Computed Tomography. *Phys. Med. Biol.*, **54**, 6337–6352.

Yukihara, E.G., Sawakuchi, G.O., Guduru, S. *et al.* (2006) Application of Optically Stimulated Luminescence (OSL) technique in space dosimetry. *Radiat. Meas.*, **41**, 1126–1135.

Yukihara, E.G., Whitley, V.H., Polf, J.C. *et al.* (2003) The effects of deep trap population on the thermoluminescence of $Al_2O_3$:C. *Radiat. Meas.*, **37**, 627–368.

Yukihara, E.G., Whitley, V.H., McKeever, S.W.S. *et al.* (2004a) Effect of high-dose irradiation on the optically stimulated luminescence of $Al_2O_3$:C. *Radiat. Meas.*, **38**, 317–330.

Yukihara, E.G., Yoshimura, E.M., Lindstrom, T.D. *et al.* (2005) High-precision dosimetry for radiotherapy using the optically stimulated luminescence technique and thin $Al_2O_3$:C dosimeters. *Phys. Med. Biol.*, **50**, 5619–5628.

Zhou, D., Semones, E., Gaza, R. and Weyland, M. (2007a) Radiation measured with passive dosimeters in low Earth orbit. *Adv. Space Res.*, **40**, 1575–1579.

Zhou, D., Semones, E., Gaza, R. *et al.* (2007b) Radiation measured for ISS-Expedition 12 with different dosimeters. *Nucl. Instrum. Methods Phys. Res. A*, **580**, 1283–1289.

Zhou, D., Semones, E., Gaza, R. *et al.* (2009a) Radiation measured with different dosimeters during STS-121 space mission. *Acta Astronaut.*, **64**, 437–447.

Zhou, D., Semones, E., Gaza, R. *et al.* (2009b) Radiation measured during ISS-Expedition 13 with different dosimeters. *Adv. Space Res.*, **43**, 1212–1219.

Zhou, D., Semones, E., O'Sullivan, D. *et al.* (2010) Radiation measured for MATROSHKA-1 experiment with passive dosimeters. *Acta Astronaut.*, **66**, 301–308.

Ziegler, J.F., Ziegler, M.D. and Biersack, J.P. (2008) SRIM (computer program), version SRIM-2008.04.

# Index

References to figures are given in italic type. Reference to tables are given in bold type.

Printed and bound by CPI Group (UK) Ltd, Croydon, CR0 4YY